Magnetism and Structure in Systems of Reduced Dimension

NATO ASI Series

Advanced Science Institutes Series

A series presenting the results of activities sponsored by the NATO Science Committee, which aims at the dissemination of advanced scientific and technological knowledge, with a view to strengthening links between scientific communities.

The series is published by an international board of publishers in conjunction with the NATO Scientific Affairs Division

A	**Life Sciences**	Plenum Publishing Corporation
B	**Physics**	New York and London
C	**Mathematical and Physical Sciences**	Kluwer Academic Publishers
D	**Behavioral and Social Sciences**	Dordrecht, Boston, and London
E	**Applied Sciences**	
F	**Computer and Systems Sciences**	Springer-Verlag
G	**Ecological Sciences**	Berlin, Heidelberg, New York, London,
H	**Cell Biology**	Paris, Tokyo, Hong Kong, and Barcelona
I	**Global Environmental Change**	

Recent Volumes in this Series

Volume 306—Ionization of Solids by Heavy Particles
edited by Raúl A. Baragiola

Volume 307—Negative Differential Resistance and Instabilities in 2-D Semiconductors
edited by N. Balkan, B. K. Ridley, and A. J. Vickers

Volume 308—Photonic Band Gaps and Localization
edited by C. M. Soukoulis

Volume 309—Magnetism and Structure in Systems of Reduced Dimension
edited by Robin F. C. Farrow, Bernard Dieny, Markus Donath, Albert Fert, and B. D. Hermsmeier

Volume 310—Integrable Quantum Field Theories
edited by L. Bonora, G. Mussardo, A. Schwimmer, L. Girardello, and M. Martellini

Volume 311—Quantitative Particle Physics: *Cargèse 1992*
edited by Maurice Lévy, Jean-Louis Basdevant, Maurice Jacob, Jean Iliopoulos, Raymond Gastmans, and Jean-Marc Gérard

Volume 312—Future Directions of Nonlinear Dynamics in Physical and Biological Systems
edited by P. L. Christiansen, J. C. Eilbeck, and R. D. Parmentier

Series B: Physics

Magnetism and Structure in Systems of Reduced Dimension

Edited by

Robin F. C. Farrow

IBM Research Division
Almaden Research Center
San Jose, California

Bernard Dieny

Commissariat à l'Energie Atomique
Grenoble, France

Markus Donath

Max-Planck-Institut für Plasmaphysik
Garching b. München, Germany

Albert Fert

Université Paris-Sud
Orsay, France

and

B. D. Hermsmeier

IBM Corporation
San Jose, California

Plenum Press
New York and London
Published in cooperation with NATO Scientific Affairs Division

Proceedings of a NATO Advanced Research Workshop on
Magnetism and Structure in Systems of Reduced Dimension,
held June 15–19, 1992,
in Cargèse, Corsica, France

NATO-PCO-DATA BASE

The electronic index to the NATO ASI Series provides full bibliographical references (with keywords and/or abstracts) to more than 30,000 contributions from international scientists published in all sections of the NATO ASI Series. Access to the NATO-PCO-DATA BASE is possible in two ways:

—via online FILE 128 (NATO-PCO-DATA BASE) hosted by ESRIN, Via Galileo Galilei, I-00044 Frascati, Italy

—via CD-ROM "NATO-PCO-DATA BASE" with user-friendly retrieval software in English, French, and German (©WTV GmbH and DATAWARE Technologies, Inc. 1989)

The CD-ROM can be ordered through any member of the Board of Publishers or through NATO-PCO, Overijse, Belgium.

Library of Congress Cataloging in Publication Data

Magnetism and structure in systems of reduced dimension / edited by Robin F. C. Farrow . . . [et al.].
p. cm.—(NATO ASI series. Series B, Physics; vol. 309)
"Published in cooperation with NATO Scientific Affairs Division."
"Proceedings of a NATO Advanced Research Workshop on Magnetism and Structure in Systems of Reduced Dimension, held June 15–19, 1992, in Cargèse, Corsica, France"—T.p. verso.
Includes bibliographical references and index.
ISBN 0-306-44529-8
1. Thin films, Multilayered—Magnetic properties—Congresses. 2. Magnetic films—Congresses. 3. Epitaxy—Congresses. 4. Superlattices as materials—Congresses. I. Farrow, R. F. C. II. North Atlantic Treaty Organization. Scientific Affairs Division. III. NATO Advanced Research Workshop on Magnetism and Structure in Systems of Reduced Dimension (1992: Cargèse, France) IV. Series: NATO ASI series. Series B, Physics; v. 309.
QC176.9.M84M33 1993 93-14193
530.4′175—dc20 CIP

ISBN 0-306-44529-8

A Division of Plenum Publishing Corporation
233 Spring Street, New York, N.Y. 10013

Printed in the United States of America

PREFACE

This volume contains the papers presented at the NATO Advanced Research Workshop on "Magnetism and Structure in Systems of Reduced Dimension", held at l'Institut d'Etudes Scientifiques de Cargèse - U.M.S. - C.N.R.S. - Université de Corte-Université de Nice Sophia - Antipolis during June 15-19, 1992.

The ordering of papers in the volume reflects the sequence of papers presented at the workshop. The aim was not to segregate the papers into rigidly defined areas but to group the papers into small clusters, each cluster having a common theme. In this way the parallel, rather than serial, development of areas such as preparation of films, magnetic and structural characterization was highlighted. Indeed the success of the field depends on such parallel development and is assisted by workshops of this nature and the international collaborations which they foster.

The organizers and participants of the NATO workshop express their thanks to Mme. Marie-France Hanseler and the staff at l'Institut d'Etudes Scientifiques de Cargèse-U.M.S. - C.N.R.S. - Université de Corte - Université de Nice Sophia - Antipolis for making the workshop and local arrangements a memorable success.

Warm thanks are also expressed to Varadachari Sadagopan and Pascal Stefanou for their encouragement and help in making the workshop a reality. We are also grateful to Kristl Hathaway, Larry Cooper and Gary Prinz for advice in developing the workshop program.

Finally we acknowledge, with thanks, the generous support of sponsors of the workshop: NATO Scientific Affairs Division, Brussels, Belgium; The Office of Naval Research; IBM Research Division, Almaden Research Center; IBM France.

Dr. Robin F.C. Farrow, IBM Research Division, Almaden Research Center, San Jose, CA, USA.

Dr. Bernard Dieny, Centre National de la Recherche Scientifique, Laboratoire de Metallurgie Physique, Grenoble, France.

Dr. Markus Donath, Max Planck Institut fur Plasmaphysik, Abteilung Oberflachenphysik, Garching, Munchen, FRG.

Prof. Albert Fert, Université Paris-Sud, Laboratoire de Physique des Solides, Orsay, France.

Dr. B.D. Hermsmeier, IBM Corporation, Storage Systems Products Division, Cottle Road, San Jose, CA, USA.

INTRODUCTION

The present volume contains the papers presented at a NATO Advanced Research Workshop on "Magnetism and Structure in Systems of Reduced Dimension", held at the Institut d'etudes Scientifiques de Cargèse, Cargese, France during June 15-20, 1992.

Magnetic systems of reduced dimension are providing a wealth of new discoveries of novel phenomena. This is a rapidly growing and active area of condensed matter physics which is driven by the need to interpret and understand the new phenomena as well as by their potential applications. Examples of recent discoveries include giant (MR) magnetoresistance in magnetic multilayers; oscillatory and long-range exchange coupling between 3d-transition metal ferromagnetic films through a variety of non-magnetic metallic interlayers; perpendicular magnetic anisotropy in magnetic films of monolayer thickness and most recently, giant MR in 2-phase heterogeneous alloy films. The purpose of the ARW was to bring together the leading groups in the field to review progress in understanding the new phenomena and especially the need to control preparation techniques to achieve reproducible results.

The workshop began with a review of the origins of the field and of the need for new magnetic materials in the information storage industry. A key conclusion was that recent developments have created an unprecedented opportunity for understanding and utilizing spin-polarized transport phenomena in low-dimensional magnetic structures. Also, since heterogeneous magnetic alloys form the basis for magnetic recording media, an improved understanding of magnetic phenomena (domain reversal, magnetotransport) in such alloys is required for next generation storage media and recording heads.

Following the overview papers the workshop proceeded with presentations on topics clustered around key areas beginning with techniques for imaging magnetic metal film nucleation and growth; a key stage in preparation of low-dimensional magnetic structures. A variety of imaging techniques were critically reviewed. Imaging of Co films grown on Cu was shown using spin-polarized low-energy electron microscopy (SPLEEM). This new technique provides in situ information on magnetic domains and their behavior as a function of film thickness. The influence of surface steps and defects on domain structure was demonstrated. Spin-polarized secondary electron microscopy confirmed the existence of micron-sized domains in single monolayer Co films which had perpendicular anisotropy. An elegant example of this technique was imaging of domains in exchange- coupled Fe (001) films through (001) Cr and Ag spacer films. This led to the discovery of short-period ($\sim$ 4.9Å) oscillatory exchange coupling through Ag (001) as well as a long period of 11.4Å.

Current theories ascribe this oscillatory coupling to an indirect exchange coupling mediated by the conduction electrons of the spacer material. Extension of the RKKY theory to a non-spherical Fermi surface predicts periods in good agreement with the experiments. Prediction of the phase and the strength of the coupling, however, requires incorporation of s-d hybridization between the ferromagnetic film and the spacer. A key prediction of extended theories is that some types of interface roughness can selectively remove short range oscillations. A full experimental description of the interfacial structure is not yet available but is being addressed by a variety of probes including X-ray diffraction, X-ray photoelectron diffraction and electron diffraction techniques. Spin-polarized electronic interface states may be a key to understanding the magnetic phenomena from the electronic structure, yet only a few experimental results are available so far.

The present understanding of giant MR in multilayers is based on spin-dependent scattering within the magnetic films and at the interfaces. However, evidence presented at the workshop showed that interface scattering is dominant in some systems. Spin-dependent scattering of the conduction electrons changes as the magnetic field reorients the magnetization in the magnetic films. In the case of giant MR in granular alloys of Co/Cu and Fe/Ag the effect may arise from spin-dependent scattering primarily at the interfaces between the magnetic particles and the non-magnetic metal. However, quantitative comparison between theory and experiment is hindered by a lack of information on magnetic particle size and distribution in the alloy.

Overall, the conclusion of the workshop was that for an understanding of the new phenomena, a full picture of the interfaces should be arrived at by continuing development of a range of structural and magnetic probes. This picture should lead to improved and more reproducible interface quality and to a test of theoretical predictions such as the absence of short period oscillations of interlayer exchange coupling for <111> oriented Cu and Ag spacer films. It will also assist the development of spin-polarized transport devices and perpendicular media films.

CONTENTS

AN HISTORICAL PERSPECTIVE OF KEY ISSUES IN THE MAGNETISM AND STRUCTURE OF SYSTEMS OF REDUCED DIMENSIONS

G.A. Prinz

Naval Research Laboratory
Washington, DC 20375

INTRODUCTION

In this introductory chapter we review some of the key issues in the field of magnetic systems of reduced dimensions in order to obtain an historical perspective for the latest work which is reported in this Volume. The field is relatively young, all of the work having been done during the professional life time of some scientists who are still working in the field. The pace of work has significantly quickened, however, in the last decade. This is due to three converging trends. First, was the development of ultrahigh vacuum (UHV) techniques which permitted the growth of films at the monolayer (ML) level, free of contamination. Second, was the concomitant development of UHV surface analysis techniques to monitor, control and analyze these films. This has more recently been followed by their spin-appended derivations to permit surface magnetism properties to be studied. This dramatic growth of the importance of surface science techniques has drawn into the magnetism community new members from the surface science community, as well as raised the surface science sophistication of the members of the more traditional magnetism world. Finally, the advent of supercomputers and the development of massive new codes has introduced sophisticated first-principals modeling of the electronic structure of magnetic films and surfaces. This not only allows one to understand much of the new experimental work on a very fundamental level, but also generates specific predictions which stimulate new experimental work.

Magnetism and Structure in Systems of Reduced Dimension
Edited by R.F.C. Farrow *et al.*, Plenum Press, New York, 1993

SMALL PARTICLES

The properties of magnetic materials in reduced dimension were first considered by Néel over forty years ago with his prediction of the fluctuation after-effect in 1949[1] . The effect was first reported observed by Bean in 1955[2] and renamed by him "super paramagnetism". The effect arises whenever the size of a magnetic system becomes sufficiently small that thermal fluctuations are sufficient to cause the magnetic moment to fluctuate between different easy directions of magnetization, as defined by the system's anisotropy. The "attempt" frequency is given by[1]

$$\frac{1}{\tau_o} = f_o e^{-(\mu K_u/M_s kT)} = f_o e^{-(4\pi r^3)K_u/3kT)} \tag{1}$$

where $f_o \approx 10^9$Hz, K_u is the anisotropy and M_S is the saturation magnetization. For spherical Fe particles at 300K, $K_u = 10^5\ J/m^3$ and $\tau_o < 100$ sec, for $r < 100$ Å.

Although this may be considered the "oldest" topic in the field, it is receiving attention for both technological and scientific reasons. Technologically, small magnetic elements can be used to store "bits" of information for data processing. As the computer industry pushes for ever higher densities of information to be stored, issues have been raised regarding the ultimate limits permitted by dimensionality. Some of the most recent work on this topic centers on lithographically produced elements of permalloy which are (1.5μm x 0.2μm x 50nm) [3]; Fe particles on a Au surface (150Å diameter x 500Å high) prepared by using STM to dissociate $Fe(CO)_5$ [4]; ferritin particles (hydrated α-Fe_2O_3) (75Å diameter containing 4500 Fe^{3+} ions) [5]; precipitated particles of Fe which are 40Å diameter [6]; and Ni islands on a Au (111) surface with the measured size of (10Å x 10Å x 2Å).[7] Scientifically, there is active discussion of the possibility of quantum tunneling between the lowest energy states defined by the anisotropy. [6]

SURFACE ANISOTROPY

The second historical prediction in this field was also made by Néel. On the basis of reduced exchange bonding coordination at the surface of a crystal, he argued in 1954[8] that there should arise a surface anisotropy. The first experimental evidence for it was reported by Gradmann[9] in 1968 using torsion magnetometry to study Ni Fe (111) on Cu (111). The next report was from ferromagnetic resonance measurements of Fe (110) on GaAs (110) in 1982.[10] The experiment which stimulated most recent work however was carried out for Fe (100) on Ag (100),[11] where results suggested that a perpendicular anisotropy large enough to overcome $4\pi M$ (20kG) may be present in films of a few ML. This result was confirmed[12] (see Fig. 1) not only for Fe on Ag, but for a large number of other

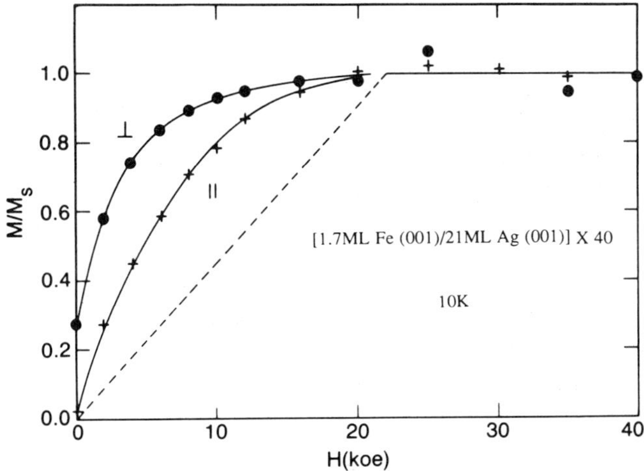

Figure 1. Normalized magnetization vs. applied field for (Fe/Ag) superlattice[8]. H is parallel (||) or perpendicular (⊥) to the film plane, and dashed line shows expected behavior for (⊥) in the absence of surface anisotropy. Note remanent magnetization for (⊥) but not for (||) orientation.

systems. It is now accepted as a commonplace phenomena in ultra thin magnetic films.

ENHANCED MOMENTS

The above experiments on the Fe (100)/Ag(100) system were actually being carried out to search for an entirely different phenomenon. They were stimulated by the theoretical predictions of an enhanced magnetic moment for a ML of Fe.[13,14] (Table I). These predictions, generated by the new supercomputer codes mentioned earlier, proved to be a powerful stimulus to many workers and are responsible for much of the recent interest in the field. It is ironic that these predictions, first published in 1985, have remained very resistant to confirmation until recently. [15]

Table I. Enhanced magnetic moments for Fe (Bohr magnetons).

		Surface	Center	Increase	
					Fe atom 4.0
	(001)	2.96	2.27	30%	
					Monolayer
bcc Fe	(110)	2.65	2.22	19%	3.2 -3.4
	(111)	2.70	2.30	17%	Adlayer Fe/Ag (001) 2.96 - 3.01

CRITICAL PHENOMENA

One of the issues that most obviously comes to mind in discussing systems of reduced dimension is that of critical phenomena. Magnetic systems have long been popular for studying critical phenomena and the predictions for critical exponents (as shown in Table II) are well known. It is interesting that all four models (Ising, XY, Heisenberg and Anisotropic Heisenberg) give the same prediction for the ordering of a magnetic surface layer. This critical exponent $\beta=0.8$ has been confirmed by photoemission experiments which shows that the surface disorders more rapidly then the bulk, although they both have the same ordering temperature T_C [16]. For a thin film, however the predictions are quite different. Recent work using the magneto-optic Kerr effect to study Fe on Pd (001) suggests quite strongly a β close to that of a 2-D Ising system (Fig. 2). T_C in this case is less than bulk and varies with thickness [17].

Table II. Critical exponent for magnetization as $T \rightarrow T_C$ from below.

$$M = M_o \, (1 - T/T_C)^{\beta}$$

3-D Heisenberg	β = 0.365	3-D Ising	β = 0.325
2-D Heisenberg	(no order)	2-D Ising	β = 0.125

Magnetic Surface Layer

$\beta \approx 0.8$

(Ising, XY, Heisenberg, Anisotropic Heisenberg)

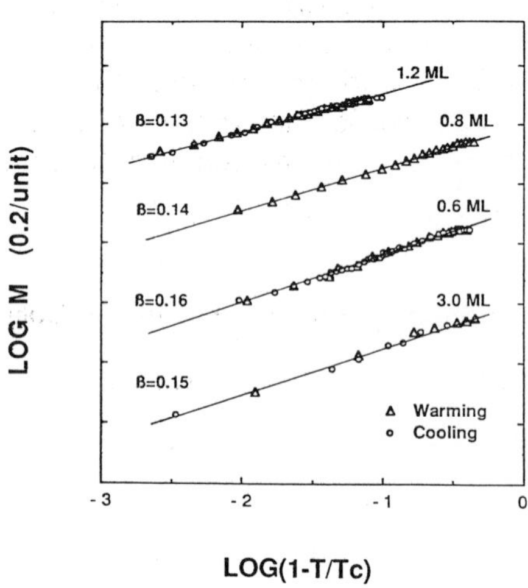

Figure 2. Temperature dependence of magnetization of Fe films on Pd(001)[17].

METASTABLE PHASES

Epitaxial growth of single crystal magnetic metal films has been one of the most fruitful results of modern UHV studies. The ability to knowledgeably prepare single crystal substrate surfaces and films was dependent upon the development of in situ electron diffraction, either at low energy (≈ 100eV) LEED or high energy (≈10keV) HEED and more recently, medium energy (≈1keV) MEED. More recent techniques, exploiting Auger or photo emitted electron forward scattering, can give elementally specific surface structure determination of epitaxial growth[18].

With these structural characterization techniques available, it became possible to witness the formation of metastable crystal structures in the deposited films[19]. The two most widely studied metastable structures are fcc Fe and fcc Co (Fig. 3). These are both high temperature phases which can be "trapped" at low temperature for study by growth upon fcc Cu substrates. Other metastable structures have also been realized, such as bcc Co and bct Mn. These metastable phases provide important extensions of the family of simple elemental magnetic crystals, especially for comparison with the predictions of first principals electronic structure calculations.

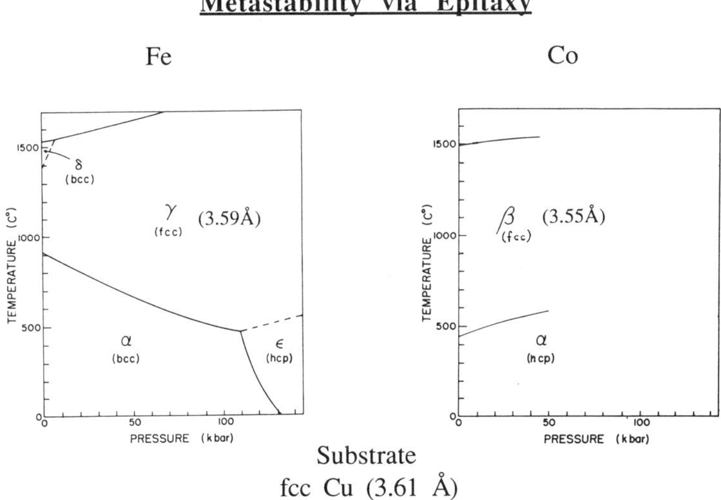

Figure 3. Bulk phase diagrams of Fe and Co with the lattice constants of their high temperature fcc phases compared with that of fcc Cu commonly used as a substrate.

COUPLED LAYERS

One of the most recent discoveries, and one that shall receive considerable attention in several articles in this Volume, stems not so much from the reduced dimensionality of the magnetic material, but rather from the reduced dimensionality of the space separating two magnetic films. This was first reported in 1986 for the coupling of rare earth layers through Y at low temperature.[20]. The discovery by Grünberg et.al [21] that two ferromagnetic Fe films, separated by a few ML of Cr would couple antiferromagnetically via the intervening metal at room temperature has stimulated a great deal of activity both experimentally as well as theoretically. The interest intensified when Parkin et.al [22] reported that the coupling strength oscillated as a function of Cr thickness. The oscillations resemble RKKY oscillations, however with a much larger length scale. Coupling has now been observed in several other magnetic/non-magnetic layered structures and there has been considerable theoretical progress in understanding its quantum mechanical basis. Indeed this coupling is the first evidence of quantum physics-based phenomena in layered magnetic metal materials. Because of the short range of exchange coupling, it did not become manifest until experimental techniques permitted the formation of carefully controlled structures at the ML level. Hence the role of reduced dimensionality is dramatically exhibited.

SPIN-POLARIZED TRANSPORT

These recent developments in multilayered structures have also provided the latest manifestation of the final topic in this overview --- spin-polarized transport. Spin-polarized transport is a phenomena unique to magnetic materials, since it implies an imbalance in the occupation of spin-states near the Fermi level. (Fig. 4) Magneto-transport studies in magnetic

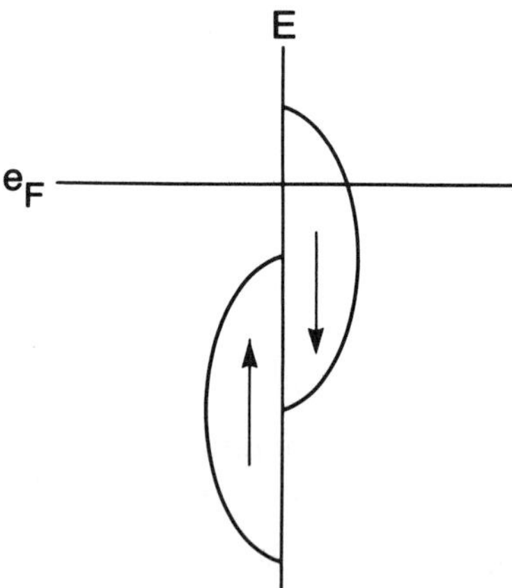

Figure 4. Idealized Stoner density of 3d states near the Fermi level.

materials are not a new topic. Dramatic results were observed in single-crystal Fe whiskers in 1964 when Reed and Fawcett observed that the resistance changed by a factor of ten at low temperature when sufficient magnetic field was applied to align all of the domains[23]. The first experiments in reduced dimension however were carried out in 1970 by Meservey et.al. when they observed spin-polarized tunneling of carriers from a magnetic film through an Al_2O_3 barrier (Fig. 5).[24] Although the results were quite dramatic (44% polarization of the current coming from an Fe film), the carriers were of the majority spins rather than the minority spins as one might have expected from the simple Stoner picture illustrated in Fig. 4. It was several years before an acceptable explanation appeared [25], based upon a detailed analysis of the spin-resolved band

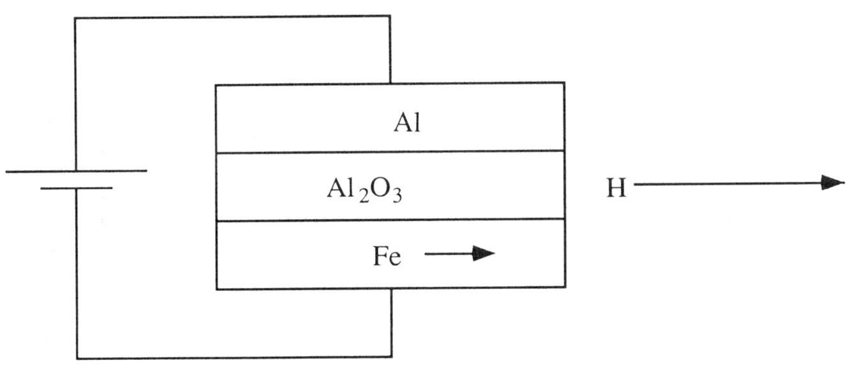

Figure 5. Tunneling of spin-polarized carriers from Fe film through insulating barrier into superconducting Al spin analyzer.[24]

structure of Fe. A few years after these initial spin-polarized tunneling experiments, an important advance was reported by Julliere in 1975 [26] and is illustrated in Fig. 6. Instead of analyzing the spin-polarization of the tunneling current using a superconducting film in an applied field, as was done by Meservey et.al., Julliere simply tunneled through a barrier into another ferromagnetic film. By choosing the two ferromagnetic films to have different coercive fields it was possible to prepare the sandwich structure to be in a state where the two magnetizations were parallel or anti-parallel. He found a 14% change in the resistance between the two configurations. This "magnetic valve" effect straight-forwardly exploited the dependence of spin-polarized transport upon the spin-dependent density of states available at the Fermi level in the two ferromagnetic metal films and demonstrated the idea of a magneto-resistance "switch".

Further experimental advance did not appear for a decade, until the challenging experiments of Johnson and Silsbee in 1985 [27]. These experiments, illustrated in Fig. 7, actually measured the spin relaxation length of carriers injected into a paramagnetic metal from a ferromagnetic film. The stunning result, that this length was 100μm at 40K in aluminum, showed that spin polarized currents could travel distances comparable to those in modern electronic device structures without losing their "memory".

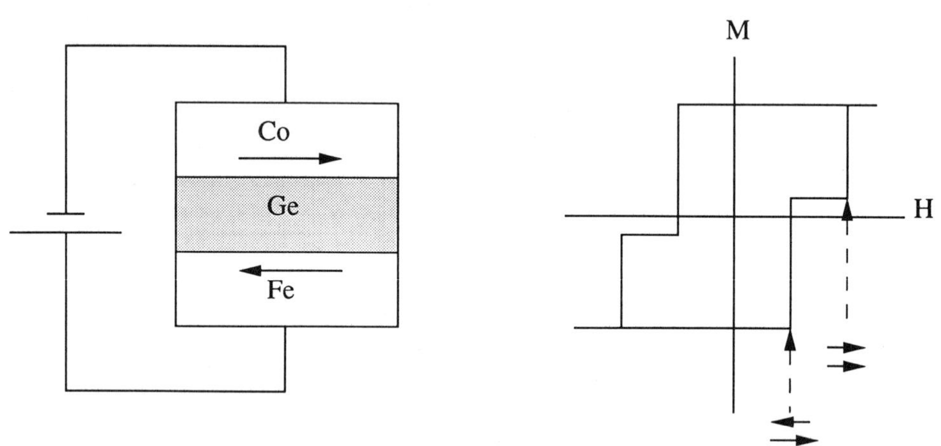

Figure 6. Tunneling of spin polarized carriers from one ferromagnetic film into another controlled by external magnetic field via differential coercivities of two films.[26]

In 1988 the well-known results of Baibich, et.al. [28] were published, which revealed the dramatic changes in the magneto-resistance of $(Fe/Cr)_n$ multilayers in an applied magnetic field sufficient to overcome the antiferromagnetic coupling between the Fe layers and align their moments. These results stimulated an enormous effort, both experimental and theoretical, which continues to the present and occupies much of the space in this Volume. This "Giant Magneto Resistance" (GMR) may be thus be viewed as the latest development in the field of spin-polarized transport. Indeed, the most recent GMR experiments do not propagate the current parallel to the layers as was done in the original (Fe/Cr)n multilayers, but propagate perpendicular to the layers [29]. These experiments are very reminiscent of the earlier tunneling experiments of Meservey and Jullierre and may prove to be more amenable to theoretical treatment than the initial parallel geometry measurements.

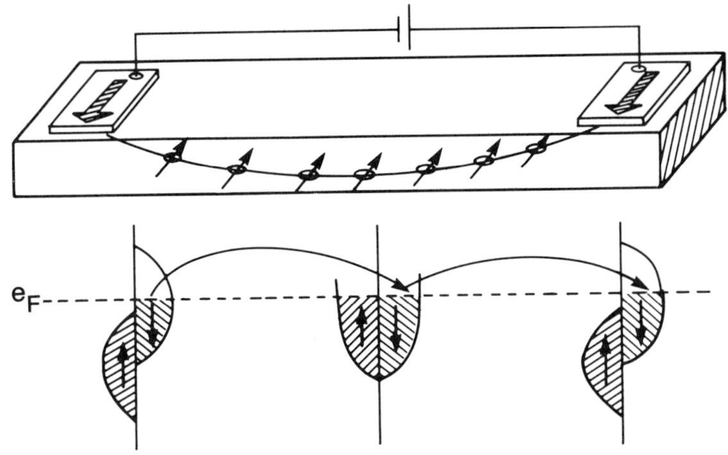

Figure 7. Injection of spin-polarized current into a paramagnetic bar from ferromagnetic pads illustrating the transfer of carriers between states at the fermi level.[28]

Finally the latest development in spin-polarized transport abandons layered structures entirely. The recently published results by Berkowitz et.al.[30] and Xiao et.al. [31] demonstrated GMR in copper films containing small particles of ferromagnetic cobalt. Here the spin-polarized transport is presumably by scattering of carriers at the copper/cobalt interface and changes in the resistance are seen to follow the alignment of the particles' moments in an applied magnetic field. We see here that we have come full circle back to small particles, but now the spaces between them are sufficiently small to be less than a spin relaxation distance.

Spin polarized transport represents the latest challenge in the field of magnetism in reduced dimensions. Since, of all magnetic phenomena, transport is most sensitive to details of structure, we should expect that both the opportunities and the understanding will depend upon levels of control over the materials ultimately at the dimensions of individual atoms.

ACKNOWLEDGMENTS

Support is acknowledged from the Office of Naval Research.

REFERENCES

1. L. Néel, *Compt. Rend.* **228**, 664 (1949).
2. C.P. Bean, *J. Appl. Phys.* **26**, 1381 (1955).
3. J.F. Smyth, S. Schultz, D.R. Fredkin, D.P. Kern, S.A. Rishton, H. Schmid, M. Cali and T.R. Koehler *J. Appl. Phys.* **69**, 5262 (1991).

4. D.D. Awschalom, M.A. McCord and G. Grinstein, *Phys. Rev. Lett.* **65**, 783 (1990).
5. D.D. Awschalom, D.P. DiVincenzo and J.F. Smyth *Science* (in press 1992).
6. F. Bodker, S. MØrup, C.A. Oxborrow, M.B. Madsen and J.W. Neiemantsverdriet *J. Magn. Magn. Mater.*, **104-107**, 1695 (1992).
7. D.D. Chambliss, R.J. Wilson and S. Chiang *Phys. Rev. Lett.* **66**, 1721 (1991).
8. L. Néel, *J. Phys. Rad.* **15**, 225 (1954).
9. U. Gradman and M. Müller, *Phys. Stat. Sol.* **27**, 313 (1968).
10. G.A. Prinz, G.T. Rado and J.J. Krebs, *J. Appl. Phys.* **53**, 2087 (1982).
11. B.T. Jonker, K.H. Walker, E. Kisker, G.A. Prinz and C. Carbone, *Phys. Rev. Lett.* **57**, 142 (1986).
12. J.J. Krebs, B.T. Jonker and G.A. Prinz, *J. Appl. Phys.* **63**, 3467 (1988).
13. C.L. Fu, A.J. Freeman and T. Oguchi *Phys. Rev. Lett.* **54**, 2700 (1985).
14. R. Richter, J.G. Gay and J.R. Smith *Phys. Rev. Lett.* **54**, 2704 (1985).
15. M. Albrecht, T. Furubayashi, U. Gradmann and W.A. Harrison, *J. Magn. Magn. Mater.* **104-107**, 1699 (1992).
16. U. Gradmann, *J. Magn. Mater.* **100**, 481 (1991).
17. C. Liu and S.D. Bader, *J. Appl. Phys.* **67**, 5758 (1990).
18. Y.U. Idzerda and D.E. Ramaker *Phys. Rev. Lett.* **69**, 1943 (1992).
19. G.A. Prinz, *J. Magn. Mater.*, **100**, 469 (1991).
20. M.B. Salamon, S. Sinha, J.J. Rhyne, J.E. Cunningham, R.E. Erwin, J. Borchers and C.P. Flynn, *Phys. Rev. Lett.* **56**, 259 (1986).
21. P. Grünberg, R. Schreiber, Y. Pang, M.B. Brodsky and H. Sowers, *Phys. Rev. Lett.* **57**, 2442 (1986).
22. S.S.P. Parkin, N. More and K.P. Roche, *Phys. Rev. Lett.* **64**, 2304 (1990).
23. W.A. Reed and E. Fawcett, *Phys. Rev.* **136**, A422 (1964).
24. R. Meservey, P.M. Tedrow and P. Fulde, *Phys. Rev. Lett.* **25**, 1270 (1970).
25. M.B. Stearns, *J. Magn. Mater.* **5**, 167 (1977).
26. M. Julliere, *Phys. Lett.*, **54A**, 225 (1975).
27. M. Johnson and R.H. Silsbee, *Phys. Rev. Lett.*, **55**, 1790 (1985).
28. M.N. Baibich, J.M. Broto, A. Fert, F. Nguyen van Dau, F. Petroff, P. Etienne, G. Creuzet, A. Friederich and J. Chazeles, *Phys. Rev. Lett.* **61**, 2472 (1988).
29. W.P. Pratt, Jr., S.F. Lee, J.M. Slaughter, R. LoLoee, P.A. Schroeder and J. Bass, *Phys. Rev. Lett.* **66**, 3060 (1991).
30. A.E. Berkowitz, J.R. Mitchell, M.J. Carey, A.P. Young, S. Zhang, F.E. Spada, F.T. Parker, A. Hutton and G. Thomas, *Phys. Rev. Lett.* **68**, 3745 (1992).
31. J.Q. Xiao, J.S. Jiang and C.L. Chien, *Phys. Rev. Lett.* **68**, 3749 (1992).

ADVANCES IN MAGNETIC DISK STORAGE TECHNOLOGY

Yoshimasa Miura

Fujitsu Laboratories Ltd.
Atsugi, 243-01 Japan

Abstract

This paper presents an overview of progress in magnetic disk drive technology, and discusses the future prospects for high-density magnetic recording.

Recent progress of magnetic materials for thin-film heads and media has greatly improved read/write performance, keeping pace with requirements of high density recording. I review the research key areas of current thin-film head and media technology. I also consider the barriers to be overcome to realise ultra-high density magnetic recording approaching 20 Mbits/mm^2.

Introduction

In the last 30 years, areal density in magnetic recording has improved almost 1,000 fold, an increase similar to the progress in semiconductor memory chips. Magnetic storage devices, such as magnetic disk drives and VCRs have become commonplace information processing devices. New technologies, such as the magnetoresistive (MR) head and perpendicular recording, will push the recording density toward 20 Mbits/mm^2.

Magnetic recording is not as simple as it seems, it involves complex meshing of a number of different key components and related techniques. Head and media technology, signal processing, and precision mechatronics, such as the spindle motor and head positioner and their control, all contribute. The head to media interface has always been critical because of the reduced flying height and the tradeoff between read/write performance and reliability.

This paper presents an overview of progress in magnetic disk drive technology, and discusses the future prospects of high density magnetic recording. This paper focuses on the progress of magnetic materials for thin-film heads and media which has improved read write to a point which meets requirements of high density recording. I finish by considering what is required to achieve a density 20 Mbits/mm^2.

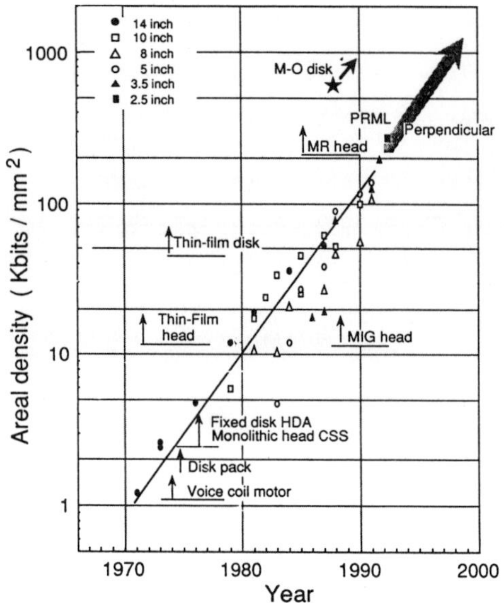

Fig. 1. Annual trend of recording density.

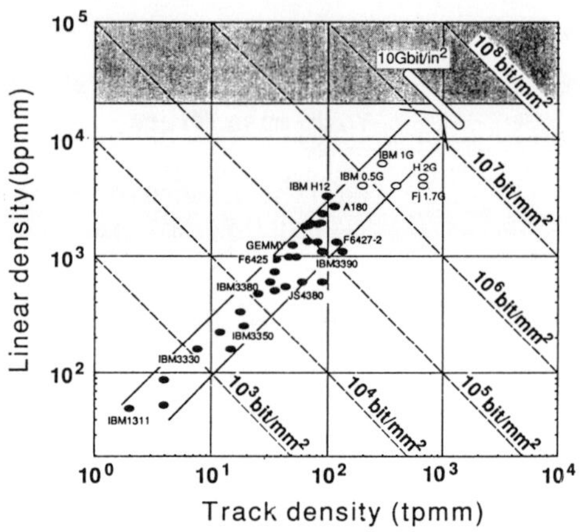

Fig. 2. Progress in areal density. -Linear density vs. track density-

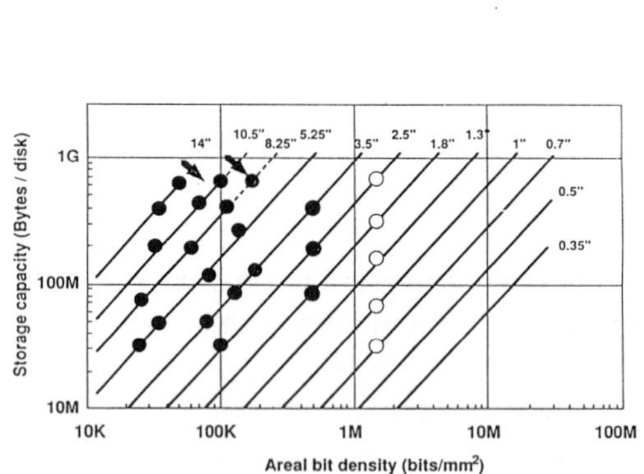

Fig. 3. Platter storage capacity versus areal bit density.

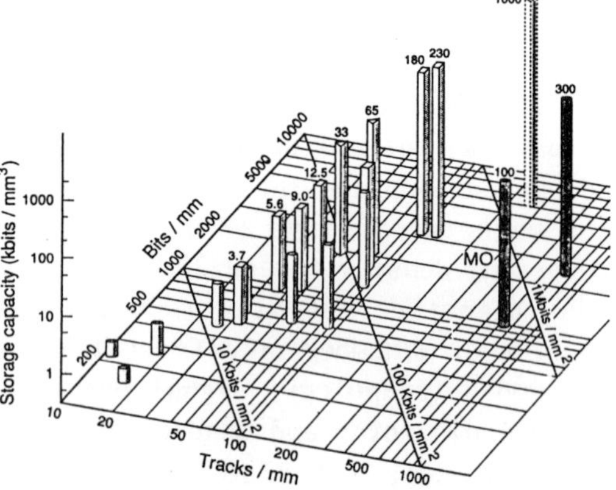

Fig. 4. Evolution of volume density.

Disk drive evolution

Figure 1 shows the trend in areal bit density of magnetic disk drives. Magnetic recording density has improved 100 fold in the last 20 years, reaching 200 kbits/mm^2.[1] The figure includes the key technologies that helped make a breakthrough possible in each development stage. The important technologies for future improvements are the MR head, partial response maximum likelihood sequential detection, and perpendicular recording.

Figure 2 shows progress in areal density as the relationship between linear density and track density. As you can see, linear density has kept up with track density. White circles plotted from 1 Mbits/mm^2 to 3 Mbits/mm^2 show recent improvements in density. Magnetic recording devices with a density over 10 Mbits/mm^2 have become a realistic target for development.[2]

The increase in density has brought about large capacity and smaller disk drives. Figure 3 shows storage capacity versus areal density estimated for different platter sizes. When 1 Mbits/mm^2 is reached , a 3.5 inch disk will store 1 gigabyte. If 10 Mbits/mm^2 becomes possible, will all disks be 1 inch or less in diameter? This is difficult to predict and the effects are not clear.

Volume density allows us to compare magnetic storage capacity with optical recording. Figure 4 shows the evolution of volume density against areal density for typical disk drives and removable optical cartridges. Volume density (bits/mm^3) was calculated as areal bit density multiplied by stacking surfaces per millimeter. The latest magnetic disk drives have a higher volume density than optical cartridges.

High-Density Magnetic Recording Technology

High-density media

The history of magnetic properties of longitudinal disk media shows increasing remanent magnetization Br and coercivity Hc. This has maintained sufficient signal strength while reducing bit cell size. A metal thin-film, such as a cobalt-base alloy, meets these requirements. As the space between the recording transitions decreases, we would expect the microstructural details of the media to become increasingly important because of their polycrystalline nature.[3] The complexity of thin-film media on a microscopic scale has prompted extensive investigation of the relationship between their microstructure (grain morphology, grain size, and crystallographic orientation), magnetic properties (coercivity Hc, saturation magnetization Bs, remanent magnetization Br, coercive squareness S *) and recording performance (noise, overwrite, and linear resolution).

Currently, the most popular materials for disks are CoCr alloys, such as CoCrTa and CoPtCr, instead of the CoNi alloy used in early disks. In longitudinal media, the micromagnetic properties of the recording medium are strongly influenced by the morphological properties of the Cr underlayer. Continuous properties of magnetic thin-films generate zigzag domains at the transition, and these domains cause modulation noise. The continuous nature

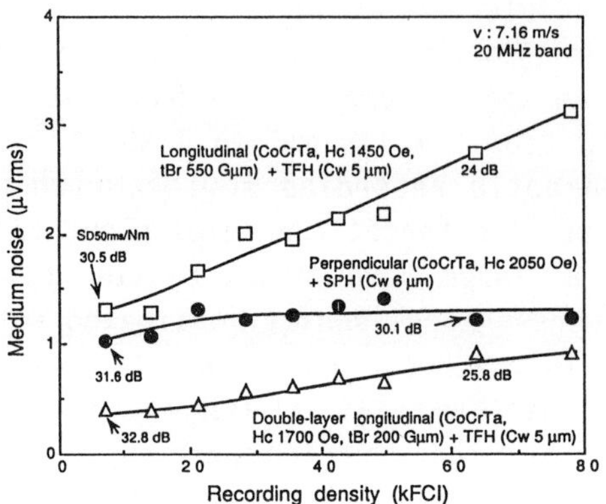

Fig. 5. Medium noise versus recording density.

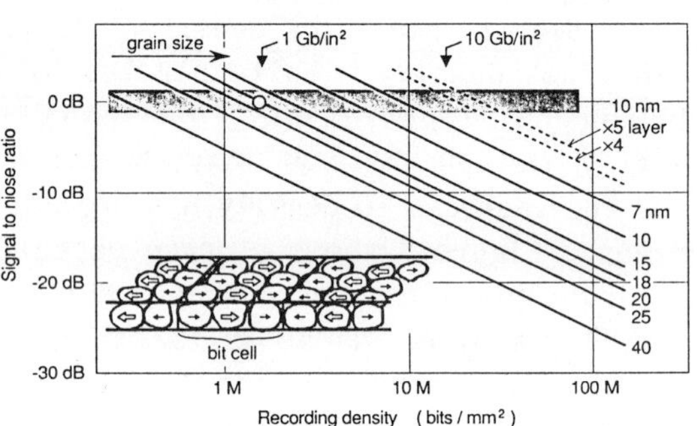

Fig. 6. Relative media S/N vs. recording density estimated for various average grain size.

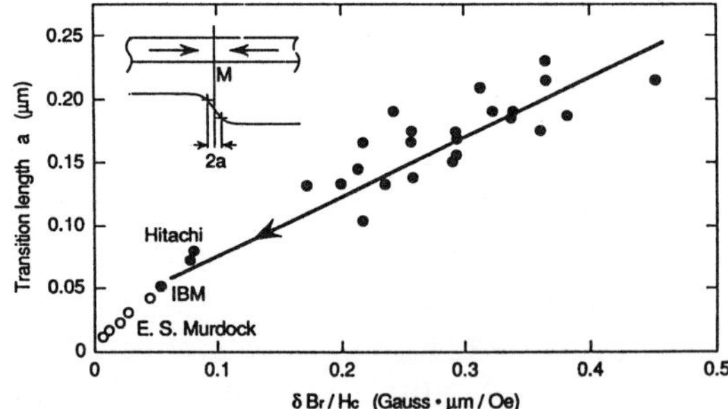

Fig. 7. Magnetic transition width, 2a, in a film medium estimated as a function of δBr/Hc.

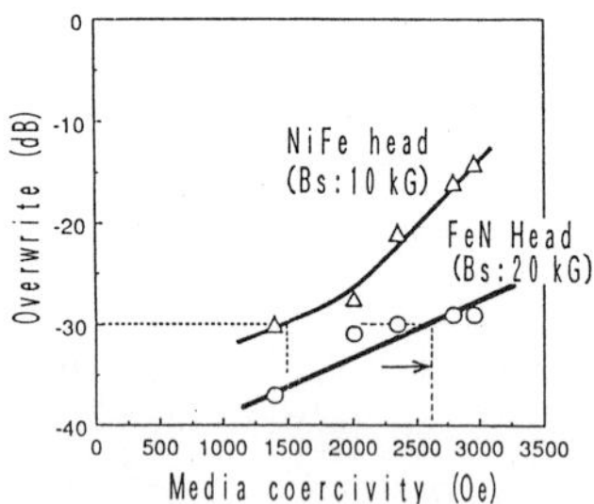

Fig. 8. Overwrite characteristics versus media coercivity.

of magnetic films originates in intergranular exchange coupling. Research into high-density low-noise media is focusing on morphological study of physical separation of grains and hence reducing intergranular exchange coupling. As figure 5 shows, the medium noise of conventional longitudinal media increases proportionally with recording density. We do not see zigzag transition noise in perpendicular recording.

In analogy to particulated media, intrinsic signal-to-noise ratio is proportional to the square root of grain numbers in a recorded bit cell. Figure 6 shows relative media S/N ratio versus recording density estimated for various average grain sizes. Future high density media will have to reduce grain size further and further while preserving their ferromagnetic nature. T. Yogi et al. achieved 1 Gbit/in^2 (1.5 Mbits/mm^2) recording with well-isolated grains with an average grain size of 18 nm.[4] According to this model, a high-density medium with 10 Mbits/mm^2 or more should have a grain size of a few nanometers.

The transition width, 2a, on the film media scale is a function of δ Br/Hc (where δ is film thickness), so a very thin film with a high coercivity is required for high-density longitudinal media. The transition width, a, in a film medium estimated as a function of δBr/Hc is plotted in Fig. 7. The 1 Gbit/in^2 and 2 Gbits/in^2 recording media demonstrated by IBM and Hitachi were made using the conventional techniques mentioned before.[4][5] E. S. Murdock et al. have attempted 10 Gbits/in^2 (15 Mbits/mm^2) recording media.[2] For recording densities of 1 Mbit/mm^2 or more, grains in film media will inevitably have a component of orientation perpendicular to the plane.

Narrow track thin-film head

High density recording requires key improvements in thin-film head technology; 1) high saturation magnetization core materials for high coercive force media, 2) suppression of magnetic domain instability which causes domain wall motion induced noise, 3) extremely narrow head-to-media spacing technology.

At present, the trend is toward increasing coercivity. To obtain good overwrite capability, the saturation magnetization of the core materials must be increased. CoZr amorphous films with Bs=13 to 14 kGauss were typical head core materials.[6] Aoyama et al. have developed sputtered FeN films with Bs=20 kGauss and used them with thin film heads.[7] Figure 8 shows overwrite characteristics versus media coercivity. FeN film heads have adequate overwrite characteristics on high-coercive media up to 3000 Oe.

We have to overcome the noise problems due to magnetic domain instability of the pole tips by reducing the track width. Write current induced instability has the following two causes.[3]

a) Write after noise: After the write operation, films which have instantaneous wall motions during domain relaxation with a relatively long time constant of several microseconds to several tens of microseconds induce sharp spike noise on read signals.

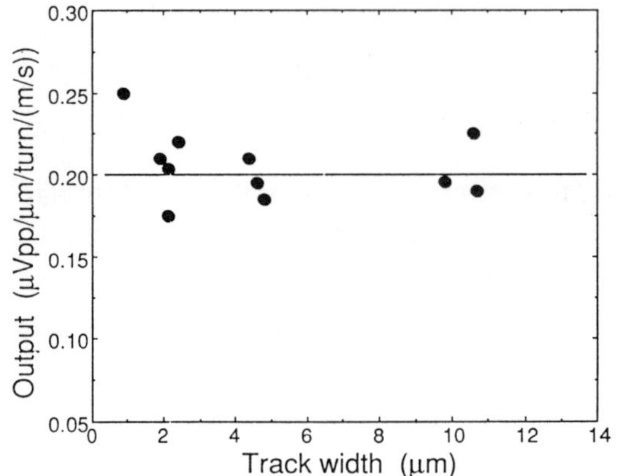

Fig. 9. Normalized output of single-pole heads vs. track width.

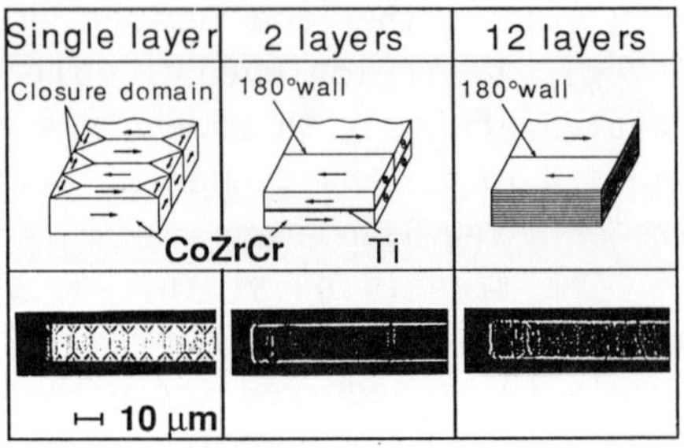

Fig. 10. Domain wall patterns for single-layer and laminated (2 and 12 layers) strip pattern elements.

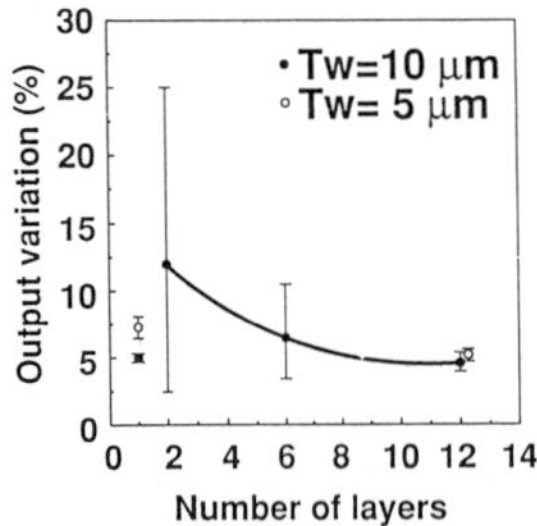

Fig. 11. Output variation dependence on number of pairs of CoZr layers.

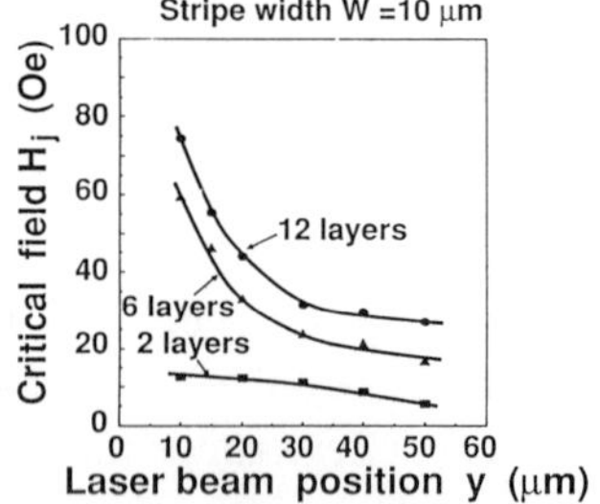

Fig. 12. Relationship between critical field Hj for wall displacement and number of pairs of CoZrCr layers.

b) Output fluctuation: This is due to instability of the domain structure configuration after a write operation. The effective permeability change causes output fluctuation.

Experimentally, a strong induced uniaxial anisotropy field Hk should be applied in the direction of track width, and internal stress and magnetostriction should be controlled to be as small as possible in order to stabilize the domains. At present, the high pearmeable potential of soft magnetic materials is limited instead of stabilizing read write characteristics.

Even in the reading process, under the influence of an external magnetic field, the domain walls move in discontinuous Barkhausen jumps and produce an irregular and noisy output. The most important task has been to eliminate Barkhausen noise. One way to eliminate domains is to couple the permalloy layer to anti-ferromagnetic layers such as FeMn.

Reducing the track width makes it difficult to keep the uniaxial easy magnetization axis to the direction of track width. A strong induced aniso tropy field, Hk, stabilizes the domain structure but reduces effective permeability of the pole tips decreasing head efficiency. We nevertheless achieved satisfactory domain control in a single-layer CoZr amorphous film down to a 1 μm track width. Figure 9 shows the normalized output versus track width.

Lamination has been studied to help to eliminate the closure domain. Figure 10 shows the domain wall patterns for single-layer and laminated (2-12 layers) stripe pattern elements. By laminating CoZrCr amorphous films with nonmagnetic Ti layers, the stripe elements appeared to have no closure domains.[8] The variation in output reproduction was measured using single-pole heads and a CoCr/NiFe double-layered medium. Figure 11 shows the output variation dependence on the number of pairs of CoZrCr layers for the main pole of single-pole heads. It appears that the output variation of the head with a two-layered main-pole film is larger than that with a single-layer main pole-film. The output variation decreased, however, as the number of pairs of CoZrCr layers increased. This is possibly because reproduction output stability depends on domain behavior.[8]

Figure 12 shows the relationship between the magnitude of critical field, Hj, for wall displacement and thè number of pairs of CoZrCr layers. Hj can be obtained by measuring the external field when the Kerr signal jumps at various laser beam spot positions. As the number of pairs of CoZrCr layers increases, Hj also increases, showing that wall displacement is suppressed and the domain structure is stabile. This domain stability agrees fairly well with the reproduction stability in perpendicular heads with multi-layered main-poles.

Magnetoresistive (MR) sensors detect magnetic flux directly, their output amplitude is independent of the relative speed between the head and the recording medium. Also, the output is proportional to the sensing current, typically producing signals higher than obtained with inductive heads. A small MR sensor, of patterned permalloy for example, is, however, typically

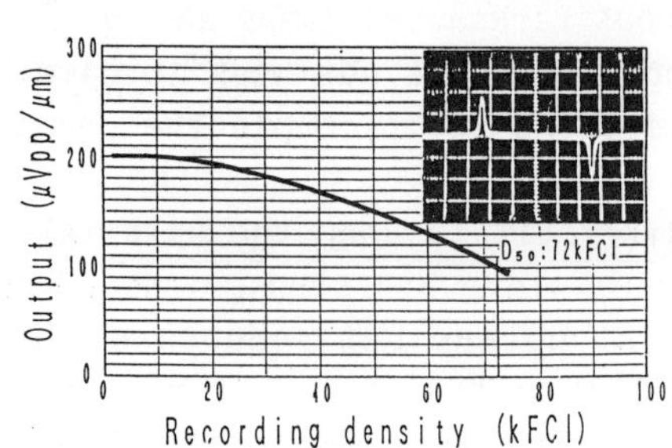

Fig. 13. Signal amplitude versus linear density

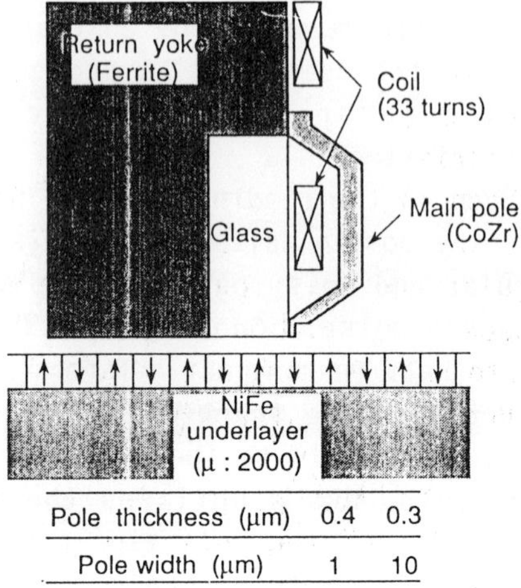

Fig. 14. Cross section of single-pole head and double-layered medium.

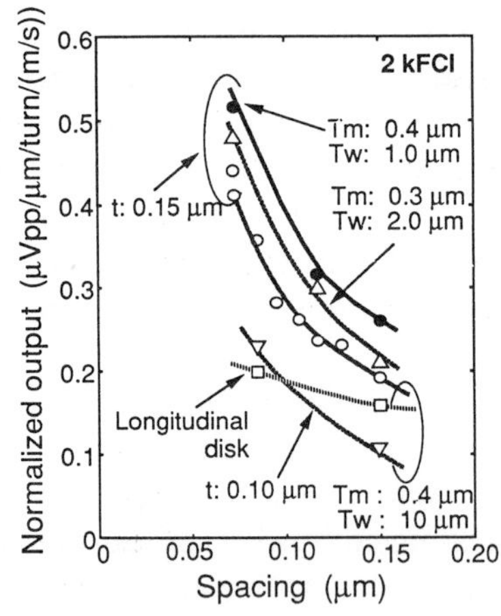

Fig. 15. Dependence of normalized output and D_{50} on spacing.

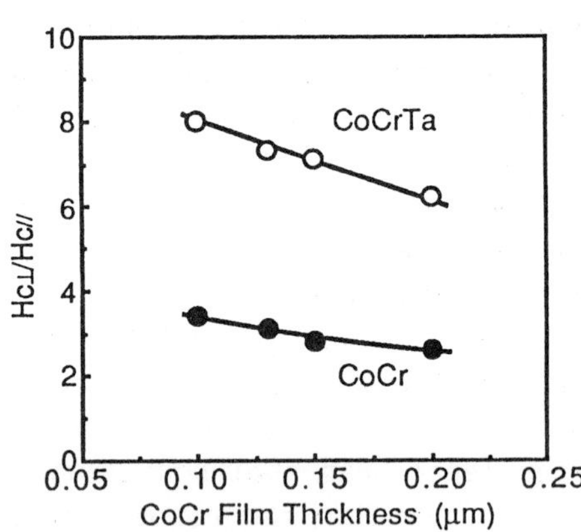

Fig. 16. Coercivity ratio Hc⊥/Hc// vs. CoCr and CoCrTa film thickness.

broken up into domains to reduce magnetostatic energy. Under the influence of an external field, the domain walls move in discontinuous Barkhausen jumps, producing irregular and noisy output.

One way to eliminate domains is by exchange coupling the permalloy layer to an antiferromagnetic (AF) layer such as FeMn. The spin structure of the AF layer can be aligned along a desired direction by heating beyond the blocking temperature of the AF layer and cooling in a magnetic field.[9] The hysteresis loop of the ferromagnetic layer is shifted from the origin by an effective exchange-bias field, H_{ua}, with a corresponding unidirectional exchange anisotropy energy.[10]

A critical magnitude of bias field was required to eliminate the domain walls completely. Too strong a coupling-field, however, reduces the effective permeability of the MR sensor and reduces reproduced signals. Various biasing methods have been suggested to control H_{ua} depending on its sensor structure; one such method is boundary control stabilization (BCS).[9]

Figure 13 shows the read characteristics of a shielded BCS-MR head. We achieved a high output of 200 $\mu V_{P-P}/\mu m$ for an isolated pulse. The MR head will be very important for reading high-density recorded signals, especially for narrow track width systems. The maximum magnetoresistance amplitude ($\Delta\rho/\rho$) of permalloy films is, however, 2 to 3 %. The multilayer structures of ultra-thin ferromagnetic films have giant-magnetoresistance with a room temperature amplitude as high as more than10%. [11] Soft magnetic multi-layer structures are therefore anticipated for future highly sensitive MR sensors with submicron track widths.

Many researchers are currently trying to reduce the flying height of heads to less than 50 nanometers, an approach sometimes referred to as contact recording. H. Hamilton proposed integrated head/flexure structures with an effective mass of about 300 μg; these structures do not crash and have the potential to cause very low wear of both head and media when operating in continuous sliding contact with rigid media.[12] The microflex head not only contributes to the trend of downsizing disk drives, but is also promising for achieving very high linear and track density.

Perpendicular recording

Perpendicular recording on rigid disks has been studied because its high-density characteristics greatly depend on the spacing between the head and medium. Figure 14 is a cross-section of the probe-type thin-film head with a ferrite return yoke and double-layer medium. Figure 15 compares longitudinal and perpendicular recording as a function of head-to-media spacing. If the head-to-media spacing is less than 0.1 μm, perpendicular recording is superior.[13]

A single-layer media with a CoCr layer on a Ti layer has satisfactory crystal orientation and magnetic properties. We therefore deposited a 20

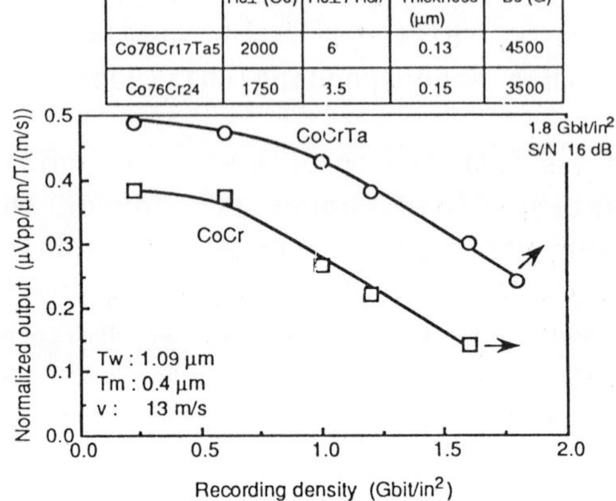

Fig. 17. Normalized output vs. recording density.

Fig. 18. Lorentz TEM microphotograph of the CoCrTa film which recorded 3 Mbits/mm^2 of information.

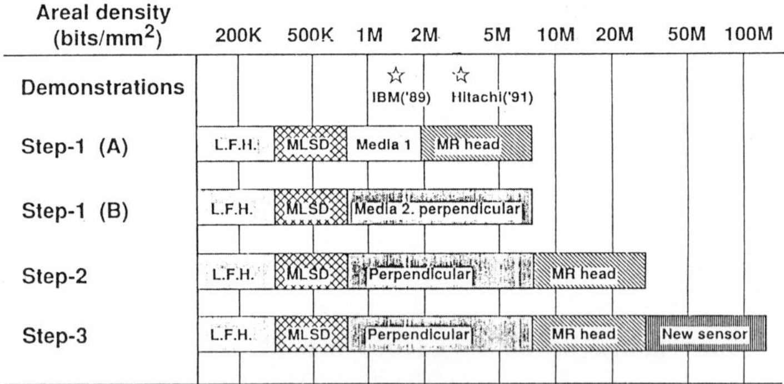

Fig. 19. Expected progress in high-density magnetic recording.

nanometer Ti interlayer between the permalloy under layer and CoCr recording layer of a conventional double-layer medium. This interlayer, however, weakend the head-to-medium interaction and worsened high-density recording at low spacing.

CoCrTa films deposited on permalloy films have good perpendicular orientation when compared with CoCr films. Using the coercivity ratio $Hc\perp / Hc//$ as a measure of crystal orientation, we observed significant differences between CoCrTa and CoCr films deposited on permalloy films (Fig. 16). Reflecting this difference, the CoCrTa/NiFe medium shows superior write/read characteristics to the CoCr/NiFe media (Fig. 17). At a recording density of 3 $Mbits/mm^2$, we obtained a C/N of 40 dB.[14] Figure 18 is a Lorentz microphotograph of the CoCrTa film which recorded 3 $Mbits/mm^2$ of information. The pattern recorded was "11000", and the track width was 1 μm. It has been suggested that the track density of perpendicular recording is potentially higher than that of optical recording.

Professor Iwasaki et al. demonstrated an ultrahigh linear bit density of 25 kbits/mm.[15] The importance of this measurement is that it demonstrated the ability of this type of medium to support magnetization reversals with a bit length, 40 nm, of about twice the diameter of the columns, 20 nm. Perpendicular recording promises to realise ultrahigh-density recording of 20 $Mbits/mm^2$.

Y. Maeda et al. investigated the compositional separation in Co-Cr base alloy film.[16] They suggested that compositional separation makes a similar contribution to reducing media noise by effectively reducing grain size.

Future prospects of high density recording

Progress in magnetic recording is not only a result of progress in heads and media. Related technologies such as signal processing, HDI reliability, head positioner, and positioning control also contribute. For further development of high-density recording, there must be several methods of synthesis a designer can choose from.

The physical limitation of conventional longitudinal recording depends mainly on magnetization transition length (δ Br/Hc). This means thinner media with a higher coercive force are required. Current research is attempting to reduce media noise by understanding the microstructure of films. The bit cell size in perpendicular recording is thought to be limited by the order of the grain size. Whenever the head-to-media spacing is less than 0.1 μm, perpendicular recording is superior.

Head-to-media spacing is one of the most important factors when trying to increase recording density. Miniaturizing magnetic disk drives has allowed the use of ultralow-flying heads and also micro-contact heads in rigid disk drives. This has made perpendicular recording preferable.

Advanced signal processing techniques, such as maximum-likelihood sequential detection (MLSD) with partial response, will improve the SNR by approximately 8 dB. An inductive write and MR read head could increase the density dramatically. The current maximum magnetoresistance sensitivity ($\Delta\rho/\rho$) of permalloy is around 2 %. Moreover, it is anticipated that giant-magnetoresistance materials will increase sensitivity by up to 10 % or more.

Figure 19 shows expected progress in high-density magnetic recording. In the near future, one of two improvements are expected.[1]

Step-1(A): Improved conventional longitudinal recording using new techniques such as MR heads, MLSD, and improved low noise media. It is estimated that recording density will reach up to a few megabits per square millimeter.

Step-1(B): Perpendicular recording combined with MLSD will have a similar high-density recording potential. Perpendicular recording will be improved 4-5 fold more by introducing MR heads as described in Step-2, and recording density will reach 30 Mbits/mm^2.

Step-3 shows the improvement expected when using new materials such as multi-layer structures. The product due to all improvement factors is 1000. This means that, with the current recording density of 150 kbits/mm^2, we can expect 150 Mbits/mm^2 recording densities within 20 to 30 years.

Theoretically, perpendicular recording media itself has the potential to record 1 Gbits/mm^2. The future of the magnetic disk recording is very promising. Realising this future, however, needs technological breakthroughs in the micromechanics of micromotors and microactuators.

Conclusions

The requirements of large storage capacity on small disk drives will become more and more demanding. Magnetic recording density has improved 1000 fold in the last three decades, a trend expected to continue at least another 20 years. High density recording accelerates disk drive downsizing both in computer systems and in integrated high-performance subsystems such as disk arrays.[17]

The access gap for magnetic disk drives is, nevertheless, very large compared to that of solid-state memory. Combining cache memory with a magnetic disk drive could greatly improve disk access performance. Rather than replacing a magnetic disk, solid-state memory will be combined with disk drives to improve their performance.

Looking the future of the disk file, downsizing the disk diameter will lead to new forms of drive, such as removable disk cartridges and chip modules that mount directly on a printed circuit board.

In conclusion, we expect the magnetic disk drive to continue to be an important storage device.

References

[1] Y. Miura, "Magnetic Disk Storage and Its Prospects," J. Magn. Soc. Jpn., vol. 15, Suppl. No. S1, p.133 (1991).

[2] E. S. Murdock, R. F. Simmons, and R. Davidson, "Roadmap for 10 Gbit/in^2 Media: Challenges," Digests of The INTERMAG Conf. JA-01 (1992).
[3] V. S. Speriosu, D. A. Herman, Jr., I. L. Sanders, and T.Yogi, "Magnetic Thin Films in Recording Technology," IBM J. Res. Develop., vol. 34, p.884 (1990).
[4] T. Yogi, C. Tsang, T. A. Nguyen, K. Ju, G. L. Gorman, and G. Gastillo, " Longitudinal Media for 1 Gb/in^2 Areal Density," IEEE Trans. Magn., vol. MAG-26, p.2271 (1989).
[5] M. Futamoto, F. Kugiya, M. Suzuki, H. Takano, Y. Matsuda, N. Inabe, Y. Miyamura, K. Akagi, T. Nakao, H. Sawaguchi, H. Fukuoka, T. Munemoto, and T. Takagaki, "Investigation of 2Gb/in^2 Magnetic Recording at a Track density of 17 kTPI," IEEE Trans. Magn., vol. MAG-27, p.5280 (1991).
[6] M. Takahashi, H. Fujimori, and T. Miyazaki, "Non-Magnetostrictive Co-bases Amorphous Alloys for Recording Head Materials," Japan Annual Reviews in Electronics, Computer and Telecommunications, vol. 10, Recent Magnetics for Electronics, ed. by Y. Sakurai, p.137 , OHM • North-Holland (1983).
[7] S. Aoyama, T. Koshikawa, Y. Uehara, E. Kanda, Y. Ohtsuka, S. Takemura, J. Toda, and K. Hosono, "Recordig Characteristics of FeN Head on High-Hc Media," Digest of the 15th Anual Conf. on Magnetics in Japan, p.17 (1991).
[8] M. Kanamine, H. Kanai, and K. Kobayashi, "Domain Stability of Multilayered CoZrCr/Ti Strips," J. Appl. Phys. vol. 69, p.5859 (1991).
[9] C. Tsang, "Unshielded MR Elements with Patterned Exchange-Biasing," IEEE Trans. Magn., vol. MAG-25, p.3692 (1989).
[10] J. Kane, H. Kanai, M. Kanamine, and K. Kobayashi, "A Study of Magnetic Properties in NiFe/FeMn Thin Films," Digest of the 15th Anual Conf. on Magnetics in Japan, p.51 (1991).
[11] T. Miyazaki, "Magnetoresistance of Alloy Films and Multilayers", J. of the Magn. Soc. Japan, vol. 16, p.615 (1992).
[12] H. Hamilton, R. Anderson, and K. Goodson, "Contact Perpendicular Recording on Rigid Media," IEEE Trans. on Magn., vol. MAG-27, p.4921 (1991).
[13] H. Wakamatsu, K. Kiuchi, M. Shinohara, and Y. Miura, "Effects of Spacing on Recording Characteristics of Perpendicular Double-Layer Disk," J. Magn. Soc. Jpn., vol. 15, Suppl. No. S1, p.875 (1991).
[14] M. Shinohara, H. Wakamatsu, Y. Miura, and K. Kiuchi, "Perpendicular Recording Media and Areal Density Characteristics of Gbit/in^2," Digests of The Mag. Rec. Conf. F5, Santa Clara (1992).
[15] S. Yamamoto, Y. Nakamura, and S. Iwasaki, "Extremely High Bit Density Recording with Single-Pole Perpendicular Head," IEEE Trans. Magn., vol. MAG-23, p.2070, (1987).
[16] Y. Maeda, and K.,Takei, "Compositional Separation in Co-Cr Based Alloy Films," J. Mag. Soc. Jpn., vol. 16, p.87 (1992).
[17] D. A. Patterson, G. Gibson, and R. H. Katz, "A Case for Redundant Arrays of Inexpensive Disks (RAID)," U. C. Berkely Report, No. UCB/CSD 87/391, Dec. (1987).

A TIME-RESOLVED SPLEEM STUDY OF MAGNETIC MICROSTRUCTURE IN ULTRATHIN Co FILMS ON W(110)

H. Pinkvos,[1] H. Poppa,[2] E. Bauer,[1] and G.-M. Kim[1]

[1] Physikalisches Institut, TU Clausthal, Leibnizstr. 4
D-3392 Clausthal-Zellerfeld, Germany

[2] IBM Research Division, Almaden Research Center
650 Harry Road, San Jose, CA 95120-6099

INTRODUCTION

Spin Polarized Low Energy Electron Microscopy (SPLEEM) is an extension of the well established Low Energy Electron Microscopy (LEEM) [1] technique with added sensitivity to surface magnetism due to the use of spin polarized electrons in the illuminating beam. SPLEEM uses parallel imaging of all image elements as LEEM does and, therefore, has the potential to study the time dependence of the magnetic surface structure. Because magnetic contrast is smaller than the usual diffraction, interference and topographic contrast and because of limitations in gun brightness and speed in data aquisition and storage, a time resolution corresponding to standard video rates as in LEEM is not possible at present. However, in this contribution we demonstrate that changes in the magnetic domain structure during growth and annealing of thin Co films can be easily resolved on times scales of seconds. This may be compared with other magnetic microscopies which provide similar resolution like special contrast modes in Scanning Transmission Electron Microscopy (STEM) [2] and, in addition, similar surface sensitivity like Scanning Electron Microscopy with Polarization Analysis (SEMPA) [3].

Thin Co films play an important role in the context of magnetic multilayers, which are often grown on noble metal substrates [4]. In systems in which the dependence on crystal orientation has been studied such as Co/Pd [5] and Co/Pt,[6] the films with the closest-packed plane parallel to the substrate generally show the largest out-of-plane anisotropy, which favours a magnetization direction perpendicular to the film plane. Co films can also be grown with the closest-packed plane parallel to the (110) surface of bcc crystals such as Mo [7] or W [8]. In contrast to the noble metal substrates mentioned, W and Mo have a larger surface free energy than Co, which leads to the formation of thermally stable monolayers and supports quasi monolayer-by-monolayer (quasi Frank-van der Merwe) growth at temperatures up to approximately 500 K. For this reason, epitaxial multilayers can be easily grown on these substrates, which makes them very suitable for systematic in-situ studies, in particular because the layers can be removed again by thermal desorption.

SPLEEM PRINCIPLES

Details of the instrument such as the gun for spin polarized electrons and the electrostatic illumination column have been described previously[9]. The magnetic images shown here have been obtained using the following procedure:

For a given spin polarization axis which is fixed to one in-plane direction in the present configuration a first image is digitized and integrated in real time with video rates using a special purpose image processing system. The integration times used in this work correspond to 16 video frames (0.64 sec) and 32 video frames (1.28 sec) per single image. The spin is then rotated by 180° and a second image is aquired. Pixelwise calculation of the normalized difference $R = (I_+ - I_-)/(I_+ + I_-)$ yields a laterally resolved map for the *total reflection asymmetry* R. R differs from the usually defined reflection asymmetry A by a factor P_0 which denotes the spin polarization of the incident beam, i.e. $R = P_0 \cdot A$. For display purposes the calculated R-values are mapped to the available grey level interval [0,255]. In this step also the black-white contrast may be adjusted for optimum visibility. The grey levels displayed in the final image correspond to the values of the dot product between the spin $\vec{S}$ and the magnetization $\vec{M}$. Maximum black and white contrast then occurs between regions, in which $\vec{M}$ and $\vec{S}$ are parallel and antiparallel. If the original images are in perfect registry with each other, the topographic details vanish completely upon subtraction. Because of image drift due to thermal motion of the sample holder during heating cycles, it has been frequently necessary in this work to shift one of the original images before subtraction.

EXPERIMENTAL PROCEDURE

Details of the preparation of the W(110) substrate crystal, the evaporation sources, the gauge and control of the evaporation rate and other experimental conditions have been described previously[10]. The standard LEEM and LEED modes of operation of the instrument are routinely used before and after each experiment to check the cleanliness of the sample, in particular to detect the formation of tungsten carbide or the adsorption of CO. The base pressure in the imaging chamber was $2 \cdot 10^{-10}$ Torr, during Co evaporation it rose to $2 \cdot 10^{-9}$ Torr.

GROWTH OF 3d ISLANDS AT 790 K

At T = 790 K Co grows on W(110) in the Stranski-Krastanov (SK) mode, i.e. all material in excess of one monolayer is incorporated into 3d islands. The growth process is documented in selected steps in Figure 1. Here the upper row of each panel shows the LEEM images taken with one spin orientation and the lower row shows the corresponding magnetic contrast. In the LEEM images the islands appear dark against the background. Thin curved lines represent step bunches as well as monoatomic steps. Some of the islands are bounded by black and white lines, which are Fresnel fringes arising from diffraction at the island boundary.

In order to avoid problems in tracing a particular feature during imaging, the crystal was allowed to thermalize for 60 min before the start of the deposition. In order to minimize the effect of CO adsorption on the clean W crystal during this phase, one monolayer of Co was deposited beforehand at room temperature. After thermal equilibrium was attained, Co was evaporated at a rate of 0.33 ML/min. SPLEEM images were taken approximately every minute and the entire sequence was recorded in parallel on video tape.

The presence of the monolayer in the regions between the islands was confirmed by LEED. Before the depositon of the second monolayer, the LEED pattern showed the (8×2) double scattering pattern characteristic for the distorted hexagonal monolayer[7,8]. After completion of the evaporation the LEED pattern had changed into the bright hexagonal pattern from the Co islands superimposed on the double scattering spots from the monolayer. This confirmed that no alloying had taken place, which causes respreading of the islands and changes in the LEED pattern [7].

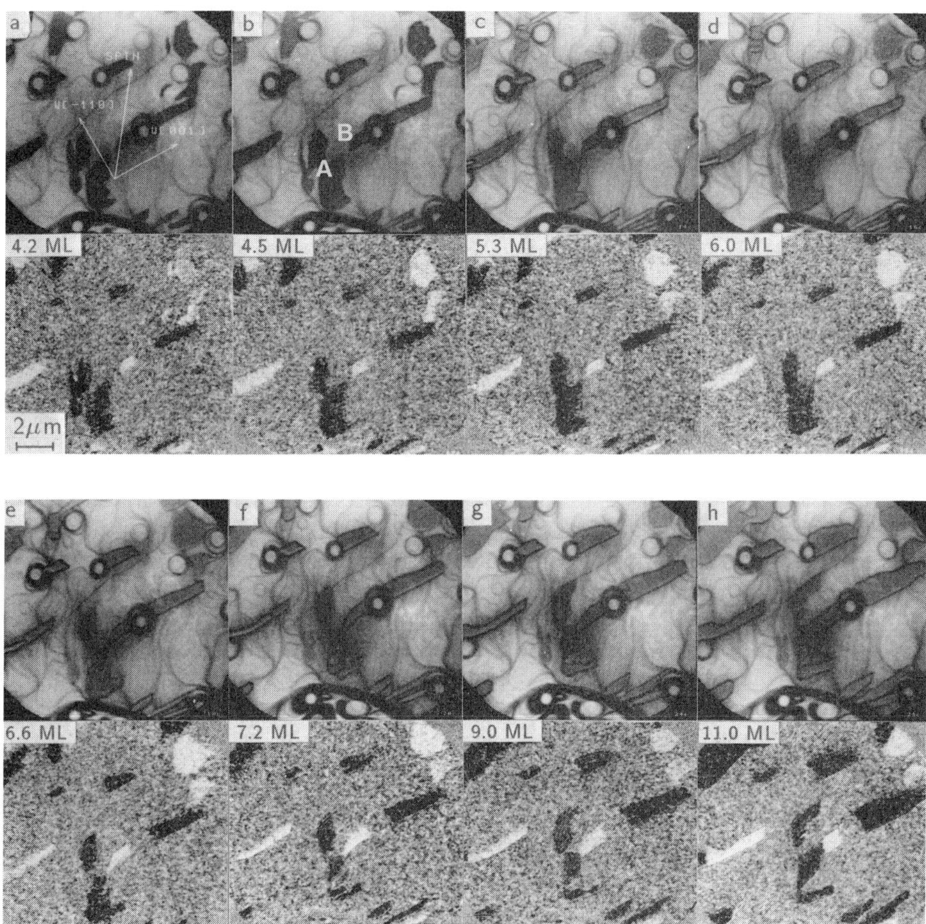

Figure 1. Time evolution of topographic (upper rows of each panel) and magnetic contrast (lower rows) imaged with SPLEEM during the evaporation of Co onto a W(110) surface at T = 790 K. All images were taken at 0.65 eV electron energy. The time elapsed between successive images corresponds to 0.75, 2.5, 2, 1.6, 2, 3.3, and 6 min respectively. Exposure times of 1.28 sec (32 video frames) were used per single topographic image. Thin white arrows in (a) indicate the main crystallographic directions of the W crystal and the spin polarization axis. The labels in the magnetic images refer to the average coverage inferred from the evaporation rate. The irregular dark line seen in all topographic images is not a surface feature but a crack in the image intensifier plates. For a detailed discussion see text.

Inspection of the LEEM images in Fig. 1 shows that the growth morphology is not simple. One type of islands nucleates at large circular W mesas which form on the substrate surface after extended use due to alloying during desorption of the layer [11] and grows continuously in a more regular shape elongated along the W[001] direction. Others, such as the island labeled by the letter A in Fig. 1b, are more irregularly shaped. Although their lateral extensions are in the range of several μm, they are not stable. Island A continuously changes its shape during observation, which is nicley documented in Fig. 1. One part even separates (Fig. 1b,c), the other then touches island B (Fig. 1c), which itself continues to grow on top of A (Fig. 1d-f). Obviously not all islands are of the same height, which may also be seen from the different Fresnel fringes.

The relative area covered by all islands increases from 11.9% in Fig. 1a to 21.4% in Fig. 1h which corresponds to an *average* island thickness of 28 ML (58 Å) and 48 ML (98 Å) respectively.

The magnetic contrast images show basically three distinct grey levels. The islands appear black and white, with asymmetry values of $R = \pm 1.5 \pm 0.3$ %. From the symmetry with respect to zero we can conclude that the magnetization points into opposite in-plane directions. It is interesting to note that one part of the two islands which grow together between Fig. 1a and Fig. 1b and form island A looses its magnetic contrast. The LEEM images show that during this rearrangement of the material this part becomes progressively thinner. At a certain crossover thickness the magnetization then switches into the out-of-plane direction and is no longer visible in Fig. 1. A second explanation involves the thickness dependence of the Curie temperature T_c. For Co on W(110) values for T_c have not been determined yet but to get a rough estimate, we note that $T_c = 434\ K$ for a monolayer of Co on Cu(111) [12], which also has hexagonal symmetry and very similar interatomic distances. Referring to the Curie temperature $T_c(\infty)$ of bulk Co, this gives $T_c(1ML) = 0.31T_c(\infty)$. This value compares well to a scaling law for the thickness dependence of T_c of densely packed Ni and Fe films, which is of the form $T_c(\infty) - T_c(D) \propto D^{-\lambda}$ [13]. Looking for the thickness D at which $T_c(D)$ corresponds to the deposition temperature, we arrive at a thickness of approximately 2.5 ML below which a Co film becomes nonmagnetic in our experiment. At the basis of this estimate also the monolayer regions between the islands are nonmagnetic. The medium grey level in these areas indeed corresponds to an asymmetry value of zero, which can be seen in the corners of the magnetic images.

The magnetization in the islands appears to be homogeneous except for those, in which domain boundaries have been created by growing-together of oppositely magnetized crystals. In particular no flux closure structures have been observed, which is very interesting since the islands in fact represent small magnetic particles with lateral dimensions between 0.5 and several μm and heights up to more than 100 Å. Depending on the actual lateral extensions such structures have been observed in small permalloy particles of comparable dimensions [14].

ANNEALING OF A 6.2 ML Co FILM TO 730 K

The observation of changes in the magnetic microstructure of a thin Co film *during* annealing turned out to be much more difficult than the growth experiment described in the previous chapter. Exposure times of 1 second or less had to be used here because of the appreciable thermal drift of the specimen. In order to increase the temperature, the heating current was incremented manually in discrete steps resulting in an average rate of change of approximately 7.5 K/min. The actual sample temperature was calculated from the measured response function of the sample. Therefore the absolute accuracy is not better than $\pm 30\ K$ but temperature differences are believed to be accurate to $\pm 10\ K$. Again SPLEEM images were taken roughly every minute and the process was continuously recorded on video tape. Figure 2 presents a subset of the time series in the same way the data were shown in Figure 1.

Flat Co films grown on refractory metals such as W or Mo at room temperature are in a metastable state. At higher temperatures the surface mobility of the Co atoms becomes sufficient to approach the equilibrium Stranski-Krastanov structure [7]. This agglomeration into 3d crystals is seen in the topographic images of Fig. 2. First deviations from a flat surface are noted at $T = 565\ K$ (Fig.2c), which compares well to the onset of agglomeration at $T \approx 500\ K$ in the system Co/Mo(110) [7]. With increasing temperature probably small holes develop (Fig. 2d), then the 3d crystals form a multiply connected network, which continuously evolves into single islands (Fig. 2e-h). Compared to the growth at 790 K, the islands are smaller and have more irregular shape.

In Fig. 2a the SPLEEM image displays a bright magnetic domain with triangular shape. The corresponding asymmetry value is $R = 9.5\%$ here, which is sufficient to make the domain visible already in the single spin LEEM image.

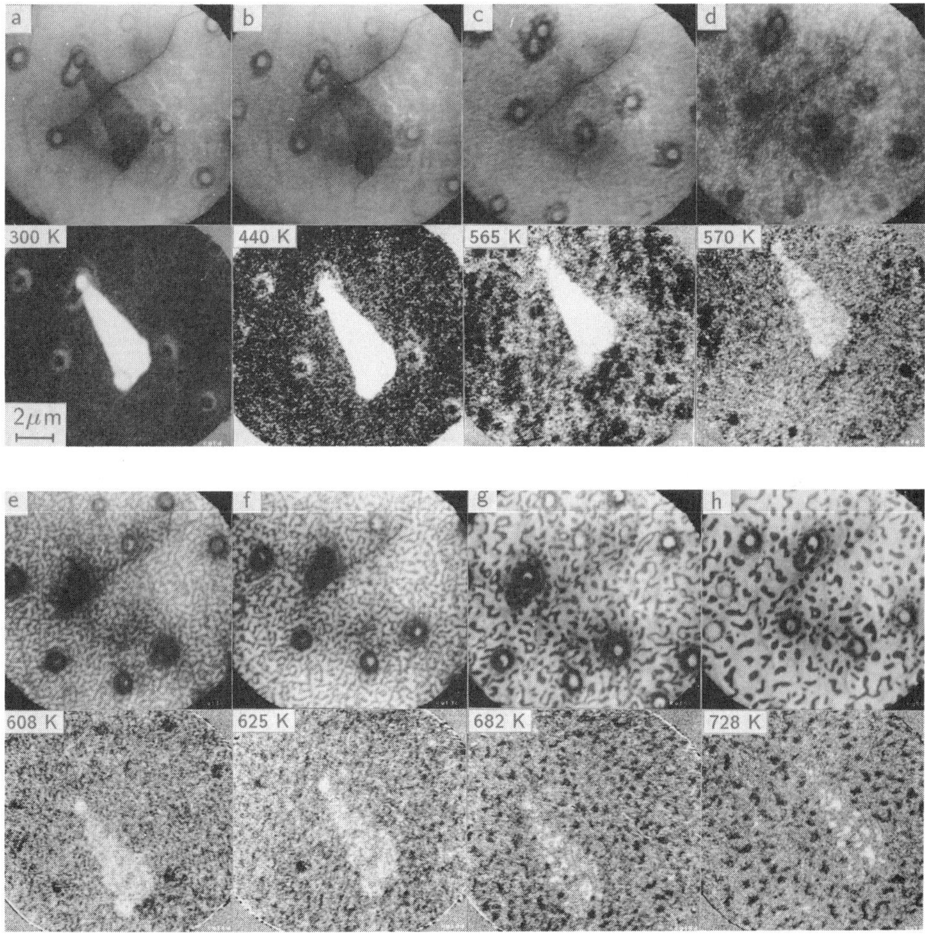

Figure 2. Agglomeration of an initially flat 6.2 ML Co film on W(110) into 3d crystals during annealing. Upper rows of the two panels show the time evolution of the surface topography, the lower rows show the corresponding magnetic contrast. The images were taken at 1.6 eV (a-d), 1.0 eV (e-g) and 0.8 eV (h) electron energy. Exposure times of 0.64 sec (16 video frames) were used per single topographic image and the time elapsed between successive images corresponds to 5.3, 8.2, 0.66, 5.3, 5.5, 15 and 10 min respectively. For a detailed discussion see text.

Thin Co films grown on W(110) at room temperature usually show very large domains with extensions of several 100μm [10]. To create domains visible in the field of view of 15 μm, an empirical preparation procedure was used here: the first monolayer was evaporated at room temperature. The heating filament was then driven by a small AC current corresponding to a maximum sample temperature of 450 K and the remaining 5.2 ML were evaporated. The recipe produced small domains which were in this case pinned by surface defects as seen in Fig. 2a. At room temperature (Fig. 2a) the measured asymmetry values are $R = +9.5\%$ for the triangular domain and $R = -5\%$ for the surrounding region, i.e. the magnetization does not point into opposite directions in this case. Up to $T = 440\ K$ the asymmetry remains constant. This is not apparent in the appearance of Figs. 2a and 2b, because the contrast has been enhanced in Fig. 2b to enable direct comparison to the following images. At $T = 565\ K$ (Fig. 2c), when first changes are seen in the single spin LEEM image, the magnetization in the outer regions has considerably rearranged into a very irregular structure, which shows no direct relation to the topography. The magnetic images become more granular at higher temperatures (Figs. 2d-f) but the network-like structure of the 3d crystals is not resolved. The islands finally also become magnetically visible (Figs. 2g-h). The asymmetry has decreased considerably to symmetric values of $R = \pm 2\%$ for the black and white patches and the regions between the islands display the zero grey level expected for the monolayer coverage. A certain decrease of the asymmetry can be expected for the islands because the magnitude of the magnetization decreases with increasing temperature. But since the change inside the triangular domain is almost twice as large than outside, it is clear that also the direction of $\vec{M}$ must have changed in at least one of the domains. In view of this fact it is very remarkable that the original domains are still visible in Fig. 2h, albeit coarsly grained, i.e. there is a memory effect for the domain shape.

SUMMARY

We have demonstrated that Spin Polarized Low Energy Electron Microscopy is well suited to study the evolution of magnetic surface structure with a time resolution of seconds. This resolution cannot be fully exploited at present because of software limitations. With improved software the time needed to write image data to the hard-disk of the processing system will be reduced soon to 4 - 5 seconds. Ultimate limits are set by the exposure time necessary to achieve a desired contrast. The exposure times used in the present experiment were of the order of one second. Because they depend sensitively on the gun brightness and the properties of the sample under investigation, improvements can be expected also here.

Simultaneous recording of nonmagnetic and magnetic contrast during evaporation showed that Co grows on W(110) in the Stranski-Krastanov mode at T = 790 K. The islands do not monotonically increase in size but instead considerable rearrangement is possible, including splitting, dissolution and changes in shape. Also very thin islands can coexist with thick ones. The regions between the islands are covered by a Co monolayer. Taking into account the thickness dependence of the Curie temperature, an estimate shows that the monolayer regions are nonmagnetic at $T = 790\ K$, which is in agreement with the observation of no in-plane contrast along the spin polarization axis probed in the experiment. At the final average coverage of 11 ML the islands then in fact represent small independent magnetic particles with lateral extensions in the range between 0.5 and 5 μm and heights up to more than 100 Å. The islands appear homogeneously magnetized in two probably opposite in-plane directions and in particular no flux closure structures have been found. The creation of domain boundaries was only observed to occur when islands with different magnetization directions grow together.

An annealing experiment for an initially flat 6.2 ML Co film on W(110) showed a remarkable memory effect for the shape of magnetic domains. Islands, which formed in a region originally belonging to the same magnetic domain, are still magnetized into a common in-plane direction, although this direction may have changed during the process of agglomeration.

ACKNOWLEDGEMENTS

We especially want to thank J. Hurst from the IBM Almaden Research Center for continuing advice and support regarding the spin polarized electron gun. We are also much indebted to N. Müller (Univ. Bielefeld) for the loan of a Pockels cell. This work was supported by the Deutsche Forschungsgemeinschaft, Bonn and by the IBM Corporation.

REFERENCES

1. E. Bauer, in: "Chemistry and Physics of Solid Surfaces **VIII**", R. Vanselow and R. Howe, ed., Springer, Berlin (1990), p. 267.
2. c.f. the contribution of J.N. Chapman in this volume and references therein.
3. c.f. the contributions of R. Allenspach and J. Unguris et al. in this volume and references therein.
4. The most recent examples for noble metal based Co multilayers can be found in various contributions to this volume.
5. B.N. Engel, C.D. England, R.A. van Leeuwen, M. Wiedmann and C.M. Falco, Phys. Rev. Lett. **67**:1910 (1991).
6. R.F.C. Farrow, C.H. Lee, R.F. Marks, G. Harp, M. Toney, T.A. Rabedeau, D. Weller and H. Brändle, this volume.
7. M. Tikhov and E. Bauer, Surf. Sci. **232**:73 (1990).
8. H. Knoppe and E. Bauer, to be published.
9. M.S. Altman, H. Pinkvos, J. Hurst, H. Poppa, G. Marx and E. Bauer, Mat. Res. Soc. Symp. Proc. **232**:125 (1990).
10. H. Pinkvos, H. Poppa, E. Bauer and J. Hurst, to be published in Ultramicroscopy.
11. G. Lilienkamp, Ph.D. Thesis, TU Clausthal 1990.
12. J. Kohlhepp, H.J. Elmers, S. Cordes and U. Gradmann, Phys. Rev. B**45**:12287 (1990).
13. U. Gradmann, in: "Handbook of Ferromagnetic Materials Vol. **7**", K.H.J. Buschow, ed., to be published.
14. J.N. Chapman, S. McVitie and I.R. McFayden, Scanning Microscopy Suppl. **1**, AMF O'Hare, Chicago (1987), p. 221.

MAGNETIC DOMAINS IN ULTRATHIN EPITAXIAL FILMS OBSERVED BY SPIN-POLARIZED SCANNING ELECTRON MICROSCOPY

R. Allenspach

IBM Research Division
Zurich Research Laboratory
8803 Rüschlikon, Switzerland

INTRODUCTION

The investigation of domain structures and domain walls is of major importance for the interpretation of numerous properties of ferromagnets, such as the balance between exchange, magnetostatic, and crystalline anisotropy energy, as well as for applications in magnetic technology. The experimental progress made in the past few years has made the growth of ultrathin magnetic layers down to a single monolayer possible.

Domain observations in these films are experimentally quite demanding and require novel techniques with high lateral resolution and high surface sensitivity. Spin-polarized scanning electron microscopy (spin-SEM[1] or SEMPA[2]) is particularly suited to these investigations owing to its ability to map the magnetization direction with a present lateral resolution of <40 nm and submonolayer surface sensitivity.[3]

This contribution will illustrate various aspects of ferromagnetism in epitaxial thin films which we have recently investigated by spin-SEM:

First the formation of domains in ultrathin layers is discussed. It is shown that out-of-plane and in-plane magnetized films behave quite differently, using Co/Au(111), Fe/Cu(100), and Co/Cu(100) films as examples.

The second topic deals with one of the fundamental relations in magnetism, the temperature dependence of the spontaneous magnetization. The model system we investigated was in-plane magnetized Co/Cu(100) films. By combining the magneto-optic Kerr effect and spin-SEM we followed the temperature dependence of remanent and spontaneous magnetization below and above the Curie temperature and find a behavior characteristic for a truly two-dimensional (2D) ferromagnetic system, i.e. very large thermal fluctuations.

The third part examines a further exciting aspect of thin film magnetism, the often large surface anisotropy which favors perpendicular magnetization. The competing term responsible for in-plane magnetization is the demagnetization energy. Since the thickness dependence of these two terms is different, the commonly observed behavior is a changeover from perpendicular to in-plane magnetization for increasing film thickness. With a spatially resolving technique like

Magnetism and Structure in Systems of Reduced Dimension
Edited by R.F.C. Farrow *et al.*, Plenum Press, New York, 1993

spin-SEM we are able to visualize this transition locally. In Co/Au(111) films we determine that the transition occurs smoothly as a rotation rather than by switching discretely. A different type of magnetization reorientation takes place in Fe/Cu(100). We find that the occurrence of domains at switching thickness is responsible for the complete disappearance of the remanent magnetization during the switching transition.

EXPERIMENTAL SETUP

Domain imaging with spin-polarized electrons has been described[1,2] in detail, so we only concentrate on the most relevant points here. A commercial scanning electron microscope modified to ultrahigh vacuum requirements is used to produce a finely focused beam of electrons which scans along the ferromagnetic sample. A huge amount of low energetic secondary electrons is ejected, collected and transferred through an energy analyzer into a spin detector, which is capable of analyzing the spin polarization $P = (N_\uparrow - N_\downarrow)/(N_\uparrow + N_\downarrow)$, where $N_\uparrow$ ($N_\downarrow$) is the number of electrons with spin parallel (antiparallel) to a given direction. The success of this spin-polarized SEM relies on the facts that the yield of low energetic (0 to 20 eV) secondary electrons is high, and that the spin polarization for these electrons is proportional to the magnetization within the topmost three to five monolayers (ML) of a surface.[4,5] Hence this is a very efficient technique to study ultrathin magnetic films of several ML thickness. By measuring the direction of P, we directly image the magnetization distribution in these thin films.

In our setup the electron microscope is equipped with a cold field emission gun of variable energy from 0.5 to 25 keV, with a minimum beam diameter of 5 nm. The typical operating energy is 2 to 5 keV, with a beam current below 1 nA. The spin detector, a high-energy (100 kV) Mott type, is capable of measuring the out-of-plane and one in-plane polarization direction simultaneously. This feature, obtained by a particular geometric arrangement of energy analyzer, spin detection axes and sample normal, is especially useful for the studies of magnetization direction reorientation. Most studies of magnetism in ultrathin films require the ability to vary temperature for detailed characterization. Our system is equipped with a home-built sample stage where the samples can be cooled and/or annealed from $T = 15$ K up to $T = 450$ K while preserving the high resolution capability (< 100 nm).

The ultrathin films are grown onto the single crystal substrates by molecular beam epitaxy from effusion cells with integrated temperature control and water cooling. Growth rates are typically 0.1 to 0.2 ML/min at a pressure of $< 2 \times 10^{-10}$ mbar. Chemical and structural characterization of the films is performed by Auger electron spectroscopy and low-energy electron diffraction (LEED). The polar and transverse magneto-optic Kerr effect is used for overall magnetic characterization of the films.

DOMAINS IN ULTRATHIN FILMS

The first topic of interest in ultrathin epitaxial films is whether domains occur at all, in particular since they have been predicted to be forbidden[6] or to occur only under very special conditions.[7] Recently we have shown that micron-sized domains exist in Co/Au(111).[8] Co/Cu(100) films, on the other hand, grow single domain.[9]

Epitaxial thin cobalt layers grown on (111)-oriented face-centered cubic (fcc) crystals display a striking similarity regardless of the particular substrate, be it Au, Cu, Pd, or Pt. Below a certain thickness they are magnetized perpendicular to the surface. Various models exist to account for this uniaxial surface anisotropy, e.g. a model based on Néel's crystalline surface anisotropy,[10] one where alloying at the interface occurs,[11] or one where strain is induced by the lattice mismatch between substrate and film.[12] In passing we note that the intrinsic crystalline bulk anisotropy of hexagonal Co also contributes to an easy axis being oriented perpendicular to the surface. However, it is too small to account for the the out-of-plane magnetization without additional anisotropy terms.

As an example of this class of materials we investigated thin Co films grown on Au(111). The out-of-plane magnetization direction has been established previously by hysteresis loop experiments.[13] With a spatially resolving technique like spin-SEM we are able to verify this perpendicular easy magnetization axis directly without applying external magnetic field by domain imaging. Fig. 1 depicts the magnetic domain image for a 3 ML Co/Au(111) film grown in a field-free configuration ($H < 1$ A/m). We find irregular shaped domains of 2 to 5 μm in size with magnetization vectors along the surface normal. The simultaneously measured in-plane magnetization vanishes completely. The fact that the domains are fully out-of-plane proves that the uniaxial anisotropy is large enough that no tilted magnetization direction occurs.[13,14] The most remarkable finding in Fig. 1 is not this presence of a vertical magnetization, but its decay into a domain structure. In fact, classical theories[6] predict a single domain for thin films, because the saving in magnetostatic energy gained by the creation of up and down domains is not sufficient to compensate for the energy required to form domain walls, the former being proportional to the film thickness d, the latter to $\sqrt{d}$. These simple arguments, however, obviously do not hold for out-of-plane magnetized Co/Au(111) films. The reason for the occurrence of these domains is not clear. We note, however, that the observation of domains is consistent with the finding that the very thin Co/Au(111) films do not exhibit square hysteresis loops but rather have reduced remanence.[15] The particularities of the growth of Co/Au(111) might be partly responsible for the small size of these domains. A recent STM study on the initial growth of Co/Au(111) showed bilayer island growth for the first two layers owing to the large lattice mismatch.[16] This structure changes when the film is annealed, which is reflected also in the domain pattern.

5 μm

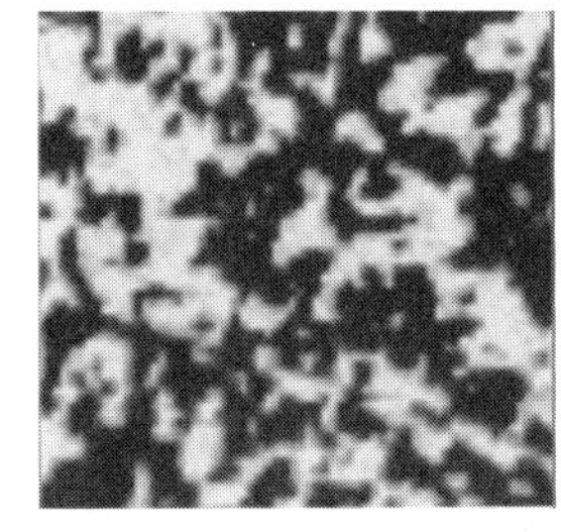

Figure 1. Magnetic domain image of a 3 ML Co/Au(111) film showing out-of-plane domains with magnetization up (white) and down (black).

5 μm

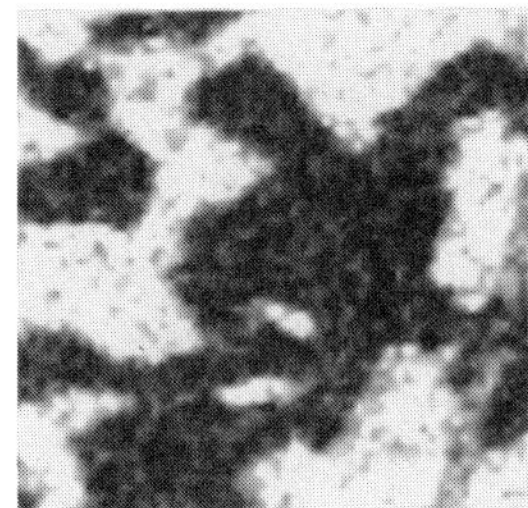

Figure 2. Magnetic domain image of a 3 ML Co/Au(111) film after annealing to $T = 620$ K showing out-of-plane domains with magnetization up (white) and down (black).

As an example, Fig. 2 displays a 3 ML Co/Au(111) film annealed to 620 K before a domain image is taken at room temperature. The domain size increases to 10 - 20 μm. In no case, however, were we able to establish a singe domain film at zero applied field.

A completely different situation occurs in in-plane magnetized films such as Co/Cu(100), again grown in zero field. These films are single domain on essentially the entire sample area of several millimeters.[9] Only at the sample edge do closure domains form.

To check the influence of the substrate on the occurrence of domains we fabricate a perpendicular film on the identical Cu(100) single-crystal, 3 ML fcc-Fe/Cu(100) grown at $T = 90$ K. Fig. 3 shows that extended defects of the substrate surface visible in topography pin thin reversed domains. We note that the typical domain size is much larger than that in Co/Au(111). Comparing Fe/Cu(100) to Co/Cu(100) we find that irregularities in the substrate are sufficient to produce pinning centers for the out-of-plane reversed domains but not for in-plane domains. We therefore conclude from our experiments that perpendicularly magnetized films generally have a strong tendency to form domains with irregular shape, whereas in-plane magnetized films are single domain even if grown in zero field. Hence we believe that the criterion for the formation of domains in perpendicularly magnetized films[7] is too restrictive to account for our observations in completely different systems like Co/Au(111) or Fe/Cu(100).

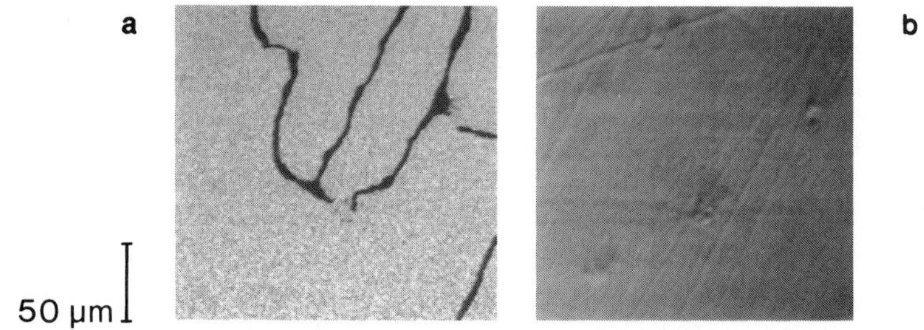

Figure 3. (a) Magnetic domain images of a 3 ML fcc-Fe/Cu(100) film displaying mainly large out-of-plane domains with magnetization up (white). (b) The corresponding topography shows substrate defects where the reversed black domains are preferentially pinned.

TWO-DIMENSIONAL MAGNETISM BELOW AND ABOVE T_C

One of the fundamental relations in magnetism is the temperature dependence of spontaneous magnetization, in particular since spontaneous magnetization as the order parameter is the quantity that characterizes the phase transition at Curie temperature T_C. A wealth of investigations exists concerning the nature of the phase transition in ultrathin epitaxial films.[17-20] Usually, however, spatially averaging techniques have been applied to determine the critical exponents and T_C. This is only justified as long as the domain configuration and hysteresis loops do not change with temperature. We combined spin-SEM and the magneto-optic Kerr effect to establish the correspondence between *microscopic* and *macroscopic* quantities, i.e. spontaneous magnetization and remanent magnetization. As a model system we investigated 1 ML Co/Cu(100) films magnetized in-plane.[21] These films grow nearly layer-by-layer, as has been verified by a recent detailed STM study.[22] We followed the temperature dependence of these quantities between 0.4 and 1.2 T/T_C by measuring hysteresis loops with the magneto-optic Kerr effect and domain images with spin-SEM. From Kerr hysteresis loops we determine the remanent magnetization $M_R \equiv M(H = 0)$, and the "saturation" magnetization $M_H \equiv M(H = 8 \text{ kA/m})$, see Fig. 4a. As a function of temperature we observe the following remarkable features. It is the *remanent* magnetization that vanishes sharply at a well-defined temperature that we tentatively identify with the Curie temperature T_C of the film. At temperatures below 0.9 T/T_C, M_R and M_H coincide, and the hysteresis loops are square. On the other hand, for $T > 0.9T_C$, the square loops deteriorate, and $M_R < M_H$. In particular M_H attains sizable values up to $T > 1.1T_C$. This observation that a large magnetic response is induced above T_C with very moderate fields of the order of several kA/m is an unfamiliar observation in 3D magnetism, except if superparamagnetic relaxation is present. We should then identify our T_C as a blocking temperature T_B, and the Curie temperature itself would be higher.

We emphasize that it is quite important to identify the Curie temperature exactly, and to pin down the influence of an external field. Numerous studies determine critical exponents and Curie temperatures, and define T_C either as the vanishing of the remanence[18] or the saturation.[20,23] From a theoretical point of

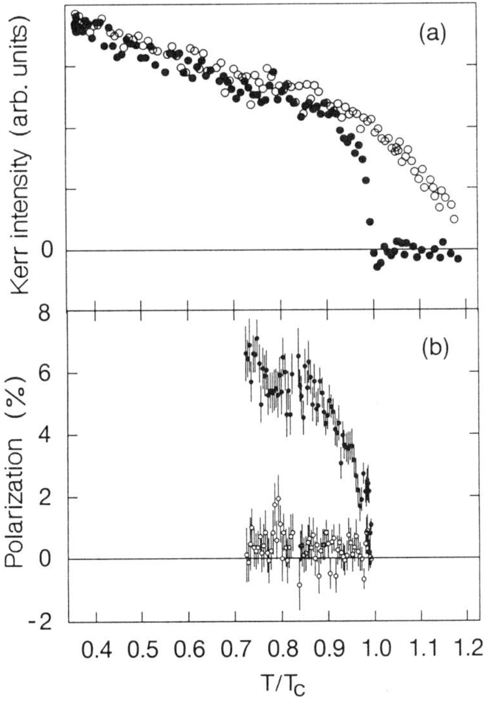

Figure 4. (a) Temperature dependence of the Kerr intensity for a Co/Cu(100) film at an applied field of $H = 0$ (M_R, solid circles) and in an applied field of $H = 8$ kA/m (M_H, open circles). (b) Temperature dependence of the spin polarization (or spontaneous magnetization M_S) deduced from scans across oppositely magnetized domains. No field has been applied. Solid circles: in-plane magnetization; open circles: out-of-plane magnetization. Note the perfect agreement with M_R as determined from the Kerr effect.

view it is clear that the *spontaneous* magnetization M_S is the order parameter for the phase transition. How the *remanent* magnetization behaves on approaching T_C is much less well understood.

We therefore set out to measure the temperature dependence of the spin polarization P with spin-SEM. We make sure that at each temperature we determine P within one domain, which by definition is M_S, apart from an irrelevant proportionality constant. The results are displayed in Fig. 4b. We find complete agreement between $M_R(T)$ and $M_S(T)$, i.e. between the macroscopic and microscopic quantities. Moreover, we observe no change in domain pattern up to 1 K below T_C,[21] nor a relaxation on the time scale in this experiment of several minutes. We therefore rule out superparamagnetism as a cause for the vanishing of M_R, and indeed identify T_C as the Curie temperature of the system. Therefore we interpret the huge response M_H of the magnetization to a moderate applied field above T_C with an intrinsic property of a truly 2D ferromagnet: large thermal fluctuations of spin blocks occur.[24] These blocks easily orient along a small field. Note that the physical concept is reminiscent of superparamagnetism, but a subtle difference exists. The spin blocks – in contrast to superparamagnetic grains – are not pinned at certain defects on the sample, but are allowed to move freely.

We expect our observations in this Co/Cu(100) ML film to be a common phenomenon in 2D ferromagnets with small anisotropy. Indeed experimental evidence exists that at least Fe/W(100),[25] Fe/W(110),[26] and Fe/Ag(100)[27] are similar. We assume in particular that the interpretation of the results for Co/Cu(111)[20] needs reexamination in view of the equivalence of remanence and spontaneous magnetization. Moreover it was recently brought to our attention that the *nuclear* magnet He on graphite[28] behaves quite analogously and is a perfect realization of a 2D magnetic system, showing once more Nature's universality among seemingly completely different fields of physics.

MAGNETIZATION DIRECTION SWITCHING

At a fixed temperature any film magnetized normal to the surface will change its magnetization direction at a certain thickness d_s owing to the predominance of shape anisotropy over crystalline surface anisotropy. Likewise, magnetization reorientation with temperature can occur for a fixed film thickness at a "switching" temperature T_s because of the different temperature dependence of shape and surface anisotropy. These two cases are completely analogous, and the relevant physics is the same. Nevertheless a difference exists: whereas the reorientation with the film thickness occurs quite generally in any system, the reorientation with increasing temperature takes place only if $T_s < T_C$.[29] Experimentally both cases have been realized. An experimentally easily accessible quantity for monitoring the transition from perpendicular to in-plane magnetization is the remanent magnetization $\mathbf{M}_R$ measured along all three orthogonal directions after applying field pulses. The first case, $\mathbf{M}_R(T = \text{const}, d)$, has been measured, for instance, for Co/Au(111)[13] and Fe/Cu(100).[30] The second case, $\mathbf{M}_R(T, d = \text{const})$, is realized in 5 to 6 ML Fe/Cu(100).[31] Quite astonishing is that these two systems show completely different behavior. For Co/Au(111), the perpendicular remanent magnetization decays smoothly near switching thickness d_s, and the in-plane remanent magnetization sets in correspondingly, thus compensating the reduction of the perpendicular component. For Fe/Cu(100), however, the perpendicular remanence decays and vanishes completely before the in-plane remanence sets in at higher thickness. Moreover, it is remarkable that at large thicknesses up to 10 ML the remanence does not recover to the value expected from an extrapolation of the thin perpendicular films. The exact mechanism of such a transition is not known *a priori*. This loss of remanent magnetization along all directions might be related to a structural phase transition at $d \simeq 6$ ML observed in LEED and EXAFS.[32] The LEED pattern changes from 4×1, 5×1 to 2×1, and the whole film structure is relaxed from distorted tetragonal to a more complicated structure. Another likely explanation would be a decay into domains in the transition region which thus reduces remanence. Or, since at d_s surface and shape anisotropy cancel (at least the short-ranged part), the symmetry arguments of Ref. 33 which predict a loss of long-range ferromagnetic order for isotropic systems might become more relevant.

With our spin-SEM we performed experiments in these two systems, hcp-Co/Au(111) and fcc-Fe/Cu(100), to monitor the magnetization direction switching locally. In the case of Co/Au(111) we are interested in whether this transition is a smooth rotation or has two discrete states, vertical and in-plane. In the case of Fe/Cu(100) we want to know whether the loss of remanence is accompanied by a loss of spontaneous magnetization or caused by the formation of domains.

We first summarize the Co/Au(111) results.[8] The experiment consists of evaporating a thin film whose thickness is subsequently increased in well-defined intervals. At each thickness, the domain image of the identical position on the sample has been probed. The results of such an experiment are displayed in Fig. 5 for a selected thickness range of 3 ML $\leq d \leq$ 6 ML, each with the out-of-plane and one in-plane component. Below 2 ML we find no magnetization at room temperature, since $T_C < 300$ K. At 3 ML, an out-of-plane domain pattern similar to that in Fig. 1 is observed. One additional layer is sufficient to let the small domains coalesce into larger domains, and some faint contrast is already visible in-plane. The crossover thickness d_s is reached by evaporating an additional 0.5 ML. Another 0.5 ML forces the film to switch essentially in-plane. For

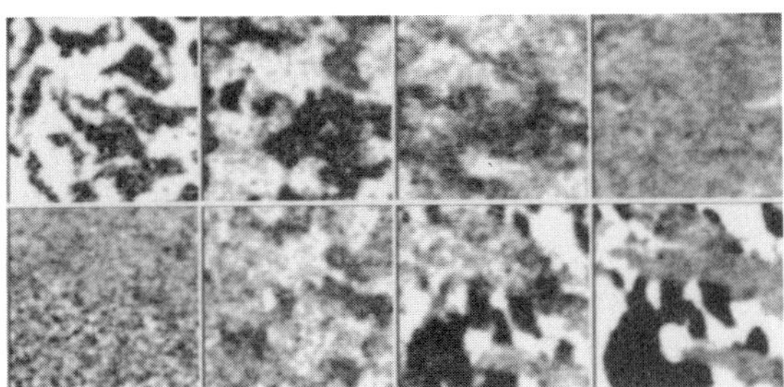

Figure 5. Magnetic domain images for Co/Au(111) thin films showing the evolution of domain size and switching behavior vs. film thickness. The pictures have been taken at identical positions on the sample between evaporation of additional layers. For each thickness, the upper image gives the out-of-plane and the lower the in-plane magnetization component (along the length of the page). The thickness from left to right is 3 ML, 4 ML, 4.5 ML and 6 ML. Gray scales from black to white indicate the magnitude of the magnetization component along the measured axis; scan area 20 μm $\times$ 20 μm.

$d \geq 6$ ML, no perpendicular component is left within the statistical error of the experiment.

From Fig. 5 the angle θ between surface normal and magnetization has been determined for each individual point. We find a *continuous* rotation from out-of-plane to in-plane magnetization in a finite thickness interval of 2 ML. In particular, a tilted magnetization direction is observed at a certain film thickness. Detailed results are given in Ref. 8. Briefly, the tilted configuration can be understood from a simple energy consideration. It is caused by a different angular dependence of the demagnetization energy and second hexagonal anisotropy constant K_2. The minima of the total energy $E = \mu_0/2 \times M^2 \cos^2\theta + K_1 \sin^2\theta + K_2 \sin^4\theta$, with M as the magnetization and K_1, K_2 as the uniaxial anisotropy constants, determine the equilibrium values for θ. Depending on the actual numbers, the solutions are $\theta = 0°$, $\theta = 90°$, and $\sin^2\theta = (\mu_0/2 \times M^2 - K_1)/2K_2$. It is this third solution that leads to a tilted magnetization direction in some thickness interval. The thickness dependence in this energy consideration is hidden in K_1 which may be split into bulk and surface contributions, $K_1 = K_{1B} + K_S/d$. From our data, we can then determine the uniaxial surface or interface anisotropy to be $K_S = 0.62 \pm 0.05$ mJ/m.

We emphasize that the phenomenological quantity K_S cannot give insight into the physical origin of the uniaxial anisotropy. We argue that strain might be responsible for the out-of-plane anisotropy. In fact, comparing the bulk lattice constants of Co and Au, the mismatch is quite large, $\simeq$14%, which leads to strained films and misfit dislocations.[34] In particular we do not think that alloying at the interface[11] plays a role in Co/Au films, since Co and Au are virtually immiscible. Since a large perpendicular surface anisotropy is quite a common phenomenon in Co films grown on fcc(111) substrates, we are tempted to expect a common cause for such a magnetic behavior, most likely strain. Experiments with unstrained hexagonal Co films would be highly desirable to resolve this ambivalence.

It might be instructive to list once more the essential ingredients that lead to this continuous magnetization rotation in a finite thickness interval. The balance between magnetization and the first anisotropy constant K_1 determines the switching thickness d_s because of the competition between demagnetization energy and anisotropy energy. The existence of a finite K_2 with a different angular dependence than M and K_1, on the other hand, determines the finite transition width responsible for the continuous rotation. To our knowledge, this is one of the first examples to show clearly the importance of the second-order anisotropy constant, a constant that can usually be neglected in most calculations and experiments.

In the following we turn to the investigation of the magnetization direction reorientation in fcc-Fe/Cu(100). For this purpose we have grown an Fe wedge on top of the Cu(100) single crystal. Wedge type samples became fashionable a year ago[35,36] after 30 years of obscurity.[37] The appealing feature is that with a spatially resolving technique like spin-SEM we are able to map out a complete vectorial thickness dependence $\mathbf{M}(T = \text{const}, d)$ in one experiment on a wedge-type sample. The results are shown in Fig. 6 for a $0 \leq d \leq 7$ ML fcc-Fe/Cu(100) sample grown at $T = 90$ K and subsequently annealed to 300 K. The actual imaging experiment takes place at $T = 200$ K. For $d < 2$ ML, no magnetization is found, neither perpendicular nor parallel to the sample surface, hence $T_C < 200$ K. Within 2 ML $< d < 6$ ML, perpendicular magnetization prevails. Above 6 ML, in-plane

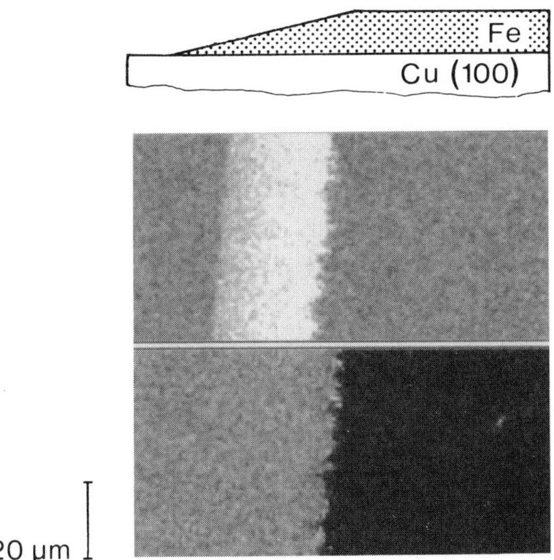

Figure 6. Magnetic domain images of an Fe wedge of $0 \leq d \leq 7$ ML atop the Cu(100) substrate grown at $T = 90$ K and measured at $T = 200$ K. A side view of the position of the wedge relative to the images is shown on top. Upper panel: magnetization component perpendicular to surface; lower panel: magnetization component parallel to surface along the [001] direction. Gray scale varies from spin polarization $P = -30\%$ (black) to $P = +30\%$ (white). Note the nonmagnetic range for thickness $d < 2$ ML, the perpendicular magnetization within 2 ML $< d <$ 6 ML, and the in-plane magnetization for $d > 6$ ML. Small in-plane domains form at crossover thickness.

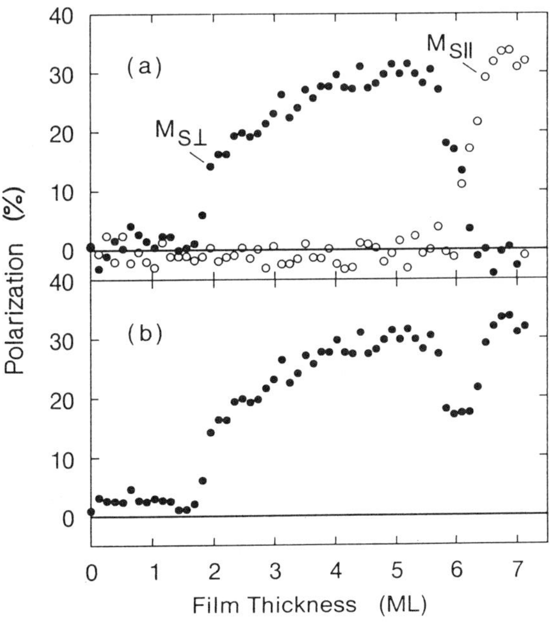

Figure 7. (a) Line scan across domain image of wedge sample in Fig. 6 displaying the perpendicular ($M_{S\perp}$, solid circles) and parallel ($M_{S\parallel}$, open circles) spin polarization or spontaneous magnetization vs. Fe film thickness, and (b) the calculated total spontaneous magnetization $M_S = [(M_{S\perp})^2 + (M_{S\parallel})^2]^{1/2}$ vs. film thickness, as deduced from (a). Note that the spontaneous magnetization increases exponentially with Fe film thickness as expected for a constant magnetic moment.

magnetization is found. Fig. 6 shows that the in-plane part of the film is magnetized along the [001] direction and forms small reversed domains in the transition region from perpendicular to in-plane magnetization. Note that the out-of-plane magnetized part of the wedge has been driven to a single domain by applying a small field pulse of $H < 4$ kA/m to facilitate the following analysis. A line scan along the wedge now gives $\mathbf{M}(d)$ directly. Fig. 7 displays both vertical and parallel magnetization components along the [001] direction, $M_{S\perp}$ and $M_{S\parallel}$, as well as the calculated magnitude of the spontaneous magnetization $M_S = [(M_{S\perp})^2 + (M_{S\parallel})^2]^{1/2}$. We clearly find no thickness range where both $M_{S\perp}$ and $M_{S\parallel}$ vanish. During the decay of $M_{S\perp}$, $M_{S\parallel}$ already sets in. Moreover the total magnetization is not reduced for the in-plane thickness range. On the other hand we see a sharp drop at crossover thickness. We point out, however, that the distance between two points, although only 0.2 ML apart, translates to $\simeq 2$ μm lateral distance on the sample surface. This is on the scale where irregularly shaped reversed in-plane domains occur at crossover thickness. The apparent reduction in M_S therefore originates from the average over this length scale and not from a truly reduced magnetization. We have verified this argument by acquiring higher resolution images in this transition region. We therefore conclude that the previously observed reduction in remanent magnetization is not caused by a diminished local magnetization (and hence a reduced magnetic moment). Instead, it displays the inability to establish a square hysteresis loop with a moderate field pulse. Thus, this magnetization reorientation experiment in Fe/Cu(100) illustrates once again that the macroscopically averaging remanent magnetization might be misleading because of the existence of domains, even in ultrathin films.

CONCLUSIONS

In this contribution, I tried to illustrate the rich variety of phenomena in thin film magnetism, where a spatially resolving technique leads to new insights. Emphasis was placed on giving illustrative examples rather than on a systematic overview of all studies existing in this field. I discussed the existence and formation of domains in ultrathin films as well as the different behavior of out-of-plane and in-plane magnetized films. The second part examined the temperature dependence of the spontaneous and remanent magnetization below and above T_C and the conclusion was drawn that the ultrathin Co/Cu(100) films behave like a model 2D Heisenberg system above T_C. The third topic dealt with the magnetization reorientation from perpendicular into plane, brought about either by a change in film thickness or by an increase of temperature. I showed that different systems behave completely differently with regard to the nature of this magnetization direction switching, and illustrate this finding with Co/Au(111) and Fe/Cu(100). All these different physical aspects show that magnetic domains play a crucial role in thin film magnetism: A single domain state, although anticipated for decades in thin films, is by no means the preferred ground state. The high spatial resolution and monolayer sensitivity of spin-SEM is therefore a very useful tool to elicit some new insight from 2D ferromagnetism.

ACKNOWLEDGMENT

Since the start of this domain imaging project several years ago, I have profited from the invaluable help of many persons, too numerous to list here. Representative of them is my technician A. Bischof, whom I take pleasure in giving special thanks for his major contributions in designing, constructing and operating the equipment.

REFERENCES

1. K. Koike, H. Matsuyama, and K. Hayakawa, *Scanning Microsc. Suppl.* 1:241 (1987).
2. G.G. Hembree, J. Unguris, R. J. Celotta, and D.T. Pierce, *Scanning Microsc. Suppl.* 1:229 (1987).
3. R. Allenspach and M. Stampanoni, *Mat. Res. Soc. Proc.* 231:17 (1992).
4. M. Taborelli, Ph.D. thesis, ETH No. 8545, Swiss Federal Institute of Technology (ETH), Zurich, Switzerland (1988).
5. D.P. Pappas, K.-P. Kämper, B.P. Miller, H. Hopster, D.E. Fowler, C.R. Brundle, A.C. Luntz, and Z.-X. Shen, *Phys. Rev. Lett.* 66:504 (1991).
6. C. Kittel, *Phys. Rev.* 70:965 (1946).
7. Y. Yafet and E.M. Gyorgy, *Phys. Rev. B* 38:9145 (1988).
8. R. Allenspach, M. Stampanoni, and A. Bischof, *Phys. Rev. Lett.* 65:3344 (1990).
9. H.P. Oepen, M. Benning, H. Ibach, C.M. Schneider, and J. Kirschner, *J. Magn. Magn. Mater.* 86:L137 (1990).
10. L. Néel, *J. Phys. Rad.* 15:376 (1954).
11. R.F.C. Farrow, C.H. Lee, R.F. Marks, G. Harp, M. Toney, T.A. Rabedeau, D. Weller, and H. Brändle, this volume.
12. G.F. Dionne, *Mat. Res. Bull.* 6:805 (1971).
13. C. Chappert, D. Renard, P. Beauvillain, J.P. Renard, and J. Seiden, *J. Magn. Magn. Mater.* 54:795 (1986).
14. G. Bayreuther, P. Bruno, G. Lugert, and C. Turtur, *Phys. Rev. B* 40:7399 (1989).
15. J. Ferré, G. Pénissard, C. Marlière, D. Renard, P. Beauvillain, and J.P. Renard, *Appl. Phys. Lett.* 56:1588 (1990).
16. B. Voigtländer, G. Meyer, and N.M. Amer, *Phys. Rev. B* 44:10354 (1991).
17. W. Dürr, M. Taborelli, O. Paul, R. Germar, W. Gudat, D. Pescia, and M. Landolt, *Phys. Rev. Lett.* 62:206 (1989).
18. C. Liu and S.D. Bader, *J. Appl. Phys.* 67:5758 (1990).
19. U. Stetter, M. Farle, K. Baberschke, and W.G. Clark, *Phys. Rev. B* 45:503 (1992).
20. J. Kohlhepp, H.J. Elmers, S. Cordes, and U. Gradmann, *Phys. Rev. B* 45:12287 (1992).
21. D. Kerkmann, D. Pescia, and R. Allenspach, *Phys. Rev. Lett.* 68:686 (1992).
22. C.M. Schneider, A.K. Schmid, H.P. Oepen, and J. Kirscher, this volume.
23. D. Pescia, G. Zampieri, M. Stampanoni, G.L. Bona, R.F. Willis, and F. Meier, *Phys. Rev. Lett.* 58:933 (1987).
24. V.L. Pokrovsky, *Advances in Physics* 28:595 (1979).

25. G.A. Mulhollan, R.L. Fink, J.L. Erskine, and G.K. Walters, *Phys. Rev. B* 43:13645 (1991).
26. U. Gradmann, M. Przybylski, H.J. Elmers, and G. Liu, *Appl. Phys. A* 49:563 (1989).
27. Z.Q. Qiu, S.H. Mayer, C.J. Gutierrez, H. Tang, and J.C. Walker, *Phys. Rev. Lett.* 63:1649 (1989).
28. H. Godfrin, R.R. Ruel, and D.D. Osheroff, *Phys. Rev. Lett.* 60:305 (1988); L.J. Friedman, A.L. Thomson, C.M. Gould, H.M. Bozler, P.B. Weichman, and M.C. Cross, *Phys. Rev. Lett.* 62:1635 (1989).
29. D. Pescia and V.L. Pokrovsky, *Phys. Rev. Lett.* 65:2599 (1990).
30. D.P. Pappas, C.R. Brundle, and H. Hopster, *Phys. Rev. B* 45:8169 (1992).
31. D.P. Pappas, K.-P. Kämper, and H. Hopster, *Phys. Rev. Lett.* 64:3179 (1990); D.P. Pappas, K.-P. Kämper, H. Hopster, D.E. Fowler, A.C. Luntz, C.R. Brundle, and Z.-X. Shen, *J. Appl. Phys.* 69:5209 (1991).
32. H. Magnan, D. Chandesris, B. Villette, O. Heckmann, and J. Lecante, *Phys. Rev. Lett.* 67:859 (1991).
33. N.D. Mermin and H. Wagner, *Phys. Rev. Lett.* 17:1133 (1966).
34. C. Chappert and P. Bruno, *J. Appl. Phys.* 64:5736 (1988).
35. P. Grünberg, S. Demokritov, A. Fuss, and J.A. Wolf, *J. Appl. Phys.* 69:4789 (1991).
36. J. Unguris, R.C. Celotta, and D.T. Pierce, *Phys. Rev. Lett.* 67:140 (1991).
37. S. Methfessel, S. Middelhoek, and H. Thomas, *J. Appl. Phys.* 31:302S (1960).

MAGNETIC-SENSITIVE SCANNING PROBE MICROSCOPY

R. Wiesendanger

University of Basel
Dept. of Physics
Klingelbergstrasse 82
CH-4056 Basel
Switzerland

INTRODUCTION

To address the important issue of the relationship between structural and magnetic properties in systems of reduced dimension, high resolution topographic and magnetic imaging techniques are required. Magnetic imaging can be performed by using a variety of experimental methods, including the Bitter technique, Kerr microscopy, electron-microscope based techniques such as Lorentz microscopy and electron holography, spin-polarized electron techniques such as scanning electron microscopy with polarization analysis (SEMPA) and spin-polarized low-energy electron microscopy (SPLEEM), and finally magnetic-sensitive scanning probe microscopies. These methods can be distinguished according to the magnetic property probed (e.g. the magnetic stray field or the sample magnetization), and the spatial resolution achieved (Fig. 1). In the following, we will first present an overview of the various novel magnetic-sensitive scanning probe methods before we will particularly focus on the spin-polarized scanning tunneling microscopy (SPSTM) technique which we have developed at the University of Basel since 1989.

MAGNETIC-SENSITIVE SCANNING PROBE METHODS

The development of scanning tunneling microscopy (STM) has triggered the invention of several different STM-related scanning probe methods, some having magnetic sensitivity. These magnetic-sensitive scanning probe techniques have already found various fields of applications. A classification of magnetic-sensitive scanning probe microscopies can be performed according to the method which is used to probe a magnetic property.

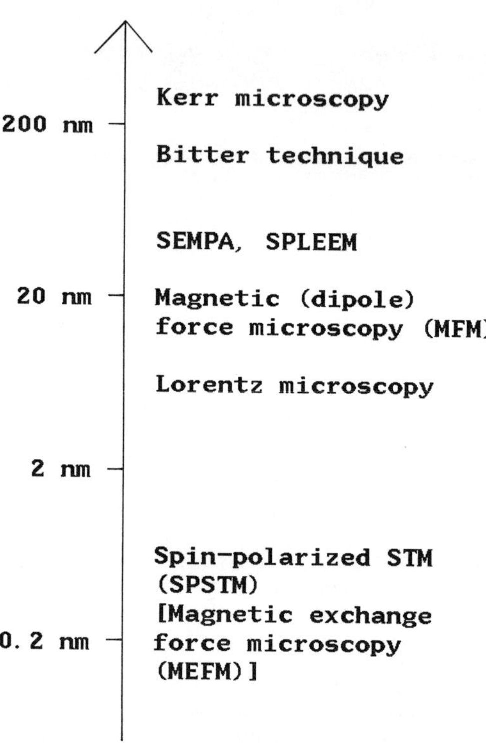

Fig. 1. Spatial resolution achieved with various magnetic imaging techniques. Magnetic exchange force microscopy still has to be developed.

1. Magnetic forces

Several scanning probe techniques are based on the measurement of magnetic forces acting between a magnetic sample and a magnetic probe tip. Magnetic (dipole) force microscopy (MFM) [1,2] and tunneling-stabilized magnetic force microscopy (TSMFM) [3,4] are outstanding for their simplicity and can be applied even in an ambient environment. They are sensitive to the magnetic stray field distribution above the sample surface and offer a spatial resolution of typically 20-100 nm, which is basically limited by the long-range nature of the magnetic dipole forces [5]. In contrast, magnetic exchange force microscopy (MEFM), based on the measurement of the short-range exchange forces at close tip-sample separations ($\lesssim$ 3 Å), is expected to offer a spatial resolution at the atomic level. However, MEFM has to be performed under ultra-high vacuum conditions on well-prepared magnetic sample surfaces.

2. Polarized light contrast

Scanning near-field optical microscopy (SNOM) combined with polarized light contrast [6] can be used to image magnetic domain structure at sub-wavelength resolution, thereby competing with conventional Kerr microscopy and scanning laser microscopy.

3. Polarization analysis of secondary particles

Various STM-related techniques involve the polarization analysis of secondary particles. For instance, the STM can be operated in the field emission mode, while monitoring the spin polarization of the emitted secondary electrons [7,8]. Another method is based on the measurement of the circular polarization of the recombination luminescence excited by electrons tunneling from a ferromagnetic tip into GaAs [9].

4. Direct analysis of the tunneling current

High-resolution magnetic-sensitive STM techniques can be based on the direct analysis of the tunneling current itself. For instance, a tunneling current modulation at the Larmor frequency due to the precession of individual paramagnetic spins has been detected [10]. Alternatively, the magnetic-valve effect [11] has been used to perform spin-polarized scanning tunneling microscopy (SPSTM) with magnetic probe tips at room temperature [12-14]. These techniques require well-defined ultra-high vacuum conditions and well-prepared sample surfaces. On the other hand, a spatial resolution down to the atomic level can be reached [15-17].

The ability of magnetic-sensitive scanning probe techniques to provide structural and magnetic information down to the atomic scale offers novel opportunities for the characterization of surfaces of magnetic materials and the growth of thin magnetic films. In the following, we will focus on STM performed with magnetic tips, leading to SPSTM which currently offers the highest-resolution magnetic imaging technique.

STM WITH MAGNETIC TIPS

The most natural way to introduce a magnetic sensitivity to STM is to use magnetic probe tips, rather than W or Pt-Ir tips. Several different magnetic tip materials have already been tried, including CrO_2, Cr, Fe, Ni, and NiMn. By using chromium dioxide thin film tips, atomic-step resolution has been achieved on a Cr(001) surface [12-14]. Atomic-step resolution has also been obtained on a Au(111) surface by using iron whisker probes [18]. However, the chromium dioxide thin film tips as well as the corners of the iron whiskers were not sufficiently sharp to achieve in-plane atomic resolution.

More recently, improvements in the in-situ preparation of atomically sharp and clean tips under UHV conditions [18,19] have allowed for routine in-plane atomic resolution studies with magnetic probes. For instance, the Si(111)7x7 surface structure has been resolved with chromium tips [18-20]. By using iron tips, atomic resolution on the Si(001)2x1 surface has been demonstrated [20,21]. The same iron tips have also successfully been applied to resolve the atomic structure of a magnetite (001) surface [15-17]. This experiment has proven for the first time that in-plane atomic resolution with a magnetic tip on a magnetic sample can be achieved. In conclusion, atomic resolution STM studies with magnetic probe tips can now routinely be performed - an important requirement for the development of atomic-resolution spin-polarized STM.

SPIN-POLARIZED SCANNING TUNNELING MICROSCOPY (SPSTM)

Spin-polarized STM (SPSTM) is based on the spin-dependence of the tunneling current flowing between two magnetic electrodes, as theoretically predicted [11] and experimentally verified for the STM by the direct observation of vacuum tunneling of spin-polarized electrons between a ferromagnetic chromium dioxide tip and a Cr(001) surface [12-14]. The chromium dioxide tips were favored for the initial SPSTM studies because a high degree of spin polarization of nearly 100% for binding energies near 2 eV below the Fermi level of chromium dioxide had been found by spin-polarized photoemission experiments [22]. The large value of the spin polarization may be explained by a spin-filter effect of chromium dioxide as a conjectured half-metallic ferromagnet [23]. The SPSTM studies on the Cr(001) surface, performed with the chromium dioxide tips, confirmed a theoretical model [24] of topological antiferromagnetism between ferromagnetic (001) terraces separated by monatomic steps [12-14]. Most recently, the alternation of the magnetization between different (001) terraces separated by monatomic steps has also been experimentally confirmed by SEMPA studies [25]. The initial SPSTM work [12,13] additionally allowed to deduce a local effective polarization of the tunnel junction on a nanometer length scale. Bias-dependent measurements of this effective polarization ('spin-polarized scanning tunneling spectroscopy' studies) have the potential to yield valuable information about the local spin-dependent electronic structure [13].

Since the chromium dioxide tips used for the initial SPSTM studies were not as sharp as required to obtain in-plane atomic resolution, a new procedure for the in-situ preparation of atomically sharp and clean magnetic tips inside the UHV system has been introduced [18,19]. Iron has been favored as an alternative tip material for high-resolution SPSTM studies because a high degree of spin polarization of iron films was found earlier in spin-polarized electron tunneling studies using planar tunnel junctions [26]. Magnetite, the best-known natural magnetic material, has been chosen as a second test system for SPSTM studies. Magnetite (Fe_3O_4) has cubic inverse spinel structure and is ferrimagnetic with a Curie temperature of 860 K (Fig. 2). There exist tetrahedrally coordinated Fe 'A-sites' as well as octahedrally coordinated Fe 'B-sites'. The A-sites are occupied by Fe(3+) whereas the B-sites are half occupied by Fe(3+) and half by Fe(2+) with spin configurations 5x(3d↑) and 5x(3d↑)/1x(3d↓), respectively. At room temperature, fluctuations of the sixth electron (3d↓) among B-sites are rapid in the bulk, explaining the relatively high electrical conductivity of about 100/(Ω cm) compared with other iron oxides. Below the so-called Verwey transition [27] at about 120 K, these charge fluctuations cease and a static periodic arrangement of Fe(3+) and Fe(2+) ions sets in, leading to a decrease in conductivity by about two orders of magnitude. The driving force is interatomic Coulomb interaction, so the Verwey transition in magnetite is considered to be an example of a Wigner crystallization [28,29] in three dimensions, albeit modified by electron-phonon coupling. Similar to the case of chromium dioxide, magnetite has also theoretically been predicted to be a half-metallic ferromagnet [30,31], thereby promising a high degree of spin polarization of the electronic states near the Fermi level which is favorable for SPSTM studies [32].

The preparation of magnetite (001) surfaces was performed by mechanical polishing followed by in-situ annealing of the single crystals up to about 1000 K in UHV. Well-ordered and clean surfaces were obtained as verified by low-energy electron diffraction (LEED) and Auger electron spectroscopy (AES) [16,20]. The correct surface stoichiometry was checked by x-ray photoelectron spectroscopy (XPS). The SPSTM experiments were performed in the same multichamber UHV system (Nanolab-I) containing the surface preparation and analysis facilities (background pressure during STM experiments: 1x10E(-11) mbar or below). Several different probe tips were used to exclude tip-specific imaging artifacts.

The topographic structure of the magnetite (001) surface was first studied by using non-magnetic electrochemically etched tungsten tips. We have mainly focused on the Fe-O (001) planes containing the octahedrally coordinated Fe B-sites which are of primary interest. In Fig. 3 a topographic STM image obtained with a tungsten tip is presented showing terraces separated by 2-Å high steps. On top of these terraces, parallel atomic rows with a spacing of 5.9 Å are visible which change their orientation by 90° from one terrace to the next. These rows have been identified with the rows of octahedrally coordinated Fe B-sites in the Fe-O planes of magnetite (Fig. 2) which are indeed separated by 2 Å. The oxygen sites are not visible in the STM images because the corresponding O-1s and p-states lie well outside the energy window accessible with the STM [17]. No atomic-scale structure is resolved along the rows of Fe B-sites by using non-magnetic tungsten tips indicating a smooth spin-averaged density-of-states corrugation along these rows.

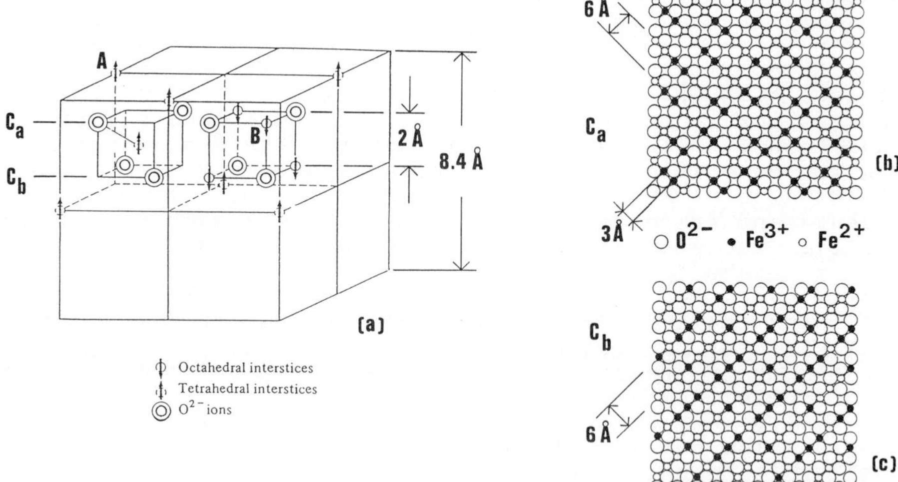

Fig. 2. a) Crystal structure of magnetite with tetrahedrally coordinated A-sites occupied by Fe(3+) and octahedrally coordinated B-sites occupied by Fe(3+) as well as Fe(2+). The Ca and Cb planes are shown in (b) and (c) respectively with an ordering of Fe(3+) and Fe(2+) according to a model [19] for the low-temperature phase of magnetite.

Fig. 3. Topographic STM image (155 Å x 165 Å) of the Fe-O (001) planes of magnetite obtained with a non-magnetic tungsten tip. The terraces are separated by 2-Å high steps. On top of these terraces atomic rows of Fe B-sites can be seen with a spacing of 5.9 Å, changing their orientation by 90° from one terrace to the next. (Tunneling current: I = 1 nA, sample bias voltage: U = +3.0 V).

In contrast, a strong corrugation along the rows of Fe B-sites is measured with ferromagnetic iron probe tips in spin-polarized STM experiments (Fig. 4). Several different periodicities are observed along these rows which are, however, always multiples of 3 Å, the nearest-neighbor distance of Fe-sites along the rows. A statistics of the observed periodicities in about twenty different STM images [33] reveals a clear dominance of a 12-Å period (Fig. 5) which is equal to the periodicity observed in the bulk for the low temperature phase of magnetite below the Verwey transition [34]. The observed static pattern with a dominant 12-Å periodicity at the Fe-O (001) surface plane of magnetite at room temperature may be explained by an increase of the Verwey transition temperature at the surface due to a band narrowing as a result of the reduced coordination at the surface [29]. The important quantity which determines whether a Wigner crystallization sets in, is the ratio between the Coulomb interaction energy V between charge carriers on nearest-neighbor sites and the electronic band-width B. A Wigner crystallization will occur if the ratio V/B is large enough, i.e. greater than a constant of order three [35]. Since the band-width B is partially determined by the co-ordination number [36], the ratio V/B will be different for the surface compared with the bulk. A reduced co-ordination number at the surface leads to a narrowed band-width B which favors electron crystallization. Therefore, the Verwey transition temperature is expected to be increased for the surface compared with the bulk value. The lack of long-range order indicates that the Wigner crystallization of the 3d↓ electrons on the Fe-O (001) surface plane takes the form of a 2D Wigner glass. At the moment, we can only speculate whether an ideal stoichiometric (001) surface plane, which never exists in nature, would exhibit a 2D Wigner crystal state. An independent proof for the non-metallic state of the Fe-O (001) surface plane at room temperature comes from local tunneling spectroscopy data from which an energy gap of about 1.5 eV has been extracted. This is in clear contrast to the metallic state of bulk magnetite at room temperature.

Besides the two examples of SPSTM studies described above, other applications of SPSTM, particularly at low temperatures [14], appear to be highly promising.

CONCLUSIONS

SPSTM is an outstanding magnetic-sensitive scanning probe method because it allows for atomic-resolution imaging of magnetic surfaces directly in real space. SPSTM can be performed at room temperature as well as at low temperatures. Half-metallic ferro- (or antiferro-) magnets are particularly suitable materials for SPSTM studies because they offer a high degree of spin polarization of the states at the Fermi level.

ACKNOWLEDGEMENTS

I would like to thank my collaborators in the SPSTM work: I.V. Shvets, D. Bürgler, G. Tarrach, T. Schaub, H.-J. Güntherodt, G. Güntherodt, J.M.D. Coey, R.J. Gambino, and R. Ruf. Financial support from the Swiss National Science Foundation and the Kommission zur Förderung der wissenschaftlichen Forschung is gratefully acknowledged.

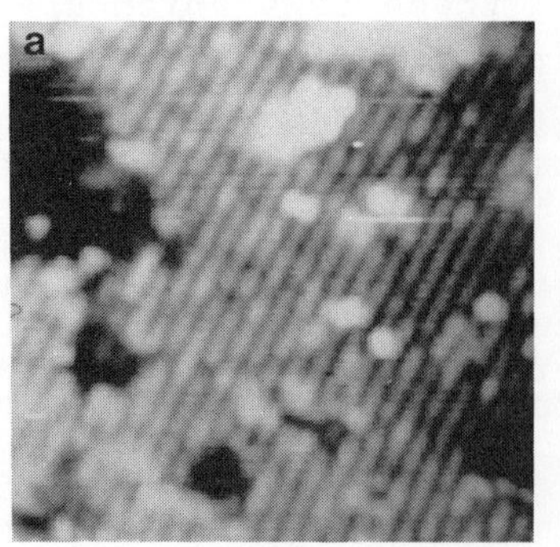

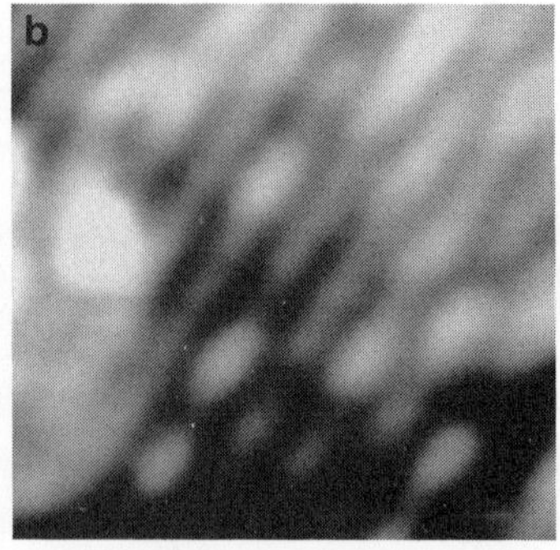

Fig. 4a. SPSTM image (155 Å x 165 Å) of the Fe-O (001) plane of magnetite obtained with a ferromagnetic iron tip. The atomic rows of Fe B-sites show a clear modulation along these rows, in contrast to the STM images obtained with tungsten tips (Fig. 3).

Fig. 4b. High-resolution SPSTM image of the Fe-O (001) plane of magnetite obtained with an iron tip. The rows of Fe B-sites with a 5.9-Å spacing and a modulation along these rows can clearly be seen. (I = 1 nA, U = +3.0 V).

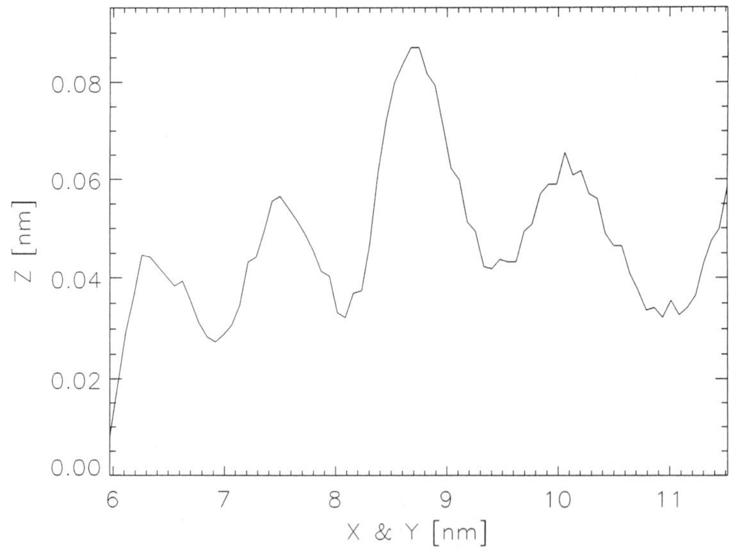

Fig. 5. A 12-Å periodicity is preferentially observed along the rows of Fe B-sites, as becomes evident from single line sections along these rows. The 12-Å periodicity corresponds to the repeat period of Fe(3+) and Fe(2+) sites in the low-temperature phase of magnetite (Figs. 2b and 2c).

REFERENCES

[1] Y. Martin and H.K. Wickramasinghe, Appl. Phys. Lett. 50, 1455 (1987).

[2] J.J. Saenz, N. Garcia, P. Grütter, E. Meyer, H. Heinzelmann, R. Wiesendanger, L. Rosenthaler, H.R. Hidber, and H.-J. Güntherodt, J. Appl. Phys. 62, 4293 (1987).

[3] R. Allenspach, H. Salemink, A. Bischof, and E. Weibel, Z. Phys. B 67, 125 (1987).

[4] J. Moreland and P. Rice, Appl. Phys. Lett. 57, 310 (1990).

[5] P. Grütter, H.J. Mamin, and D. Rugar, in: Scanning Tunneling Microscopy II, eds. R. Wiesendanger and H.-J. Güntherodt, Springer Series in Surface Sciences Vol. 28, p. 151. Springer, Berlin/Heidelberg (1992).

[6] J.K. Trautman, E. Betzig, J.S. Weiner, D.J. DiGiovanni, T.D. Harris, F. Hellman, and E.M. Gyorgy, J. Appl. Phys. 71, 4659 (1992).

[7] R. Allenspach and A. Bischof, Appl. Phys. Lett. 54, 587 (1989).

[8] P.N. First, J.A. Stroscio, D.T. Pierce, R.A. Dragoset, and R.J. Celotta, J. Vac. Sci. Technol. B 9, 531 (1991).

[9] S.F. Alvarado and Ph. Renaud, Phys. Rev. Lett. 68, 1387 (1992).

[10] Y. Manassen, R.J. Hamers, J.E. Demuth, and A.J. Castellano, Jr., Phys. Rev. Lett. 62, 2531 (1989).

[11] J.C. Slonczewski, Phys. Rev. B 39, 6995 (1989).

[12] R. Wiesendanger, H.-J. Güntherodt, G. Güntherodt, R.J. Gambino, and R. Ruf, Phys. Rev. Lett. 65, 247 (1990).

[13] R. Wiesendanger, D. Bürgler, G. Tarrach, A. Wadas, D. Brodbeck, H.-J. Güntherodt, G. Güntherodt, R.J. Gambino, and R. Ruf, J. Vac. Sci. Technol. B 9, 519 (1991).

[14] R. Wiesendanger, D. Bürgler, G. Tarrach, H.-J. Güntherodt, and G. Güntherodt, in: Scanned Probe Microscopy, ed. H.K. Wickramasinghe, AIP Conf. Proc. Vol. 241, p. 504. AIP, New York (1992).

[15] R. Wiesendanger, I.V. Shvets, D. Bürgler, G. Tarrach, H.-J. Güntherodt, and J.M.D. Coey, Z. Phys. B 86, 1 (1992).

[16] R. Wiesendanger, I.V. Shvets, D. Bürgler, G. Tarrach, H.-J. Güntherodt, J.M.D. Coey, and S. Gräser, Science 255, 583 (1992).

[17] R. Wiesendanger, I.V. Shvets, D. Bürgler, G. Tarrach, H.-J. Güntherodt, and J.M.D. Coey, Europhys. Lett. 19, 141 (1992).

[18] R. Wiesendanger, D. Bürgler, G. Tarrach, T. Schaub, U. Hartmann, H.-J. Güntherodt, I.V. Shvets, and J.M.D. Coey, Appl. Phys. A 53, 349 (1991).

[19] R. Wiesendanger, D. Bürgler, G. Tarrach, I.V. Shvets, and H.-J. Güntherodt, Mat. Res. Soc. Symp. Proc. Vol. 231, 37 (1992).

[20] I.V. Shvets, R. Wiesendanger, D. Bürgler, G. Tarrach, H.-J. Güntherodt, and J.M.D. Coey, J. Appl. Phys. 71, 5489 (1992).

[21] R. Wiesendanger, I.V. Shvets, D. Bürgler, G. Tarrach, H.-J. Güntherodt, and J.M.D. Coey, Ultramicroscopy (1992).

[22] K.P. Kämper, W. Schmitt, G. Güntherodt, R.J. Gambino, and R. Ruf, Phys. Rev. Lett. 59, 2788 (1987).

[23] K. Schwarz, J. Phys. F 16, L211 (1986).
[24] S. Blügel, D. Pescia, and P.H. Dederichs, Phys. Rev. B 39, 1392 (1989).
[25] J. Unguris, R.J. Celotta, and D.T. Pierce, paper presented at the NATO workshop on Magnetism and Structure in Systems of Reduced Dimension, Cargèse (France), June 15-19 (1992).
[26] P.M. Tedrow and R. Meservey, Phys. Rev. B 7, 318 (1973).
[27] E.J.W. Verwey, Nature 144, 327 (1939).
[28] E. Wigner, Trans. Far. Soc. 34, 678 (1938).
[29] N.F. Mott, Metal-Insulator Transitions, Taylor & Francis Ltd., London, 1974.
[30] A. Yanase and K. Siratori, J. Phys. Soc. Jpn. 53, 312 (1984).
[31] Z. Zhang and S. Satpathy, Phys. Rev. B 44, 13319 (1991).
[32] R.A. de Groot, Physica B 172, 45 (1991).
[33] R. Wiesendanger, in: New Concepts for Low-Dimensional Electronic Systems, ed. G. Bauer, Springer Series in Solid State Sciences, Springer, Berlin/Heidelberg (1992).
[34] S. Iida, K. Mizushima, M. Mizoguchi, K. Kose, K. Kato, K. Yanai, N. Goto, and S. Yumoto, J. Appl. Phys. 53, 2164 (1982).
[35] J.R. Cullen and E.R. Callen, Phys. Rev. B 7, 397 (1973).
[36] N.F. Mott, in: Festkörperprobleme XIX, ed. J. Treusch, p. 331. Vieweg, Braunschweig (1979).

MORPHOLOGY AND STRUCTURE OF ULTRATHIN Co- and Au-FILMS GROWN ON Ru(0001) SUBSTRATES

J. Vrijmoeth, C. Günther, J. Schröder, R.Q. Hwang*, and R.J. Behm

Abteilung für Oberflächenchemie und Katalyse
Universität Ulm
Postfach 4066
D-7900 Ulm
Germany

We have studied the morphology, growth mechanisms, and atomic structure of ultrathin Co- and Au-films on Ru(0001) substrates using Scanning Tunneling Microscopy (STM). During growth at room temperature, kinetic limitations dominate the film morphology rather than thermodynamics. As a consequence, both systems follow a quasi layer-by-layer growth mechanism at room temperature. It proceeds via nucleation and subsequent growth of two-dimensional islands. The crucial role of adatom mobility in island growth is illustrated.

The thermodynamically stable film morphology is reached after annealing. Both Co and Au follow a Stranski-Krastanov growth mode on Ru(0001), with one and two closed layers, respectively.

The systems show relaxation of the lattice strain. Au films are relaxed from the first layer on. For Co, the first layer is pseudomorphic with the substrate. Thicker Co-films are strain-relaxed. The relaxation observed may be described in a model assuming an almost fully relaxed bulklike Co lattice which is commensurate with the substrate.

1. STM STUDY OF ULTRATHIN MAGNETIC FILMS

In recent years, there has been an increasing interest in the magnetic behavior of ultrathin films and multilayers.[1] The unique magnetic properties exhibited by these quasi-twodimensional systems, such as the giant magnetoresistance,[2] present intriguing fundamental and technological challenges.[3] A more complete understanding of the specific magnetic effects in thin-film systems requires detailed knowledge of their

*Present address: Sandia National Laboratories, Livermore, CA 94551, USA.

Magnetism and Structure in Systems of Reduced Dimension
Edited by R.F.C. Farrow *et al.*, Plenum Press, New York, 1993

structure and morphology, on an atomic scale. This is most easily achieved for model systems consisting of atomically clean films deposited on single crystalline, flat substrates, which are increasingly studied for these purposes.

In this contribution we present results of a comparative study on the growth mechanism, which determines the film morphology, and the atomic structure of ultrathin Co and Au films on a Ru(0001) substrate. Both systems form atomically abrupt interfaces. Because of the large surface free energy of the substrate, both elements are expected to wet a Ru(0001) surface, resulting in either a flat layer-by-layer (Frank- van der Merwe) or a layer-plus-cluster (Stranski-Krastanov) growth mode rather than in the formation of three-dimensional (3-D) clusters (Vollmer-Weber growth mode).[4] The growth behavior and atomic structures at the interfaces differ significantly for both systems, and demonstrate the strong, system specific phenomena which are to be anticipated. We will show that at room temperature, kinetic limitations for mass transport dominate the film development so that the energetically most stable configuration cannot be reached. To this end, we subsequently discuss the mechanisms for the nucleation of the first layer, the growth of the first layer, and the nucleation of the second layer, as characteristic stages of film development.

Structure and morphology were characterized mainly by Scanning Tunneling Microscopy (STM) measurements. Because of the high resolution and the local nature of the technique, STM gives direct insight into mechanistic details of film growth.[5,6] Further details of Au film growth on the Ru(0001) substrate have been published earlier.[5,7]

2. EXPERIMENT

The studies were performed in an ultra-high vacuum (UHV) system (base pressure less than 1 x 10^{-10} mbar),[8] which contains facilities for Low Energy Electron Diffraction (LEED), Auger Electron Spectroscopy (AES), and Thermal Desorption Spectroscopy (TDS), as well as a STM. The Ru(0001) crystal was heated by electron bombardment. Temperatures were monitored using a Pt-PtRh thermocouple spot-welded to the crystal. Co and Au were evaporated by direct-current heating of a thoroughly outgassed Co wire and a W filament supporting a Au ball. Film thicknesses are given in monolayers (ML), where 1 ML corresponds to 1 adatom per substrate atom.

The sample was cleaned using standard methods involving Ar^+-sputtering and cycles of O_2-exposure and annealing to remove residual traces of C.[9] The well-prepared surface exhibited large terraces (up to ~2 μm in size), separated by steps of monoatomic height. Areas with large step densities were encountered as well.

STM-images were acquired with a louse-type microscope. The data were taken in the constant-current mode with tunneling voltages ranging from 0.01 V to 0.2 V. For preparation of the W tunneling tip voltages of up to 10 V were occasionally applied. Tunneling currents ranged between 1 nA for large-scale topographs and 30 nA for high-resolution images. Data processing exclusively involved background subtraction and correction of scan lines containing tip instabilities. Images are shown in a gray-scale representation, with darker areas representing lower regions. Fig. 5 is presented as an illuminated landscape (light source from the left side of the image).

3. GROWTH AND MORPHOLOGY

3.1. Nucleation of the First Layer

Upon room-temperature deposition of either Co or Au onto a Ru(0001) substrate, two-dimensional islands are observed to form even at the lowest coverages investigated (0.03 ML). This is illustrated in the STM images reproduced in Fig. 1 (a) and (b) of submonolayer Co and Au films respectively. The measured island heights of 1.85 Å (Co) and 2.90 Å (Au) correspond with single layers of deposited material. The formation of islands provides direct proof that the metal adatoms are sufficiently mobile to reach the nearest island from their adsorption site, consistent with the high mobility of metal atoms on metal substrates derived from Field Ion Microscopy (FIM) measurements.[10] As a consequence, individual metal adatoms have never been observed in STM experiments performed at room temperature.

The island density was found to increase with increasing coverage up to a certain critical coverage beyond which it remained constant. This observation indicates a homogeneous nucleation mechanism, in which stable two-dimensional (2-D) nuclei are formed by condensation of mobile adatoms diffusing in a 2-D lattice gas.[11,12] Once the nuclei have formed, they grow rapidly by trapping the adatoms impinging at their edges. Thus the adatom density is reduced within a certain critical radius, so that the probability for forming new nuclei in the vicinity of existing islands is small. The critical radius reflects the effective mean free path (EMFP) of the adatoms, i.e., the distance an adatom can travel before being captured in a nucleation process or being

(a)

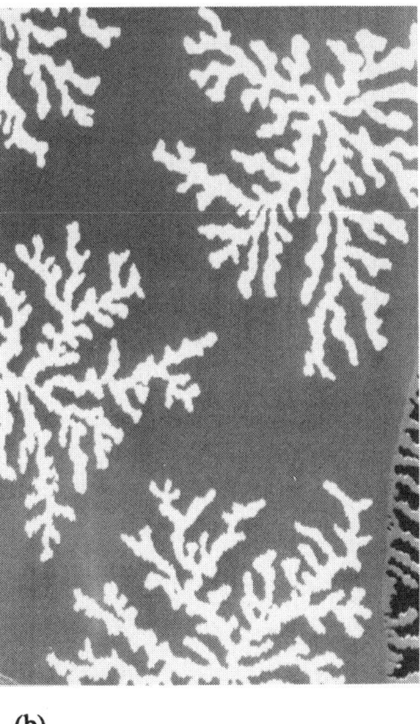

(b)

Figure 1. STM micrographs of metal films grown on Ru(0001) at room temperature. (a) Co, coverage 0.19 ML, evaporation rate 0.15 ML/min, image dimensions 9000 Å x 7300 Å; (b) Au, coverage 0.3 ML, evaporation rate 0.20 ML/min, image dimensions 8000 Å x 6500 Å.

trapped at the edge of an existing island, in the initial stage of growth.[13] As a consequence, the island density saturates beyond a critical coverage. In this model the island density and EMFP depend on preparation variables such as deposition rate and substrate temperature as well as on intrinsic parameters of the growth system, in particular the adatom mobility and cluster energies. The influence of adatom mobility is directly illustrated by the pronounced differences in island density both for different evaporants and for different substrates. For similar deposition conditions (300 K, ~0.2 ML/min.) we find that the Au island density on Ru(0001) saturates at ~3 x 10^8 cm^{-2}, whereas for Co the highest island density was ~3 x 10^{10} cm^{-2}. On a chemically modified, oxygen-precovered, Ru(0001) substrate (Θ_O = 0.5 ML) the Au island density rises by a factor of ~10^5 to 3 x 10^{13} cm^{-2}. For Co, this effect is less pronounced and the island density only increases to 1 x 10^{12} cm^{-2} on an oxygen-preadsorbed surface.

Additional experiments further confirm the tendencies in island density as predicted by these models, namely an increase in island density at lower substrate temperature or, equivalently, at higher deposition rates. Raising the deposition rate by a factor of ten led to an increase in the Au island density by about a factor of four.[14]

In conclusion, our data are consistent with current theories for homogeneous nucleation.[11,12] Based on these models the observed difference in island density for Au/Ru(0001) and Co/Ru(0001) implies that the mobility of individual diffusing Au adatoms on Ru(0001) is significantly larger than that of Co adatoms, by up to a factor of 10^6 if the critical nucleus size would be equal to one in both cases.[15]

Apart from being involved in homogeneous nucleation, adatoms may interact with defects such as steps. Ascending steps act as heterogeneous nucleation sites by trapping the adatoms, as is evident from the small amounts of metal wetting the steps (Fig. 1 (a) and (b)). As a consequence, the adatom density is reduced in a depletion zone along the step, where the probability for island nucleation is small. Indeed, for both Co (Fig. 1 (a)) and Au (not shown) the island density is reduced in an area next to the ascending steps. As expected, the width of the depletion zone is of the same order of magnitude as the EMFP.[16]

By contrast, the island density near a descending step is unaffected by the presence of the step (see Fig. 1(a)), both for Co and for Au. Apparently the adatoms arriving at a descending Ru step edge experience a diffusion barrier and are not trapped but rather reflected. On the other hand, since atoms cannot pass from the lower to the upper terrace the adatom density is not reduced in the vicinity of the descending step. In other cases, in which adatoms are not reflected by the descending step edge but fall down and get trapped at the lower step edge, the island density on both sides is reduced.[16] Analogous behavior is also observed for second-layer Au adatoms, as discussed in section 3.3.

It is important to realize that nucleation is a purely kinetic phenomenon. In thermodynamic equilibrium, each terrace would only support a single island.

3.2. Growth of the First Layer

After nucleation of the film, continued metal deposition results in the growth of the existing first-layer nuclei. In both of the present systems, growth is purely two-dimensional, i.e. it takes place within the first layer, for a certain limited coverage range (see section 3.3). The actual 2-D growth behaviors in the two systems, however,

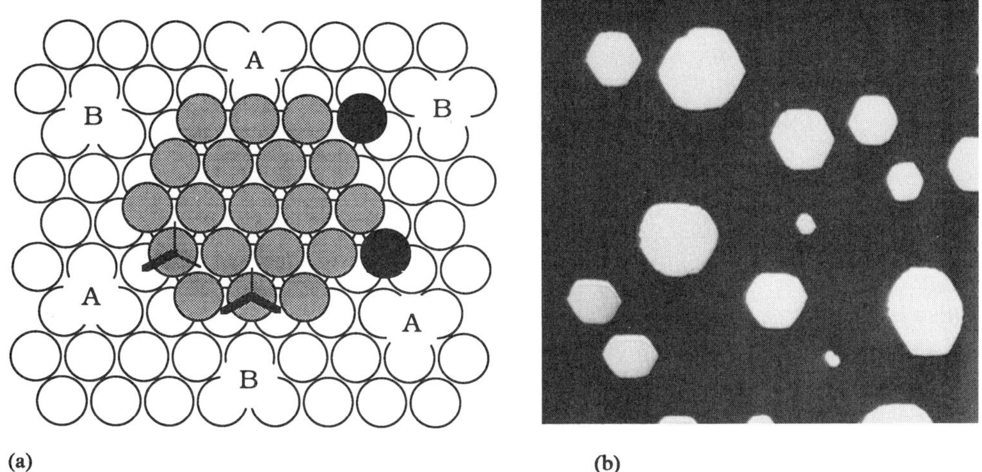

(a) (b)

Figure 2. (a) Different classes A and B, with different stacking geometries, of edges at an hexagonal island on Ru(0001). To guide the eye, bonds to the substrate for edge atoms in each of the classes are indicated. Energies and mobilities of edge atoms and adsorbed atoms (black) may be different for edges A and B. (b) STM image showing a Co-film grown on Ru(0001) at room temperature and annealed afterwards at 600 K. The originally triangular islands obtain a hexagonal shape upon annealing (see text). Coverage 0.19 ML, evaporation rate 0.2 ML/min. image dimensions 5600 Å x 5800 Å.

shows pronounced differences as evidenced by the images shown in Fig. 1. Au islands exhibit a strongly dendritic shape, with overall diameters of up to 4000 Å while the arms are only 100 Å wide. These dendritic shapes were explained to originate from a limited mobility of Au atoms along the perimeter existing islands, in contrast to the high mobility of individual free adatoms. As a consequence, the most stable island shape cannot be reached (Fig. 1(b)).[5] That conclusion has been confirmed by comparison of the data with simulations assuming zero mobility along the island edges (Diffusion-Limited Aggregation, DLA).[17] The dendrites obtained from that model have a fractal dimension of $D = 1.70 \pm 0.02$. Analysis of measured dendrites yields $D = 1.72 \pm 0.07$, in good agreement with the simulations mentioned above.[5] That the dendrites originate from a limited edge atom mobility is further evident from annealing the room-temperature deposited film to 500 K (not shown). At that temperature the island density does not change, which shows that no appreciable mass transport between islands takes place. However, the now increased mobility of the edge atoms allows the islands to assume an energetically more favorable morphology, as is evident from their collapse into non-dendritic, more compact forms.[5]

In the case of Co, the mobility of the atoms along the island perimeter is sufficiently large for a more compact shape to be reached already at room temperature. Fig. 1(a) reveals that the morphology of the first-layer islands at room temperature is triangular. This island shape arises from the difference in atomic structure in the two inequivalent edge classes A and B on the threefold symmetric hcp Ru(0001) substrate (see Fig. 2(a)). The triangles change orientation by 180° from terrace to terrace because of the ABAB stacking of subsequent (0001) substrate layers (Fig. 1 (a)). Interestingly, after a flash-anneal (~3 sec) to 600 K the triangular island shapes have transformed into hexagonal ones, with edge classes A and B about equally populated (see Fig. 2(b)).

Two different schemes may be proposed to explain these observations, dependent on whether the hexagonal or the triangular island shape is thermodynamically more stable, i.e., lower in free energy. The observation of hexagons at elevated temperature seems to suggest that this compact shape is energetically favored over the triangular one. If that is the case, the observed formation of triangles at room temperature should, similarly to the case of Au, arise from kinetic limitations, such as different mobilities for adatoms diffusing along edges A and B. The annealing then would allow for the formation of the favored compact morphology.

Alternatively, the triangular shapes may represent the equilibrium configuration resulting from a significant difference in energy between edges of type A and B. In this interpretation, the occurrence of the hexagon after annealing would in turn be induced by kinetic barriers. It would result from the quenching to room temperature of islands in a disordered 'droplet' state existing at elevated temperature. If the latter interpretation is valid, it is expected that upon cooling down more slowly a more triangular island shape might be obtained. Experiments are planned to clarify this issue.

The data illustrate that, similarly to the case of film nucleation, thermodynamics and kinetics compete in determining the 2-D growth behavior and the resulting island shape of the film. A film tends to minimize its edge energy, which would result in the formation of compact islands. However, our results confirm that kinetic limitations such as possible diffusion barriers strongly affect the first-layer island shapes as well.

3.3. Second-layer Nucleation and Film Morphology

The nucleation of the second layer represents the transition from two- to three-dimensional film growth. It may be regarded as a representative case for the growth of higher layers as well as for the role of nucleation in determining the effective three-dimensional growth behavior of the system. For the often desired layer-by-layer like growth to occur it is important that subsequent-layer nucleation begins only when the preceding layer is almost complete.

Again we find significant differences between the two systems. This is demonstrated in comparing the STM images in Fig. 3, which show a Co film just beyond the onset of second-layer nucleation (Fig. 3(a), $\Theta_{Co} = 0.45$) and two Au films just below and above that point (Fig. 3(b), $\Theta_{Au} = 0.69$; Fig. 3(c), $\Theta_{Au} = 0.8$). Second-layer nuclei form for integral coverages larger than 0.4 ML and 0.8 ML of Co and Au, respectively. For the higher-coverage Au film (Fig. 3(c)) we see several small second-layer nuclei, which have formed randomly on the large first-layer island, except for a depletion zone of about 200 Å width along the island perimeter. In Fig. 3(a) the number of second-layer islands is too small to yield a representative picture, but from these and other data we find similar trends for Co/Ru(0001). (Note that the second-layer islands in the left part of the image, which have nucleated at or close to the (first-layer) island edge, are all located above substrate steps. Apparently the latter act as heterogeneous nucleation sites.)

These observations indicate that homogeneous nucleation in higher layers, on top of existing islands, may be described in a picture similar to that used before for the description for first-layer nucleation. In that picture second-layer adatoms impinging on top of existing first-layer islands will form stable nuclei once a critical density of (second-layer) adatoms is reached. In the vicinity of steps the adatom density is reduced if adatoms can pass the descending step edge, and atoms from the lower terrace level

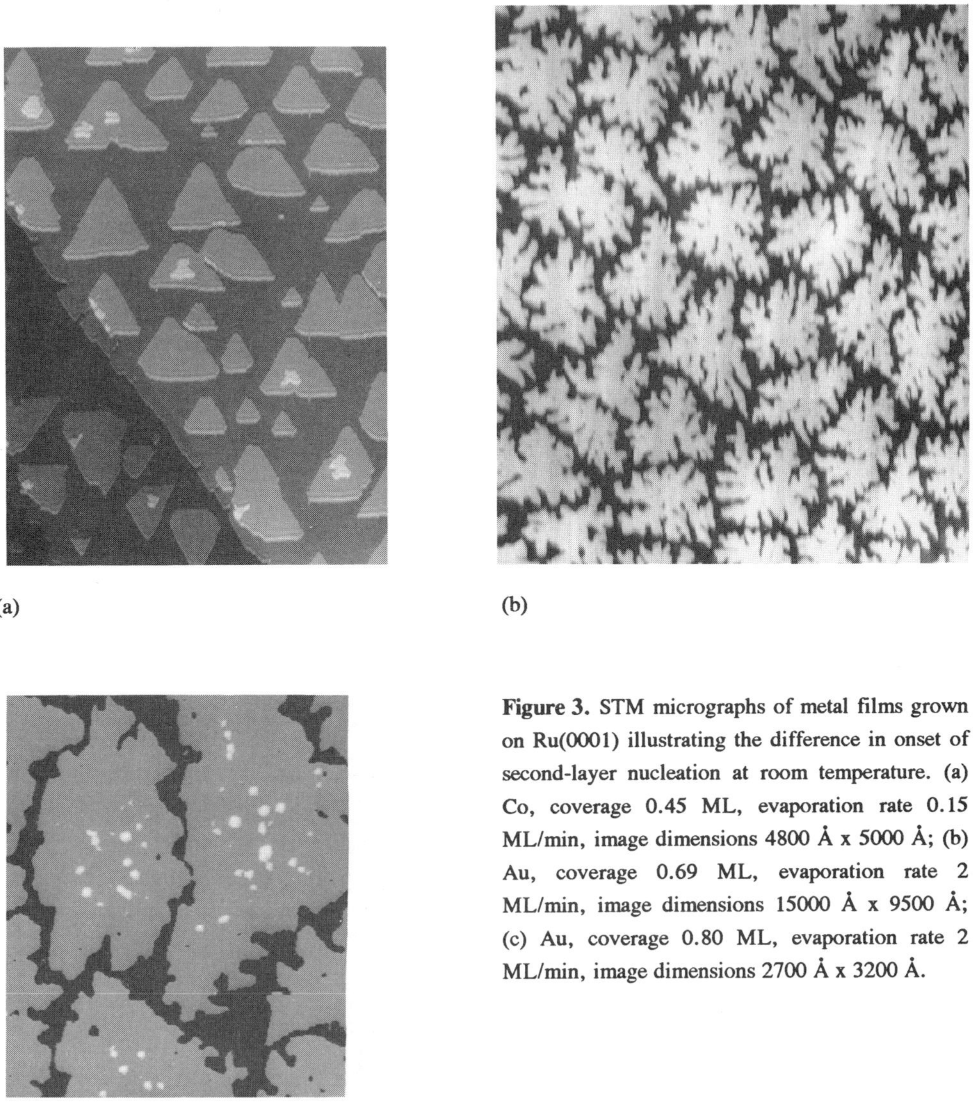

(a)

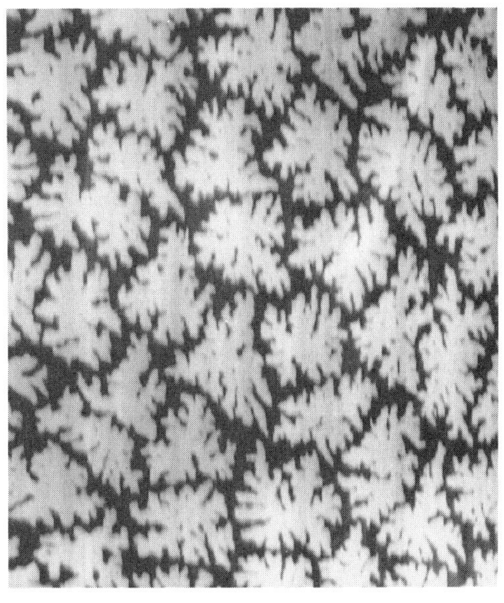

(b)

(c)

Figure 3. STM micrographs of metal films grown on Ru(0001) illustrating the difference in onset of second-layer nucleation at room temperature. (a) Co, coverage 0.45 ML, evaporation rate 0.15 ML/min, image dimensions 4800 Å x 5000 Å; (b) Au, coverage 0.69 ML, evaporation rate 2 ML/min, image dimensions 15000 Å x 9500 Å; (c) Au, coverage 0.80 ML, evaporation rate 2 ML/min, image dimensions 2700 Å x 3200 Å.

cannot pass the ascending step edge.[5,14,18] This mechanism reduces the density of second-layer adatoms on an area along the first-layer island edges and thereby inhibits nucleation of second-layer islands. In contrast, if adatoms would be efficiently reflected at the descending edges the adatom density and nucleation probability would remain constant up to the step edge, as was observed for the clean Ru(0001) surface (section 3.1).

In the present systems, only if the first-layer islands form a 'substrate' of sufficient size they may support second-layer growth. In the case of Co, the 0.4 ML integral coverage suffices to allow second-layer Co island nucleation on top of the compact first-layer islands (Fig. 3(a)). However, for a Au-coverage as large as 0.69 ML (Fig. 3(b)), the absence of compact areas on the dendritic islands still prevents nucleation of the

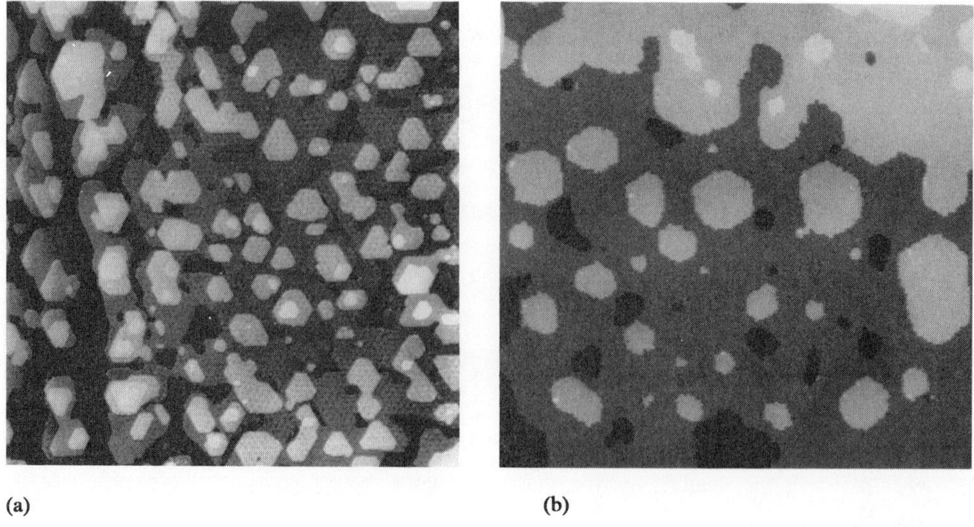

(a) (b)

Figure 4. STM micrographs of metal films grown on Ru(0001) demonstrating the quasi layer-by-layer mode at room temperature. (a) Co, coverage 4 ML, evaporation rate 0.15 ML/min, image dimensions 2800 Å x 2800 Å; (b) Au, coverage 3 ML, evaporation rate 2 ML/min, image dimensions 4000 Å x 2500 Å.

second Au layer. The latter only occurs at an integral Au coverage of 0.8 ML (Fig. 3 (c)), when the dendrites have filled in and the compact regions on the first-layer islands have become larger than ~1000 Å. The 'delayed' nucleation of Au on an almost closed first layer (Fig. 3(c)) leads to a rather flat morphology approaching that of the ideal layer-by-layer growth mode. By contrast, the 'early' nucleation of the second Co layer yields a rougher film morphology (Fig. 3(a)).

The latter trends are further evidenced during continued growth. Fig. 4 (a) and (b) show images recorded from Co- and Au-films with coverages of 4 ML and 3 ML, respectively, which give a direct impression of the effective growth scheme at room temperature. A *quasi* layer-by-layer mechanism occurs for both systems, with more than two layers exposed simultaneously. The difference in growth behavior as discussed above for second-layer growth continues during subsequent metal deposition, with the earlier nucleation of new Co layers leading to a rougher morphology of the Co films (Fig. 4 (a)), as compared to the Au films (Fig. 4 (b)).

The quasi layer-by-layer growth mechanism does not yield the thermodynamically most stable configuration.[4] In both systems the morphology changes drastically upon annealing. The Co film imaged in Fig. 5 has been deposited at room temperature and annealed to 600 K. In striking contrast to the morphology of the films shown in Fig. 4, the film now consists of a closed layer with a height of 1 ML, on top of which three-dimensional clusters have formed. The latter preferentially nucleate along steps in the underlying substrate. Assuming that this configuration represents the thermodynamically most stable morphology it is concluded that Co follows a Stranski-Krastanov growth mode on Ru(0001), with a critical thickness of one layer.

Similar behavior, i.e. a Stranski-Krastanov growth mode, was also observed for Au/Ru(0001). In this case, however, the critical thickness preceding the onset of 3-D cluster formation amounts to two layers.[13]

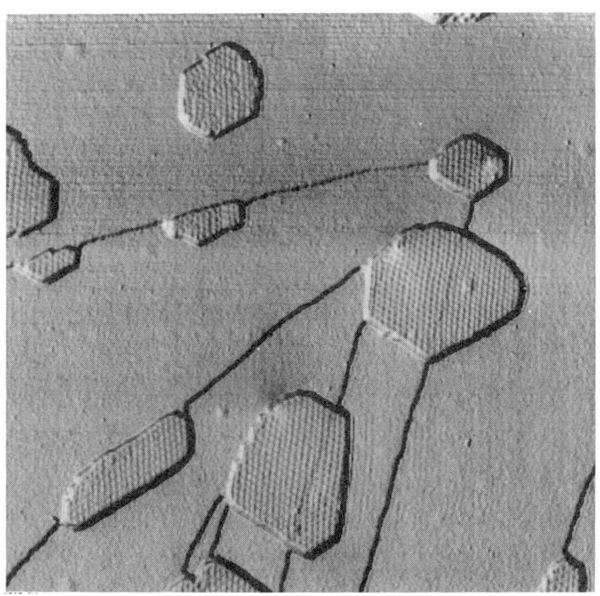

Figure 5. STM image showing a Co-film grown on Ru(0001) at room temperature and annealed afterwards to 600 K. This image shows both the energetically most favorable film morphology and the mechanism for strain relaxation. Coverage 1.8 ML, image dimensions 5600 Å x 5800 Å.

4. ATOMIC STRUCTURE

The atomic structure of the films is determined by the lattice geometry of both substrate and evaporant. For the present systems the substrate and film exhibit relatively large misfits of -7% for Co and +6.6 % for Au. Since in fully pseudomorphic films the total strain energy resulting from the lattice mismatch increases linearly with their thickness, both systems are expected to relax towards their bulk geometry already at relatively low thicknesses.

Au films indeed show strain relaxation beginning from the first layer. STM images show a corrugation pattern consistent with a unidirectionally expanded, hexagonal layer.[19] A variety of corrugation patterns is resolved for larger local film thicknesses. The underlying film structures result from the tendency to transform from the twofold symmetric first-layer structure - with three rotational domains - to a bulklike, threefold symmetric, structure.[7]

In contrast to the behavior of Au, the first Co atomic layer is pseudomorphic with the substrate, as is evident from the absence of any additional spots in the LEED pattern. In addition, no longe-range corrugation, other than the atomic corrugation, was resolved with STM. This applies both to the first-layer islands (Figs. 1(a) and 2(b)) and to the closed first layer obtained after annealing (Fig. 5).

We do observe relaxation of the lattice strain for thicker Co films with (local) thicknesses of 2 layers and more, both for films deposited at room temperature (Fig. 4(a)) and for the three-dimensional clusters obtained after annealing (Fig. 5). The relaxation gives rise to a threefold symmetric corrugation pattern with an amplitude of ~0.4 Å, which indicates a uniform relaxation of the strain. The corrugation has a

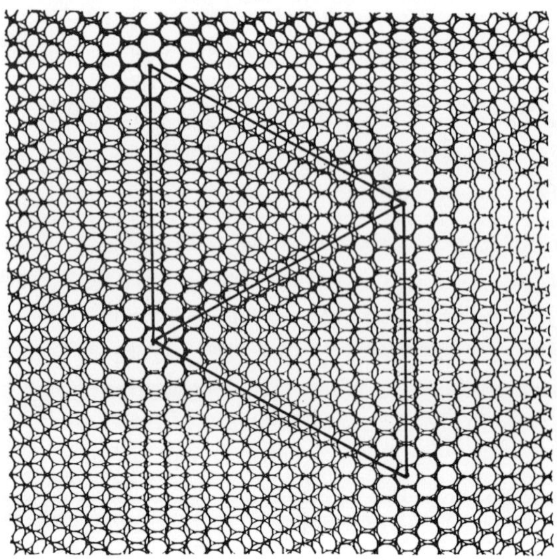

Figure 6. Strain relaxation through formation of a commensurate Co overlayer. A hard-sphere model for the relaxed system is shown, containing two unstrained lattices with a mismatch of 7 %. The (15 x 15)Co/(14 x 14)Ru unit cell is indicated.

wavelength of ~ 38 Å, which corresponds to ~ 14 Ru interatomic distances. Such a corrugation pattern is expected for a commensurate interface between two hexagonal layers with ~ 7 % lattice mismatch, i.e. with a 14/15 ratio in lattice spacings. Fig. 6 shows a hard-sphere model for the relaxed film. Hence the strain relaxation occurs via formation of an essentially unstrained hexagonal Co overlayer,[20] with nearly the bulk Co lattice constant, which is commensurate with the Ru(0001) substrate. In that case, 15 overlayer interatomic distances coincide with 14 interatomic distances of the substrate. High resolution images confirm this structural assignment. In addition they reveal a slight tetragonal distortion of the Co lattice.[21] Such a distortion may arise from lattice defects or imperfections in the 3-D Co clusters. In fact, closer examination of Fig. 5 shows that most of the 3-D clusters are distorted, which is evident from the numerous irregularities in their corrugation patterns. Imperfections in the Co lattice may, for example, arise from steps in the underlying substrate. Indeed in some of the islands shown in Fig. 5 the substrate step may still be identified in the corrugation of the cluster surface.

Both for the epitaxy of Co and of Au on Ru(0001), strain relaxation already occurs within the first few layers. This may either proceed via formation of misfit dislocations, as discussed in detail in the literature[22] or formation of periodic commensurate or incommensurate overlayer structures, as observed in these two examples. In fact, similar structures have been observed for other epitaxial metal on metal systems[19,23,24] and for clean reconstructed metal surfaces where the topmost layer is contracted with respect to the bulk.[25,26] The difference between the two epitaxial systems discussed, namely strain relaxation as opposed to pseudomorphic growth for the first layers of Au and Co, may be rationalized in a simple picture. The asymmetry of the interatomic interaction potential makes it energetically more costly to compress a 2-D layer (by ~7 % in the case of Au) than to expand a layer (by ~7 % in the case of Co), always compared to their respective bulk lattice spacing. On the other hand, the lateral corrugation of the adsorption potential for adatoms, as reflected by their mobility, allows qualitative conclusions regarding the likelyhood for pseudomorphic growth. In the case of very mobile adatoms, such as for Au/Ru(0001), the energy price to be paid for the occupation of non ideal adsorption sites is less than in the case of less mobile adatoms, such as Co/Ru(0001). Hence pseudomorphic growth is more probable in the latter case.

5. CONCLUSIONS

A comparative STM study, pertinent for the understanding of ultrathin magnetic film growth, has been performed on the nucleation, growth and structure of thin Co- and Au-films on atomically flat Ru(0001)-substrates. In both cases kinetic limitations, rather than thermodynamics, dominate growth at room temperature.

Growth proceeds via homogeneous nucleation and subsequent growth of 2-D islands, complemented by adatom condensation at step edges. The nucleation behavior is consistent with existing theories on homogeneous nucleation. The mobility of individual adatoms plays a decisive role in determining the nucleation density. The 2-D shape of islands, which is equally determined by kinetic effects, has been shown to be important for the onset of nucleation of higher layers. For both systems a quasi layer-by-layer growth mechanism results, with several layers exposed at the same time. The thermodynamically most stable film morphology is only reached upon heating to elevated temperatures. Co and Au follow a Stranski-Krastanov growth mode with one and two critical layers, respectively.

The atomic structure of the films is dominated by the tendency for strain relaxation. For Au this already begins in the first layer. The relaxation leads to complex periodic superstructures. Co grows pseudomorphically in the first layer. Strain relaxation occurs for Co film thicknesses from two layers on. In the relaxed films, the Co lattice forms a commensurate interface with the substrate.

6. ACKNOWLEDGMENTS

The authors thank J. Bohr for discussions. Financial support for this work came from the Deutsche Forschungsgemeinschaft via Sonderforschungsbereich 338 and from IBM Corporation within the Shared University Research (SUR) Program (contract No. 0861). We gratefully acknowledge fellowships from the Alexander-von-Humboldt Foundation (JV and RQH).

REFERENCES

1. 'Magnetism in Thin Films', edited by D. Pescia, in Appl. Phys. **A49**, 437 - 526 and 547 - 629 (1989).
2. S.S.P. Parkin, N. More, and K.P. Roche, Phys. Rev. Lett. **67**, 2304 (1990).
3. This workshop.
4. E. Bauer, Z. Kristall. **110**, 372 (1958).
5. R.Q. Hwang, J. Schröder, C. Günther, and R.J. Behm, Phys. Rev. Lett. **67**, 3279 (1991).
6. M. Bott, Th. Michely, and G. Comsa, Surf. Sci. **272**, 161 (1992).
7. J. Schröder, C. Günther, J. Vrijmoeth, R.Q. Hwang, and R.J. Behm, to be published.

8. W. Hösler, R.J. Behm, and E. Ritter, IBM J. Res. Development **30**, 403 (1986).
9. G. Pötschke, J. Schröder, C. Günther, R.Q. Hwang, and R.J. Behm, Surf. Sci. **251/252**, 592 (1991).
10. G. Ehrlich, Surf. Sci. **246**, 1 (1991).
11. J.A. Venables, G.D.T. Spiller, and H.M. Hanbrücken, Rep. Prog. Phys. **47**, 399 (1984).
12. Y.-W. Mo, J. Kleiner, M.B. Webb, and M.G. Lagally, Phys. Rev. Lett. **66**, 1998 (1991); and Y.-W. Mo, J. Kleiner, M.B. Webb, and M.G. Lagally, Surf. Sci. **268**, 275 (1992).
13. R.Q. Hwang, C. Günther, J. Schröder, S. Günther, E. Kopatzki, and R.J. Behm, J. Vac. Sci. Technol. **A10**, 1970 (1992).
14. R.Q. Hwang, and R.J. Behm, J. Vac. Sci. Technol. **B10**, 256 (1992).
15. This number is derived by using the equations given in ref. 11 and $i=1$.
16. E. Kopatzki, S. Günther, W. Nichtl-Pecher, and R.J. Behm, submitted for publication.
17. T.A. Witten, Jr., and L.M. Sander, Phys. Rev. Lett. 47, 1400 (1981).
18. R. Kunkel, B. Poelsema, L.K. Verheij, and G. Comsa, Phys. Rev. Lett. **65**, 733 (1990); and B. Poelsema, R. Kunkel, N. Nagel, A.F. Becker, G. Rosenfeld, L.K. Verheij, and G. Comsa, Appl. Phys. **A53**, 369 (1991).
19. G.O. Pötschke, and R.J. Behm, Phys. Rev. **B44**, 1442 (1991).
20. The Co film lattice may either have hcp or fcc stacking, exposing the (0001) or (111) surface orientations, respectively. From the data, we cannot distinguish between either of both possibilities.
21. C. Günther, J. Vrijmoeth, R.Q. Hwang, and R.J. Behm, to be published.
22. F.C. Frank, amd J.H. van der Merwe, Proc. Roy. Soc. **A198**, 205 (1949).
23. K. Yagi, K. Takayanagi, K. Kobayashi, and G. Honjo, J. Cryst. Growth 9, **84** (1971).
24. D.L. Doering, and S. Semancik, Phys. Rev. Lett. **53**, 66 (1984).
25. Ch. Wöll, S. Chiang, R.J. Wilson, and P.H. Lippel, Phys. Rev. B **39**, 7988 (1989).
26. J.V. Barth, H. Brune, G. Ertl, and R.J. Behm, Phys. Rev. B **42**, 9307 (1990).

STABILIZATION OF METASTABLE PHASES BY EPITAXY

Michel Piecuch[1], Stéphane Andrieu[1], Jean François Bobo[1], and Phillipe Bauer[2]

[1]Laboratoire Mixte CNRS-Saint Gobain CRPAM BP109
54704 Pont-à-Mousson, France
[2]Laboratoire de Physique des Solides, Université de Nancy I BP239
54506 Vandoeuvre-lés-Nancy, France

INTRODUCTION

The interplay between magnetism and crystalline structure is illustrated in the stable structure of the transition metals. While in the second and in the third series of transition metals ranging from yttrium to palladium and lutetium to platinum respectively, there is a regular succesion of the stable phases (HCP BCC HCP FCC), the existence of spin polarized electronic ground states perturbs dramatically this simple scheme in the first series. Manganese has, thus, a complicated cubic structure, iron is BCC and cobalt HCP in their low temperature states instead of being HCP, HCP and FCC respectively. It is, then, very attractive to obtain these first series transition metals in new crystalline structures to modify their magnetic properties and understand the interplay between crystalline and electronic properties. On the other hand, a simple band structure argument predicts the enhancement of magnetic moments with the augmentation of interatomic distances. It is, then, also interesting to obtain these metals with large interatomic distances. Recent progress in molecular beam epitaxy of metals offers new ways to realise such a program. The pseudomorphic epitaxy may constrain deposited metals in non equilibrium structures and with interatomic distances very different from those of the usual stables phases. Since 1988, we have demonstrated the possibility of growing several pseudomorphic iron and manganese planes on various single crystalline surfaces of ruthenium or iridium. Single crystal superlattices with quasi ideal interfaces have also been realized. This paper will give some examples of our recent work in this field with two main different basic starting points. In the two first parts, we will describe the structure and magnetic properties of iron grown on dense plane of ruthenium and iridium. The goal, here, is to obtain iron in close packed phases (FCC or HCP) but with expanded

Magnetism and Structure in Systems of Reduced Dimension
Edited by R.F.C. Farrow *et al.*, Plenum Press, New York, 1993

interatomic distances in order to recover some magnetic moment. In the third part, we will show how we have explored another way, the obtention of tetragonaly distorted iron phases starting with the epitaxy of iron on (100) iridium.

IRON ON (0001) RUTHENIUM

Single crystals of ruthenium with (0001) face have been obtained by epitaxial growth of ruthenium on (11$\bar{2}$0) sapphire[1]. The surface of ruthenium is reasonably flat as observed by electron diffraction (RHEED). We, then, tried to grow iron on this surface. Depending on the deposition rates, several planes of iron can grow pseudomorphically on (0001) ruthenium. Three pseudomorphic planes are obtained at low deposition rates (around 1Å/mn) while six ones are obtained at larger deposition rates (up to 1Å/s). At low deposition rates, we have obtained also, RHEED intensity oscillations, showing the layer by layer mode of growth during the deposition of the first third atomic layers[2].

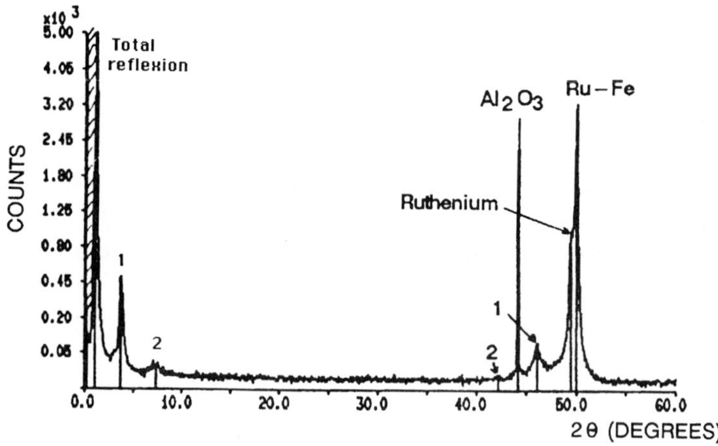

Figure 1. X ray spectra of a Fe/Ru superlattice with 8Å of iron and 20Å of ruthenium.

After the initial pseudomorphic growth, iron undergoes a phase transition to the so called Kurdjumov Sachs epitaxy of a BCC phase on a HCP (FCC) dense plane[3]. This epitaxy is relatively complicated since we have the coexistence of six different domains of BCC iron. The distances between iron rows relaxe slowly to the bulk values[3]. In superlattices we have obtained pseudomorphic flat surfaces up to six monolayers of iron[1]. The pseudormophy of iron on (0001) ruthenium implies in plane interatomic distances of the order of 2.71Å. To measure iron interplane distances in the superlattices, we have used Xray diffraction. The Xray spectra of this kind of superlattices (Figure 1) are relatively simple: one main peak corresponding to average interplane distances and several satellites corresponding to the modulation. Then we have analysed the spectra using[4]:

$$\frac{\sin\theta}{\lambda} = \frac{t_{Fe}}{\Lambda}\left(\frac{1}{d_{Fe}} - \frac{1}{d_{Ru}}\right) + \frac{1}{d_{Ru}} \qquad (1)$$

Where t_{Fe} is the iron thickness, Λ is the period of the superlattice, d_{Fe} and d_{Ru} are the unknown interplane distances of iron and ruthenium in the superlattices and θ is the position of the main Xray peak in a θ, 2θ scan.. This formula is fitted to the experiments in figure 2 and gives d_{Ru}= 2.142 Å as in bulk HCP ruthenium, but d_{Fe} is anomalously dilated along the c-axis (modulation axis): d_{Fe} = 2.073 Å. We were surprized by this result since a simple elasticity argument predict that, since iron is dilated in plane, it has to be contracted between planes. Then, as compared to the iron ε HCP phase (11.3Å^3/at.) or to the γ FCC phase (11.4Å^3/at.), iron in the Fe/Ru superlattices has a volume expansion of more than 11% (12.8Å^3/at.)[5].

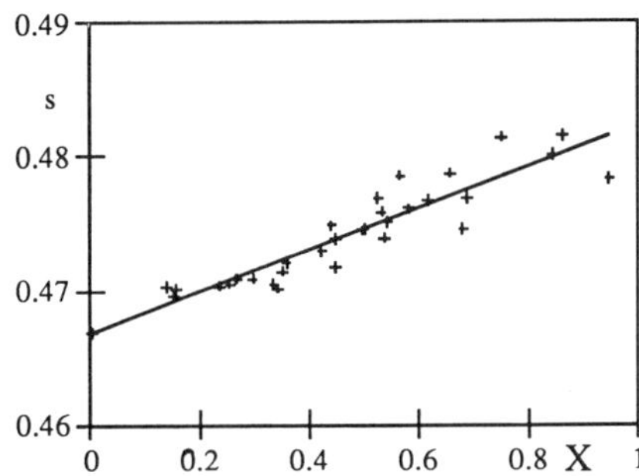

Figure 2. Plot of s= $\sin\theta/\lambda$ versus X= t_{Fe}/Λ where θ is the position of the main Xray peak, λ the X ray wavelength, t_{Fe} the iron layer thickness and Λ the modulation period of the superlattice. The experimental points corresponds to 30 superlattices.

X-ray Absorption Spectroscopy experiments were performed at LURE in Orsay. Two sets of measurements have been done using the X-ray linear polarization of the synchrotron beam. One set was measured with the electric field in the plane and the second one with the electric field perpendicular to the plane of the film[6]. This kind of geometry can probe the atomic environment and interatomic distances selectively in plane or out of plane. The main results are the anisotropic near edge spectra (XANES) which are HCP (or FCC) like in plane but BCC like out of plane. These results can be interpreted with a BCC like stacking of the iron hexagonal planes. This kind of stacking explains not only the observed anisotropic XANES and the distances and neighbours numbers determined by extended X-ray absorption spectra (EXAFS) but also provides a natural explanation of the large distances between hexagonal planes. It consists in the translation of each dense (0001) plane with respect to the hexagonal stacking positions: the atoms in the second plane are not above the center of gravity of the triangles of atoms in the first plane but they move to the center of these triangles sides. The atoms in the third plane are, like in a normal HCP metal, above the atoms in the first one.

Concerning magnetic properties, the volume expansion was expected to increase the iron local magnetic moment as a consequence of band narrowing. Surprisingly Mössbauer measurements[7] revealed two-planes thick magnetically dead layers (4 planes of atoms in a superlattice have a zero hyperfine field). The magnetic atoms have a slighty lower hyperfine field than the BCC one: 330 kOe instead of 342 kOe at 4.2 K. The Mössbauer results were confirmed by magnetization measurements at liquid Helium temperature[8]. We, again, interpreted the results with two atomic dead layers: using the following formula:

$$\mu = \mu_0 - \frac{\mu_0 \, n_0}{n} \qquad (2)$$

Where μ is the measured magnetic moment (in Bohr magnetons per iron atoms), n the number of iron layers, μ_0 is the expected magnetic moment of the magnetic iron atoms and n_0 is the number of dead layers. The comparison with experiment is made in figure 3.

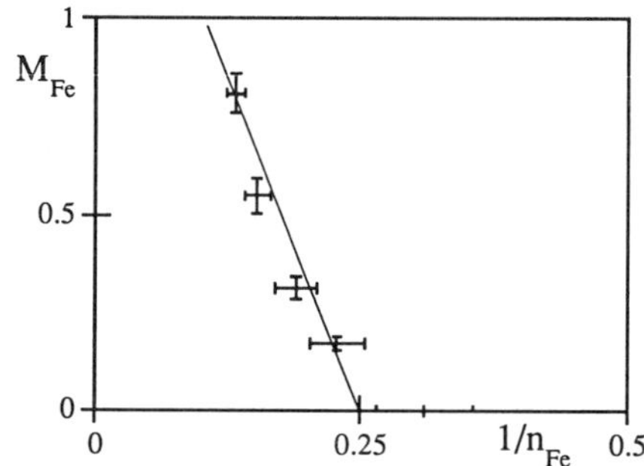

Figure 3. iron magnetization in Bohr magnetons per atoms versus $1/n_{Fe}$ in Fe/Ru (0001) superlattices. n_{Fe} is the number of iron atomic planes. The error bars are calculated using a constant thickness error of about 1Å.

One can see that equation (2) is close to the experiment with $\mu_0 = 1.7\mu_B$ and $n_0 = 4$ (2 dead layers at each interfaces). The bulk iron atomic moment is significantly lower than in the BCC phase[9].

IRON ON (111) IRIDIUM

To understand the influence of the substrate on the stacking of the dense planes and to increase the iron magnetic moment, we substitued HCP ruthenium by FCC iridium. Iridium has the same interatomic distances than ruthenium in its close packed planes but iron is much more magnetic in IrFe FCC solid solutions than in the HCP FeRu ones. The initial single crystals where 200Å thick layers of either (0001) ruthenium or (111) iridium epitaxialy grown on (11$\bar{2}$0) sapphire. RHEED revealed pseudomorphic iron growth up to three atomic layers and RHEED oscillation can also be observed. Xray diffraction experiments on Fe/Ir superlattices were analyzed with the same method than for the Fe/Ru superlattices. The distance between iron hexagonal planes was found to be 2.10Å , that is an even bigger atomic volume of 13.3$Å^3$/at. for iron than in the Fe/Ru system (iridium keeps its bulk value of 2.217Å). But, Xray diffraction on powdered superlattices have shown an FCC stacking[2]. The XANES spectra on Fe/Ir superlattices are the fingerprints of a classical FCC or HCP stacking. Conversely to the Fe/Ru case, the Fe/Ir XANES spectra are perfectly isotropic[2]. This result confirms the perfect FCC structure of iron. This isotropy in the Fe/Ir superlattices reveals that the c/a ratio (equal to 1.55 and far from the ideal hcp ratio, equal to 1.63) was not responsible to the XANES anisotropy in the Fe/Ru ones. This observation provides another argument in favor of a non simple crystal structure in Fe/Ru. However, the large interplane distances in Fe/Ir superlattices cannot be explained by the same argument as in Fe/Ru. These distances can be related, may be, to the fact that the FCC lattices are robust against a trigonal distortion (expansion or contraction along a <111> axis).

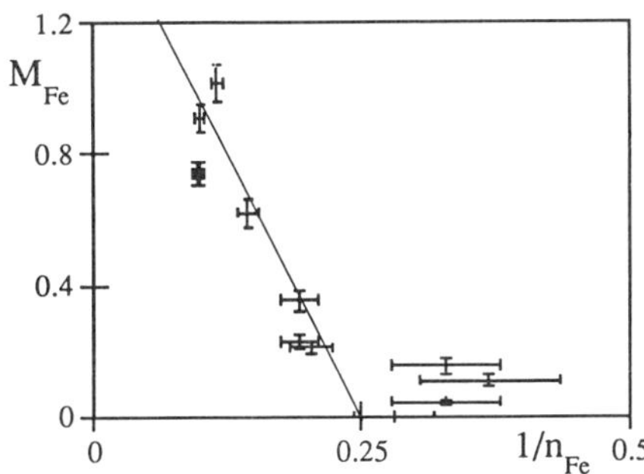

Figure 4. iron magnetization in Bohr magnetons per atoms versus $1/n_{Fe}$ for Fe/Ir (111) superlattices. n_{Fe} is the number of iron atomic planes. The error bars are calculated using a constant thickness error of about 1Å.

These results show that long range interaction with the non magnetic spacer not only produces long range magnetic coupling but also leads to different crystalline structures. One may quote[10] that the interplane pair energy in close packed phases which is responsible of the various stacking sequences of the dense planes has exactly the same oscillatory dependance with the distance between planes than the Bruno and Chappert[11] version of the RKKY theory.

The magnetic properties of these superlattices have been investigated at room temperature with a gradient force magnetometer. The magnetization results are plotted in figure 4 in the same way than in figure 3 for the Fe/Ru system. We again found two magnetically dead layers. The iron magnetic moment corresponds to about 1.5μ_B. However, preliminary measurements at 4.2K reveals only one (or zero) dead layer[9] and a larger magnetic moment.

IRON ON (100) IRIDIUM

Finally, we have started the study of iron grown on (100) iridium. The goal of this study is to test the magnetic and structural properties of BCT (body centered tetragonal) iron with c/a values intermediate between the BCC iron and FCC iron. A BCC phase is evidently a BCT phase with c/a=1 but a FCC phase can also be described (by a 45° rotation of its two axes in a (100) plane) as a BCT phase with c/a=$\sqrt{2}$. Since the distances between atoms in the (100) iridium plane (2.71Å) are just intemediate between the distances in (100) BCC iron (2.87Å) and (100) FCC iron (2.52Å), we can expect the obtention of a new BCT iron phase by pseudomorphic epitaxy of iron on (100) iridium[§].

(100) iridium buffer layers of about 200Å are prepared by epitaxial growth of iridium on (100) MgO. These iridium layers show high quality RHEED patterns and are fully relaxed at about 100Å. RHEED oscillations are observed at the beginning of the growth of iron on this (100) iridium surface up to 10 atomic planes at around 270K and 4 planes at 400K. We do not observe any alteration of the initial (100) iridium RHEED patterns for unlimited iron thicknesses. The surface lattice of iron thus keeps the square symmetry.

The measurement of the distances in the plane of growth at room temperature shows that the iron interatomic distance does not vary from the iridium one (2.715Å) up to four deposited planes, as relaxation to BCC interatomic distance thus occurs for thicker deposits[14]. In the superlattices, the iridium overlayer stabilize the pseudomorphic phase and we have obtained pseudomorphic superlattices up to six monolayers of iron.

The determination by X-Ray diffraction of the distances between planes is more complicated to undertake than in the dense plane superlattices since the interplane distances are, now, very different and leads to relatively complicated Xray diffraction patterns. We have tried to determine these distances by fitting the Xray patterns with a model of a quasi ideal superlattice. This model fits both the positions and intensities of the lines. We obtain always qualitative agreement between the calculated lines intensities and positions and the experimental ones. In pseudomorphic superlattices we find with this procedure: 1.92Å (equal to bulk (200) iridium distance) for iridium and 1.71 $\pm$ 0.01 Å for iron, in a series of samples with four iron planes per layer[14]. Since we find that iridium is thus not strained, we conclude that iron is in a BCT structure with a c/a ratio equal to 1.26. The atomic volume of iron is equal to 12.6 Å^3/at and is then also expanded. The in plane nearest neighbours of an iron atom are obviously at 2.71Å like in the (100) iridium plane, but the out of plane nearest neighbours are at 2.57Å which is also a larger distance than in usual FCC iron (2.52Å).

[§] One can quote here that the BCT and FCT (face centered tetragonal) phases are two equivalent description of the same phase but the c/a ratios are reversed: the FCC phase is FCT with c/a=1 and the BCC is FCT with c/a= $\sqrt{2}/2$ (this point needs to be quoted since the two words are randomly used in the litterature). We have followed here an old historical tradition: the BCC-FCC transition in iron was described as a BCT deformation by Bain[12] long time ago (the so called Bain transformation). We may recall also that the international convention[13] for the tetragonal lattice admits only primitive and body centered tetragonal Bravais lattices.

For thicker iron deposits, iron seems to abruptly transit in its regular BCC phase. However, details about this relaxation along the growth axis still remain unclear. For thick iron layers (15Å) and thin iridium ones (< 3 planes), iron and iridium distances along the growth axis are found to be equal to 1.45 and 1.85 ± 0.05Å respectively, which shows that iron is nearly relaxed in its regular BCC phase (c/a= 1.01) and that iridium is strained. Assuming that the surface lattice is relaxed in the BCC (100) iron one, an elastic calculation gives an iridium distance between planes in the direction of growth equal to 1.845Å. In these superlattices iridium is in a BCT phase of c/a=1.29 and has an atomic volume of 15.2 $Å^3$/at instead of 14.2 $Å^3$/at. Then that is iridium which is expanded and in a metastable phase. For the intermediate regime (iron thickness > 6 planes and iridium thickness > 3 planes), iridium deposits are sufficiently thick to relaxe the strain imposed by iron. The situation is thus more complex but seems to corresponds to elastic equilibrium between BCC iron and FCC iridium. Both elements are then in a BCT phase, iron with a c/a between 1 and 1.1 and Ir with c/a between 1.3 and $\sqrt{2}$. However, we cannot rule out a solution with variable interplane distances in the two elements.

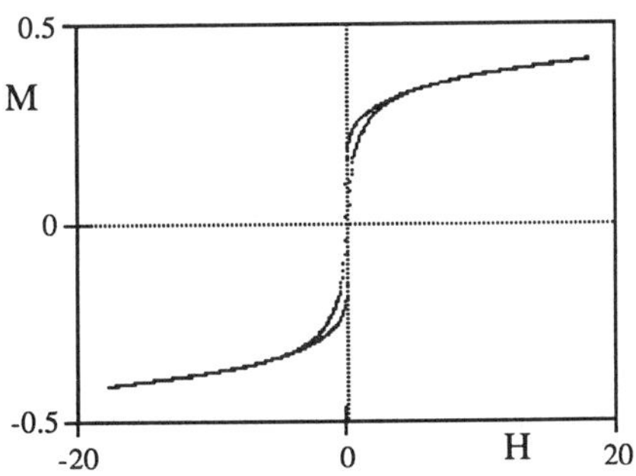

Figure 5. magnetization curve at room temperature for a sample with 7.4Å of iron (4.5 atomic planes) and 5.6Å of iridium, M is in Bohr magnetons per iron atoms and H in kOe.

In conclusion, we have demonstrated that depending on the ratio of the iron and iridium thicknesses we obtain three situations: pseudomorphic BCT iron (named BCT 1 in the following), fully relaxed BCC iron and an intermediate BCT phase close to BCC iron (named BCT 2 in the following). The situation for the non pseudomorphic phases is simpler than in the case of epitaxy on dense plane since the symmetry of non pseudomorphic iron is here the same as the pseudomorphic one and then we obtain only one domain in that case and good epitaxial superlattices.

As this paper concerns the metastable pseudomorphic phases, we will concentrate on the BCT 1 phase. The magnetic properties of the samples were determined at room temperature with a gradient force magnetometer. The results can be divided in two different families, the first one for low iridium interlayer thicknesses and the second one for the larger iridium thicknesses. Fig. 5 shows hysteresis loops observed on BCT 1 structure for a superlattice with low iridium thickness. The iron atoms seem to be clearly magnetic. We have tried to determine the saturation moment by plotting the magnetization results as M^2 versus H/M (Arrot plot). The Arrot plots are reasonably linear at high field and we can determine the saturation magnetization by extrapolation at 0 field. We have found a magnetic moment of about 0.31 μ_B per atoms. However for the other ferromagnetic samples the magnetic moment varies and seems to be a complicated function of iridium thickness. The magnetic properties of the superlattices with Ir thickness greater than about 12Å have also been analyzed by Arrot plots. They are all paramagnetic in that sense (negative intercept with the M^2 axis) but they can have an overall magnetic signal at 18 kOe which is of the same order of magnitude as for the preceding samples and even greater.

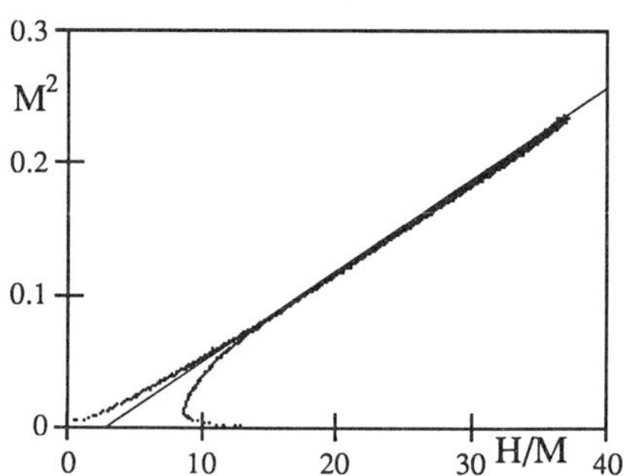

Figure 6. Arrot plot (M^2 versus H/M) for a sample with 9Å of iron (5.6 atomic layers) and 30Å of iridium. M is in Bohr magnetons per iron atoms and H in kOe, the straight line is a fit.

One typical Arrot plot for such a sample is shown on figure 6. One can see that the curve is almost perfectly linear, except in very low field. Again the parameters deduced from the Arrot plots, the zero field susceptibilities and the high field slopes are non monotonous fonctions of the iridium thickness. One simple interpretation of this behavior is that the long range magnetic order is very sensitive to the interlayer coupling, when this coupling is high and ferromagnetic we have a weak ferromagnet, when it is low or antiferromagnetic we loose the long range order and since this coupling oscillate with iridium thickness, as demonstrated by neutron scattering on the more magnetic BCC and BCT 2 samples[15], the behavior of the magnetization is an oscillating function of the iridium thickness. This seems to be an entirely new phenomenon of metamagnetism induced not by the field but by the interlayer coupling. This interpretation is supported by the plot of the Landau parameter a deduced from the Arrot plots versus the iridium thickness (figure 7), a is defined as:

$$a = A/B \tag{3}$$

Where A is the intercept of the Arrot plot with the y axis (M^2 axis) and B is the slope. In a conventional mean field description of the magnetic phase transition, a is given by:

$$a = a_0(T_C - T) \tag{4}$$

Where T_C is the Curie temperature. If a is positive we have an ordered magnetic phase and when a is negative we are in a paramagnetic phase. As in mean field theory, T_C is given by:

$$T_C = n\,J \tag{5}$$

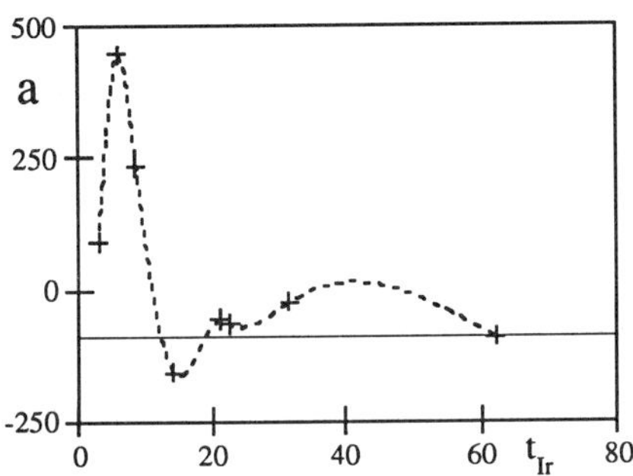

Figure7. Landau parameter (arbitrary units) versus iridium thickness(Å). The horizontal line shows the estimated asymptotic value. The line between experimental points is only a guide for the eyes.

Where J is the ferromagnetic exchange integral and n a constant related to the number of neighbours, here we can write:

$$T_C = n_1 J_1 + n_2 J_2 \tag{6}$$

Where J_1 is the intralayer coupling and J_2 the interlayer coupling. Then the variation of a with the iridium thickness is a picture of the variation of J_2, the interlayer magnetic coupling. This picture shows a clearly defined ferromagnetic maximum at $t_{Ir} \approx 6$Å, a more diffuse one at 30Å and a clearly defined antiferromagnetic maximum at $t_{Ir} \approx 14$Å. While the neutrons results show three well defined antiferromagnetic peaks at $t_{Ir} \approx 4$Å, 14Å and 18Å[15]. A more precise picture of the magnetic properties needs, however, the determination of the magnetic isotherm at various temperature. This will be made in the near future.

DISCUSSION

We have determined in this paper the structure and magnetic properties of various pseudomorphic phases of iron on ruthenium or iridium. We will discuss now these properties according to current theoretical predictions. We first, will emphazise that the obtained phases are, in the three case studied here, novel phases and not simple elastic modifications of known phases. The most exotic case is the iron on ruthenium phase which seems to be an hexagonal modification of BCC iron but with a large volume. The other two are non trivial modification of the FCC phase, a very large volume sligthy trigonaly deformed one in the case of iron on (111) iridium and a largely tetragonalized one with a rather large volume, also, in the case of iron on (100) iridium. The large strain imposed by the epitaxy are certainly out of the application of a linear elasticity theory but the epitaxy is not simply a strain process and the long range interactions with the substrate are clearly shown by the different behavior for the iron on (0001) ruthenium and the iron on (111) iridium where the strains are almost the same.

The magnetic properties of HCP or FCC iron have been calculated recently by several groups[16,17]. For FCC iron, the situation can be summarized as follow[16]: several different solutions have been tested, non magnetic (NM), antiferromagnetic with (100) planes having alternating moment (AF), high spin ferromagnet (HS) and low spin ferromagnet (LS). The AF solution is stable for atomic volume between 10.2 $Å^3$/at. and 12.3 $Å^3$/at, the NM phase is stable at small volume and the HS at high volume. Then, our pseudomorphic iron on (111) iridium should be in the HS phase. This seems to be the case although the experimental magnetic moment (at room temperature) seems to be low (1.5μ_B instead of 2.9μ_B). This may be due to the trigonal distortion along the (111) direction which can lower the magnetic moments (six atoms are at 2.62Å while the six other first neighbours are at 2.71Å). This is, however, only a second order effect on the band width. A more plausible explanation will be that the order temperature is rather low and that, at low temperature the magnetic moment are bigger. This last point is preliminary confirmed by a SQUID measurement at low field (100 Oe) which shows a significant increase of the magnetization around 20K.

In HCP iron the theoretical predictions are almost the same[17], the ferromagnetic phase is, however, only stable against the antiferromagnetic one at a very large volume of the order of 13$Å^3$/at. At our experimental volume the ferromagnetic phase is only metastable and its theoretical magnetic moment is very large (2.5μ_B). Motivated by our experimental results and by the close ones of Liu and Bader on epitaxial iron on (0001) ruthenium[18], several groups have tried to calculate the magnetic properties for iron on ruthenium or in superlattices[19,20,21]. The main problem was the existence of two magnetically dead layers and also the possible existence of a very peculiar phase intermediate between HCP and BCC iron. Concerning this second point, Koenig and Knab[19] have shown that this structure stabilizes the ferromagnetic structure against the antiferromagnetic one and that one obtains no dead layers and a high iron magnetic moment (although the iron atoms close to the interface have a lower moment than those inside the layers). Concerning the dead layer problem, Wu and Freeman[20] have proposed an elegant solution: they found that the first overlayer is, in plane, antiferromagnetically coupled and then cannot be seen by magnetization measurement. They also found a significant decrease of the value of the magnetic moment by hybridization with ruthenium. However, they predict a small but measurable hyperfine field of 48kOe. In our Mössbauer measurements[6] we have not detected any hyperfine field and we have even shown, by application of an external field, that iron atoms bear a zero magnetic moment (absolute zero susceptibility). In conclusion, on hexagonal iron, it seems that the experimentally determined hybridization between iron and ruthenium is bigger than the theoretical one (this hybridization plays a central role since iron lacks its moment as a dilute impurity in ruthenium). This can be due to the experimentally inevitable interface imperfections. But, if this is possible for the first iron layer, this seems more difficult in the second one since we have demonstrated the relatively high perfection of the interfaces in our iron/ruthenium superlattices[5]. Again, the magnetic moment of iron is rather low ($1.7\mu_B$ at 4.2 K) inside the layers and this seems to be a general discrepancy between theory and experiments.

For the tetragonaly deformed phase on (100) iridium, the atomic volume is just slighty larger than the theoretical critical volume for the transition between AF and HS solution in FCC iron[16]. Then, the situation can be complicated. Experimentally, a lot of work has been done on (100) iron on (100) copper. In that case, the deformation is expected to be small and iron close to a only slighty expanded FCC iron. However, it has been demonstrated recently[22] that the iron layers grown at low temperature and with few monolayers of iron are in a distorded FCC structure and that fully FCC iron on copper is only obtained at high temperature or (and) at relatively large thickness. Then, the results of Macado and Keune[23] are the most reliable for the magnetic properties of (100) FCC iron on copper. They found that iron is antiferromagnetic with a Neel temperature of 65K[23]. Several groups[24,25] have recently attempted to calculate the properties of tetragonaly deformed iron. The main result of the calculations is that the two stable (or metastable) phases are the NM FCC or the HS BCC, the other values of c/a seem to be instable. But, the calculations have not tested so far the antiferromagnetic solution which is the most stable one for FCC iron both in the experiments[23] and in the first principle calculations[16]. For instance at our atomic volume and c/a value the predicted magnetic moment is around $2.2\mu_B$[24]. Two possible explanation of our results can then be conjectured. The first one is that we are at a temperature bigger than the order temperature of our phase. This cannot be ruled out since the order temperature may decrease with the volume and that the order temperature of FCC iron on copper is only 65K[23]. Then, what we see can be magnetic properties induced by the interlayer coupling between paramagnetic phases close to the order temperature. The second possible explanation is that we have an antiferromagnetic phase, but, due to a non even number of planes (4.5 or 5.4 in our samples) we see a weak magnetization (corresponding to 1/9 or 1/11 full magnetic signal), in that case the real in plane magnetic moment will be relatively huge ($2.8\mu_B$). Again, in that case the interlayer coupling is crucial to propagate the magnetic signal throughout the sample and weak or antiferromagnetic coupling between the layers should decrease the magnetic signal. This second solution is however improbable since conversion electron Mössbauer measurements at room temperature show a very small broadening of the lines, only compatible with an hyperfine field smaller than 15kOe.

CONCLUSION

In this paper, we have demonstrated the possibility of growing thin films of iron with metastable structures very far from the equilibrium BCC one. The obtained phases depends critically, not only of the epitaxial constraint but also of the interaction with the substrate, leading for instance at the same epitaxial constraint (iron on (0001) ruthenium or (111) iridium) to very different phases ("simple" FCC or complicated hexagonal). The obtained phases can usually be maintained only for a few atomic planes. But, the realization of epitaxial superlattices permits the obtention of more important samples.

The magnetic properties of these phases cannot be described by simple arguments like the volume expansion. The interplay between the influence of the substrate or interlayer (interlayer coupling and hybridization) and the structure leads to unusual magnetic properties which hardly compare to the theoretical results although some general trends can be extracted, the large volume seems to favorize the high spin phases both in experiment and in theory and the hybridization is crucial and tends to diminish the magnetic moments as in solid solutions. This last point, however, needs certainly more progress in both sides: more realistic structures for the theoretician (problem of computing time) and better samples for the experimentalist since it remains relatively hard to discriminate between the effect of hybridization and the influence of defects at the interfaces.

Acknowledgements

The authors and particularly M.P. acknowledge all the peoples which have participed at this program during its five years of accomplishement and particularly Marc Maurer, Marie Françoise Ravet and Jean claude Ousset which were at the begining of this work.

REFERENCES

1. M.Maurer, J.C Ousset, M.F. Ravet, M. Piecuch, Europhys. Lett. **9**: 803 (1989)
2. S Andrieu, M.F.Ravet, O. Lenoble, V. Dupuis, M. Piecuch, Europhys. Lett. **18**: 529 (1992)
3. S Andrieu, J.F. Bobo, M. Piecuch, Phys Rev. B to appear (August 1992)
4. S Andrieu, M.F.Ravet, M. Maurer, V. Dupuis, M. Piecuch, Mat. Res. Soc. Symp. Proc **230**: 365 (1991)
5. M. Piecuch, M. Maurer, M.F. Ravet , Mat. Res. Soc. Symp. Proc **187:** 231 (1990)
6. M. Maurer, M. Piecuch, M.F. Ravet, J.C. Ousset, J.P. Sanchez, C. Aaron, J. Dekoster, D. Raoux, A. De Andres, M. De Santis, A. Fontaine, F. Baudelet, J.L. Rouvière, B. Dieny, Journal of Magn.Magn. Mat. **93**: 15 (1991)
7. J.P.Sanchez, M.F. Ravet, M. Piecuch, M. Maurer, Hyperfine Interactions **57**: 2077 (1990)
8. M.Piecuch, M. Maurer, M.F. Ravet, J.C. Ousset, J.M. Broto, B. Dieny, Nato ASI series **259**: 195 (1990)
9. M. Piecuch, S Andrieu, J.F. Bobo, M. Maurer, M.F.Ravet, V. Dupuis, J.C. Ousset, B. Dieny, Mat. Res. Soc. Symp. Proc **232**: 165 (1991)
10. A. Blandin, J. Friedel, G. Saada, J. Phys. (Paris) **27**: C3, 128, (1966)
11. P· Bruno, C. Chappert, Phys. Rev. Lett. **67**: 1602 (1991)
12. E.C. Bain:,Trans. Am. Inst. Min. Metall. Pet. Eng. **70**: 25 (1924)
13. International tables for crystallography, Volume A, Th. Hahn editor Reidel Dordrecht (1987)
14. S. Andrieu et Al, To be published
15. M. Piecuch et al, To be published
16. V.L. Moruzzi, P. M. Marcus, J. Kübler, Phys Rev B **39**: 6957 (1989)
17. J. Kübler, Solid State Com. **72**: 631 (1989)
18. C.Liu, S.D. Bader, Phys Rev B **41**: 553 (1990)
19. D. Knab, C. Koenig, Phys. Rev. B **43**: 8370 (1991)
20. R. Wu, A. J. Freeman, Phys. Rev. B **44**: 4449 (1991)
21. M. O. Selme, P. Pecheur, J. Mag. Mag. Mat. **93**: 285 (1991)
22. H. Magnan, D. Chandesris, B. Villette, O. Heckmann, J. Lecante, Phys. Rev. Let. **67**: 859 (1991)
23. W. A. A. Macedo, W. Keune, Phys. Rev. Let. **61**: 475 (1988)
24. S. Peng, H. J. F. Jansen, J. Appl. Phys. **67**: 4567 (1990)
25. G. L. Krasko, G. B. Olson, J. Appl. Phys. **67**: 4570 (1990)

REAL-TIME RHEED STUDIES OF EPITAXIAL Co-Cr SUPERLATTICES

D. Barlett*, W. Vavra, S. Elagoz, C. Uher and Roy Clarke

*Applied Physics Program
Department of Physics
Randall Laboratory
University of Michigan
Ann Arbor, MI 48109-1120

INTRODUCTION

The epitaxial growth[1,2] of a bcc form of cobalt, on (110) GaAs, has stimulated a great deal of interest in metastable phases of the magnetic transition metal elements. An issue of general importance in this kind of system concerns the accommodation of lattice mismatch, including possible differences in symmetry, at the heterostructure interfaces. The resulting interface structure is known to play a crucial role in the magnetic properties of such materials, especially with regard to anisotropy[3], interlayer coupling,[4,5] and giant magnetoresistance effects[6].

We report here the results of a time-resolved, quantitative reflection high-energy electron diffraction (RHEED) study on the growth of Co-Cr superlattices. Using a newly developed CCD-based detection and analysis system[7], lattice spacing and domain size were recorded simultaneously as a function of Cr-layer thickness. The data reveal the existence of a new close-packed form of Cr oriented in the (111) direction. For thicknesses greater than ~ 6Å we observe a transition to the bulk bcc structure in the classic Kurdjumov-Sachs (KS) and Nishiyama-Wasserman (NW) epitaxial geometries.

SAMPLE GROWTH

Samples were grown on (110) GaAs substrates in a Vacuum Generators V80 molecular beam epitaxy chamber at 1-5 x 10^{-9} mbar. Cr and Co were evaporated from electron beam hearths at rates that ranged from 0.35 to 0.5 Å/s. Cr rates lower than this led to poor sample quality indicating that supersaturation growth conditions are beneficial. To establish a smooth (0001) Co surface on which to begin superlattice growth the following buffer layers were grown on the GaAs: first, 500 Å of (110)Ge, then 25 Å bcc Co, and finally 20 Å (111)Au. Superlattice bilayers consisting of [20Å Co/*x* Å Cr] with *x*=4,6,8,10 and 12 Å were then grown on the Au. Lastly, a protective 10 Å Au cap was deposited on the samples. The Ge was grown at 550°C, all metals at 50°C, and the total thickness of each superlattice was 1000 Å. X-ray photoemission showed that the surface contamination (e.g., oxygen and carbon) was less than the detectable limit (<2%).

Magnetism and Structure in Systems of Reduced Dimension
Edited by R.F.C. Farrow *et al.*, Plenum Press, New York, 1993

RESULTS

Figure 1 shows an x-ray χ scan, in which the diffraction vector lies in the sample plane[8], for a [20Å Co/4 Å Cr]$_{42}$ superlattice. The diffraction peaks with 60° symmetry indicate a highly coherent close-packed structure. A c* scan[8] [inset (b)], which probes the stacking of the layers, indicates the Co is (0001)-oriented hcp, and a low angle scan [inset (a)] shows well-formed superlattice peaks. Although the Co is hcp, we use fcc notation to describe all surface directions. This is done to make comparisons with previous literature more straightforward, as well as to simplify the notation.

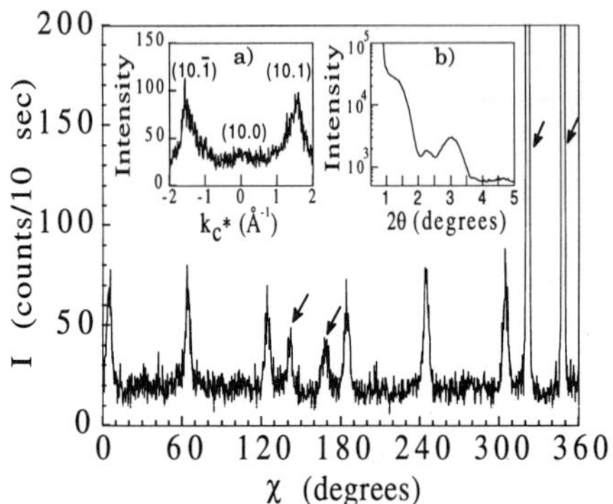

Figure 1. X-ray χ-scan for [20Å Co/4 Å Cr]$_{42}$ superlattice, showing diffraction peaks with 60° symmetry. Inset (a): x-ray c* scan indicating hcp stacking of Co layers. Inset (b): low-angle x-ray scan showing superlattice peaks. Arrows indicate GaAs (112) and (420) peaks.

We now turn our attention to the RHEED images of Figure 2. These are diffraction patterns from Co and Cr layers in samples with 4Å and 10Å Cr thickness. The electron beam is parallel to Co<110>. One can see that the superlattice is well ordered for a sample with 4Å Cr layers, but the RHEED pattern indicates a structural transition has occurred for the 10Å Cr sample. Two separate regions of the RHEED patterns were monitored simultaneously during growth. These two regions are labeled "NW" and "KS" in Figure 2, to be explained below. Note the same regions were scanned for all the superlattices, but these regions become visibly distinct only after the structural transition has occurred. Streak separation and width (giving respectively the lattice spacing and domain size) were measured in these two regions at a rate of 15 data points per Å during deposition.

The same measurements were also performed along the other major azimuth, Co<112>. Absolute values of lattice spacings are calibrated with respect to the Au buffer layer. Domain sizes (see Figure 3 insets) are lower limits and strictly speaking are the average in-plane structural coherence lengths perpendicular to the incident electron beam. The contribution to the widths from instrument broadening has been deconvoluted from the data.

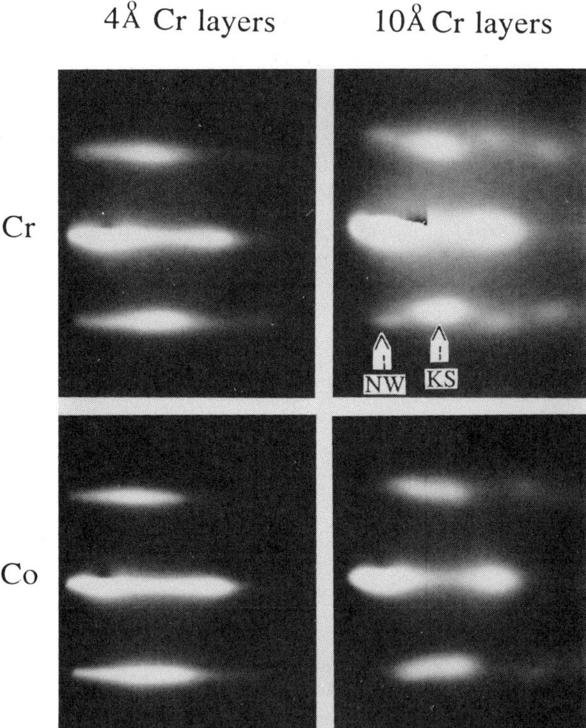

Figure 2. Representative RHEED images along the Co<110> direction. Sharp RHEED streaks are seen for the Co and Cr layers of the [20Å Co/4 Å Cr]$_{42}$ superlattice. In contrast, the data for Co and Cr layers of the [20Å Co/10Å Cr]$_{33}$ superlattice indicate a structural transition has occurred.

In Figure 3(a) we have plotted lattice spacing versus thickness of the same Cr layers shown in Figure 2. Again, the RHEED beam is parallel to Co<110>. First, we see that the epitaxial strain of the overlayer begins relaxing as soon as growth is started. Therefore, there is no critical thickness up to which true pseudomorphic growth occurs without the formation of misfit dislocations [Figure 3(a) inset]. This is in contrast to strained-layer semiconductor growth[9], and is due in part to the isotropic nature of the metallic bond compared with the more directional semiconductor bond.

The most important feature in Figure 3 is that the Cr d-spacing initially *increases* along *both* azimuths during growth. We will show presently that this is inconsistent with bcc Cr growth. The 4Å Cr layers (dashed lines) show an increase in d-spacing throughout the growth of the layer, whereas the 10Å Cr layers (solid lines) undergo uniform expansion only until 4-5Å Cr thickness. Beyond this thickness the d-spacing along Co<110> [Figure 3(a)] reaches a maximum in the NW region then begins decreasing. This maximum in d spacing coincides with a minimum in domain size [Figure 3(a) inset]. In other words the misfit dislocation density increases with Cr thickness until there is a transition to a lower strain energy form of epitaxy, then the sample begins healing.

Along Co<112> [Figure 3(b)], the Cr transition is even more pronounced: at 4-6Å an abrupt 15% change in the d-spacing occurs. Correspondingly, the domain size (inset) shows a deep minimum at this thickness and then the surface becomes more coherent. This again clearly shows evidence for a structural transition.

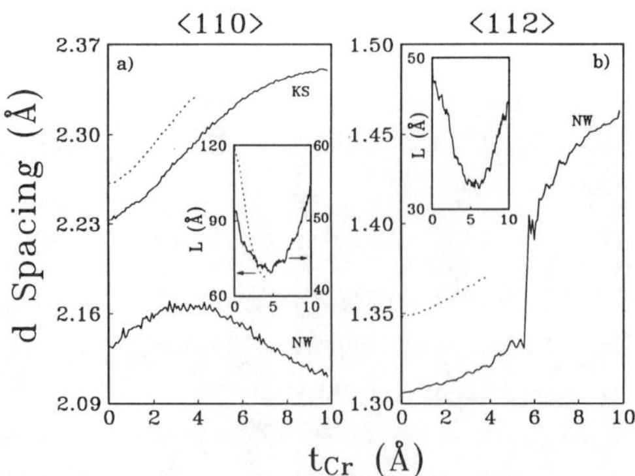

Figure 3. a) d spacing profiles during growth for the Cr data shown in Figure 2. The RHEED beam is parallel to Co<110>, while the dashed and solid lines refer to the 4Å and 10Å Cr layers, respectively. b) RHEED beam parallel to Co<112>. Insets are domain size (L) profiles with arrows indicating the appropriate y-axis. X-axes have the same scale as in the d spacing plots.

DISCUSSION

The analysis of the transition is based on extensive work found in the literature on bcc(110)/fcc(111) (or hcp 0001)) metal epitaxy. Enlightening theoretical predictions of the orientation and growth for this kind of heteroepitaxy have been performed by several authors[10-13]. The results of these studies show that minima in the total overlayer free energy occur for orientations such that the most dense packed rows are parallel. This is easily seen by considering the potential energy of a monolayer deposited on a substrate[13],

$$U = \phi_\circ + \sum_{G} \sum_{G'} \phi_G \delta_{G,G'} \qquad (1)$$

Here **G** and **G'** are the surface reciprocal lattice vectors (rlv's) of the substrate and overlayer, respectively, and ϕ_G are unknown *negative* Fourier coefficients whose absolute values increase rapidly with decreasing $|G|$. This expression shows that the total potential energy is lowered by matching overlayer and substrate rlvs, i.e. by atomic row matching. The lowest energy configuration is where *all* rlv's match and the overlayer assumes the structure of the substrate. There is, however, an increase in strain energy as the thickness of this pseudomorphic layer increases. When these two competing terms offset each other there is a transition to the next lowest energy configuration. In this next configuration the overlayer takes its preferred (bulk) phase and aligns its *shortest* rlv's with those of the substrate, thus making the most densely packed rows of atoms parallel. These orientations correspond to the Kurdjumov-Sachs (Cr<111>//Co<110>) and Nishiyama-Wasserman (Cr<001>//Co<110>) orientations. The KS and NW orientations for Co-Cr are illustrated in Figure 4, with row spacing mismatches of $(d^{Cr} - d^{Co})/d^{Co}$ = 8.5% and -6.1%, respectively. It is interesting to note that similar behavior[14] has been observed in Fe or Ir.

Our d-spacing data for the five samples are summarized in Figure 5. The underlying (bulk) Co(111) d-spacings are also plotted for reference. Measured d-spacings are taken from the top surface of individual layers, and these were typically averaged over 3-4 layers at various stages of each superlattice. We emphasize that the results in Figure 5 do not correspond to a growth front profile as in Figure 3, but what is shown is a plot of the average Cr layer d-spacing for each of the superlattices. Figure 5 shows that after the transition, the measured d-spacings for both directions approach those expected for the KS and NW orientations. In contrast, we see isomorphic Cr expansion for the 4Å and 6Å Cr samples, which is inconsistent with bcc growth in the NW or KS orientations. This isomorphic expansion, plus the strong 60° in-plane symmetry and the large in-plane coherence length for these samples (from RHEED and X-ray analysis), indicates that the 4Å and 6Å Cr superlattices are strained coherently into the close-packed structure of the underlying Co.

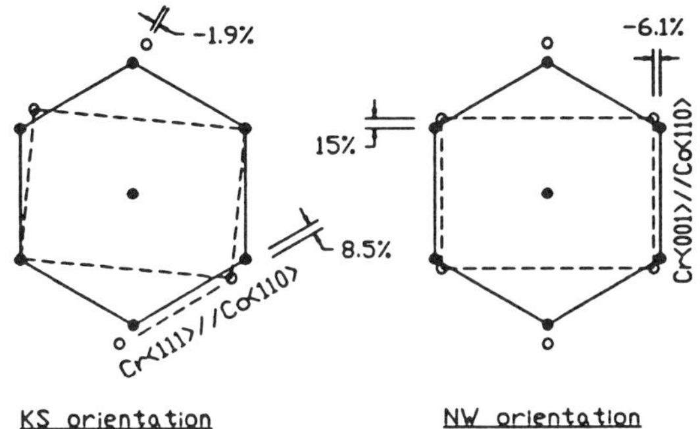

Figure 4. Schematic diagram of bcc (110) Cr mesh (open circles and dashed line) on hcp (0001) Co mesh in the KS and NW orientations. As in the text, Co directions are given in fcc notation. Row spacing mismatches were determined using the bulk lattice constants a(Co)=2.507Å and a(Cr)=2.884Å. We have also labeled the misfit along Co<112> for the two orientations: for KS, Cr<113> is within 0.5° of being parallel to Co<112> with a mismatch of 1.9%; and for NW, Cr<110> is parallel to Co<112> and the mismatch is 15%. Our measured strains along these directions are given in Figures 3(b) and 5(b).

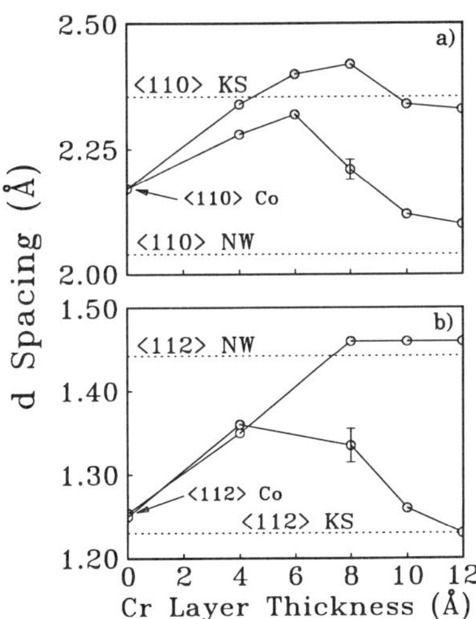

Figure 5. Average Cr d-spacings for the series of Co-Cr superlattices. Measured d-spacings are taken from the top surface of individual layers and these were typically averaged over 3-4 layers at various stages of each superlattice. The dashed lines show the expected d-spacings for bcc (110) Cr in the indicated orientations. In (a) the data suggest that the 4Å and 6Å Cr samples are in the KS orientation, however the *increase* in d-spacing along Co<112> is inconsistent with KS. Data along Co<112> were not recorded for 6Å Cr sample.

Figure 6 shows the evolution of the Cr first-order RHEED streak during the growth of a 10Å Cr layer. The electron beam is along Co<112>. Initially the Cr strains to the underlying close-packed Co d-spacing, but an abrupt 15% change and a final d-spacing value of 1.46 Å are indicative of bcc (110) Cr in the NW orientation (see Figure 4). It is important to point out that we see diffraction from both orientations along a given azimuthal direction: hence the labeling "NW" region and "KS" region in the images of Figure 2. This is due to the 6-fold symmetry of the hcp (0001) surface allowing three equivalent alignments for a given orientation.

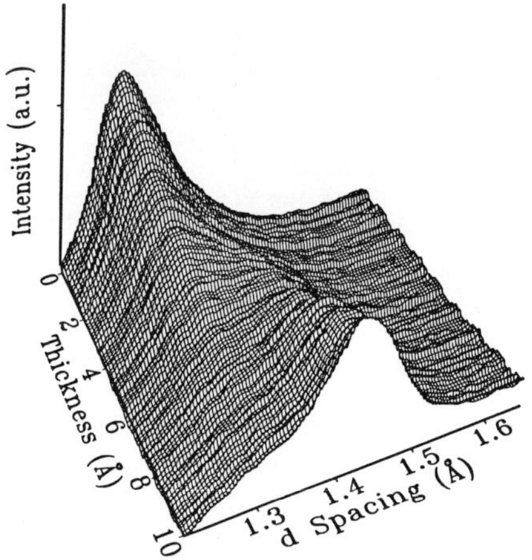

Figure 6. Evolution of Cr RHEED streak along Co<112> during the growth of the 5th Cr layer in the [20Å Co/10Å Cr]$_{33}$ superlattice. The plot shows the growth of 4-5Å of close-packed Cr on (0001) Co, followed by a transition to bcc(110) Cr. The 15% change in d-spacing and the final d-spacing value are both consistent with bcc(110)Cr in the NW orientation.

CONCLUSIONS

In summary, we have grown a series of epitaxial Co-Cr superlattices in which the first 5-6Å of each Cr layer is close-packed. Beyond this thickness there is an abrupt transition to the Kurdjumov-Sachs and Nishiyama-Wasserman bcc orientations. In contrast to Fe-Cr superlattices[15] we find very weak coupling across the Cr layers. Given the altered crystal structure of the Cr it is reasonable to expect its electronic structure to be significantly different than bulk Cr. This is also expected based on band structure calculations of fcc[16] and hcp[17] Cr which predict a density of states at the Fermi energy roughly three times higher than bcc Cr. These changes in the electronic structure can inhibit the formation of spin density waves which have been identified with the interlayer coupling in Fe-Cr structures[18].

ACKNOWLEDGEMENTS

This work is supported by Office of Naval Research grant number N00014-92-1335.

REFERENCES

1. G. A. Prinz, Phys. Rev. Lett. **54**, 1051 (1985).
2. Y. U. Idzerda, W. T. Elam, B. T. Jonker, and G. A. Prinz, Phys. Rev. Lett. **62**, 2480 (1989).
3. U. Gradmann, J. Mag. Mater. **54-57**, 733 (1986).
4. P. Grunberg, J. Appl. Phys. **57**, 3673 (1985).
5. J. C. Slonzewski, Phys. Rev. Lett. **67**, 3172 (1991).
6. M. N. Baibich, J. M. Broto, A. Fert, F. Nguyen Van Dau, F. Petroff, P. Etienne, G. Creuzet, A. Friederich, and J. Chazelas, Phys. Rev. Lett. **61**, 2472 (1988).
7. D. Barlett, C. W. Snyder, B. G. Orr, and Roy Clarke, Rev. Sci. Instrum. **62**, 1263 (1991).
8. See R. Clarke, F. J. Lamelas and S. Elagoz, in this volume.
9. C. W. Snyder, D. Barlett, B. .G. Orr, P. K. Bhattacharya, and J. Singh, J. Vac. Sci. Techol. B **9**, 2189 (1991).
10. L. A. Bruce and H. Jaeger, Philos. Mag. A **38**, 223 (1978).
11. E. Bauer and J. H. van der Merwe, Phys. Rev. B **33**, 3657 (1986).
12. R. Ramirez, A. Rahman, and I. K. Schuller, Phys. Rev. B **30**, 6208 (1984).
13. S. Stoyanov, Surface Sci. **172**, 198 (1986).
14. S. Andrieu, M. F. Ravet, O. Lenoble, V. Depuis, M. Piecuch, S. Pizzini, F. Baudelet and A. Fontaine, Europhys. Lett. **18**, 529 (1992); see also M. Piecuch et. al. in this proceedings.
15. P. Grunberg, R. Schreiber, Y. Pang, M. B. Brodsky, and H. Sowers, Phys. Rev. Lett. **57**, 2442 (1986).
16. Jian-hua Xu, A. J. Freeman, T. Jarlborg, and M. B. Brodsky, Phys. Rev. B **29**, 1250 (1984).
17. D. A. Papaconstantopoulos, J. L. Fry, and N. E. Brener, Phys. Rev. B **39**, 2526 (1989).
18. Y. Wang, P. M. Levy, and J. L. Fry, Phys. Rev. Lett. **65**, 2732 (1990).

INTERLAYER COUPLING AND ITS RELATION TO GROWTH AND STRUCTURE

P.A. Grünberg, A. Fuss, Q. Leng, R. Schreiber, and J.A. Wolf

Forschungszentrum Jülich, IFF
Postfach 1913
D-5170 Jülich, Germany

INTRODUCTION

It is now well established that interlayer coupling is characterized by oscillations between ferro (F)- and antiferro- (AF) magnetic type coupling with long- and short periods[1-5] The antialigned state due to AF coupling is often associated with a giant magnetoresistance (GMR)-effect[6,7]. Also a contribution to the interaction has been found which wants to align the magnetic films on both sides of the interlayer perpendicular to each other[8,9]. We refer to this in the following as 90° - or biquadratic coupling. Although a strong effect of the growth and structure of the samples on the coupling is expected so far the situation is rather unclear. It appears that AF type coupling and long period oscillations are easier to obtain for polycrystalline samples obtained by sputtering. Recently the relation between the GMR and interface roughness has been studied and established[10]. However whereas a relation between roughness and GMR is reasonable one would intuitively think that the strongest coupling should be obtained for samples with the smoothest interfaces. The observation that these should be the sputtered polycrystalline and not the epitaxial ones comes somewhat as a surprise.

The short periods on the other hand have so far only been observed in epitaxial samples fabricated by thermal evaporation which fits into the usual cliché that they have smoother interfaces. Biquadratic coupling has been ascribed to the existence of short period oscillations in the presence of interface roughness[11]. It has also only been observed in epitaxial samples so there seems to be again a contradiction.

It is clear that in this situation reliable results from samples with the best possible quality are highly desirable. If we choose Fe as magnetic material than Cr, Al, Au and Ag are in this context particularly interesting interlayer materials because of the good match of the crystal lattices if (100) type growth is chosen. Cr has bcc structure like Fe but Al, Au and Ag have fcc structure. However if we consider (100)-planes than in all cases a good match is obtained after a rotation of the inplane [100]-directions of the bcc- and fcc strutures with respect to each other. This is due to the good match of the nearest neighbour distances within the (100)-planes which are 0.287 nm for Fe, 0.288 nm for Cr and Au, 0.286 nm for Al and 0.289 nm for Ag. In this way

Magnetism and Structure in Systems of Reduced Dimension
Edited by R.F.C. Farrow *et al.*, Plenum Press, New York, 1993

one can obtain well grown epitaxial layered structures with pseudomorphic growth of the interlayer and little strain. We exploit these properties by growing the Fe/X/Fe structures on an Ag surface and by chosing X=Cr, Au, Al.

For Fe/Cr-layered structures the best possible growth known so far is on Fe-whisker substrates. It has been shown that in this case in addition to long period- also short period oscillations can be observed[4,5]. The peculiar whisker geometry on the other hand has the disadvantage that many experiments are not possible or difficult. We want to show here results from samples with very good growth on single crystalline Ag-surfaces, obtained by thermal evaporation on GaAs substrates.

SAMPLE PREPARATION

In order to obtain good epitaxial growth the preparation was performed in UHV (better than $2 \cdot 10^{-8}$ Pa) by e-gun evaporation. The samples were made on (100)-type GaAs substrates. These were first cleaned in an ultrasonic bath and then for a few minutes in boiling propanol. After transfer to the UHV system they were baked out at 580°C for one hour in order to remove an oxide layer without destroying the structure. This was verified by means of LEED.

Deposition started with a 1 nm thick seed layer of Fe[12] at a substrate temperature T_s = 373 K and a rate r = 0.01 nm/s. Following this a 150nm Ag buffer layer was evaporated with r= 0.1nm/s and T_s = 373 K. The growth was monitored by RHEED and SPALEED, some results will be shown in the next paragraph. After deposition the buffer was annealed for 1h at 573 K which significantly improved its structure. On the Ag buffer the Fe/X/Fe structure to be investigated was deposited. For the various interlayer materials X, optimum results were obtained with different preparation conditions. We have summarized these conditions in table 1. On top of the Fe/X/Fe structure we deposited a 50nm thick film of ZnS. It served as protection against air humidity and as antireflection coating in the optical experiments.

Most samples were made with an interlayer X of variable thickness d_o. This was achieved by moving the substrate during the deposition of the interlayer behind a shutter. The d_o-variation was between 0 at the front- and 3 to 10nm at the rear end. In this way the d_o-dependence of the interlayer coupling could be investigated for a whole range of d_o on the same sample and for the different values of d_o the magnetic films remained the same. A further advantage of this method therefore is that scatter due to variations in the properties of the magnetic films is strongly suppressed.

Table 1. Preparation conditions for the Fe/X/Fe- structures. Rates r for the condensation of Fe films were always r = 0.01 ... 0.03 nm/s and for the interlayers X always r = 0.02 nm/s. The table quotes materials and substrate temperatures T_s.

	1	2	3	4
1.Fe	293K	293 K (2nm) 523 K (3nm)	293 K	293 K
X	X = Cr 293 K	X = Cr 523 K	X = Au 353 K	X = Al 353 K
2.Fe	293 K	523 K	353 K	353 K

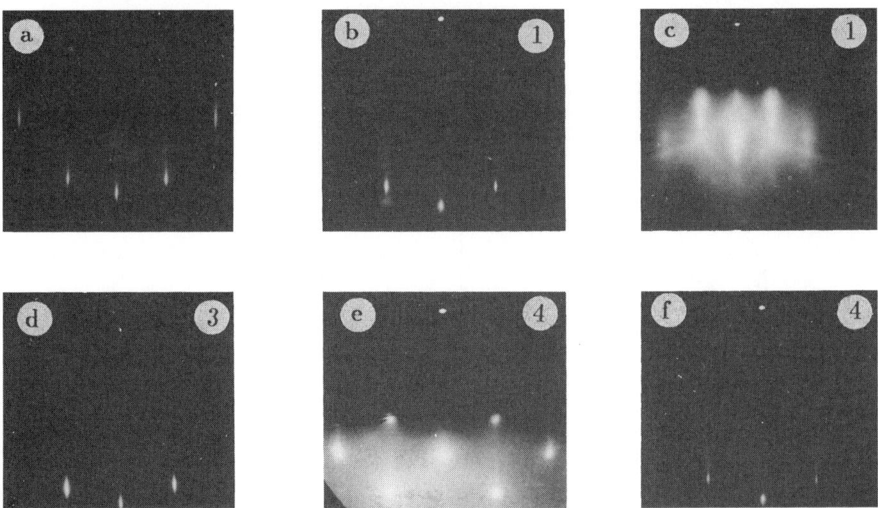

Figure 1. RHEED patterns from surface of: (a) Ag buffer layer ,(b) first Fe film, (c) Cr interlayer, (d) Au interlayer, (e) Al interlayer, (f) second Fe film. The numbers in circles give the column number in table 1 where the related preparation conditions are quoted.

STRUCTURAL CHARACTERIZATION

As mentioned common to all samples was a thick Ag buffer layer on the GaAs substrate and the good quality of its top surface was crucial to the results on the coupling reported here. Without the buffer for example we were never able to observe oscillations of the coupling-neither with long nor with short periods. (Only the first AF coupling range for the coupling across Cr seemed to be rather robust and was practically always observable).

Fig.1a shows a RHEED pattern of the Ag surface after annealing. The spots are somewhat streaky but lie clearly on an arc which proves the good epitaxial structure and flatness of this surface. (For an interpretation of such patterns see ref.13). The additional features are Kikuchi lines. On the Ag buffer the Fe/X/Fe structure (X=interlayer material) to be investigated was deposited. In fig.1b-d we display RHEED patterns from the surface of the first Fe film, the surfaces of the Cr, Au, and Al interlayers and of the second Fe-layer. The related preparation conditions from table 1 are quoted by the encircled numbers which refer to the corresponding column of table 1. The patterns show that the Fe layers in all cases are well structured and flat. Good results are also obtained for Au-interlayers whereas interlayers of Cr show some- and of Al appreciable roughness. Note that the RHEED pattern of Al displays spots on a two dimensional lattice and not on an arc. It originates from electrons which are transmitted through the material at edges which are abundant in a rough surface.

We performed also some RHEED investigations of samples prepared according to column 2 of table 1 but we do not want to display them here because we did to few to judge whether they are representative.

During the growth of the various materials we were also able to observe RHEED intensity oscillations some of which are displayed in fig.2. As expected the growth of further Ag on the annealed Ag surface gives the best result in that the oscillations not only show the smallest damping but have also a cusp like behaviour, indicative of good layer by layer growth[14].

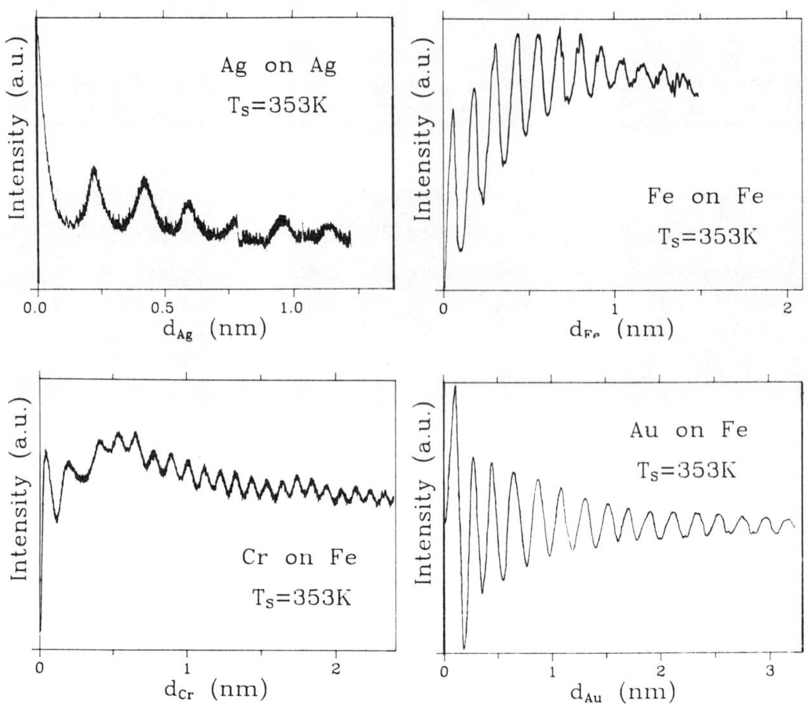

Figure 2. RHEED intensity vs. film thickness d during growth of various materials. Substrate temperatures T_s as indicated. d was independently determined with a quartz crystal monitor.

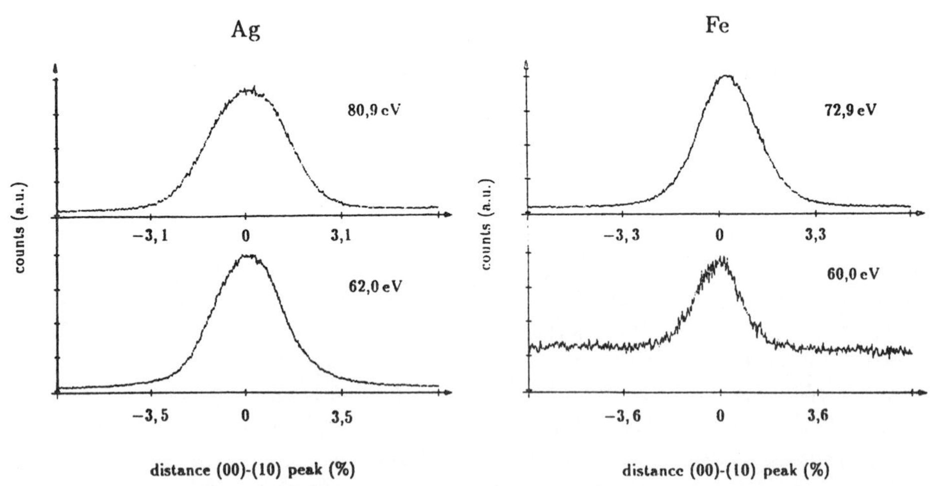

Figure 3. Profile of the (00) LEED spot of annealed Ag surface and of Fe film grown with T_s = 293 K on the Ag. Upper traces: Bragg condition, lower traces: close to anti-Bragg condition.

Fig.3 shows results of SPALEED (Spot Profile Analysis of Low Energy Electron Diffraction) from the annealed Ag surface and for the Fe surface for different energies of the incident electron beam. The measured width of $\approx$ 800 meV is by a factor of 4 larger than the instrumental transfer width of 200 meV and corresponds to a coherence length of $\approx$ 20 nm. The shape of the spot does not change when going from the Bragg- to the anti-Bragg condition. Therefore a "Henzler structure" like reported for Fe on an Ag single crystal[15] does not appear here.

The films were also characterized by means of RBS (Rutherford Back- Scattering) of He atoms. The results of these investigations are reported elsewhere[16].

EVALUATION OF THE INTERLAYER COUPLING

The phenomenological description of the interlayer coupling is made with an expression of the form

$$E_s = -J_1 \cdot cos\Delta\varphi - J_2 \cdot (cos\Delta\varphi)^2 \tag{1}$$

Here $\Delta\ \varphi$ is the angle between the magnetizations of the two films and E_s denotes the energy per surface unit of the interlayer. (In previous work we used other parameters $A_{12} = J_1/2$, $B_{12} = J_2/2$. Meanwhile J_1 seems to be more widely accepted and then J_2 is the logical alternative to B_{12}).

The first term of (1) is called "bilinear"- and the second term "biquadratic" coupling. Bilinear coupling is of ferro- and antiferromagnetic nature for positive and negative J_1 respectively. Biquadratic coupling produces for negative J_1 an alignment where the magnetizations of the magnetic films are perpendicular to each other. Positive J_2 produces like positive J_1 ferromagnetic type coupling.

In the following we consider films with magnetization M and cubic anisotropy with anisotropy constant K. For films parallel to a (100) plane there are two inplane easy axes, perpendicular to each other. Suppose the external field is applied along one of the two, than the total magnetic energy of the bilayer in the external field H can be written in the form

$$E = E_s - \mu_o HM(d_1 cos\varphi_1 + d_2 cos\varphi_2) + K[d_1(sin2\varphi_1)^2 + d_2(sin2\varphi_2)^2]/4 \tag{2}$$

Here φ_1 and φ_2 are the angles between the magnetizations of the individual films and the external field and d_1 and d_2 are the corresponding film thicknesses The minima of E can be found in the usual way by evaluation of equ.(2) and its derivatives with respect to φ_1, φ_2, where $\Delta\varphi=\varphi_1-\varphi_2$. In this way the orientation of the magnetizations as a function of the external field can be found for a certain set of the parameters and compared with the experiment. The parameters are obtained from the best fit. In the limit of strong AF coupling (J_2 neglected, K small) one can deduce J_1 simply from the saturation field H_s. In this case together with $d_1=d_2=d$ we have

$$J_1 = -\frac{1}{2}M\mu_o Hd - Kd \tag{3}$$

This formula is only valid for a double layer. In the limit of an infinite multi layer the factor of 1/2 on the r.h.s. has to be replaced by 1/4, which is due to the fact that each magnetic layer has then interlayers on both sides.

As mentioned for the investigation of the coupling phenomena we used samples with wedge type interlayers. This has the advantage that a continuous range of interlayer thicknesses can be studied on one sample with identical Fe films. The different interlayer thicknesses can be investigated by scanning a focussed laser beam across the

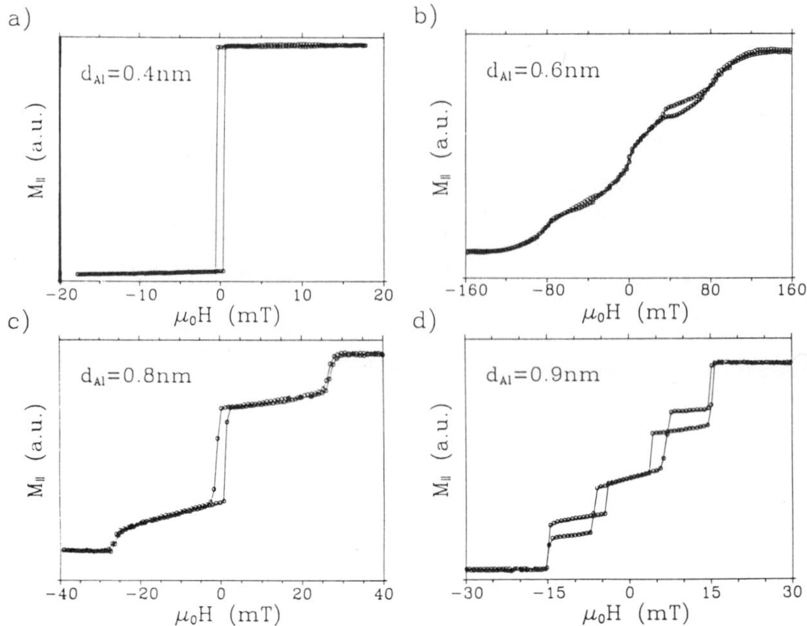

Figure 4. M(H) curves from Fe/Al/Fe structures displaying different types of coupling.

sample. In this way we obtain point by point the thickness dependence of the M(H)-curves by means of the magnetooptical Kerr effect (SMOKE) and of the spinwaves via a frequency analysis of the inelastically scattered light. The analysis of these data yields the thickness dependence of the total coupling J_1+J_2. In favorable cases J_1 and J_2 can be determined separately from the shape of hysteresis curves. More details on the method are contained in references 3,8 and 17.

In order to give a better feeling how these couplings show up during magnetization reversal we display in fig.4 M(H) curves taken from Fe/Al/Fe structures with different Al thicknesses. $M_{||}$ represents here the component of the total sample magnetization parallel to the external field H. Panel (a) represents ferromagnetic type coupling. This type of hysteresis curve is also observed from single Fe-films or from Fe bilayers if they are decoupled. Hence by the M(H) method apart from a trick called "spin engineering"[18] ferromagnetic type coupling can not be determined quantitatively. Panel (b) is representative of strong AF type coupling. Between the saturated states for large positive and negative values of H the individual magnetization of the two films perform rotations in opposite sense, in order to be antialigned for small H. Panel (c) is typical of 90°-coupling. The plateaus at about half the saturation value of $M_{||}$ are due to a perpendicular alignment of the magnetizations of the individual Fe-films. AF type coupling and 90°-type coupling do not exclude each other but can occur simultanuously. This is demonstrated by the M(H) curve of panel (d) where the plateaus at about half saturation are due to perpendicular alignment and those at $M_{||} \approx 0$ due to antialignment.

RESULTS FOR THE COUPLING OF Fe ACROSS Cr

Fig. 5a shows the interlayer coupling for Fe/Cr/Fe- structures prepared according to column 1 of table 1, i.e. with T_s = 293 K. There is a strongly damped oscillation

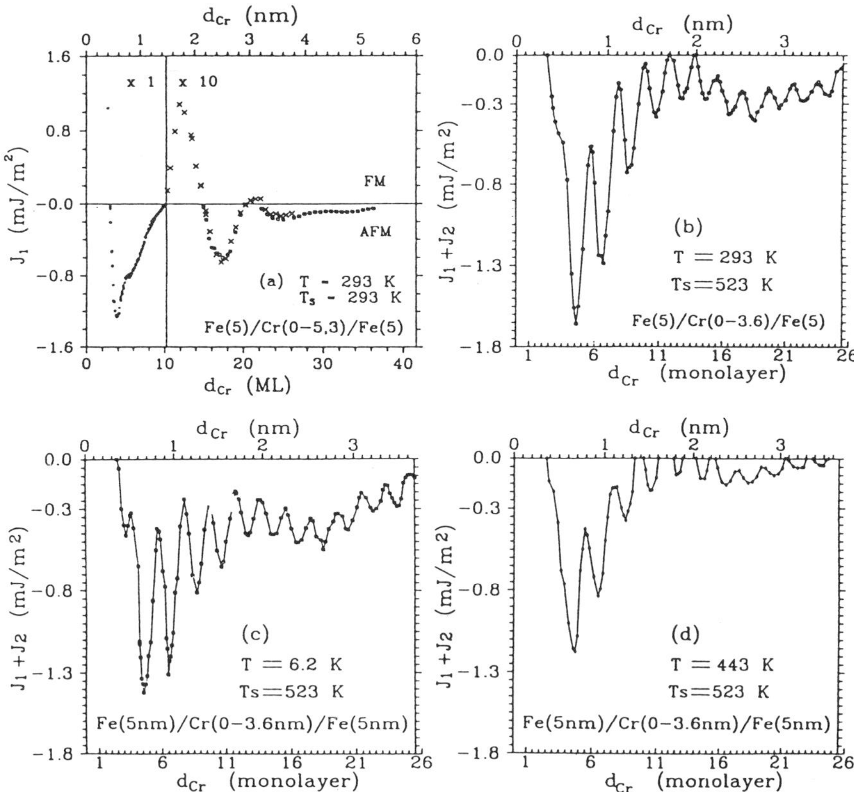

Figure 5. Interlayer exchange as a function of Cr-thickness measured at the indicated temperatures T for samples made at T_s=293 K in panel (a) and at T_s = 523 K in panels (b)-(d).

with a long period of $\lambda_{l.p} \approx 1.8$ nm. (Here and in the following we take $\lambda_{\ l.p}$ as the distance between the first two maxima of the long period oscillation). The contribution of 90° -type coupling in this case was sufficiently small to be negligible therefore only J_1 is plotted.

The first range of AF type coupling for small d_{Cr} in Fig.5a shows a finestructure. It indicates the existence of further maxima of AF type coupling within that range. From theory[19-23] such short period oscillations are expected to appear with a period $\lambda_{s.p.}$ approximately equal to two monolayers (ML), i.e. 0.288 nm of Cr and very flat interfaces are a neccessary precondition for their occurence. In fig.5 b we see the effect of a higher substrate temperature during the fabrication of the sample. Here after the deposition of the first 2nm of the first Fe film the substrate temperature was raised from T_s = 293 K to T_s = 523K (see column 2 of table 1). The first 2nm were made at T_s = 293 K in order to avoid diffusion of the Ag. In addition to the long periods there are many short period oscillations with $\lambda_{\ s.p} \approx$ 2ML. Plotted here is $J_1 + J_2$ because there is now an appreciable contribution of 90°-coupling which cannot generally be separated from AF type coupling. Obviously at the higher T_s flatter interfaces form.

Figs.5 c,d show the coupling curve from the same sample as in fig.5b only taken at low and high temperature, as indicated. At low temperatures the whole coupling curve seems to be shifted to negative values. At the higher temperature this background is reduced. As will be seen below a similar behaviour is also found in the coupling of Fe across Au.

RESULTS FOR THE COUPLING OF Fe ACROSS Au AND Al

Fig.6 shows the coupling curve measured at low temperatures for Fe interspaced by Au. The important parameters for the growth of these samples are summarized in column 3 of table 1. Since there is an appreciable J_2 contribution which cannot in general be separated from J_1 we display again $J_1 + J_2$. There is a short range oscillation with $\lambda_{s.p} \approx 2$ML and a long range oscillation with $\lambda_{l.p} \approx 8$ ML.

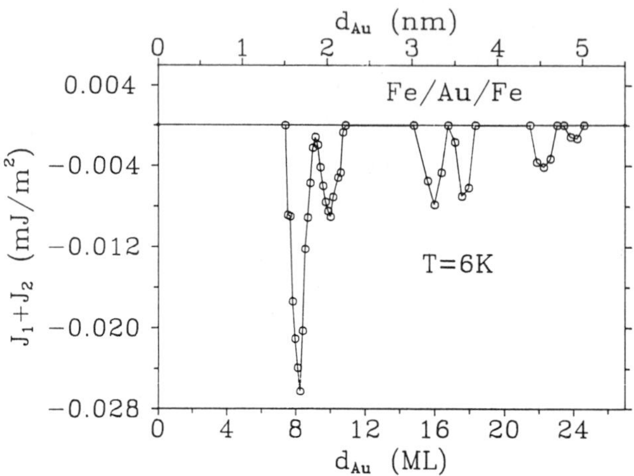

Figure 6. Interlayer exchange in Fe/Au/Fe

More detailed investigations for Au-interlayers revealed a strong temperature dependence of the J_2 contribution. In fig.7a we display the first two short period oscillations measured at 6K and at 300K. The curve for 300K seems to be upshifted as compared to the one for 6K. At the same time one finds that at 300K along the whole curve the J_2 contribution has dissappeared where it plays an important role at 6K. The vanishing of J_2 towards higher temperatures is displayed in more detail in fig.7b. The values of J_1 and J_2 are shown as a function of temperature for the first maximum of $J_1 + J_2$ at d_{Au}=1.68 nm. Unfortunately a separate determination of J_1 and J_2 appreciably away from this point is not possible due to the particular shape of the corresponding hysteresis curves. Nevertheless this result indicates that the differences in the two curve s in fig.7a are mainly due to the vanishing of J_2 at 300 K, with a slight reduction of J_1. Similar results have recently been obtained by Gutierrez et al.[24] from Fe/Al/Fe structures.

For the Fe/Al/Fe system the coupling curve is displayed in the main part of fig.8. Clearly there are long period oscillations and the maximum coupling strength is $J_1 + J_2 \approx 0.6$ mJ/m^2 - about half as strong as in Cr. In addition we see a fine structure. However these fine features had poor reproducibility and changed from sample to sample.

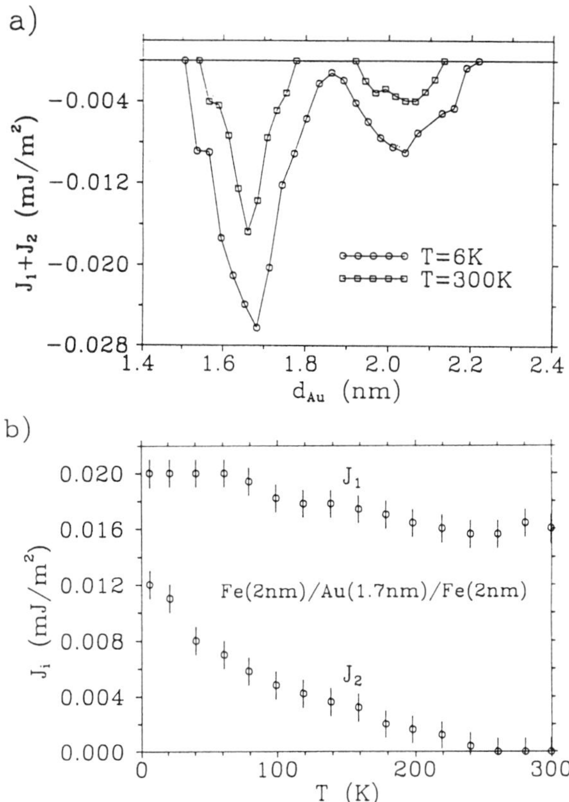

Figure 7. The first two short period oscillations of $J_1 + J_2$ in Fe/Au/Fe at two different temperatures (a) and the temperature dependence of J_1 and J_2 at $d_{Au} = 1.68$nm (b).

An obvious explanation comes from inspection of fig.1e which indicates that the surface of the Al interlayers and therefore also of the corresponding Al/Fe interface is rough. Hence the Al thickness is not well defined and there can be strong local fluctuations which can wash out any possible short period oscillation in a non-reproducible way. From theory on the other hand such fine oscillations are not expected[25].

There is also a strong J_2 contribution in the case of the Fe-coupling across Al. It dominates in the whole first range of negative $J_1 + J_2$ as can be seen in the insert of fig.8. Displayed is the domain structure of the sample with the Al wedge at room temperature and the correspondence to the coupling curve is indicated. Let us first consider the range on the left hand side where $J_1 + J_2$ is positive, hence the coupling is ferromagnetic. The magnetizations in the lower and upper Fe film are parallel and along the two easy axis which are vertical and horizontal. These are the resulting magnetization directions within the domains. Between them we find 90°-Bloch walls in the diagonal directions. In the neighbouring range of negative $J_1 + J_2$ biquadratic coupling dominates so the upper and lower film are magnetized perpendicular to each other. Now the resulting moments in the domains are along the diagonals and the directions of the walls are vertical and horizontal (this was a different sample than the one of fig. 4b).

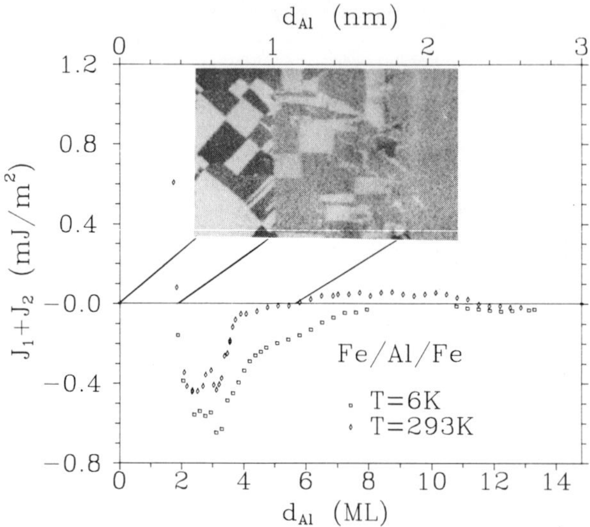

Figure 8. Interlayer exchange and magnetic domains in a Fe/Al/Fe sample with an Al wedge.

DISCUSSION, CONCLUSIONS

Long And Short Period Oscillations

The properties of interlayer coupling of Fe across Cr reported here should be compared with the results by Unguris et al.[4] In their case the information is obtained by means of SEMPA. These investigations were made on samples grown on Fe-whiskers which have supposedly the best growth known so far for these systems and therefore the results can serve as a guideline. By means of SEMPA one detects magnetic domains which are here a result of the coupling. Therefore the technique gives only a qualitative information on the kind of the coupling which dominates (F,AF , or 90°-type) and does not provide e.g. the strength of the coupling.

Our result displayed in fig.5b is at variance with the result by Unguris et al. in the following details. Most fine oscillations are not associated with a change of the sign like in the case of Unguris et al. There is no phase slip at 24 ML of Cr. On the other hand the fact that 90°-coupling is observed in the experiment of fig.5b but not in the SEMPA experiment does not neccessarily mean that it is not present in the samples on the whisker. In zero external field as is the case in the SEMPA experiment the direction of the static magnetization is determined by the interaction which dominates, hence a possible 90°-contribution might here simply escape detection. Altogether more experiments are needed before a final comparison between the data from the whiskers and the films on GaAs is possible.

According to the current theoretical understanding the oscillations of the interlayer coupling are related to certain distances at the Fermi surface of the interlayer material (see e.g. ref.26 and the article of P. Bruno in this volume) . For example $\lambda_{s.p.}$ of the coupling of Fe across Cr is thought to be associated with the famous "nesting vector" of Cr. For $\lambda_{l.p}$ of fig.5a an associated distance has not been identified. The two periods visible in fig.6 have been interpreted as due to the two calipers of the "dogs bone" orbit of Au in the [100] direction . For Al only a long period is expected[25] which has been observed (fig.8).

Table 2. Results of experiments where short period oscillations have been observed.

System	$\lambda_{s.p.}$ [ML]	$\lambda_{l.p.}$	J_1^{max} [mJ/m^2]	J_2^{max} [mJ/m^2]	substrate	ref.
Fe/Cr/Fe	2	1.8 nm	-1.8^a	0.4	GaAs/Ag	this work
Fe/Au/Fe	2	8ML			GaAs/Ag	this work
Fe/Al/Fe		1.8 nm	-0.8		GaAs/Ag	this work
Fe/Cr/Fe	2		-1.2		Fe whisker	5
Fe/Cr/Fe	2^b				Fe whisker	4
Fe/Mn/Fe	2		-0.14		Fe whisker	27
Fe/Cu/Fe	≈ 2		-0.1		Fe whisker	28
Co/Cu/Co	2.6	8ML	-0.4		Cu crystal	28
Fe/Mo/Fe	3		-0.2		Mo crystal	29

a) at low temperatures

b) with phase slips

Short periods are only expected to be found in the experiments if the interface roughness is small enough- i.e. if the interfaces are atomically flat over large areas. The fact that only in epitaxial samples of the best quality the short periods have so far been found supports this. We have collected in table 2 data from the literature for the cases where both long and short periods have been observed. In all cases growth was epitaxial on especially prepared substrates

While the relation between growth and observation of short periods is rather clear and as expected this is not so in the case of the long periods. For many systems the long periods have been first seen on polycrystalline samples made by sputtering and only after special efforts also on epitaxial samples made by thermal evaporation. The case of coupling of Co across (111)-type Cu is a famous example (see the related articles in this volume). There has been the suspicion that this is due to more magnetic "shorts" or "bridges" in the epitaxial samples which is probably correct.

90°-Coupling

Fig.7b can be compared with a theoretical prediction by Slonczewski[11]. In his theory 90°-coupling is the result of a competition between F- and AF- type coupling which can occur in the presence of interface roughness. Suppose that due to interface roughness areas of the sample with F- and with AF- type coupling are very close then there will be a competition between F- and AF- alignment in these areas. Under such conditions an energy minimum can be found by a 90°-alignment. Hence the J_2-parameter is not independent of J_1 and under certain conditions Slonczewski predicts $J_2 \sim (\Delta\ J_1)^2$ where $\Delta\ J_1$ is the change of J_1 between F- and AF-type areas for a corresponding smooth interface. Suppose now $\Delta\ J_1$ decreases linearly with temperature- as approximately in fig.7b- then J_2 should decrease quadratically. Since the decrease of J_2 in fig.7b is clearly stronger than with T^2 one could conclude that the Slonczewski mechanism does not explain the experimental results. We should keep in mind however that the T -dependence was derived under certain conditions and might not generally apply.

Alternatively 90°-coupling has been interpreted as a higher order effect in a series expansion in a similar fashion as suggested by equ.(2) (see the article by D.Edwards et al. in this volume). Here a strong temperature dependence -namely exponential- is predicted.

Further explanations emphasize the fact that 90°-coupling is equivalent to a dependence of J_1 on $\Delta\varphi$ (equ.1) as was pointed out first by B. Heinrich et al.[9] or to the assumption that not only the atomic planes of the ferromagnetic films adjacent to the interlayer interact but also further planes[30].

At the moment an experimental decision on the true mechanism can not be made. For example for the coupling of Fe across Cr we see an increase of the 90° coupling in the sample which displays also the fine oscillations. At a first glance this seems to be in disagreement with the Sloczewski mechanism. However according to a discussion given by B. Heinrich (see the article in this volume) a stronger 90°-coupling from the samples with the better interfaces is in fact in agreement with this mechanism. From the measured temperature dependences which show the same trends for Cr interlayers (fig.5) and Au interlayers (fig.7) one would also not like to make final conclusions. Interesting is the case of Al-interlayers. As mentioned one expects for this type of coupling only a long period oscillation[25]. Therefore and important ingredient for the Slonczewski mechanism - the short period oscillation - is here missing. As is seen from fig.8 the 90°-coupling can here be particularly strong.

FINAL REMARK AND ACKNOWLEDGEMENT

The intention of this article was to indicate some relations between interlayer coupling and the growth and structure of the samples. In most cases (in particular for the 90°-coupling) final conclusions are not yet possible. An important side effect of these investigations should also be mentioned. With the coupling there is not only a new criterion but also driving force towards better magnetic films. In many cases this has already led to an appreciable improvement .

We would like to thank W.Zinn for continuous interest and support of this work.

REFERENCES

1. P. Gruenberg, R. Schreiber, Y. Pang, M.B. Brodsky, H. Sowers, Phys. Rev. Lett. 57:2442 (1986).

2. S.P. Parkin, N. More, K.P. Roche, Phys. Rev. Lett. 64: 2304 (1990).

3. S. Demokritov, J.A. Wolf, P. Gruenberg, Europhysics Letters 15: 881 (1991) and S. Demokritov, J.A. Wolf, P. Gruenberg, W. Zinn, Mat. Res. Soc. Symp. Proc. vol.231:133 (1992).

4. J .Unguris, R.J. Celotta, D.T. Pierce, Phys. Rev. Lett.67:140 (1991) and submitted. See also the article by Unguris et al. in this volume.

5. S.T. Purcell, W.Folkerts, M.T. Johnson, N.W.E. McGee, K.Jager, J.aan de Stegge, W.B. Zeper, W.Hoving, P.Gruenberg, Phys. Rev. Lett.67:903 (1991).

6. G. Binasch, P. Gruenberg, F. Saurenbach, W. Zinn, Phys. Rev. B39:482 (1989).

7. M.N. Baibich, J.M. Broto, A. Fert, F. Nguyen van Dau, F. Petroff, P. Etienne, G. Creuzet, A. Friederich, J. Chazelas, Phys. Rev. Lett. 61:2472 (1988).

8. M. Ruehrig, R. Schaefer, A. Hubert, R. Mosler, J.A. Wolf, S. Demokritov, P. Gruenberg, phys. stat. sol. (a) 125:635 (1991).

9. B. Heinrich, J.F. Cochran, M. Kowalewski, J. Kirschner, Z. Celinski, A.S. Arrott, and K. Myrtle, Phys.Rev. B44:9348 (1991).

10. E.E. Fullerton, D.M. Kelly, J. Guimpel, I.K. Schuller, Phys.Rev. Lett.68:859 (1992).

11. J.C. Slonczewski, Phys. Rev. Lett. 67:3172 (1991).

12. R.F.C. Farrow, U.S. Speriosu, S.S.P. Parkin, C. Chien, J.C. Bravman, R.F. Marks, P.D. Kirchner, Mater. Res. Soc. Symp. Proc. 130 (1989) and P. Etienne, J. Massies, F. Nguyen-Van-Dau, A. Barthelemy, A. Fert, Appl. Phys. Lett. 55:2239 (1989).

13. see the figure on p.162 of the article by M.G. Lagally et al. in "RHEED and Reflection Electron Imaging of Surfaces" Plenum Press 1988, NATO ASI series.

14. A.S. Arrott, B. Heinrich, S.T. Purcell, "Kinetics of Ordering at Surfaces" ed. by M. Lagally, Plenum Press 1990, p. 321.

15. B. Heinrich, A.S. Arrott to be published.

16. J.A. Wolf, Q. Leng, P. Gruenberg, W. Zinn, Symposium on Magnetic Ultrathin Films, Mulitlayers and Surfaces, 7.-10 sept. 1992, Lyon, France.

17. A. Fuss, S. Demokritov, P. Gruenberg, W.Zinn, JMMM 103:L221 (1992).

18. S.S.P. Parkin, see contribution in this volume.

19. C.L. Fu, A.J. Freeman, JMMM 54-57:777 (1986).

20. Y. Wang, P.M. Levy, J.L. Frey, Phys.Rev.Lett.65:2732 (1990).

21. D. Stoeffler, K. Ounadjela, F. Gautier, JMMM 93:386 (1991).

22. H. Hasegawa, Phys.Rev.B42:2368 (1990).

23. G. Mathon, D.M. Edwards,, R.B.Muniz, M.S.Phan, JMMM 104-107:1734 (1992).

24. C.J. Gutierrez, J.J. Krebs, M.E. Filipkowski, G.A. Prinz, submitted.

25. G. Mathon, private communication.

26. P. Bruno, C. Chappert, Phys. Rev. Lett. 67:1682 (1991); 67(E):2592 (1991).

27. S.T. Purcell, M.T. Johnson, N.W.E. Mc Gee, R. Coehoorn, W. Hoving, Phys. Rev.B45:13064 (1992).

28. M.T. Johnson, S.T. Purcell, N.W.E. Mc Gee, R. Coehoorn, J.aan de Stegge, W. Hoving, Phys.Rev.Lett.68:2688 (1992).

29. Z.Q. Qiu, J.Pearson, A.Berger, S.D. Bader, Phys.Rev.Lett.68:1398 (1992).

30. J.Barnas, P. Gruenberg, Symposium on Magnetic Ultrathin Films, Multilayers and Surfaces, 7.-10 sept. 1992 , Lyon, France.

SEMPA STUDIES OF OSCILLATORY EXCHANGE COUPLING

John Unguris, Daniel T. Pierce, Robert J. Celotta, and
Joseph A. Stroscio

National Institute of Standards and Technology
Gaithersburg, MD 20899

ABSTRACT

We have used scanning electron microscopy with polarization analysis (SEMPA) to investigate the magnetic exchange coupling between an Fe film and an Fe whisker separated by either a Cr or Ag spacer layer. The thickness dependence of the oscillatory coupling in these atomically well ordered, epitaxial films was precisely measured. The periodicity of the exchange coupling is consistent with predictions based on Fermi surface spanning vectors of the interlayer material.

INTRODUCTION

The exchange coupling between magnetic layers separated by nonmagnetic layers oscillates between ferromagnetic (FM) and antiferromagnetic (AFM) coupling as a function of the spacer layer thickness in a wide variety of systems.[1] The strength and periodicity of this oscillatory exchange coupling depends sensitively upon the atomic order in the films. While most polycrystalline exchange coupled films seem to show an oscillation period of about 1 nm,[2] well-ordered, single crystal films exhibit more complex, sometimes multi-periodic behaviors.[3-10] The origin of the coupling and the factors that determine the period of the oscillations in the coupling are subjects of intense interest, as is the technologically important spin-valve magnetoresistance observed in multilayers of these materials. The goal of this work is to provide precise measurements of the oscillation periods in high quality films in order to better couple theory with experiment.

In this article, we describe measurements for two different sandwich structures where the spacer materials are either a transition metal, Cr, or a noble metal, Ag. The spacer films are deposited on nearly perfect, single crystal Fe whisker substrates and covered with a thin Fe film. The atomic structure of the films is characterized by reflection high energy electron diffraction (RHEED) and scanning tunneling microscopy (STM). The magnetization of the films is measured with scanning electron microscopy with polarization analysis (SEMPA). For both spacer layers, the oscillatory exchange coupling consists of both short and long period contributions, and the oscillations persist over at least 10 nm of spacer material. In each case the observed periodicities are consistent with models of

Magnetism and Structure in Systems of Reduced Dimension
Edited by R.F.C. Farrow *et al.*, Plenum Press, New York, 1993

the exchange coupling in which spanning vectors of the spacer layer's Fermi surface dominate the magnetic response of the multilayer.

EXPERIMENT

The experimental procedures used to investigate the Fe/Cr/Fe and Fe/Ag/Fe structure have been previously described.[3,4] The magnetization, crystalline order, and chemical composition are measured using SEMPA,[11] RHEED, and scanning Auger microscopy, respectively. All of this analysis is done *in situ* and, when necessary, during film evaporation. In addition, STM measurements of the atomic scale order were made using similar Fe whisker substrates in a separate experiment. The use of high spatial resolution analytical techniques allows us to study films that are grown on Fe whisker substrates that are only a few tenths of a mm across by one cm long. The Fe whiskers are extremely high quality, strain free crystals with very low dislocation densities.[12] The whisker surfaces are naturally flat and can be prepared by well established ion sputtering and heating procedures.[13] STM measurements of Fe(100) whisker surfaces show atomically flat terraces that are separated by single atom high steps that are at least 1 μm apart.[14] RHEED patterns from the whiskers show sharp diffraction spots distributed along Laue rings.

In order to measure the exchange coupling as a continuous function of spacer layer thickness, a wedge shaped interlayer was grown on the Fe substrate. A schematic of the wedge is shown in Fig. 1. We obtained the wedge shape by moving a precision, piezo-controlled shutter during the evaporation. The slope of the wedge is very small, between 0.001° and 0.0001°. The thickness of the spacer layer is determined by measuring RHEED intensity oscillations as the focused electron beam is scanned along the wedge. Using spatially resolved RHEED to measure the thickness eliminates any errors due to drifting evaporation rates or local thickness variations arising from microscopic defects in the shutter.

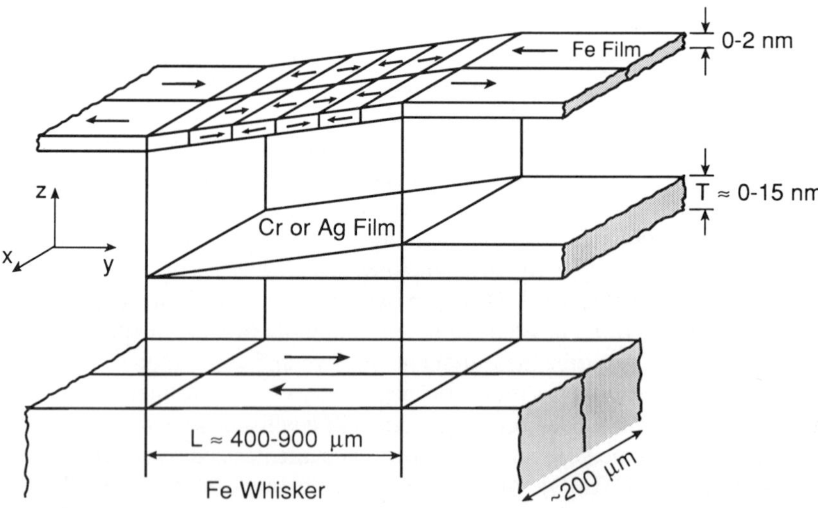

Fig. 1. A schematic exploded view of the sample structure showing the Fe(001) single-crystal whisker substrate, the evaporated Cr wedge, and the Fe overlayer. The arrows in the Fe show the magnetization direction in each domain. The z-scale is expanded approximately 5000 times.

The magnetic coupling between the Fe(100) substrate and Fe film overcoat are determined using SEMPA.[3] Specifically, SEMPA measures the spin polarization of the secondary electrons which, for a simple transition metal ferromagnet, is directly proportional to the magnetization. A coupling measurement therefore simply involves measuring the direction of the magnetization in the Fe overcoat and comparing it to the magnetization of the bare Fe substrate. In these measurements we assume that the magnetization of the Fe whisker is unchanged. The escape depth of the secondary electrons is about 1 nm so that only the magnetization of the outermost layers is measured. Possible instrumental offsets in the SEMPA measurements are determined and eliminated by only using Fe whiskers that are divided lengthwise into two oppositely magnetized domains. Finally, it should be noted that SEMPA does not directly measure the magnitude of the exchange coupling, but it is extremely sensitive to the direction of the magnetic coupling in these multilayer systems.

Cr RESULTS

Both Cr and Fe are body centered cubic crystals with only a 0.7% lattice mismatch, and the surface free energy of Cr is less than that of Fe. Therefore one should expect Cr to grow in a layer by layer mode on Fe. Experimentally, we find that the growth is strongly temperature dependent so that temperatures in the range of 250°C to 350°C are required in order to achieve true layer by layer growth. The temperature dependence of the growth is illustrated in Fig. 2 in which STM images of bare Fe(100) and of Cr films grown at various temperatures are shown. The STM image of the Cr film grown at 100°C, Fig. 2(b) shows that the growth front extends over at least four layers. The corresponding RHEED pattern from the 100°C Cr film shows weak diffraction spots, enhanced diffuse scattering and transmission diffraction features characteristic of a rough sample. In contrast, the RHEED pattern from the Cr film grown at 300°C is almost as sharp as the bare Fe RHEED pattern and the STM image, Fig. 2(c), shows a nearly ideal growth front with only two layers participating.

The temperature dependence of the Cr growth has a dramatic effect on the Fe/Cr/Fe magnetic exchange coupling.[3] Fig. 3 shows SEMPA images of the magnetic domain structure in the bare Fe whisker and in an Fe film deposited on top of Cr wedges that were grown at 30°C, Fig. 3(b), and 350°C, Fig. 3(c). These SEMPA images show the magnetization component along the whisker; white corresponds to magnetization pointing towards the right and black to the left. The oscillatory exchange coupling is clearly present for both Cr wedges and in both cases the oscillations persist for at least 10 nm, but the periodicity of the coupling is very different. While the coupling through Cr grown at the lower temperature oscillates with a period of 11-13 layers (1.6-1.9 nm), the coupling through Cr grown at higher temperature changes, after the initial 5 layer FM coupling region, with each layer of Cr giving an oscillation period of nearly two layers. We say "nearly" because at 24-25, 44-45, and 64-65 layers, indicated by the arrows at the top of the figure, no reversal takes place. This corresponds to a phase slip resulting from the accumulation of a phase difference owing to the incommensurability of the exchange period and the lattice constant. The actual period of the short period oscillations at room temperature is therefore 2.11 ± 0.03 Cr layers (0.304 ± 0.004 nm).

Further evidence for the correlation between short period oscillations and Cr film quality can be obtained by comparing the coupling measurements with RHEED intensity oscillations measurements. An example of the correlation between RHEED intensity oscillations and short period oscillations is shown in Fig. 4. The RHEED image is obtained by measuring the intensity of the specular RHEED beam as the incident electron

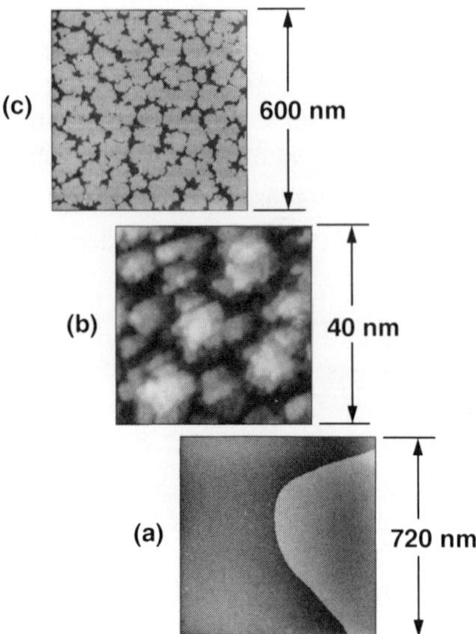

Fig. 2 STM images of (a) a clean Fe whisker substrate with a single, one Cr atom high step, (b) a Cr film deposited on an Fe whisker at 100°C showing the rough, multilevel surface, and (c) a Cr film deposited at 300°C showing a nearly ideal 2 layer growth front. The size of the scanned area is given to the right of each image.

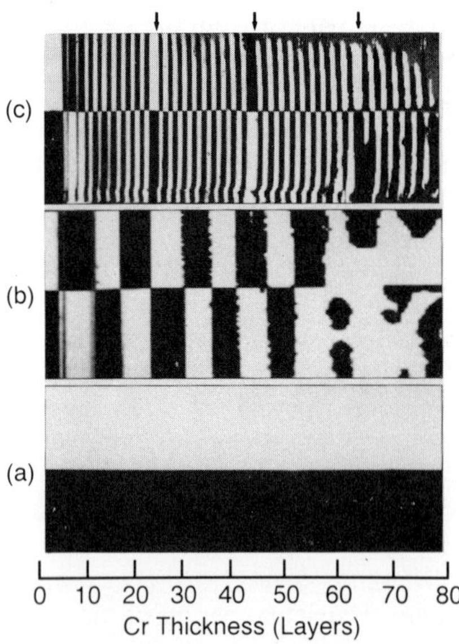

Fig. 3. SEMPA image of the magnetization M_y (axes as in Fig. 7) showing domains in (a) the clean Fe whisker, (b) the Fe layer covering the Cr spacer layer evaporated at 30°C, and (c) the Fe layer covering a Cr spacer evaporated on the Fe whisker held at 350°C. The scale at the bottom shows the increase in the thickness of the Cr wedge in (b) and (c). The region of the whisker imaged is about 0.5 mm long. The Cr thicknesses at which exchange coupling phase slips occur are indicated by the arrows at the top of the figure.

beam is rastered over an uncovered Cr wedge that was grown at 250°C. We attribute the absence of RHEED oscillations in the thicker part of the wedge to roughness of the Cr film, possibly caused by the slightly less than optimal growth temperature and damage in the Fe substrate from too many sputtering and annealing cycles. The important point to note, however, is that in the magnetization image of the Fe overlayer, M_y, the coupling reverts to long period oscillatory coupling at the same thickness that the RHEED intensity oscillations disappear.

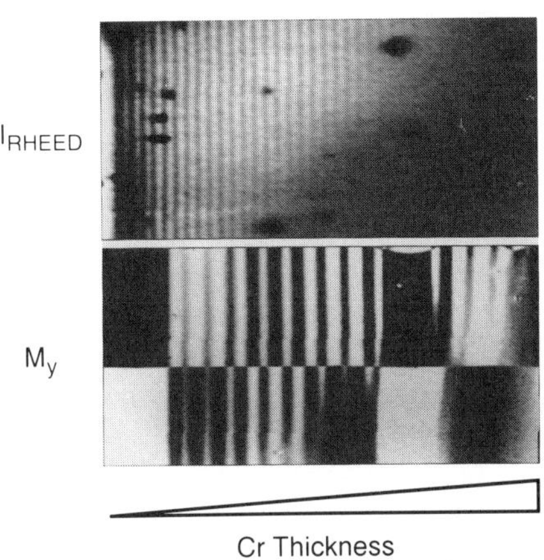

Fig. 4. The effect of roughness on the interlayer exchange coupling is shown by a comparison of the oscillations of the RHEED intensity along the bare Cr wedge with the SEMPA M_y magnetization image over the same part of the wedge.

It is noteworthy that a calculation of the exchange coupling in Fe/Cr/Fe predicted short period oscillations, in addition to the long period oscillations, before short period oscillations had been observed experimentally. Wang et al[15] pointed out that the apparent discrepancy with earlier experiments, which showed no short period oscillations, could be accounted for by interfacial roughness corresponding to the displacement of one quarter of the atoms in the interface by one layer. For a Cr film n layers thick,this roughness corresponds to 25% of the surface being at n-1 layers, 50% at n layers, and 25% at n+1 layers. This is equivalent to a three layer growth front with 0.10 nm rms roughness. Therefore, in order to observe the short period oscillations, the Cr must grow in nearly an ideal layer by layer manner where one layer is completed before the next layer begins.

Data from the SEMPA and scanned RHEED images can be used to take a more quantitative look at the Cr thickness dependence of the exchange coupling. Fig. 5 shows the polarization of the bare Cr, P(Cr), compared with the polarization of the same wedge after coating it with a thin (~1 nm thick) Fe film. The Cr thickness scale is calibrated using the RHEED intensity oscillations shown at the top of Fig 5. The RHEED measurements were made near the out of phase diffraction condition where peaks in the specular intensity correspond to filled layers. The RHEED and SEMPA measurements are registered by aligning defects, usually found near the edge of the whisker, that appear in both the RHEED and SEMPA intensity images. The build-up of disorder and roughness with increasing Cr thickness is indicated by the corresponding decrease in the amplitude of the RHEED intensity oscillations.

The measured spin polarization of secondary electrons from the bare Cr, before and after removing the background polarization from the Fe substrate, is shown in the middle of Fig. 5. The high polarization of electrons from the Fe at the start of the wedge decreases exponentially as the Fe electrons are attenuated by the Cr film of increasing thickness. A fit to the exponential gives a 1/e sampling depth for SEMPA in Cr of 0.55 ± 0.05 nm. Subtracting the exponential background leaves the Cr polarization, P(Cr), which is shown magnified by a factor of 5. Because of the attenuation of the electrons coming from layers below the surface, P(Cr) is dominated by the polarization of the surface layer which is seen to reverse nearly every layer.

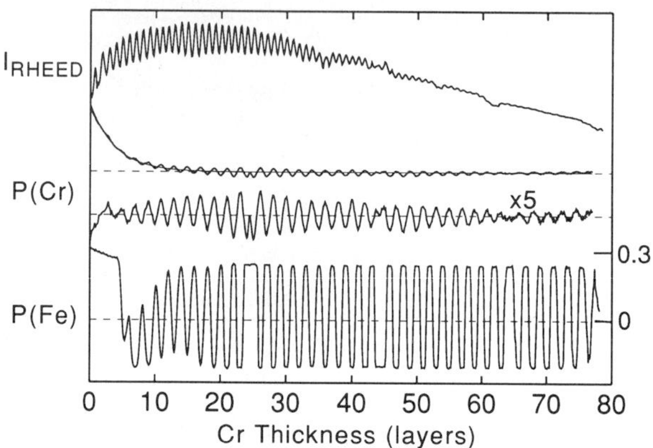

Fig. 5. Scanned RHEED and SEMPA measurements from the same Cr wedge showing the thickness dependence of the RHEED intensity oscillations (top), the spin polarization of secondary electrons emitted from the bare Cr, P(Cr), before and after removing the Fe substrate polarization (middle), and the M_y polarization component from an Fe layer, P(Fe), deposited on top of the Cr (bottom).

After coating this Cr wedge with a 5 layer thick Fe film a SEMPA measurement yields the thickness dependence of the magnetic coupling shown in the bottom of Fig. 5. The polarization of the Fe overlayer, P(Fe), is seen to be opposite to that of the top layer of the bare Cr. This observation is consistent with spin-polarized photoemission[16] and electron energy loss[17] measurements which have shown that the Fe-Cr interface is antiferromagnetically coupled. In addition, the phase slips at 24-25, 44-45, and 64-65 layers are seen in both P(Cr) and P(Fe). With AFM coupling at each interface and if the Cr orders antiferromagnetically with successive planes of reversed moments in the [100] direction, one expects the Fe/Cr/Fe coupling for even numbered layers of Cr to be AFM, and FM for odd numbers of Cr layers. However, looking at P(Fe) we see that Fe separated by 7 layers of Cr is antiferromagnetically coupled, opposite to expectations. A close examination of P(Cr) reveals that there is a defect in the AFM layer stacking of Cr between 1 and 4 layers. That is, at a thickness of less than four layers, adjacent layers of Cr must have parallel moments.

One might argue that the observation of short period oscillatory coupling through Cr interlayers is not surprising, since bulk Cr is a spin density wave (SDW) antiferromagnet so that the coupling is simply an extension of the antiferromagnetic stacking. However, the SDW and the oscillatory exchange coupling have the same origin. The SDW arises from an enhanced susceptibility at a wave vector that spans extended parallel regions of the Cr Fermi surface. Cr is special in that there is a strong "nesting" of the Fermi surface.[18] This same enhanced susceptibility at a particular spanning wave vector is also responsible for the short period oscillations in Ruderman-Kittel-Kasuya-Yosida (RKKY)-like models of the Fe/Cr/Fe exchange coupling through paramagnetic Cr interlayers. This makes it hard to distinguish between an RKKY-like coupling and an explanation where the Fe magnetization is locked to the antiferromagnetism of the Cr. In an attempt to differentiate between these two mechanisms we have measured the temperature dependence of P(Cr) and P(Fe).[19] We find that the oscillatory coupling, along with the phase slips associated with the SDW, persist from just below the Néel temperature, T_N=38 °C, to 1.8 T_N. Above T_N bulk Cr is paramagnetic which suggests that either the coupling takes place through paramagnetic Cr or that the presence of the Fe substrate stabilizes the antiferromagnetism in Cr. We therefore have two closely related ways to view this response in the Cr film. In one view, since even above T_N the magnetic susceptibility is enhanced at the nesting wavevector, an antiferromagnetic response can be induced in the Cr by the presence of the Fe. Alternatively, if the Cr is viewed as a paramagnet, RKKY-like oscillations would be established which would be quite similar to the antiferromagnetic order because both derive from the same strong Fermi surface nesting.

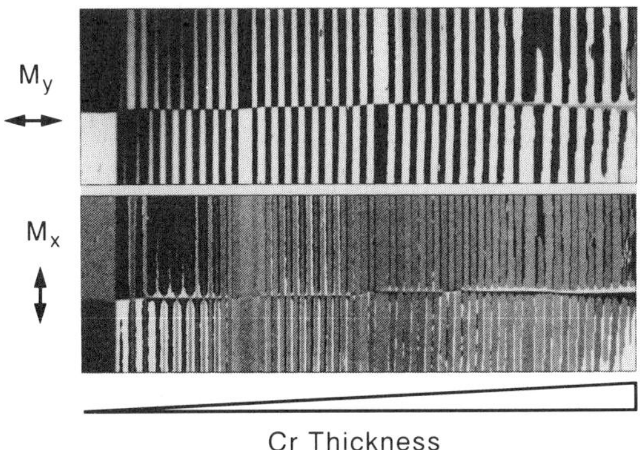

Fig. 6. SEMPA images from a Fe/Cr/Fe(100) wedge showing the M_y and M_x magnetization components. The presence of biquadratic coupling can be seen by the increased contrast of M_x at the FM/AFM transitions.

Finally, one should note that the exchange coupling in Fe/Cr/Fe is not purely FM or AFM, but there can also be an in-plane component that is not parallel or antiparallel to the substrate magnetization. Phenomenologically this additional coupling may be treated as a higher order term in an expansion of the exchange coupling energy per unit area,[20]

$$E = A_{12}[1-\mathbf{m}_1\bullet\mathbf{m}_2] + \tfrac{1}{2}B_{12}[1-(\mathbf{m}_1\bullet\mathbf{m}_2)^2]$$

where $\mathbf{m}_1$ and $\mathbf{m}_2$ are unit vectors in the direction of magnetization in the two Fe layers,

A_{12} is the coefficient of the bilinear coupling term which yields either FM or AFM alignment, and B_{12} is coefficient of biquadratic coupling which favors an orthogonal magnetization. Biquadratic coupling in Fe/Cr/Fe was initially observed in a magneto-optic Kerr microscopy investigation.[20] In the SEMPA measurements, the biquadratic coupling can be most clearly seen in images of the in-plane magnetization component that is orthogonal to the FM/AFM axis. Fig. 6 shows images of both in-plane magnetization components for an Fe/Cr/Fe(100) wedge. M_y is the magnetization component along the whisker magnetization axis, while M_x is the simultaneously measured orthogonal component. The large contrast in M_x at the FM/AFM transitions shows that the biquadratic coupling dominates the bilinear at these thicknesses. The width of these transition zones is proportional to the strength of the biquadratic coupling relative to the bilinear.[20] Note that the transition zones are not simply domain walls in the Fe film, since they are much wider than the domain walls in Fe and they scale inversely with the slope of the wedge. The physical origin of the biquadratic exchange coupling is still not well understood. Both intrinsic[21] and extrinsic[22] origins for the coupling have been proposed.

Ag RESULTS

Unlike Cr with its complex Fermi surface and its SDW antiferromagnetism, Ag is diamagnetic and has a relatively simple Fermi surface. In this regard Ag is a better interlayer material for testing models of the oscillatory exchange coupling. Unfortunately Ag is not as close a perfect epitaxial match to Fe as Cr is. The larger Ag fcc unit cell, when rotated by 45° in order to match the bcc Fe substrate, results in a lateral mismatch of only 0.8%, but, perpendicular to the surface, a single atom high Ag step is 42% larger than an Fe step. Ag only grows well over a narrow range of substrate temperatures and Ag evaporation rates, and Ag does not grow as well as Cr on Fe. Nevertheless our investigations of the exchange coupling in Fe/Ag/Fe(100) show oscillatory coupling for spacer films that are up to at least 50 Ag layers (>10nm) thick.[4]

Fig. 7 shows a scanned RHEED image obtained from a bare Ag wedge and SEMPA images from the same sample after coating it with Fe. As in the Cr measurements, the thickness of the Ag wedge is measured using the intensity oscillations in the bare Ag RHEED image. M_y and M_x are the magnetization components along and orthogonal to the Fe whisker axis, respectively. The SEMPA images show that both long and short period oscillations contribute to the exchange coupling, although the short period contribution is much less pronounced when compared with that of the Cr interlayers. The M_x image shows that there is also a significant biquadratic coupling contribution to the exchange coupling. The relative strength of the biquadratic coupling appears to increase with increasing Ag thickness, since the widths of the FM/AFM transitions increase in the thicker part of the Ag wedge. Fig. 8 shows averaged line scans of M_y and M_x and the derived angle of the in-plane magnetization, Θ, obtained from the images in Fig. 7 and plotted as a function of the Ag thickness.

In order to extract the coupling periods from the Fe/Ag/Fe(100) SEMPA data, we model the magnetization by adding together two sine waves with adjustable periods, phases, and amplitudes. This coupling function is discretized at the Ag lattice and then all positive coupling values are set to the same magnetization and all negative values are set to the equal but opposite magnetization. The coupling periods determined by varying the parameters to achieve the best fit are 5.73 ± 0.05 Ag layers (1.17 ± 0.01 nm) and 2.37 ± 0.07 Ag layers (0.485 ± 0.014 nm).

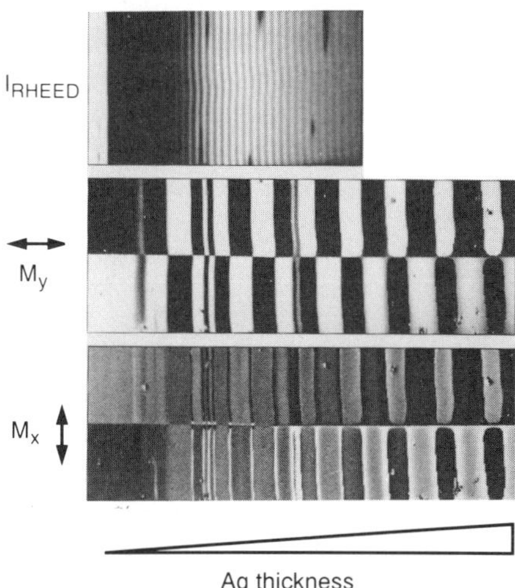

Fig. 7. A scanned RHEED image of a bare Ag wedge, and SEMPA images, showing the in-plane magnetization components, M_y and M_x, of the same wedge after coating it with three layers of Fe. The RHEED image was used to calibrate the Ag thickness.

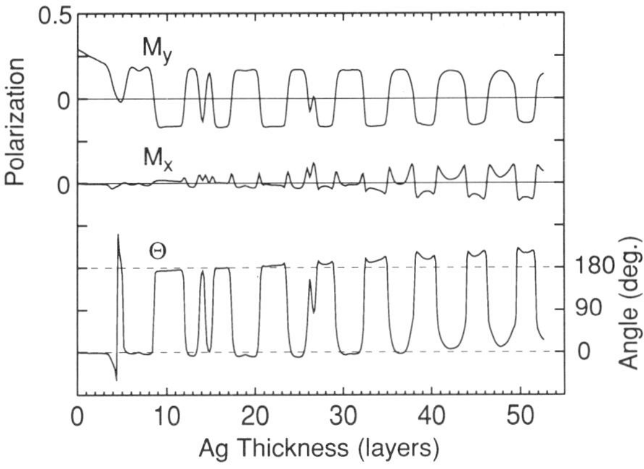

Fig. 8. The Ag thickness dependence of M_y, M_x and the in-plane magnetization angle, Θ, derived from the images in Fig. 7. The angle is plotted so that 0° (180°) corresponds to FM (AFM) coupling.

DISCUSSION AND CONCLUSION

Measurements of the oscillatory exchange coupling provide a test of various exchange coupling theories. In particular, for theories that are based on RKKY-type interactions, the coupling periodicities are closely related to the size and shape of the Fermi surface of the interlayer material. The periodicity of the oscillatory coupling is dominated by reciprocal lattice vectors that are perpendicular to the interface and span nearly parallel sections of the Fermi surface. The enhanced susceptibility of the spacer layer at these interface Fermi surface spanning vectors results in the oscillatory exchange coupling. A simple test of various coupling models, therefore, is to see how well they predict the observed periods of the oscillatory exchange coupling.

Table 1

Exchange Coupling Period (layers)				
	Fe/Cr/Fe(100 (0.144nm/layer)		Fe/Ag/Fe(100) (0.204nm/layer)	
Experiment	2.11±0.03	12±1	2.37±0.07	5.73±0.05
Theory Bruno, Chappert[23]			2.38	5.58
Stiles[24]	2.10	11.3		

Table I summarizes the Cr and Ag interlayer coupling periods and compares them with the results of two different derivations of Fermi surface spanning vectors. Bruno and Chappert[23] use De Haas-van Alphen and cyclotron resonance measurements of Fermi surface extremals as the basis for their Fermi surface calculations of the noble metals. Stiles[24] uses a tight-binding approximation to fit the band structure in order to calculate the Ag and Cr spanning vectors. In the case of Cr, it is clear that the short period oscillations are related to the spanning vector that bridges the unusually strong nesting feature of the Cr Fermi surface. This is the same Fermi surface nesting feature that leads to the enhanced susceptibility and the related SDW antiferromagnetism in Cr. The origin of the long period oscillation in Cr is less clear, however, since the complicated Cr Fermi surface has 10 other spanning vectors which also correspond to significant Fermi surface nesting. While one of these spanning vectors results in a period that is close to what we observe, it is not clear why we do not observe oscillatory coupling associated with the others. Roughness may play a role, as we have shown in the case of low temperature Cr growth. In addition, Fermi surface nesting only reveals which periodicities are possible, while the matrix elements, which have not as yet been calculated, determine the relative amplitudes of the various periodic contributions to the exchange coupling.

For Ag the comparison is easier since the Ag Fermi surface is less complex and only two spanning vectors are found which can lead to coupling in the [001] direction. Table I shows that the measured coupling periods are in excellent agreement with the theoretical values, considering possible uncertainties in the theory and the possibility of some slight tetragonal distortion of the strained Ag interlayer. The SEMPA measurements show that the oscillations of the interlayer coupling in Fe/Ag/Fe(100) are consistent with theories in which oscillation periods are derived from Fermi surface spanning vectors.

A common feature of both the Fe/Cr/Fe and Fe/Ag/Fe multilayer coupling investigations is the sensitivity of the exchange coupling to the quality of film growth. This work emphasizes that, in order to make meaningful and quantitative comparisons with theoretical predictions, coupling measurements must be made on interlayer films that are atomically well ordered and characterized. In this regard, SEMPA is especially well suited for measuring the coupling periodicities, because the high resolution of SEMPA permits the use of small, nearly perfect, single crystal Fe whisker substrates. SEMPA also has the surface sensitivity to interrogate only the top layer of the trilayer structures and it can be used in situ, along with RHEED and Auger spectroscopy, to study the multilayer structures as they are made. SEMPA measurements using Fe whisker substrates have given the most precise determination of the periods of oscillatory exchange coupling in Fe/Cr/Fe(100) and Fe/Ag/Fe(100), the two systems that have been studied by SEMPA to date, and the experimental results on Cr and Ag interlayers support theories of interlayer exchange coupling based on Fermi surface properties.

ACKNOWLEDGEMENTS

We wish to thank M. H. Kelley and M. D. Stiles for many helpful discussions. The Fe whiskers were grown at Simon Fraser University under an operating grant from the National Science and Engineering Research Council of Canada. This work is supported by the Office of Technology Administration of the Department of Commerce and by the Office of Naval Research.

REFERENCES

1. See review articles in "Ultrathin Magnetic Structures", (B. Heinrich and A. Bland, eds.), Springer Verlag, 1992.
2. S.S.P. Parkin, Phys. Rev. Lett. **67**, 3598 (1991).
3. J. Unguris, R.J. Celotta, and D.T. Pierce, Phys. Rev. Lett. **67**, 140 (1991).
4. J. Unguris, R.J. Celotta, and D.T. Pierce, to be published.
5. S. Demokritov, J.A. Wolf, P. Grünberg, and W. Zinn, in Proc. Matls. Res. Soc. Symp. **231**, 133 (1992).
6. S.T. Purcell, W. Folkerts, M.T. Johnson, N.W.E. McGee, K. Jager, J. aan de Stegge, W.B. Zeper, W. Hoving, and P. Grünberg, Phys. Rev. Lett. **67**, 903 (1991).
7. M.T. Johnson, S.T. Purcell, N.W.E. McGee, R. Coehoorn, J. aan de Stegge, and W. Hoving, Phys. Rev. Lett. **68**, 2688 (1992).
8. A. Fuss, S. Demokritov, P. Grünberg, and W. Zinn, J. Mag. and Mag. Mat. **103**, L221 (1992).
9. Z.Q. Qiu, J. Pearson, A. Berger, and S.D. Bader, Phys. Rev. Lett. **68**, 1398 (1992).
10. S.T. Purcell, M. T. Johnson, N.W.E. McGee, R. Coehoorn, and W. Hoving, Phys. Rev. **B45**, 13064 (1992).
11. M.R. Scheinfein, J. Unguris, M.H. Kelley, D.T. Pierce, and R.J. Celotta, Rev. Sci. Instrum. **61**, 2501 (1990).
12. P.D. Gorsuch, J. Appl. Phys. **30**, 837 (1959).
13. S.T. Purcell, A.S. Arrott, and B. Heinrich, J. Vac. Sci. Technol. **B6**, 794 (1988).
14. J.Stroscio, D.T.Pierce, J.Unguris, R.J.Celotta, and R.A.Dragoset, to be published.

15. Y. Wang, P.M. Levy, and J.L. Fry, Phys. Rev. Lett. **65**, 2732 (1990).
16. F.U. Hillebrecht, C. Roth, R. Jungblut, E. Kisker and A. Bringer, Europhys. Lett. **19**, 711 (1992).
17. T.G.Walker, A.W.Pang, H.Hopster, and S.F.Alvarado, Phys. Rev. Lett. **69**, 1121 (1992).
18. E. Fawcett, Rev. Mod. Phys. **60**, 209 (1988).
19. J. Unguris, R.J. Celotta, and D.T. Pierce, Phys. Rev. Lett. **69**, 1125(1992).
20. M. Rührig, R. Schäfer, A. Hubert, R. Mosler, J.A. Wolf, S. Demokritov and P. Grünberg, phys. stat. sol. (a) **125**, 635 (1991).
21. R.P. Erickson, K.B. Hathaway, and J.R. Cullen, to be published.
22. J.C. Slonczewski, Phys. Rev. Lett. **67**, 3172 (1991).
23. P. Bruno and C. Chappert, Phys. Rev. Lett. **67**, 1602 (1991); Phys. Rev. **B46**, 261 (1992).
24. M.D. Stiles, to be published.

GIANT MAGNETORESISTANCE IN TRANSITION METAL MULTILAYERS

S.S.P. Parkin, R.F. Marks, R.F.C. Farrow, and G.R. Harp

IBM Research Division
Almaden Research Division
650 Harry Road
San Jose, California 95120-6099

INTRODUCTION

The indirect magnetic exchange coupling of magnetic moments via non-magnetic metals has attracted considerable interest over the past few decades. Much work has been carried out on systems comprised of low concentrations of magnetic atoms randomly distributed in metallic hosts, for example, Mn or Fe atoms dissolved in Cu or Au. It was found that the magnetic atoms are exchange-coupled via a spin polarization of the conduction electrons of the host metal[1-3]. The spin polarization has been inferred from, for example, Cu^{63} nuclear magnetic resonance (NMR) measurements in which satellites are observed surrounding the main NMR line[3]. The satellites, corresponding to successive spherical shells of Cu atoms surrounding the magnetic impurities, are shifted alternately to higher and lower magnetic resonance fields resulting from oscillations in the spin polarization of the Cu conduction electrons. For higher concentrations of magnetic impurities the oscillating spin polarization is manifested as an oscillating exchange interaction, alternating between ferromagnetic and antiferromagnetic coupling depending on the separation of the magnetic impurities. This coupling is of the well-known Ruderman-Kittel-Kasuya-Yosida (RKKY) form (see, for example, [4]).

At first sight a simpler system in which to study the magnetic coupling through non-magnetic metals is that of metal multilayers comprised of thin magnetic layers separated by thin metallic layers (see, for example, [5]). However from the 1960's until about 1988 the coupling of such magnetic layers through most metals, including Cu and Au, was found to be ferromagnetic in sign with a strength that typically decayed exponentially with increasing separation of the magnetic layers[6, 7]. These experiments were however in contradiction with most theoretical models which predicted an oscillating exchange interaction analogous to that found in the dilute magnetic alloys[8]. Whilst an oscillatory magnetic coupling of the RKKY form was observed in multilayers composed of the rare-earth metals Gd and Y[9] only recently has an oscillating magnetic exchange coupling been found in transition metal multilayers. The first observations were made in Fe/Cr and Co/Ru multilayers[10], and subsequently in Co/Cu[11] and later in the majority of transition metal multilayered systems[12]. Surprisingly, oscillatory coupling was first found in transition metal (TM) multilayers grown by conventional sputter deposition techniques and only later in single crystalline multilayers prepared in ultra high vacuum (uhv) deposition systems using electron beam or thermal evaporation cells. The latter are often referred to as *molecular beam epitaxy* (MBE) systems.

Whilst oscillatory interlayer exchange coupling was first discovered in sputtered multilayers, antiferromagnetic coupling via a non-magnetic transition metal had previously been found in crystalline bcc (100)Fe/9ÅCr/Fe sandwiches[13] (and later in fcc (100) 9ÅCo/9ÅCu [14] multilayers). Both (100) Fe/Cr/Fe sandwiches[15] and (100) Fe/Cr multilayers[16] were found to show very large changes in resistance with the application of a magnetic field. Indeed the changes were so large that the phenomenon has been termed *giant magnetoresistance (GMR)*. A similar phenomenon was subsequently observed in sputtered Fe/Cr and Co/Ru multilayers[10]. The largest GMR effects have, however, been found in antiferromagnetically coupled polycrystalline Co/Cu multilayers[17]. The GMR effect derives from a change in the magnetic structure of the magnetic layers from an antiparallel configuration of neighboring magnetic layers in small fields to a parallel configuration with application of a sufficiently large magnetic field, H_S. In this paper we briefly review the giant magnetoresistive properties of Cu based multilayered structures, prepared by both sputtering and MBE techniques.

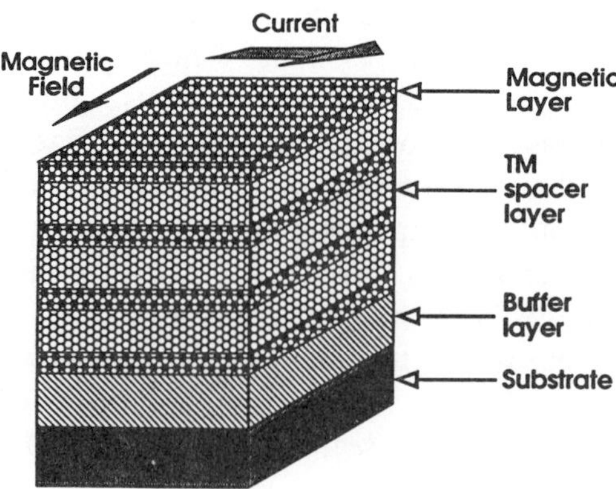

Figure 1. Schematic diagram of a multilayer comprised of alternating magnetic and non-magnetic layers, grown on a buffer layer. The magnetoresistance measurements described here were taken with the current and magnetic field in the plane of the layers with the magnetic field either parallel or orthogonal to one another.

GIANT MAGNETORESISTANCE OF SPUTTERED Cu BASED MULTILAYERS

A schematic figure of a multilayer is shown in Figure 1. The properties of such structures are often very sensitive to growth conditions including, for example, the deposition method, the temperature of growth, the substrate material and the buffer layer, if any, between the substrate and multilayer. The structure and thus the physical properties including GMR of multilayers containing Cu layers has been found to be particularly sensitive to deposition conditions, both for sputtered and MBE prepared multilayers (ML). This is demonstrated in figure 2 which

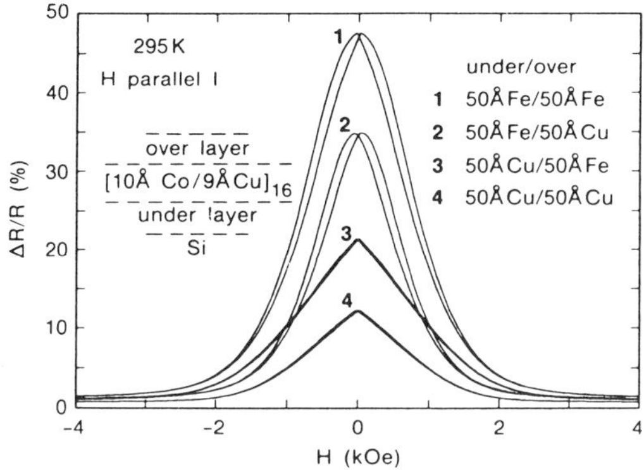

Figure 2. Room temperature resistance versus field curves for four samples of the form Si(100)/buffer layer/$[10ÅCo/9ÅCu]_{16}$/capping layer with 50Å thick buffer and capping layers of respectively: (1) Fe and Fe, (2) Fe and Cu, (3) Cu and Fe and (4) Cu and Cu.

shows room temperature magnetoresistance data for several nominally identical magnetron sputtered Co/Cu multilayers deposited on 50Å thick Fe or Cu buffer layers. (The growth of the structures is described elsewhere[11]). The change in resistance, ΔR, is normalized to the resistance of the multilayer at high field, R. The magnitude of the saturation magnetoresistance, ΔR/R, is very large and is almost 50% in the structure grown on an Fe buffer layer with an Fe capping layer. As can be seen by comparing structures with the same underlayer, changing the capping layer from Fe to Cu considerably reduces the MR. However, the capping layer is not expected to significantly alter the structure or properties of the multilayer. The reduction in MR can be simply accounted for by the higher electrical conductivity of Cu compared to Fe which results in a significant shunting of the sensing current through the capping layer. This reduces the proportion of current passing through the multilayer itself and so reduces the MR. In contrast, changing the underlayer material significantly alters the MR, for the same degree of shunting of the sensing current. The latter is realized, to a first approximation, by preparing structures with the same net thicknesses of the Fe, Co and Cu layers in the structure. For example, by comparing the MR data for structures 2 and 3 in Figure 2 which have underlayers/ capping layers of 50ÅFe/50ÅCu and 50ÅCu/50ÅFe respectively, it is clear that the structure grown using an Fe buffer layer displays a significantly higher MR. The reason for this is evident from both magnetic and structural characterization of these samples. Magnetic studies show incomplete antiferromagnetic coupling of the Co layers for the multilayer grown on Copper buffer layers. Since the magnitude of the GMR is directly related to the degree of AF coupling this accounts for the reduced MR. Both cross-section transmission electron microscopy (XTEM) images as well as auger sputter depth profiling show the presence of Cu more than 100Å beneath the silicon surface. These studies suggest that the Cu underlayer reacts with the silicon substrate. The XTEM studies clearly show that the reaction of the Cu with the silicon results in rumpled Co and Cu layers as compared with growth on Fe buffer layers. Indeed varying the buffer layer is an important method to control the structural morphology of the multilayer and thus influence the magnitude of the GMR.

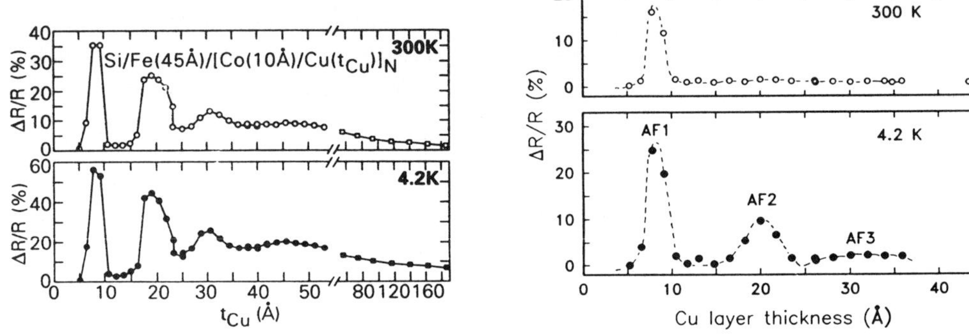

Figure 3. (left) Dependence of saturation transverse magnetoresistance on Cu spacer layer thickness for a family of related superlattice structures of the form Si/Fe(40Å)/[Co(10Å)/Cu(t_{Cu})]$_N$. An additional Cu layer was deposited on each film structure such that the uppermost Cu layer was ≃55Å thick. The number of bilayers in the superlattice, N, is 16 for t_{Cu} below 55 Å (●, ○) and 8 for t_{Cu} above 55 Å (□, ■). (right) Dependence of saturation magnetoresistance versus Cu spacer layer thickness for multilayers of the form, Si(100)/50Å$Ni_{81}Fe_{19}$/[15Å$Ni_{81}Fe_{19}$/Cu(t_{Cu})]$_{14}$/25ÅRu, at (upper) 300 K and (lower) 4.2 K. Ranges of Cu spacer layer thicknesses for which the $Ni_{81}Fe_{19}$ layers are coupled antiferromagnetically are shown as AF1, AF2 and AF3.

The dependence of saturation magnetoresistance on Cu layer thickness is shown in Figure 3 for Co/Cu and permalloy ($Ni_{81}Fe_{19}$)/Cu multilayers[11, 18]. The former are prepared on Fe and the latter on $Ni_{18}Fe_{19}$ buffer layers. In both cases, at 4.2 K well defined oscillations in the MR are found as a function of Cu spacer layer thickness. It was first shown for Fe/Cr and Co/Ru multilayers that the oscillations in MR are clearly tied to oscillations in the exchange coupling mediated by the non-magnetic layer[10]. In these systems enhanced MR, larger than the anisotropic MR of the magnetic material itself, is found only in the antiferromagnetically coupled multilayers whereas no significant enhancement of the MR is observed in multilayers with substantial ferromagnetic interlayer coupling. In the latter case the relative orientations of the local magnetizations in adjacent magnetic layers are unaffected by the magnetic field.

For the Co/Cu system similar oscillations are found at all temperatures from below 4.2 K to above 400 K. At still higher temperatures the structures are unstable and the Co and Cu layers interdiffuse into one another. For the permalloy/Cu multilayers only a single oscillation is observed at room temperature for magnetron sputtered multilayers. This can be readily explained as follows. For the permalloy/Cu system the exchange coupling is significantly weaker than in Co/Cu. At 4.2 K the antiferromagnetic coupling, J_{AF}, for $t_{Cu} = \simeq 8$Å, is about 3-6 times smaller than that in Co/Cu. Note that J_{AF} is related to the saturation field, H_S by $J_{AF} \simeq - H_S M_S t_F/4$, where M_S and t_F are the saturation magnetization and thickness, respectively of the magnetic layers. If we assume that the structures are not perfect and that there are defects in the system, for example, pin-holes of magnetic material extending across the Cu layers from one magnetic layer to the next, or similarly, necks in the Cu spacer layers where the layer is locally thin, such defects may give rise to ferromagnetic coupling of the magnetic layers. Not only is the magnetic coupling weak in the permalloy/Cu system but it has a much stronger temperature dependence than in Co/Cu. In the latter system the AF coupling at the first AF peak, J^1_{AF}, weakens by only ≃25% between 4.2 and 300 K, whereas in the former J^1_{AF} changes by a factor of 2.5. Thus it seems reasonable to argue that in NiFe/Cu multilayers at low temperatures, where the AF coupling is considerably stronger, it is likely that more oscillations in coupling will be observed than at higher temperatures where the coupling may be weak compared to direct ferromagnetic coupling via defects.

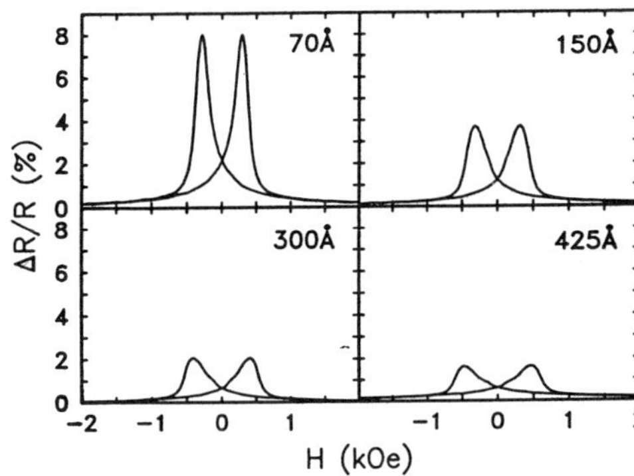

Figure 4. Resistance versus field curves for four Co/Cu multilayers of the form, Si(111)/Ru(50Å)/[Co(11Å)/Cu(t_{Cu})$]_6$ /Ru(15Å) with Cu spacer layer thicknesses, t_{Cu}, of 70, 150, 300 and 435 Å.

Figure 3 shows the variation of magnetoresistance in Co/Cu multilayers for Cu layer thicknesses extending out to 180Å. Actually substantial magnetoresistance is observed for Cu layers with thicknesses even several times thicker. Typical resistance versus field curves for Co/Cu multilayers are shown in Figure 4 for Cu layers ranging up to 425Å thick. The shape of these curves is distinct from that of the bell shaped curves shown in Figure 2. Note that the permalloy/Cu structures display characteristic triangularly-shaped resistance versus field curves often indicative of incomplete antiferromagnetic coupling of the magnetic layers. This is also the case for incompletely coupled Co/Cu multilayers as, for example, illustrated by samples 3 and 4 in Figure 2. The structures shown in Figure 4 exhibit a double peaked MR curve with maximum resistance at fields corresponding to the $\pm H_C$, where H_C is the coercive or switching field at which the magnetization passes through zero. The interlayer exchange coupling decreases with increasing Cu thickness much faster than the GMR effect, such that the exchange coupling fields are much weaker than H_C for the structures shown in Figure 4. Thus the MR in figure 4 results from the random arrangement of magnetic domains in successive magnetic layers. A schematic diagram of the likely magnetic domain structure is shown in Figure 5.

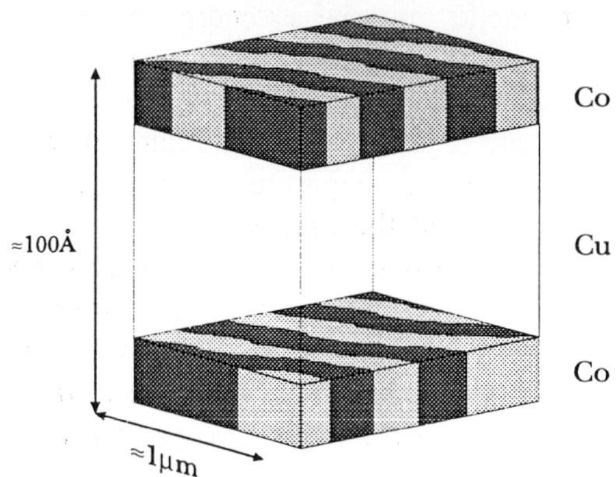

Figure 5. A schematic diagram of the arrangement of the magnetic domains in the remanent magnetic state of a Co/Cu multilayer. The darker and lighter shaded regions correspond to longitudinal magnetic domains aligned parallel and antiparallel to the magnetic field direction.

Lorentz electron microscopy of the remanent magnetic state of polycrystalline Co films a few hundred angstroms thick shows longitudinal magnetic domains aligned along the magnetizing field direction[19]. For Co/Cu multilayers in which the interlayer coupling is weak it is likely that there will be domains in neighboring magnetic layers with their magnetization axes aligned non-parallel to one another. The magnetic configuration with the highest degree of antiparallelism of magnetic domains will arise when the net magnetic moment is zero i.e. at $\pm H_C$. The degree of antiparallelism will depend on the nature, (i.e. the symmetry and the strength) of the magnetic anisotropy within the magnetic layers themselves, as well as any magnetic coupling (exchange or magnetostatic) between the magnetic layers. Perhaps it is surprising that the giant MR effect was not observed in magnetic multilayer systems long before the observation of antiferromagnetic coupling since the data in Figure 4 suggest that strong antiferromagnetic coupling is not a necessary requirement for the observation of giant MR. Indeed enhanced MR, although small (1-3 %), was reported in uncoupled single crystalline Co/Au/Co sandwiches[20] prior to the observation of giant MR in antiferromagnetically coupled Fe/Cr multilayers and sandwiches. It was only later recognized that this was most probably a manifestation of the same GMR phenomenon[21]. In any case it appears that the magnitude of the giant MR effect is tied to that of the interlayer coupling. Within the two families of multilayered systems comprised of Co magnetic layers with respectively, non-ferromagnetic transition-metal spacer layers and noble-metal spacer layers,

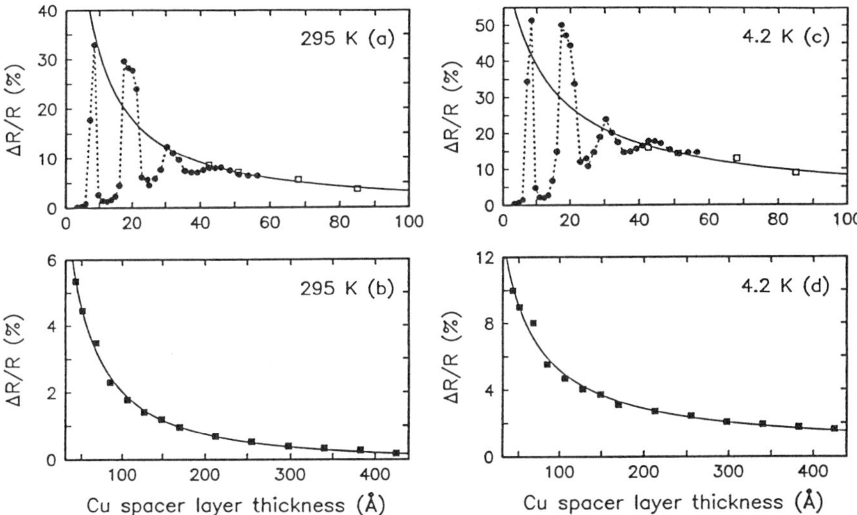

Figure 6. Saturation magnetoresistance versus Cu spacer layer thickness for several series of structures of the form, Si(111)/Ru(50Å)/[Co(11Å)/Cu(t_{Cu})]$_N$/Ru(15Å). The number of Co/Cu periods, N, is 20 (solid circles) and 6 (open and closed squares). Data are shown for temperatures of (a) and (b) 295 K, and (c) and (d) 4.2 K. Since the MR increases with N, in figures (a) and (c) the data for the structures with N=6 has been scaled by a factor of 1.6 to make comparison with the N=20 data easier. The curves through the data are of the form $1/t_{Cu} \exp -(t_{Cu}/\lambda_{Cu})$ at 295 K and $1/t_{Cu}$ at 4.2 K. Note the actual curves shown in the figure are of the exact form, $\Delta R/R = 289/(4.3 + t_{Cu}) \exp - (t_{Cu}/318)$ and $\Delta R/R = 0.28 + 554/(13 + t_{Cu})$ at 295 and 4.2 K respectively, and are also scaled by a factor of 1.6 in (a) and (c).

those systems which exhibit the largest interlayer coupling also exhibit the largest giant MR effect[22].

The dependence of saturation magnetoresistance on Cu thickness, t_{Cu}, for thick Cu spacer layers is shown in detail in Figure 6 for Co/Cu multilayers grown on Ru buffer layers[23]. Well defined oscillations in MR are observed for thinner Cu layers. For Cu layers thicker than ≃60Å the MR decays. The dependence of MR on t_{Cu} is straightforward. At 4.2 K (see fig 6(d)), the MR decays approximately as $1/t_{Cu}$. Since the GMR phenomenon is usually discussed in terms of spin-dependent scattering within the interior of the magnetic layers (*bulk* scattering) or at the interfaces between the magnetic and non-magnetic layers (*interfacial* scattering), this can be readily understood[23] as *dilution* of the spin-dependent scattering regions as the measuring current, which is parallel to the layers, is shunted away from these regions through the Cu layers. Furthermore, since GMR is found only in systems comprised of at least two magnetic layers separated by a non-magnetic layer, the effect requires the flow of electrons from one magnetic layer to neighboring layers. Scattering within the spacer layers will diminish the flow of electrons and so reduce the magnitude of the GMR. Such scattering should be related to *volume* scattering within the interior of the spacer layers. Therefore it can be described by a scattering length,

λ, which should be simply related to the mean free path of thick Cu layers where surface scattering is small compared to volume scattering. Taking into account both volume scattering and *dilution* one expects[23] the GMR to decay as $\simeq 1/t_{Cu} \exp - (t_{Cu}/\lambda_{Cu})$. Figure 6(b) shows that such a functional form well describes the dependence of MR on Cu layer thickness at 295 K in Co/Cu for Cu layer thicknesses ranging from 50 to more than 500Å. The value of λ of $\simeq$320Å compares with $\simeq$390Å in single crystalline Cu. From measurements of the dependence of MR on Cu thickness at various temperatures λ_{Cu} is found to have a strong temperature dependence increasing as the temperature is reduced. In particular, at 4.2 K, the exponential factor differs so little from unity for the range of Cu thickness included in Figure 6, that the MR decays simply as $1/t_{Cu}$ (see Figure 6(d)).

MBE (111) ORIENTED Co-Cu MULTILAYERS

The sputtered Co/Cu multilayers discussed above are polycrystalline and are textured along the (111) orientation. It was thus surprising when several groups, prompted by the studies on polycrystalline (111) textured Co/Cu, reported no evidence for antiferromagnetic coupling or GMR in crystalline (111) Co/Cu multilayers[24, 25]. In contrast oscillatory coupling was observed in crystalline (100) Fe/Cu/Fe and (100) Co/Cu/Co sandwiches grown on copper single crystals[26, 27]. These results led to speculation[25] that the magnetic coupling in the polycrystalline films arose not from (111) oriented crystallites but from crystallites

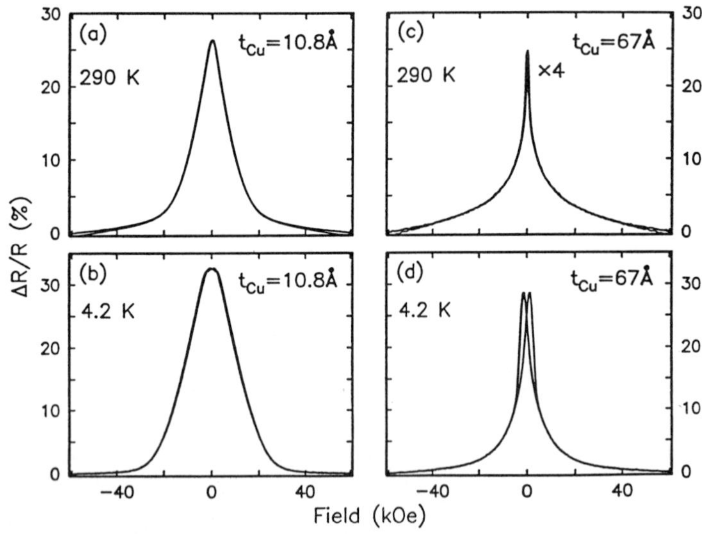

Figure 7. Magnetoresistance versus applied field for two superlattices deposited at 0°C The structure of the superlattices is sapphire(0001)/Pt(60Å)/[Co(8Å)/Cu(t_{Cu})]$_{30}$/Pt(30Å) Figures (a) and (b) are the room temperature and $\simeq$4.2K MR measurements from a superlattice having Cu layers 10.8Å thick, (c) and (d) are the analogous measurements made on a superlattice having Cu layers 67Å.

oriented along other directions, in particular, (100). Consistent with such speculations x-ray diffraction[28] and transmission electron microscopy studies[29] on sputtered Co/Cu multilayers do show evidence for the presence of (200) oriented grains. However, for typical Co/Cu multilayers containing thin Co and Cu layers ($\simeq$10 – 20Å) x-ray $\theta - 2\theta$ measurements give ratios of the integrated intensity in the (200) to the (111) peaks ranging from 0.06 to 0.13 depending on the substrate, the buffer material and thickness and the growth conditions. For much thicker Cu layers, higher (200)/(111) intensity ratios (> 1) are possible. In contrast to these early MBE studies, several groups later reported evidence for both antiferromagnetic coupling and enhanced MR in highly oriented fcc (111) Co/Cu multilayers, at least for Cu layer thicknesses corresponding to the first antiferromagnetic peak in Figure 3. These groups used very different growth procedures. Two groups used thin Au buffer layers: Au layers $\simeq$250Å thick on float glass[30], and Au layers $\simeq$10Å thick on 15Å bcc (110) Co layers on 500Å Ge buffer layers on GaAs(110) substrates [31], respectively. Since metallic buffer layers shunt the sensing current these layers need to be as thin as possible for MR studies. A third group studied only the coupling but not the MR in (111) Co/Cu/Co sandwiches grown on thick ($\simeq$2500Å) (111) Cu layers on $11\bar{2}0$ sapphire substrates[32].

We have considered crystalline (111) Co/Cu structures grown on a variety of substrates using different buffer layers[33, 34]. The best results were obtained for Co/Cu multilayers grown on thin ($\simeq$20Å) Pt buffer layers on sapphire (0001) substrates. Typical MR results for such structures are shown in Figure 7. Very large MR, although substantially smaller than that observed in sputtered films, is found at room temperature of up to $\simeq$35%. A comparison of Figure 7 with that of Figure 2 shows that much larger saturation fields are found in the MBE compared to sputtered Co/Cu multilayers. Indeed values of J^1_{AF} for the MBE structures are about four to five times larger than those for comparable sputtered films. However magnetization data show evidence for incomplete antiferromagnetic coupling in the MBE films with remanent magnetic moments in low fields of at least 70% of the saturation magnetization. This may account for the comparatively low MR values in such structures, since, as mentioned earlier, the magnitude of the MR is related to the degree of antiparallelism of the magnetic layers. The most likely explanation[11] for the incomplete antiferromagnetic coupling in the MBE structures is that they contain subtle defects, such as pin-holes between the magnetic layers, giving rise to direct exchange coupling of the magnetic layers which overwhelms the indirect magnetic coupling mediated by the Cu itself. A similar explanation probably accounts for the numerous reports prior to 1988, as mentioned above, of ferromagnetic coupling in many magnetic multilayered systems. The properties of magnetic multilayers are clearly extremely sensitive to structural defects which can give rise to ferromagnetic bridging of the magnetic layers. Indeed MBE grown highly crystalline (111) Co/Cu multilayers of very similar structural perfection, as evidenced by x-ray scattering studies, can exhibit widely varying properties. For example, whereas Co/Cu structures containing thin Cu layers prepared on Pt buffer layers on sapphire substrates show substantial GMR, similar structures of

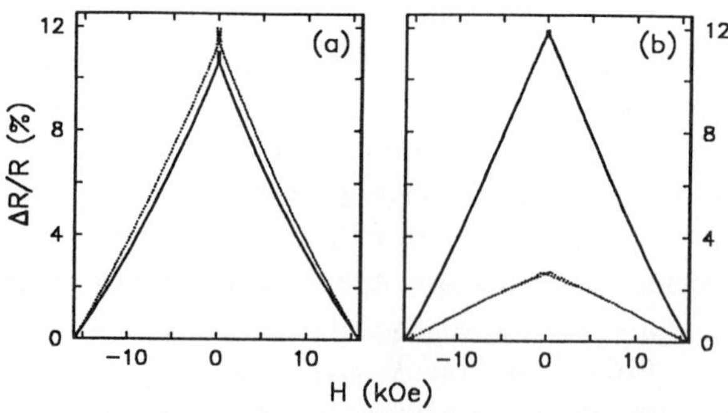

Figure 8. Resistance versus field curves at 295 K (full lines) and 4.2 K (dotted curves) for two samples of the form sapphire(0001)/Pt(t_{buffer})/[Co(16Å)/Cu(7Å)]$_{13}$/Co(16Å)/Pt(30Å) with t_{buffer} = (a) 30 Å and (b) 300 Å.

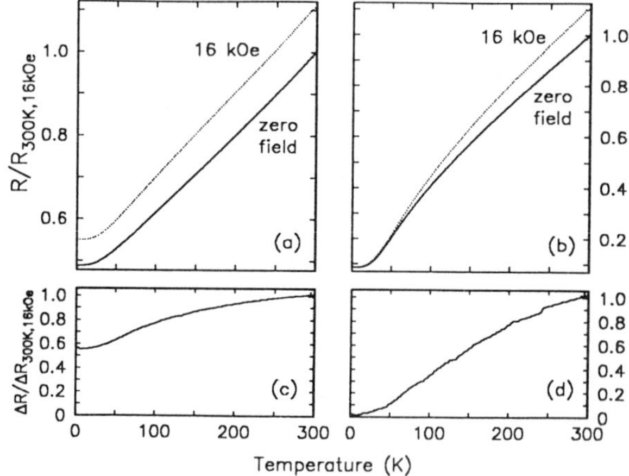

Figure 9. (a) and (b) Resistance vs temperature in zero field (full lines) and 16 kOe (dotted lines) for the two samples of Figure 8. (c) and (d) Difference between the two curves in (a) and (b) respectively, normalized to 1 at 295 K. (a) and (c) correspond to the sample with t_{buffer} = 30Å and (b) and (d) correspond to the sample with t_{buffer} = 300Å.

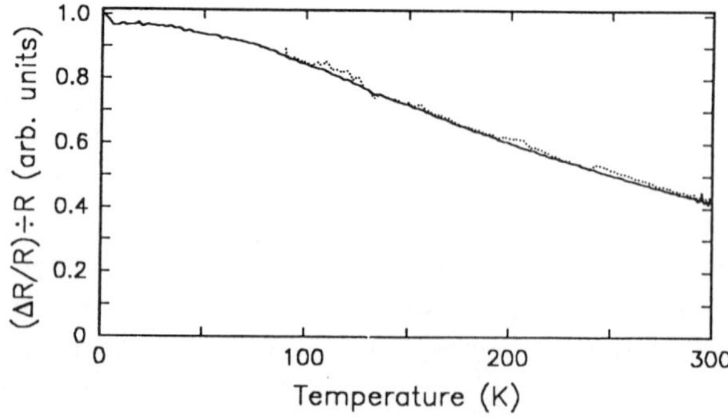

Figure 10. Comparison of the normalized quantity MR/R vs temperature for the two samples of Figures 8 and 9. The full line and dotted lines correspond to the samples with t_{buffer} = 30Å and 300Å respectively.

comparable structural quality prepared on Co buffer layers on GaAs(110) substrates show negligible MR[33].

For the MBE films the buffer layer plays a particularly important role in that it controls the crystallographic orientation of the multilayer. Moreover the structural perfection of the multilayer is influenced by that of the buffer layer. For these reasons it is often necessary to use relatively thick buffer layers. For the Co/Cu structures grown on Pt buffer layers mentioned above, seed layers 20Å thick give rise to high quality structures. Nevertheless such layers are sufficiently conducting to influence the magnitude of the MR. Figure 8 compares MR data on two identical multilayers of the form, sapphire(0001)/Pt(t_{buffer})/[Co(16Å)/Cu(7Å)$]_{13}$/Co(16Å)/Pt(30Å) prepared on Pt buffer layer thicknesses, t_{buffer} of (a) 30Å and (b) 300 Å respectively. Data are shown at 300 K (full curves) and 4.2 K (dotted curves). Note that the resistance is not saturated at the maximum field strengths used of 16 kOe. However data taken at higher fields shows firstly that the resistance change in $\pm$16kOe is about 80% of the saturation change in resistance and secondly the saturation field only increases weakly as the temperature is decreased. Figure 9 shows the temperature dependent resistance of the same samples in zero field and at 16 kOe as well as the difference, $\Delta R(T)$, between these two curves for each sample. The magnetoresistance of the sample with $t_{buffer} = 30$Å varied little from 11% at 300 K to 12% at 4.2 K, whereas for $t_{buffer} = 300$Å the MR varies substantially from 11.9 % at 300 K to just 2.7% at 4.2 K. The most likely explanation is shunting of the current through the relatively more conducting Pt buffer and capping layers which increases as the temperature is decreased. This is very similar to the effect of dilution discussed above for the sputtered Co/Cu multilayers. Indeed assuming the Pt layers are highly conducting the shunting of the current through the Pt layers can be taken into account as a first approximation by the factor $1/R(T)$ where $R(T)$ is the high field temperature dependent resistance of the structure. This is illustrated in Figure 10 which shows a normalized plot of $(\Delta R/R) \times 1/R$ for the two samples of figure 8 and 9 as a function of temperature. The temperature dependence of this quantity is essentially identical for the two samples demonstrating that the temperature dependence of the MR is indeed controlled by shunting through the Pt buffer layers.

Although substantial MR is exhibited by (111) oriented Co/Cu the dependence of MR on Cu spacer layer thickness is distinct from that exhibited by sputtered Co/Cu multilayers. This can be seen by comparing the dependence of MR on Cu layer thickness for crystalline (111) Co/Cu multilayers shown in Figure 11 with that illustrated in Figure 3 for polycrystalline Co/Cu multilayers. Data in Figure 11 are shown at 4.2 K and 290 K for two sets of Co/Cu multilayers deposited at 0°C and 150°C. Only a single oscillation in MR is found centered at approximately 9Å, the same position of the first antiferromagnetic peak in the polycrystalline multilayers. For thicker Cu layers the MR varies monotonically with increasing Cu layer thickness. The variation depends on growth temperature. For samples grown at 0°C the MR is almost independent of Cu layer thickness for t_{Cu} ranging up to >70Å but at 4.2 K the MR steadily increases over the same thickness range. Typical data from this series of samples

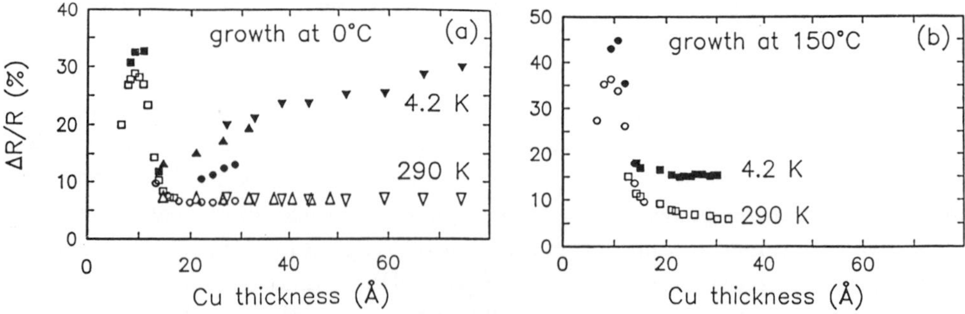

Figure 11. Magnetoresistance at room temperature (open symbols) and 4.2K (filled symbols) for samples deposited at (a) 0°C and (b) 150°C. Data are displayed from a total of 6 wedged samples, with data from a single wedge plotted using the same symbol. The maximum applied field for each measurement was 60 kOe. In all cases, only a single MR peak is observed, near 10Å. For Cu thicknesses greater than 15Å, the variation of the MR is monotonic with Cu layer thickness for all samples, and displays no further MR maxima.

for Cu layer thicknesses of 10.8Å and 67Å are shown in Figure 6. One possible explanation of these data is that, as mentioned earlier, there is significant direct ferromagnetic coupling of the Co layers as suggested by the large remanent moments in the magnetization loops. The ferromagnetic coupling may be so large that it dominates the antiferromagnetic coupling at the second and subsequent AF peaks. Nevertheless as t_{Cu} increases the ferromagnetic coupling will eventually become so weak that the magnetization in neighboring Co layers will switch independently of each other. The MR will then presumably derive from the mis-alignment of magnetic domains in successive layers. Eventually it would be expected that dilution would cause the MR to decrease, as observed for (111) Co/Cu multilayers prepared on GaAs(110) substrates[33].

Note that the data in Figure 11 is from series of samples prepared as multilayer wedges. The wedges contains Co layers of uniform thickness but the thickness of the Cu layers is varied linearly across the substrate by taking advantage of the oblique incidence of the Cu beam on the sample. Since deposition of a single multilayer by MBE takes about 20 times longer than for sputter deposition this makes it difficult to precisely vary the Cu layer thickness from sample to sample. This leads to considerable scatter in the dependence of MR on t_{Cu} measured in a series of multilayered samples. However, by depositing a wedge, a series of samples can be prepared in which the Cu layer thickness can be monotonically varied over a certain thickness range. The data in Figure 11 correspond to a compilation of data measured on six different wedge samples. The substrates were rectangular in shape, $\simeq 11 \times 55\text{mm}^2$ in size, and were pre-scribed at 2mm intervals along the back of the substrates. After deposition the substrates could readily be broken into small strips suitable for MR measurements. These studies are described in more detail elsewhere[34].

INTERFACIAL ORIGIN OF GIANT MAGNETORESISTANCE

The origin of the giant magnetoresistance effect is of great interest. A variety of models have been proposed to account for the origin of the giant MR effect in magnetic multilayers[35-40]. Many of these models assume the current is carried by two conduction-electron channels, associated with spin-up and spin-down electrons. Scattering within these channels will be different because of the different electronic states into which the spin-up and spin-down electrons can be scattered. The magnitude of the giant MR will be related to the ratio of the scattering rates within the two conduction channels of electrons flowing from one magnetic layer to adjacent magnetic layers. The spin dependent scattering obviously derives from the magnetic layers, but of particular importance is whether such scattering occurs within the interior of the magnetic layers (*bulk* scattering) or predominantly at the interfaces between the magnetic and spacer layers (*interfacial* scattering). A simple method[41, 42] to examine the importance of interfacial scattering is to modify the magnetic/non-magnetic interfaces by adding a third material. It is important to consider cases where the MR might be expected to increase since it is very easy to reduce the MR of a structure by inserting material at the interfaces which gives rise to increased scattering or equivalently resistivity. This includes almost any non-magnetic material.

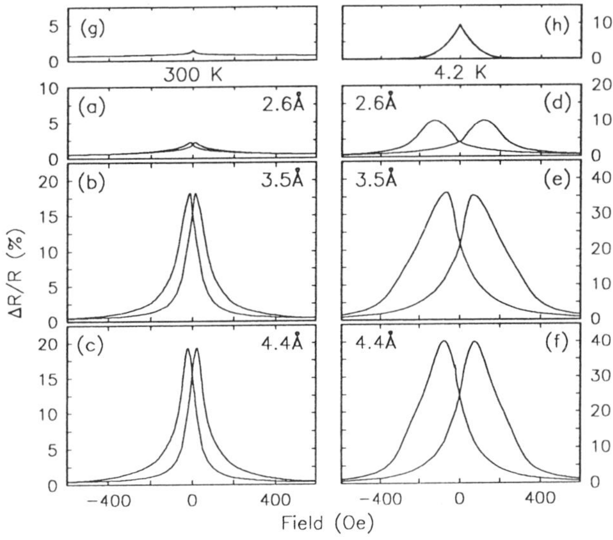

Figure 12. Resistance versus in-plane field for structures of the form (a) and (b), Si/$Ni_{81}Fe_{19}$(50Å)/[15Å$Ni_{81}Fe_{19}$(15Å)/Cu(20Å)$]_{14}$/Ru(25Å) and (c), (d), (e), (f), (g) and (h), Si/Ru(34Å)/$Ni_{81}Fe_{19}$(10Å)/Co(t_i)/[Cu(19Å)/Co(t_i)/$Ni_{81}Fe_{19}$(10Å)/Co(t_i)$]_{19}$/Ru(14Å) with t_i = (c) & (d) 2.6Å, (e) & (f) 3.5Å, and (g) & (h) 4.4Å. Data in the left hand column of the figure are taken at 295 K and data in the right hand column, are measured at 4.2 K.

A particularly interesting case is that of $Ni_{81}Fe_{19}$/Cu multilayers in which thin layers of Co are inserted at each $Ni_{81}Fe_{19}$ – Cu interface[42]. Since the magnetoresistance of $Ni_{81}Fe_{19}$/Cu multilayers is about half that of similar Co/Cu multilayers with layers of comparable thickness we might expect a significant increase in the MR of the permalloy/Cu multilayers if interfacial scattering is significant. Figure 12 demonstrates that the properties of $Ni_{81}Fe_{19}$/Cu multilayers are indeed dramatically modified by inserting thin Co layers at the $Ni_{81}Fe_{19}$/Cu interfaces. The Figure includes resistance versus field curves at temperatures of 300 and 4.2 K for three $Ni_{81}Fe_{19}$/Cu structures, Si/Ru(34Å)/$Ni_{81}Fe_{19}$(10Å)/Co(t_i)/[Cu(19Å)/Co(t_i)/$Ni_{81}Fe_{19}$(10Å)/Co(t_i)$]_{19}$/Ru(14Å) , with Co interface layer thicknesses, t_i, of 2.6 , 3.5 and 4.4 Å. For comparison resistance versus field data are also shown in Figure 12 for two $Ni_{81}Fe_{19}$/Cu multilayers without Co interface layers of the form, Si/$Ni_{81}Fe_{19}$(50Å)/[$Ni_{81}Fe_{19}$(15Å)/Cu(20Å)$]_{14}$/Ru(25Å). All these structures contain Cu spacer layers which correspond to the second antiferromagnetic oscillation in Figure 3. The addition of the thin Co layers restores the antiferromagnetic coupling, absent at room temperature in the pure $Ni_{81}Fe_{19}$/Cu structures at this Cu layer thickness, and, perhaps more remarkably, dramatically increases the MR of the multilayers. The room temperature data in Figure 12 clearly show that the interlayer exchange coupling is determined by the character of the ferromagnetic / spacer layer interfaces. However, since the MR is present only for antiferromagnetically coupled structures, perhaps it is not surprising that the reestablishment of antiferromagnetic coupling in these structures is accompanied by a giant MR effect. In contrast, the low temperature data, for which there is antiferromagnetic coupling even in the pure $Ni_{81}Fe_{19}$/Cu structures unambiguously demonstrates that the addition of Co layers just 3-4 Å thick dramatically increases the magnetoresistance, almost quadrupling it. By using exchange biased sandwich structures instead of multilayers the dependence of the MR on the thickness of the inserted interfacial layer can be examined in detail[41]. From such data a characteristic length can be ascribed to the thickness of the interfacial layer required to establish the character of the interface and the magnitude of the MR. This length is extremely short at just 2 to 3Å at room temperature for a wide range of magnetic/ non-magnetic material combinations. Such experiments seem to unambiguously demonstrate the interfacial origin of the giant MR effect. This conclusion is supported by a large number of experiments but there is no space to discuss them here.

SUMMARY

We have presented a brief review of the properties of polycrystalline (111) textured and crystalline (111) oriented Co/Cu multilayers. Such structures show interesting properties including, in particular, very large magnetoresistance values at room temperature. In sputtered structures grown on Fe buffer layers saturation magnetoresistance values exceed 65% although the fields required are large, $\simeq$10kOe. In MBE multilayers MR values exceeding 35% have been reported at room temperature but the saturation fields are even higher than those found in sputtered multilayers, exceeding 40 kOe. The giant MR phenomenon in these structures results from the strong antiferromagnetic coupling of the Co layers mediated by the Cu spacer layers. Whereas beautiful oscillations in MR resulting from oscillations in the magnetic coupling between ferromagnetic and antiferromagnetic are found in sputtered Co/Cu multilayers as the Cu layer thickness is varied, only one oscillation in MR is found in the MBE (111) oriented structures. The observation of MR in the crystalline multilayers appears to resolve a controversy as to whether there is an intrinsic oscillatory coupling along the (111) direction. The origin of the giant magnetoresistance effect is predominantly derived from scattering at the interfaces between the magnetic and non-magnetic layers. This can be clearly demonstrated by engineering structures in which the interfaces are systematically modified by inserting thin layers of different magnetic materials.

ACKNOWLEDGEMENTS

We thank K.P. Roche for technical support.

REFERENCES

1. R.E. Walstedt and J.H. Wernick, Phys. Rev. Lett. **20**, 856 (1968).
2. J.B. Boyce, and C.P. Slichter, Phys. Rev. B **13**, 379 (1976).
3. L.R. Walker and R.E. Walstedt, Phys. Rev. B **22**, 3816 (1980).
4. R.M. White, *Quantum Theory of Magnetism* (Spinger-Verlag, 1983).
5. T. Shinjo and T. Takada, *Metallic Superlattices* (Elsevier, 1987).
6. J-C. Bruyere, O. Massenet, R. Montmory and L. Neel, C.R. Acad Sc. Paris **258**, 1423 (1964).
7. P. Grunberg and F. Saurenbach, MRS Int'l. Mtg. Adv. Mats. **10**, 255 (1989).
8. A. Bardasis, D.S. Falk, R.A. Ferrell, M.S. Fullenbaum, R.E. Prange, and D.S. Mills, Phys. Rev. Lett. **14**, 298 (1965).
9. C.F. Majkrzak, J.W. Cable, J. Kwo, M. Hong, D.B. McWhan, Y. Yafet, J.V. Waszczak, and C. Vettier, Phys. Rev. Lett. **56**, 2700 (1986).
10. S.S.P. Parkin, N. More and K.P. Roche, Phys. Rev. Lett. **64**, 2304 (1990).
11. S.S.P. Parkin, R. Bhadra and K.P. Roche, Phys. Rev. Lett. **66**, 2152 (1991).
12. S.S.P. Parkin, Phys. Rev. Lett. **67**, 3598 (1991).
13. P. Grunberg, R. Schreiber, Y. Pang, M.B. Brodsky and H. Sowers, Phys. Rev. Lett. **57**, 2442 (1986).
14. A. Cebollada, J.L. Martinez, J.M. Gallego, J.J. de Miguel, R. Miranda, S. Ferrer, F. Batallan, G. Fillion, and J.P. Rebouillat, Phys. Rev. B **39**, 9726 (1989).
15. G. Binasch, P. Grunberg, F. Saurenbach and W. Zinn, Phys. Rev. B **39**, 4828 (1989).
16. M.N. Baibich, J.M. Broto, A. Fert, F. Nguyen van Dau, F. Petroff, P. Etienne, G. Creuzet, A. Friederich, and J. Chazelas, Phys. Rev. Lett. **61**, 2472 (1988).

17. S.S.P. Parkin, Z.G. Li and D.J. Smith, Appl. Phys. Lett. **58**, 2710 (1991).
18. S.S.P. Parkin, Appl. Phys. Lett. **60**, 512 (1992).
19. I.R. McFadyen and P.S. Alexopoulous, in *Science and Technology of Nanostructured Magnetic Materials*, edited by G.C. Hadjipanayis and G.A. Prinz (Plenum Press, New York, 1991) p. 99.
20. E. Velu, C. Dupas, D. Renard, J.P. Renard and J. Seiden, Phys. Rev. B **37**, 668 (1988).
21. C. Dupas, P. Beauvillain, C. Chappert, J.P. Renard, F. Trigui, P. Veillet, E. Velu, and D. Renard, J. Appl. Phys. **67**, 5680 (1990).
22. S.S.P. Parkin, (unpublished).
23. S.S.P. Parkin, A. Modak and D.J. Smith, Phys. Rev. B. Rap. Comm. (to be published).
24. R.F. Marks, R.F.C. Farrow, S.S.P. Parkin, C.H. Lee, B.D. Hermsmeier, C.J. Chien, and S.B. Hagstrom, in *Heteroepitaxy of Dissimilar Materials*, edited by R.F.C. Farrow, J.P. Harbison, P.S. Peercy and A. Zangwill (Mat. Res. Soc. Sym. Proc., 1991) Vol. 221, p. 15.
25. W.F. Egelhoff and M.T. Kief, Phys. Rev. B **45**, 7795 (1992).
26. W.R. Bennett, W. Schwarzacher and W.F. Egelhoff, Phys. Rev. Lett. **65**, 3169 (1990).
27. A. Cebollada, R. Miranda, C.M. Schneider, P. Schuster and J. Kirschner, J. Mag. Mag. Mat. **102**, 25 (1991).
28. S.S.P. Parkin, in *Magnetic Surfaces, Thin Films and Multilayers*, edited by S.S.P. Parkin, H. Hopster, J-P. Renard, T, Shinjo and W. Zinn (Mat. Res. Soc. Sym. Proc., 1992) Vol. 231, p. 211.
29. S.S.P. Partin, K.P. Roche and T. Suzuki, Jap. J. Appl. Phys. **31**, L1246 (1992).
30. J.P. Renard, P. Beauvillain, C. Dupas, K. Le Dang, P. Veillet, E. Velu, C. Marliere, and D. Renard, J. Mag. Mag. Mat. **115**, L147 (1992).
31. D. Greig, M.J. Hall, C. Hammond, B.J. Hickey, H.P. Ho, M.A. Howson, M.J. Walker, N. Wiser, and D.G. Wright, J. Mag. Mag. Mat. **110**, L239 (1992).
32. J. Kohlepp, S. Cordes, H.J. Elmers and U. Gradmann, J. Mag. Mag. Mat. **111**, L231 (1992).
33. S.S.P. Parkin, R.F. Marks, R.F.C. Farrow, G.R. Harp, Q.H. Lam, and R.J. Savoy, Phys. Rev. B **46**, RC9262 (1992).
34. G. R. Harp, S. S. P. Parkin, R. F. C. Farrow, R. F. Marks, M. F. Toney, Q. H. Lam, T. A. Rabedeau, and R. J. Savoy, (preprint).
35. R.E. Camley and J. Barnas, Phys. Rev. Lett. **63**, 664 (1989).
36. P.M. Levy, K. Ounadjela, S. Zhang, Y. Wang, C.B. Sommers and A. Fert, J. Appl. Phys. **67**, 5914 (1990).
37. D.M. Edwards, J. Mathon, R.B. Muniz and M.S. Phan, Phys. Rev. Lett. **67**, 493 (1991).
38. J-I. Inoue, A. Oguri and S. Maekawa, J. Phys. Soc. Jap. **60**, 376 (1991).
39. F. Trigui, E. Velu and C. Dupas, J. Mag. Mag. Mat. **93**, 421 (1991).
40. S. Zhang, P.M. Levy and A. Fert, Phys. Rev. B. **45**, 8689 (1992).
41. S.S.P. Parkin, (preprint).
42. S.S.P. Parkin, Appl. Phys. Lett. **61**, 1358 (1992).

PERPENDICULAR MAGNETORESISTANCE IN Ag/Co AND Cu/Co MULTILAYERS

P. A. Schroeder, J. Bass, P. Holody, S-F. Lee,
R. Loloee, W. P. Pratt Jr., and Q. Yang

Department of Physics and Astronomy and
Center For Fundamental Materials Science
Michigan State University
East Lansing, MI 48823-1116

INTRODUCTION

In this paper we present measurements and analyse data on the magnetoresistance measured with the current perpendicular to the layer planes (CPP-MR) of magnetic multilayers of Ag and Co, and of Cu and Co. Because of the small resistances involved (~ $10^{-7}\Omega$) the measurements are considerably more difficult than those for the more usual orientation with current in the layer planes, (CIP-MR). However as we have recently shown[1-3], the interpretation is much simpler, and permits quantitative determination of relative magnitudes of interface and bulk spin dependent scatterings that are fundamental to current theories of MR[4].

We first describe our experimental method and our samples. We then show that the CPP-MR for the Cu/Co system oscillates with the non-magnetic metal (N) layer thickness, t_N, as already observed in CIP-MR[5,6]. For large t_N, for which exchange coupling is negligible, we have previously shown that a simple series resistor model, in which the resistances of the N and ferromagnetic (F) layers and the N/F interface resistances are added in series, provides a good description of data on Ag/Co multilayers, and permits a clear separation of the important bulk and interface scattering contributions to the MR[2]. We showed more recently[3] that a more complex 2-spin model provided a satisfactory alternative description of the Ag/Co multilayer data, with the advantage of allowing the determination of the spin resistivity parameters $\alpha_F = \rho_F^{\downarrow}/\rho_F^{\uparrow}$ and $\alpha_{F/N} = R^{\downarrow}_{F/N} / R^{\uparrow}_{F/N}$. Here $\rho_F^{\downarrow}(\rho_F^{\uparrow})$ is the resistivity when electron spin and local magnetization are in opposite (same) directions. A similar definition holds for the interface resistance $R_{F/N}$. For AgSn/Co multilayer data, in contrast, the 2-band model was essential[3]. In this paper, we are mainly concerned with the application of both models to Cu/Co multilayers, and for the first time we derive values of and α_{Co} and $\alpha_{Co/Cu}$. We do include some data on Ag/Co multilayers and compare the results on both systems.

A more detailed justification for the 2-spin model is given in the following paper by Dr. A. Fert.

EXPERIMENTAL

To measure the CPP-MR, one has a choice of using samples of macroscopic cross-section with very small resistance, in conjunction with a very sensitive voltage measuring device, or of making samples of very small cross-section and using a conventional digital voltmeter to measure the voltages. Because of our previous experience in high precision measurements of very small resistances using a SQUID system[7], we chose the former method.

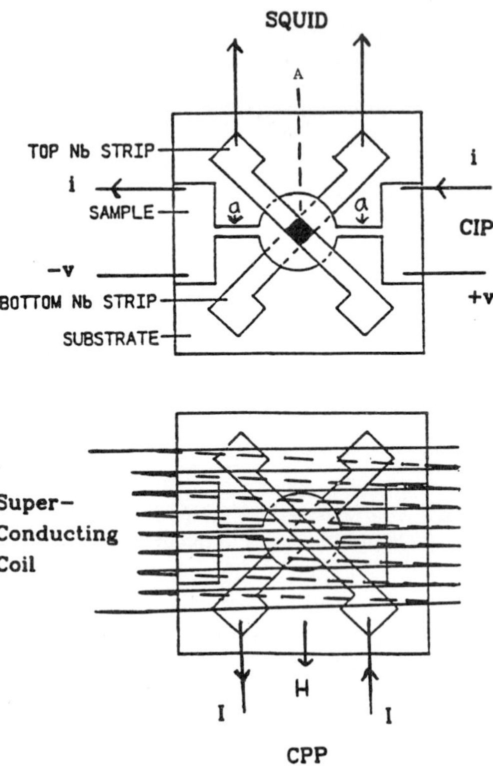

Figure 1. The geometry of the experimental samples. The black region A in the upper diagram is the area of the CPP-MR sample. The strips labelled "a" represent the CIP-MR sample. The lower diagram shows the superconducting solenoid fitting around the sample.

The geometry of the samples is shown in Fig. 1. The samples and connecting leads are all formed by D.C. magnetron sputtering. First the bottom strip of Nb is sputtered through a mask onto a substrate. Then the multilayer of the shape shown is sputtered through a second mask. Finally, the top Nb strip is laid down through a third mask. To measure the CPP resistance, current is passed through one leg of the lower superconducting strip along the path of least resistance, through the small square (Area A = 1.2 mm^2 ± 4%) where the Nb strips overlap, and out of one leg of the top Nb strip. The other legs of the Nb strips are connected to the SQUID measuring system via superconducting wire. CIP measurements are made on the thin sample sections labelled "a" in Fig. 1. The resistance of these sections is ~ 1 Ω and can easily be measured by conventional methods. In the lower part of Fig. 1 we show the sample inserted into an oval cross-section superconducting coil, so that the resistance can be measured with magnetic fields in the plane of the sample, which corresponds with the plane of easy magnetization. For CPP measurements we used a current I = 0.05 A so that the voltages to be measured were ~ 5×10^{-9} V. A few samples showed current dependence, indicative of superconductivity by proximity effect in the multilayers, presumably due to pin holes. These samples were discarded. All measurements were performed at 4.2 K.

Our samples were all prepared by sputtering in a cryopumped ultra high vacuum compatible system[8] containing an essential in-situ mask changing device. The base pressure before sputtering was < 10^{-8} torr and the sputtering pressure was 2.5 mtorr. Deposition rates were ~ 0.9 nm/s for Ag and Cu and ~ 0.6 nm/s for Co. Substrates were sapphire for Ag/Co and silicon, with or without a 5 nm buffer layer of Fe, for Cu/Co.

At the moment we can only claim that the results we obtain apply to the samples prepared under the above circumstances. To what extent these results are representative of differently prepared Cu/Co and Ag/Co samples, and to what extent the data depend on the detailed structure of the samples, has yet to be determined.

We note the following characteristics of our samples. The X-ray θ -2θ spectra are similar to those published elsewhere[9] and indicate a coherence length of 2 - 4 bilayers. Rocking curve widths about <111> are ~ 10°. This is confirmed in pole figures which show textured growth in the <111> direction. <200> reflections from crystallites with (200) in the plane of the crystal were not observed. Transmission electron microscopy indicates that the initial layers are nicely parallel, but after ~ 10 layers columnar growth and disruptive layering becomes apparent.

NMR measurements on our Cu/Co samples have been made by Mény and Panissod[10]. Multilayers with low t_{Co}(~ 1.5 nm) show purely FCC cobalt stacking. With increasing t_{Co} the more stable HCP phase begins to appear, but as t_{Cu} increases the proportion of HCP Co decreases. The detailed study involved fitting the data to various models and searching for the model with the lowest χ^2. In samples with thin Co layers, if bulk defects in the Co are excluded, then the Co and Cu are intermixed over 3 monolayers. However such a model does not fit well for samples with thick Co layers. There the existence of bulk defects is strongly suggested, with the best fit giving sharp Co/Cu interfaces (~ 1 mixed layer) and ~ 0.7% defects. If the latter are caused by columnar growth, then the in-plane diameter of the columns is ~ 9.0 nm, a value in good agreement with the 10 - 20 nm thickness of the columns observed by electron microscopy.

RESULTS

In Fig. 2 we show the CPP-MR, CIP-MR, and magnetization of a Ag(6nm)/Co(6nm) sample[1]. We define MR by

$$MR(H) = \frac{R(H) - R(H_{sat})}{R(H_{sat})} \qquad (1)$$

where R(H) and $R(H_{sat})$ are the resistances at magnetic field H and H_{sat} (where the resistance saturates) respectively. The shapes of the MR curves are very similar, but note that the vertical scales are different and the CPP-MR is ~ 4 times the CIP-MR for this particular sample. H_P is the field at which the peak MR occurs after saturation, and

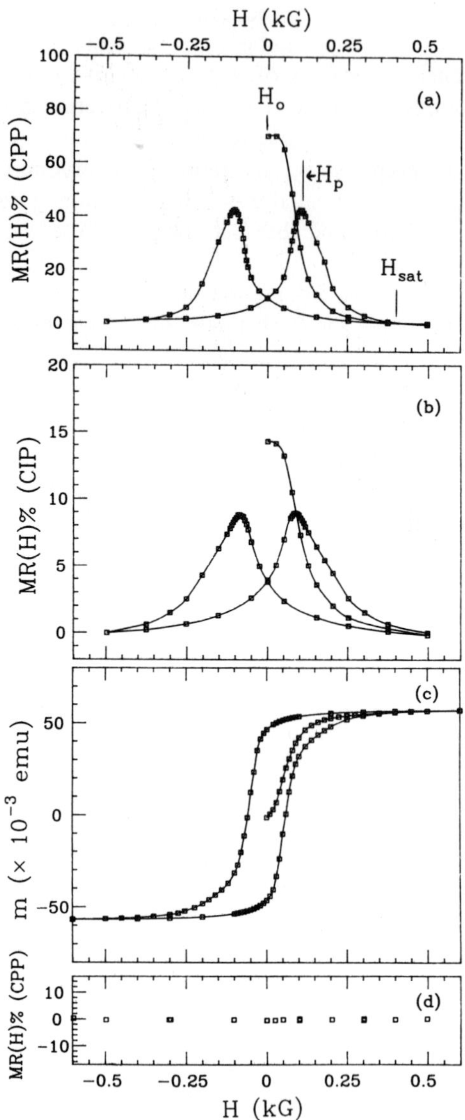

Figure 2. (a) CPP-MR, (b) CIP-MR, and (c) magnetization m for a Ag(6nm)/Co(6nm) sample. In (a), H_0, H_P and H_{sat} are defined. (d) shows the CPP-MR for a Nb/Co/Nb sandwich on the same scale as (a).

comparison with the magnetization curve indicates that H_P is somewhat above the coercive field (H_C). [Measurements over a large number of samples indicate that, the Cu/Co samples nearly always have H_C < H_P(CPP) < H_P(CIP), while the Ag/Co samples usually have H_P (CPP) > H_P(CIP) > H_C]. The CPP-MR for a Nb/Co/Nb sandwich is included (Fig. 2d) for comparison and to show that the MR of the Ag/Co sample cannot possibly be ascribed to the MR of bulk Co and the two Nb/Co interfaces alone.

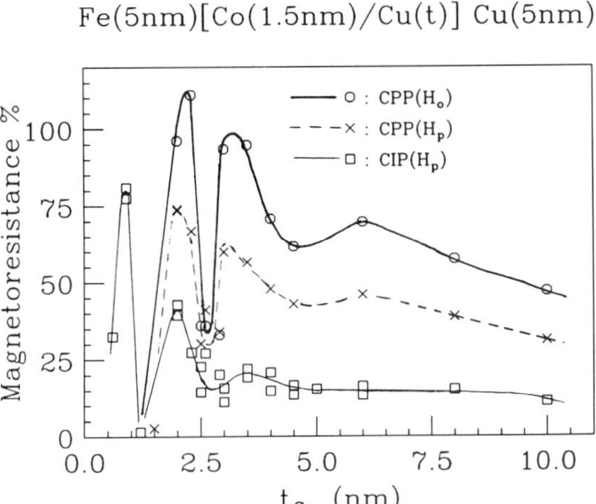

Figure 3. CPP-MR for H_0, and H_P, and CIP-MR for H_P plotted as a function of copper layer thickness for the Co(1.5nm)/Cu(t_{Cu}) system. Samples are deposited on an Fe buffer layer and have a 5 nm overlayer of Cu. Lines are guides to the eye.

A surprising aspect of these curves is that the MR for the initial virgin H = 0 state-- we will call this the H_0 state-- is considerably greater than the MR at $H = H_P$. Furthermore, the H_0 magnetic state is not an irreproducible, badly defined state; we will see that it varies systematically with N and F layer thickness. With our present limited knowledge of the micromagnetics of our samples, all we can say is that there must be more antiferromagnetic alignment of neighboring Co layers at H_0 than at H_P. In what follows we tentatively refer to the initial H_0 state as having antiferromagnetic (AF) alignment. We will present some justification for this later.

In Fig. 3 we show the CPP-MR at $H = H_0$ and $H = H_P$, along with new and previously measured[6] CIP-MR data, as a function of Cu thickness (t_{Cu}) for the Cu/Co system. All the samples in this figure were grown on a 5 nm buffer layer of Fe[5]. The CPP-MR data faithfully follow the CIP-MR oscillations, and we note especially the systematic variation of the H_0 data.

We have searched for similar oscillations in the Ag/Co system, but, in agreement with Araki et al.[11], we have only found them when t_{Co} is very small. These data will be the subject of another paper[12].

For the remainder of this paper we will limit ourselves mostly to the region where $t_{Cu} > 6.0$ nm, where the oscillations and the exchange coupling have died out. Experimentally the first simple systematic relation we observed was that, for films of fixed total thickness, t, the total measured resistance, $R_T(H)$, was proportional to the number of bilayers, M This behaviour is illustrated in Figures 5, 6, and 7 for Cu/Co samples with $t_{Co} = t_{Cu}$, 1.5 nm and 6 nm respectively. Unless specifically stated otherwise, these samples and those that follow do not have an Fe buffer.

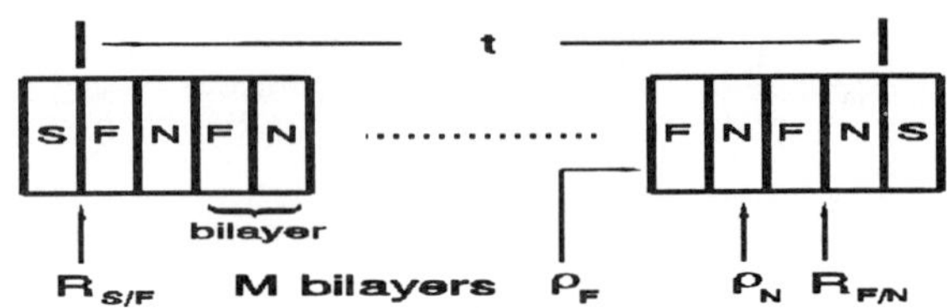

Figure 4. Details of the series resistor model.

SIMPLE SERIES RESISTOR MODEL

The behaviour in Figs. 5, 6 and 7 can be easily explained by the simple series resistor model illustrated in Fig. 4. Here S refers to the superconducting Nb contacts. We add the resistances of each layer and the boundary resistances $R_{F/N}$ and $R_{S/F}$ in series to give

$$AR_T = AR_{sample} + 2AR_{S/F} \tag{2}$$

where A is the sample cross-section and AR_T is the appropriate quantity for interface phenomena. Taking the resistors that make up R_{sample} in series, we get

$$\begin{aligned} AR_T(H) = 2\,AR_{S/F} + (\mathrm{M}-1)\rho_N t_N + \mathrm{M}\rho_F(H)t_F \\ +2\,AR_{F/N}(H)(\mathrm{M}-1) \end{aligned} \tag{3}$$

where R_T, ρ_F and $R_{N/F}$ are regarded as field dependent entities. The factor (M - 1) occurs because the last N layer on the right of Fig. 5 is turned superconducting by proximity effect. For fixed t, and M-1 approximated by M (later calculations will always use the correct M - 1 value), we find that

$$AR_T(H) = Intercept + Slope \times \mathrm{M} \tag{4}$$

The intercepts and slopes for two independent conditions, t_F = constant, and $t_F = t_N$, are as given in Table 1.

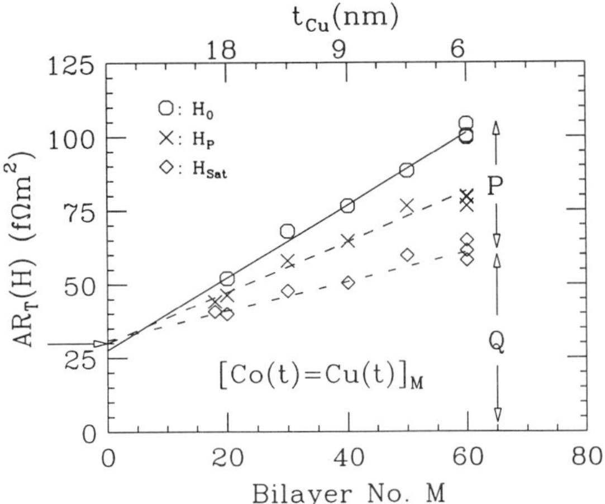

Figure 5. Sample area times total sample resistance measured perpendicular to the layers plotted against the number of bilayers, M, in the sample, for total thickness t = 720 nm and $t_{Cu} = t_{Co}$. The straight lines are least square fits. The arrow on the vertical axis is the intercept expected from Eq. 5(a-d)

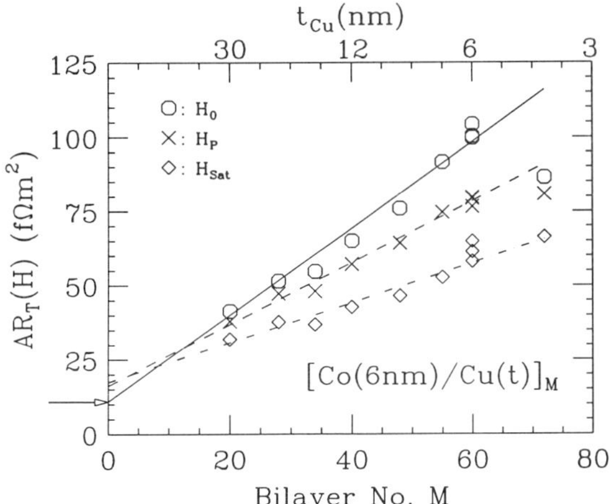

Figure 6. As for Fig. 5 but t = 360 nm and t_{Co}= 1.5 nm.

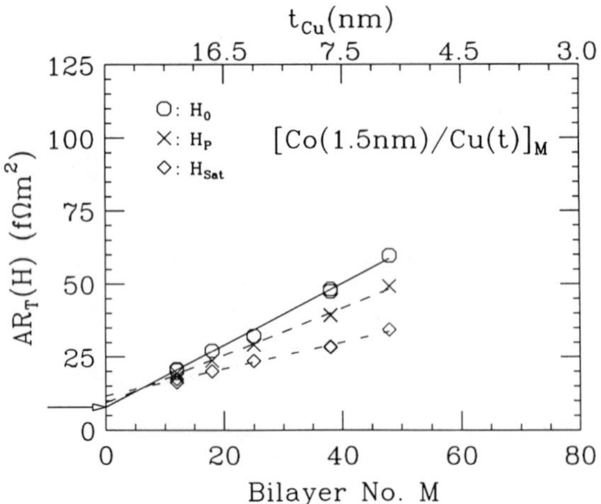

Figure 7. As for Fig. 5 but t_{Co} = 6 nm.

Table 1. Intercepts and Slopes for Eq. 4.

t_F, t_N relations	Intercept	Slope
t_F = constant	$2AR_{S/F} + \rho_N t$	$[\rho_F(H) - \rho_N]t_F + 2AR_{F/N}(H)$
$t_F = t_N$	$2AR_{S/F} + [\rho_N + \rho_F(H)]t/2$	$2AR_{F/N}(H)$

The quantities $\rho_F(0)$, ρ_N, and $R_{Nb/Co}$ have been measured independently. Their values are[13]:

$$\rho_{Cu} = (0.6 \pm 0.1) \times 10^{-8}\,\Omega m \tag{5a}$$

$$\rho_{Ag} = (1.0 \pm 0.2) \times 10^{-8}\,\Omega m \tag{5b}$$

$$2AR_{Nb/Co} = (6.1 \pm 1.0)\, f\Omega m^2 \tag{5c}$$

$$\rho_{Co} = (6.0 \pm 1.0) \times 10^{-8}\,\Omega m \tag{5d}$$

(f = femto = 10^{-15}). The first three values allow a prediction of the common (H independent) intercept for t_F.= constant; the arrows in Figs. 6 and 7 show that this prediction agrees reasonably well with the data for Cu/Co, especially for H = 0. The slope of the $t_F = t_N$ graph gives $R_{F/N}(H)$ directly. The relationship between the field dependent quantity, $\rho_{Co}(H)$, and the field independent quantity, ρ_{Co}, is not uniquely specified in this model. Interestingly, for Cu/Co, the arrow in Fig. 5 shows that the intercepts for $t_F = t_N$ all lie fairly close to what is predicted simply by inserting the independently measured ρ_{Co}.

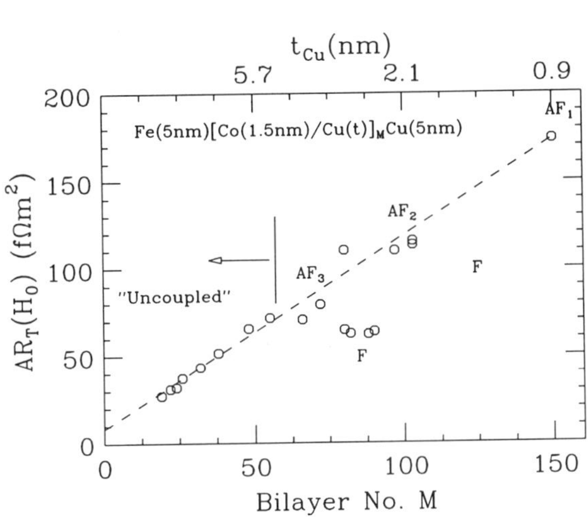

Figure 8. Plot of $AR_T(H_0)$ against M for the Co(1.5nm)/Cu(t_{Cu}) series relating measurements in the exchange coupling region with those in the "uncoupled" region. AF_1, AF_2, AF_3 correspond with the 1st → 3rd antiferromagnetic peaks in Figure 3. F indicates Cu thicknesses for ferromagnetic exchange.

In Fig. 8 we plot $AR_T(H_0)$ versus the number of bilayers (M) for the Co(1.5nm)/Cu(t_{Cu}) system with a 5 nm buffer layer of Fe. The straight line is drawn through the highest point (t_{Cu} = 0.9 nm) and the theoretical intercept. The fact that the "uncoupled" data lie so close to this line can be interpreted in two ways. Firstly, of all samples, one might expect the t_{Cu} = 0.9 nm sample, with strong exchange coupling, to be nearest to complete antiferromagnetic alignment; this line then suggests that for the uncoupled samples the magnetizations in successive layers ($\vec{M}_i$) are also close to complete antiferromagnetic alignment. (If we believe that exchange coupling is negligible at these t_{Cu}s, then presumably the alignment would be magnetostatic in origin.) The second interpretation is that random $\vec{M}_i$'s (with $\Sigma\vec{M}_i$'s = 0) and antiferromagnetic $\vec{M}_i$'s produce the same MR[14].

Fig. 5 provides visually semi-quantitative information on the relative importance of interface and bulk scattering. For M = 60, for example, the MR = P/Q. According to Table 1, P is determined primarily by the difference in slopes for H_0 and H_{sat}, and only minimally by the difference in the intercepts. The slopes are a direct measure of $R_{Co/Cu}(H)$, while the difference in the intercepts is determined by $\rho_{Co}(H)$. Therefore we can say that for M > ~ 20, (i.e. t_{Cu} < 18 nm), the field dependence of $R_{Co/Cu}$ is the main factor determining the MR. Also by comparing $R_{F/N}$ obtained from the slope for the H_0 data with ρ_{Co}, we find that the resistance at H_0 of one Co/Cu interface is about equal to that of 10 nm of Co.

Although the simple series resistor model is in broad agreement with experiment, there are some discrepancies. The H_P and H_{sat} data for t_{Co} fixed (Figs. 6 and 7) give intercepts which are greater than the H = 0 intercepts. Pronounced disagreement occurs when a high resistivity N material is used[3]. Finally, according to the simple series resistor model, the plot of (AR_{AF} - AR_{FM}) (where R_{AF} and R_{FM} are the total resistances for antiferromagnetic and ferromagnetic alignment respectively) against M for t_{Co} fixed should pass through the origin, but experimentally these plots make a small but perceptible negative intercept on the Y axis. A two-spin model[15] largely eliminates these discrepancies.

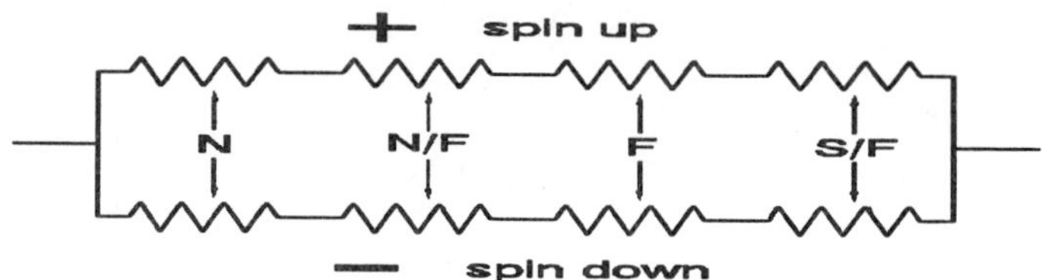

Figure 9. Equivalent circuit for the 2-spin model.

TWO-SPIN MODEL

The equivalent circuit for this model is shown in Fig. 9. We assume that the scattering length for spin flip scattering >> $t_F + t_N$. We follow the notation of Valet and Fert[16] that $^+$ and $^-$ refer to absolute spin directions. Then we can use a series resistance model to obtain R^+ and R^- for each spin channel. Taking these two resistances in parallel we have $R = R^+ R^-/(R^+ + R^-)$. We define the following quantities. $\rho_F^{\uparrow}$ and $R^{\uparrow}{}_{F/N}$ are used when spin and magnetization are in the same direction. $\rho_F^{\downarrow}$ and $R^{\downarrow}{}_{F/N}$ are used when spin and magnetization are in opposite directions. We assume that $R^{\uparrow}{}_{S/F} = R^{\downarrow}{}_{S/F}$.

$$\alpha_F = \frac{\rho_F^{\downarrow}}{\rho_F^{\uparrow}} = \frac{(1+\beta)}{(1-\beta)} \quad \text{and} \quad \alpha_{F/N} = \frac{R^{\downarrow}{}_{F/N}}{R^{\uparrow}{}_{F/N}} = \frac{(1+\gamma)}{(1-\gamma)} \tag{6}$$

When the magnetizations in consecutive F layers are antiferromagnetically aligned, $R^+ = R^-$, and

$$AR_{AF} = 2AR_{S/F} + \mathrm{M}\left[\rho_F^{\,*} t_F + \rho_N t_N + 2AR^*{}_{F/N}\right] \tag{7}$$

$$\text{where } \rho_F^{\,*} = \frac{\rho_F}{\left(1-\beta^2\right)} \quad \text{and} \quad \mathrm{AR}^*{}_{F/N} = \frac{AR_{F/N}}{\left(1-\gamma^2\right)} \tag{8}$$

As in Eq. 4, again we obtain a linear relation between AR_{AF} and M. The slopes and intercepts for the special cases, t_F = constant, and $t_N = t_F$ are the same as those in Table 1 with ρ_F replaced by ρ^*_F, and $R_{F/N}$ by $R^*_{F/N}$. (We note that, in his paper in these proceedings, Dr. Fert uses $r_b{}^*$ in place of our $AR^*_{F/N}$).

Both ρ^*_F and $R^*_{F/N}$ can be obtained from slopes and intercepts of the AR_{AF} vs M plots and $\beta$and γ can then be estimated, as follows.

When the magnetizations in consecutive layers are ferromagnetically aligned, $R^+ \neq R^-$.

$$AR^+{}_{FM} = \left[\rho_F{}^{\uparrow} t_F + 2\rho_N t_N + AR^{\uparrow}{}_{F/N}\right]\mathrm{M} + 4AR_{S/F}$$

$$AR^-{}_{FM} = \left[\rho_F{}^{\downarrow} t_F + 2\rho_N t_N + AR^{\downarrow}{}_{F/N}\right]\mathrm{M} + 4AR_{S/F}$$

Adding these terms in parallel gives

$$AR_{FM} = AR_{AF} - \frac{\mathrm{M}^2\left[\beta\rho_F{}^* t_F + 2\gamma AR^*{}_{F/N}\right]^2}{AR_{AF}} \tag{9}$$

Clearly the linear relation between AR_{FM} and M breaks down. However, if we take the case of t_{Co}= constant, when Eq. 7 becomes

$$AR_{AF} = \left[2AR_{S/F} + \rho_N t\right] + \mathrm{M}\left[\left(\rho_F{}^* - \rho_N\right)t_F + 2AR^*{}_{F/N}\right], \tag{10}$$

then we see that for M small or ρ_N large, the first term in Eq. 10 predominates and

$$AR_{AF} - AR_{FM} \propto \mathrm{M}^2 \tag{11}$$

But for M large, the second term in Eq. 10 predominates and

$$AR_{AF} - AR_{FM} \propto \mathrm{M} \tag{12}$$

The latter situation exists over most of the range of our data, but simple extrapolation of the high-M AR_{FM} data to zero gives an incorrectly high value for the intercept.

We analyse our data in terms of the two band model as follows. First ρ^*_F, ρ_N, and $R^*_{F/N}$ can all be directly obtained from the linear relation between R_{AF} and M obtained for H_0. With data from t_{Co}= 1.5 and 6.0 nm, and t_{Co}= t_{Cu} for the Cu/Co system, we have much more than the minimum data necessary to extract these quantities. We therefore make a "global" fit as follows. First we make a linear fit and compute χ^2 for each of the sets of $AR(H_0)$ data. We then minimize $\Sigma\chi^2$ using $\rho_F{}^*$, ρ_N, and $R^*_{F/N}$ as variable parameters. With the resulting best fit values of these parameters, we calculate AR_{AF}- AR_{FM} using Eq. 9. We repeat these calculations for many combinations of β and γ, and obtain χ^2 for each combination. The combination which gives a minimum χ^2 gives the best values for β and γ. An identical analysis has been applied to the Ag/Co data.

RESULTS

For the Cu/Co system:-

$$\rho^*_{Co} = (8.60 \pm 0.45) \times 10^{-8}\,\Omega m, \qquad \beta = 0.50 \pm 0.10, \qquad \gamma = 0.76 \pm 0.05$$

$$AR^*_{F/N} = (0.50 \pm 0.02)\, f\Omega m^2 \qquad \alpha_{Co} = 3.0 \pm 1, \qquad \alpha_{Co/Cu} = 7.5 \pm 2.0$$

$$\rho_{Cu} = (0.67 \pm 0.20) \times 10^{-8}\,\Omega m$$

(Experiments on both Ag/Co and Cu/Co are continuing. Error bars may become smaller, but we do not anticipate major changes in these values).

From the above value of ρ_{Co}^* and Eq. 8 we obtain ρ_{Co} = (6.5 ± 0.3) x 10^{-8} Ωm. This value of ρ_{Co} with the above value for ρ_{Cu}, fall well within the range of the independently measured values in Eq. 5.

Our preliminary value for α_{Co} for the Ag/Co system[3] is ≈ 3 and it is thus satisfying to see that α_{Co} is the same within experimental error for the Ag/Co and Cu/Co systems. It appears that $\alpha_{Co/Ag}$ and $\alpha_{Co/Cu}$ have not been independently measured in dilute alloys, but $\alpha_{Co/Ag} \approx 13$ and $\alpha_{Co/Cu}$ are large compared with α values for dilute Co alloys[17] with non magnetic impurities. Thus either Ag and Cu must be unusual as solutes or alloying at interfaces is not the prime cause of the $\alpha_{F/N}$ values.

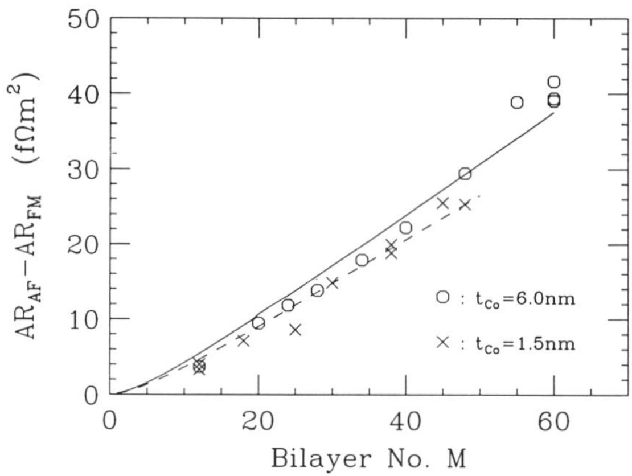

Figure 10. Plot of AR_{AF} - AR_{FM} vs M for t_{Co} = 1.5 nm and 6 nm. The solid line represents the values expected from Eq. 9 for t_{Co} = 6 nm; the dashed line for t_{Co}= 1.5 nm.

In Fig. 10 we show (AR_{AF} - AR_{FM}) as a function of M, both for the experimental data and from Eq. 9 (lines). A straight-line extrapolation of the experimental points gives a negative intercept on the vertical axis. The theoretical curve passes through the origin. It is clear that only for M < ~ 15 does the nonlinearity become significant in these samples. Similar curves are found for the Ag/Co system. We show[3] elsewhere that high resistivity samples, in which the N layers consist of a 4% alloy of Sn in Ag, show the strongly pronounced curvature predicted by Eqs. 9 and 11.

CONCLUSIONS

We have studied CPP-MR in Ag/Co and Cu/Co multilayers. A simple series resistor model provides clear evidence that the interfaces play a vital role in the spin dependent scattering essential to GMR. Our modification of the series resistor model theory leads to the more complex 2-spin model, which is verified in a more sophisticated manner in the following paper by Dr. Fert. Our experiments not only verify the predictions of this model but also permit us to evaluate α_N and $\alpha_{F/N}$. Future experiment will determine to what extent the values we obtain depend on the detailed structure of the samples.

ACKNOWLEDGEMENTS

This research was supported in part by the US NSF under Grants DMR-88-013287 and DMR-91-22614, the Michigan State University Center for Fundamental Materials Research, and the Ford Motor Company. The authors acknowledge helpful discussions with A. Fert, and P. Levy. High resolution transmission electron micrographs were kindly provided by Yuanda Cheng and Mary Beth Stearns. The electron microscopy was performed at the Center for High Resolution Microscopy, at Arizona State University, supported by US NSF under Grant DMR-91-15680.

REFERENCES

1. W.P. Pratt Jr., S-F. Lee, J.M. Slaughter, R. Loloee, P.A. Schroeder, and J. Bass, Perpendicular giant magnetoresistance of Ag/Co layers,Phys. Rev. Lett. 66:3060(1991).
2. S-F. Lee, W.P. Pratt Jr., R. Loloee, P.A. Schroeder, and J. Bass, The field dependent interface resistance of Ag/Co multilayers, Phys. Rev. B 46:548 (1992).
3. S-F. Lee, W.P. Pratt Jr., Q. Yang, P. Holody, R. Loloee, P.A. Schroeder, and J. Bass, Two spin analysis of CPP-MR data for Ag/Co and AgSn/Co multilayers. Submitted to J. Mag. Magn. Mat.
4. M.N. Baibich, J.M. Broto, A. Fert, F. Nguyen Van Dau, F. Petroff, P. Etienne, G. Creuzet, A. Friederch, and J.Chazelas, Giant magnetoresistance of (001)Fe/(001)Cr magnetic superlattices, Phys. Rev. Lett. 61:2472(1988).
5. S.S.P. Parkin, Z.G. Li and D.J. Smith, Giant magnetoresistance in antiferromagnetic Co/Cu multilayers, App. Phys. Lett. 58:2710 (1991): S.S.P. Parkin, Systematic variation of the strength and oscillation period of indirect magnetic exchange coupling through the 3d, 4d, and 5d transition metals, Phys. Rev. Lett. 67:3598 (1991).
6. D.H. Mosca, F. Petroff, A. Fert, P.A. Schroeder, W.P. Pratt,Jr., R. Loloee, and E. Leqien, Oscillatory interlayer coupling and giant magnetoresistance in Co/Cu Multilayers, J. Mag. Mag. Mat. 94:L1(1991).
7. M.L. Haerle, W.P. Pratt, Jr., and P.A. Schroeder, Electron transport properties of deformed potassium, J. Low Temp. Phys. 62:397 (1986): J. Bass, W.P. Pratt, Jr., and P.A. Schroeder, The temperature dependent electrical resistivities of the alkali metals, Rev. Mod. Phys. 62:645 (1990).
8. J.M. Slaughter, W.P.Pratt Jr., and P.A. Schroeder, Fabrication of layered metallic systems for perpendicular resistance measurements, Rev. Sci. Instrum. 60:127(1989).
9. H. Sato, P.A. Schroeder, J.Slaughter, W.P. Pratt Jr., and W. Abdul-Razzaq, Galvanomagnetic properties of Ag/M (M=Fe,Ni,Co) layered metallic films. Superlattices and Microstructures, 4:45(1988).
10. C. Meny, P. Panissod, and R. Loloee, Structural study of cobalt copper multilayers by NMR, Phys. Rev. 45:12269(1992).

11. S. Araki, K.Yasui, and Y.Narumiya,Giant magnetoresistance and oscillatory magnetic coupling of evaporated and MBE-grown Co/Ag multilayers, J. Phys. Soc. Japan 60:2827(1991).
12. R. Loloee, J. Bass, W.P. Pratt Jr., P. A. Schroeder and A. Fert, Oscillatory behaviour of giant magnetoresistance in Ag/Co multilayers, submitted to J. Mag. Mag.Mat.
13. C. Fierz, S.-F. Lee, J. Bass, W.P. Pratt, Jr., and P.A. Schroeder, Superconductor/ferromagnetic boundary resistances, J. Phys., Cond. Matt. 2:9701 (1990); C. Fierz, S.-F. Lee, J. Bass, W.P. Pratt, Jr., and P.A. Schroeder, Perpendicular resistances of thin Co films in contact with superconduction Nb, Physica B165-166:453 (1990).
14. S. Zhang and P.M. Levy, Interpretation of the magnetoresistance of multilayered structures, submitted to Phys. Rev.
15. S. Zhang and P.M. Levy, Conductivity perpendicular to the plane of multilayered structures, J. Appl. Phys. 69:4786 (1991).
16. T. Valet and A. Fert, Perpendicular magnetoresistance in magnetic multilayers: theory and comparison with experiments. Submitted for publication.
17. I.A. Campbell, and A. Fert,Transport properties of ferromagnets, in "Ferromagnetic Materials", vol 3.E. P. Wohlfarth, ed.,North-Holland (1982).

THEORY OF THE PERPENDICULAR MAGNETORESISTANCE IN MAGNETIC MULTILAYERS

A. Fert[1] and T. Valet[2]

[1] Laboratoire de Physique des Solides, Université Paris-Sud 91405 Orsay, France

[2] Laboratoire Central de Recherche, Thomson-CSF, 91404 Orsay, France

INTRODUCTION

In the talk presented by one of us (AF) at Cargèse, the mechanisms of the giant magnetoresistance (MR) in magnetic multilayers were discussed very generally for both the CIP (current in plane) and CPP (current perpendicular to the planes) geometries. A so general discussion would be too long for these proceedings and the present paper will be restricted to the CPP case. We present the theoretical model that we have recently worked out [1,2] and discuss its application to experimental data obtained for the Ag/Co and Cu/Co systems at Michigan State University [3,4,5] and presented at Cargèse by Pr. P.A. Schroeder [6]. We describe the specific fundamental problems related to the spin accumulation effects occuring in the CPP geometry and we calculate the magnetoresistance. The expressions of the MR become relatively simple in the limit where the layer thicknesses are much smaller than the spin diffusion length and we justify the analysis of experimental results developed at Michigan State University [3-6]. We also relate our theory to those of Johnson et al [7,8], van Son et al [9] and Zhang and Levy [10]. Of course, it is not in the scope of the present paper to develop calculations presented elsewhere in detail [1] and we will focus on the presentation of the basis of the model and the discussion of our results.

DIFFERENCES BETWEEN THE EXISTING MODELS AND OUR THEORY

The **spin accumulation effects** at interfaces between ferromagnetic and non-ferromagnetic metals have been described by Johnson et al [7,8] and Van

Son et al [9] for **isolated** interfaces separating two **semi-infinite media**. The point is as follows. If, in the ferromagnet, the current is spin polarized, there will be spin accumulation around the interfaces with the non-magnetic metal. To balance spin accumulation by spin relaxation, the chemical potentials of the spin↑ and spin↓ direction are shifted in opposite directions, which gives rise to spin dependent pseudo-electric fields. The final result is an additional potential drop, $\Delta V_I = r_{SI}J$, where J is the current density and r_{SI} the "**spin coupled interface resistance**"[7]. The **extension to the case of multilayers** proposed by Johnson [8] consists in a sort of addition of **interface resistances calculated for isolated interfaces.**

The calculations of Johnson et al [7,8] and van Son et al [9] are based on a "**macroscopic approach**" in which each medium is characterized by macroscopic parameters (resistivity, spin diffusion length) and macroscopic equations (Ohm's law and diffusion equation for the spin accumulation).

Our approach departs from that of Johnson et al [7,8] and van Son et al [9] in two points.

1) Since, in principle, the "macroscopic approach" of Ref[7-9] is appropriate only when the characteristic length of the inhomogeneities in a multilayered structure is much longer than the electron mean free path (MFP), we adopt a "microscopic approach" based on the Boltzmann equation. However, we show that this "microscopic approach" reduces to the "macroscopic one" when the MFP λ is much shorter than the spin diffusion length l_{sf}. According to the ESR data, this condition is fulfilled at low temperature for multilayers based on 3d and noble metals [11]. These points are developed in our paragraph "From Boltzmann equation to the macroscopic model ".

2) To describe the spin accumulation effects in a multilayer, we use the macroscopic equations justified in 1) and similar to those already used by Johnson et al [7,8] or van Son et al [9] for **isolated interfaces**. We show that the spin accumulations induced by **successive interfaces** partly balance each other and that Johnson's calculations [9] based on the interface resistances calculated for **isolated interfaces** is generally not valid. Our expression of the CPP-MR is definitely different from that of Johnson [9].

Another model of the CPP-MR model has been proposed by Zhang and Levy [10]. This model, developed in the Kubo formalism, is based on the assumption of current conservation in each spin channel (no spin flip scattering or, in our language, no spin relaxation and infinite spin diffusion length). Our calculation shows that this assumption is justified in the limit where the layer thicknesses are much smaller than the spin diffusion length l_{sf}. Our model is more general and can also describe the CPP-MR out of this limit. It also presents the advantage of giving analytical expressions of the CPP-MR.

FROM BOLTZMANN'S EQUATION TO THE MACROSCOPIC MODEL

We consider a structure with ferromagnetic (F) layers alternating with normal (N) metal layers. The current density $\vec{J}$ is along the z axis perpendicular to the layers. The electrons are free electrons. Our notation is ± for the absolute spin direction ($s_x = \pm 1/2$) and ↑, ↓ for majority and minority spins in the F layers. The electron distribution function for the spin s is written as :

$$f_s(z,\vec{v}) = f^0(v) + \frac{\partial f^0}{\partial \varepsilon}\{[\mu_0 - \mu_s(z)] + g_s(z,\vec{v})\} \qquad (1)$$

where $g_s(z,\vec{v})$ is the conventional term associated with the electron flow and $\mu_s(z)$ the spin and position dependent chemical potential introduced to express the spin accumulation (μ_0 is the equilibrium chemical potential). The Boltzmann equation in a layer is written as

$$v_s\frac{\partial g_s}{\partial z}(z,\vec{v}) + \left[\frac{1}{\tau_s} + \frac{1}{\tau_{sf}}\right]g_s(z,\vec{v}) = \left[v_z\frac{\partial \overline{\mu}_s}{\partial z} + \frac{\overline{\mu}_s(z) - \overline{\mu}_{-s}(z)}{\tau_{sf}}\right] \qquad (2)$$

where $\overline{\mu}_s = \mu_s(z) - eV(z)$ is the electro-chemical potential ($V(z)$ is the electrical potential), τ_s is the momentum relaxation time for spin s and τ_{sf} is the spin relaxation time. At low temperature the spin flip scattering by electron-magnon collisions is frozen out and the spin relaxation is due only to spin flip scattering by the spin-orbit part of the scattering potentials. Data on the ratio τ_{sf}/τ_s derived from ESR have been reviewed by Monod and Schultz [11] and it turns out that, for spin orbit scattering involving 3d or noble metals potentials, this ratio is generally larger than 10^2. This means that, for the multilayers investigated up to now, the expansion of the solution of Eq(2) can be limited to the lowest order in τ_s/τ_{sf}. We have shown in Ref [1] that, at its lowest order, Eq.(2) reduces to the following macroscopic equations.

$$J_s = \frac{\sigma_s}{e}\frac{d\overline{\mu}_s}{dz} \qquad (3)$$

$$\frac{e}{\sigma_s}\frac{dJ_s}{dz} = \frac{\overline{\mu}_s - \overline{\mu}_{-s}}{l_s^2} \qquad (4)$$

where $\sigma_s = \frac{ne^2}{2m}\tau_s$ and $l_s = v_F\tau_{sf}\left[\tau_s/3(\tau_s + \tau_{sf})\right]^{1/2}$ are the conductivity and the spin diffusion length for the spin s channel in a given layer.

Eq.(3) is the Ohm law in which the electric field is replaced by a spin dependent pseudo-electric field ~ $d\overline{\mu}_s/dz$. Eq.(4) is the equation expressing that the spin accumulation generated by the dependence of J_s on z is balanced by the spin relaxation due to spin flip scattering. The same macroscopic equations, in slightly different forms, are at the basis of the calculations of Johnson et [7,8] and van Son et al [9].

As shown in the next paragraph, Eqs (3-4) can be used to calculate the z-dependence of the electro-chemical potential and the current for each spin direction in a multilayer. Then one derives the potential drop accross the multilayer, and consequently the resistivity. As clearly seen in Eqs (3-4), the only remaining scaling length of the problem is the spin diffusion length (SDL), whereas the mean free path λ is no longer involved. The replacement of the MFP by the SDL as scaling length is an important aspect of the CPP-MR problem. The CPP-MR will be a function of the ratio of the individual thicknesses to the SDL. Most experiments are for thicknesses much smaller than the SDL and, as we see below, the expression of the CPP-MR is very simple in this limit.

MACROSCOPIC MODEL

We write the spin dependent potential $\overline{\mu}_{\pm} = \overline{\mu} \pm \Delta\mu$ where $\Delta\mu$ is the term related to the spin accumulation (in a free electron model, $\Delta\mu$ is related to the out of equilibrium magnetization by $|\Delta\mu| = 2\mu_0 |\Delta M| / 3n\mu_B$). We call F(z) the spin independent pseudo-electric field related to the spin independent term $\overline{\mu}$ by

$$F(z) = -\frac{1}{e}\, d\overline{\mu}/dz \tag{5}$$

Eqs (3-4) transform into

$$J_{\pm}(z) = \sigma_{\pm}\left[F(z) \pm \frac{1}{e}\frac{\partial \Delta\mu}{\partial z}\right] \tag{6}$$

$$\frac{e}{\sigma_{\pm}}\frac{\partial J_{\pm}}{\partial z} = \pm\frac{\Delta\mu}{l_s^2} \tag{7}$$

This leads to a diffusion equation for $\Delta\mu$

$$d^2\Delta\mu/dz^2 = \Delta\mu/l_{sf}^2 \tag{8}$$

$$\text{where } (1/l_{sf}^2)^2 = (1/l_{\uparrow})^2 + (1/l_{\downarrow})^2 \tag{9}$$

and to a charge conservation equation

$$\frac{\partial^2}{\partial z^2}\left[\sigma_+\overline{\mu}_+ + \sigma_-\overline{\mu}_-\right] = 0 \tag{10}$$

The general solution for $\Delta\mu$ has the form

$$\Delta\mu = A \exp\left[z/l_{sf}\right] + B \exp\left[-z/l_{sf}\right] \tag{11}$$

where A and B are fixed by boundary conditions at the interfaces. Similar forms are found for $J_\pm(z)$ and F(z).

Equations of the type of Eqs (8-11) have already been applied by Johnson and Silsbee [7] or van Son et al [9] in the case of an isolated interface between two semi-infinite materials. In our model we apply these equations to the case of a multilayer composed of non-magnetic layers (thickness=t_N, resistivity = ρ_N^*, which means that the resistivity for each spin channel is $2\rho_N^*$) and magnetic layers (thickness = t_F, resistivity of the spin $\uparrow(\downarrow)$ channel = $\rho\uparrow(\downarrow) = 2\rho_F^*[1-(+)\beta]$. The scattering by the interfaces is supposed to occur within an infinitisimally thin interfacial layer, so that it can be described by a spin dependent interface resistance, $r\uparrow(\downarrow)=2r_b^*[1-(+)\gamma]$. These interface resistances fix the following boundary condition for spin s at an interface located at $z = z_0$.

$$\overline{\mu}_s(z=z_0^+) - \overline{\mu}_s(z=z_0^-) = r_s\, J_s\,(z=z_0)/e \tag{12}$$

A second boundary condition expresses the conservation of current $J_s(z=z_0^+) = J_s(z=z_0^-)$

The solutions for $\Delta\mu(z)$, F(z), $J\pm(z)$ in the multilayer are obtained by imposing the above boundary conditions to the general solution for $\Delta\mu(z)$, Eq(11), and to the corresponding general solutions for F(z) and $J\pm(z)$. For simplicity, we first write down these solutions for r_s=0 (bulk scattering only). For an antiferromagnetically ordered multilayer and a total current density J we obtain :

$$\Delta\mu(z) = 2\mu_F^{AP} \sinh\left[\frac{z-z_A}{l_{sf}}\right] \tag{13}$$

$$F(z) = (1-\beta)^2\, \rho_F^* J + \frac{2\beta\mu_F^{AP}}{e l_{sf}} \cosh\left[\frac{z-z_A}{l_{sf}}\right] \tag{14}$$

$$J_{\pm}(z) = (1 \mp \beta)\frac{J}{2} \pm \frac{\mu_F^{AP}}{e\rho_F^* l_{sf}} \cosh\left[\frac{z-z_A}{l_{sf}}\right] \quad (15)$$

in a magnetic layer (z_A is its center) and

$$\Delta\mu(z) = 2\,\mu_N^{AP} \cosh\left[\frac{z-z_B}{l_{sf}}\right] \quad (16)$$

$$F(z) = \rho_N^* J \quad (17)$$

$$J_{\pm}(z) = \frac{J}{2} \pm \frac{\mu_N^{AP}}{e\rho_N^* l_{sf}} \sinh\left[\frac{z-z_B}{l_{sf}}\right] \quad (18)$$

in a non-magnetic layer centered at z_B.

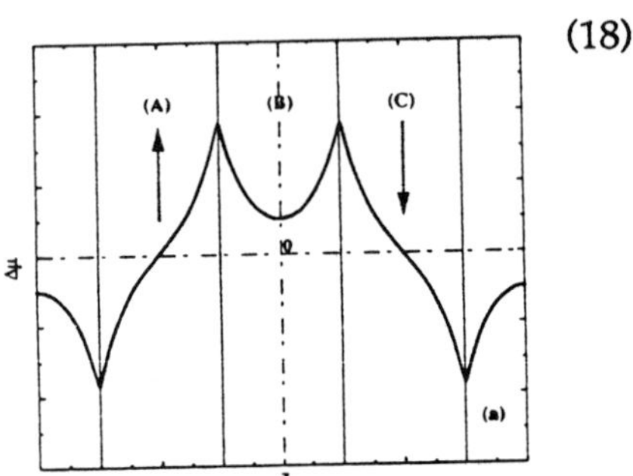

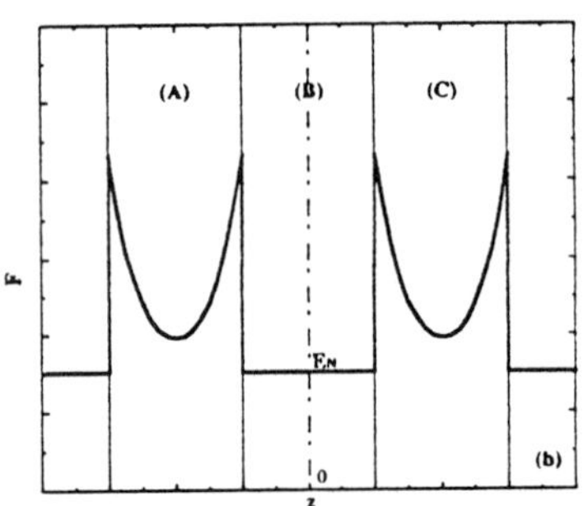

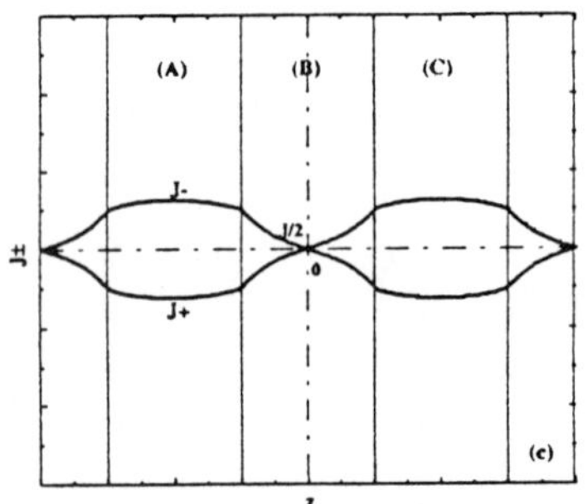

Figure 1 . Variation of $\Delta\mu$ (spin dependent part of the electro-chemical potential), F (spin independent part of the pseudo-electric field), $J_{\pm}$ (current density in the spin+ and spin- channels) in a multilayer. The curves are derived from Eqs(13-18), that is for only bulk scattering and for an antiferromagnetically ordered multilayer (the arrows indicate the majority spin direction in the magnetic layer). They have been calculated for faster minority electrons (β<0) and for a current flowing from left to right (check that the minority electrons acccumulate on the left side of the magnetic layers). The figure represents the case with the same order of magnitude for the thicknesses and the spin diffusion length. For $t \ll l_{sf}$, the sinh- and cosh- like curves become oblique and horizontal straight lines respectively, see Eqs (19-22).

The coefficients μ_P^{AP} and μ_N^{AP} are determined by continuity conditions and their expression is useless here. In Fig. 1, we show the oscillatory variation of $\Delta\mu$, F(z) and $J\pm$ (z) with z. $\Delta\mu$, F(z) and $J\pm$ (z) could also be calculated in the same way for ferromagnetic ordering and also in the general case with bulk *and* interface scattering.

The potential drop across the multilayer has no contribution from the oscillatory part $\Delta\mu$ of the electrochemical potential and arises only from the integration of F(z). The global resistivity is derived from this potential drop. The general expressions of the resistivity for AP (antiparallel) and P (parallel) ordering are given in Ref [1,2] and will not reproduced in this paper. Here we focus on the limit $t_{F(N)} \ll l_{sf}$ which correspond to most experiments [3-6].

In this limit $t_{F(N)} \ll l_{sf}$ and, again, in the case with only bulk scattering for AP ordering, the development of Eq(13-18) in first order in t_F/l_{sf} and t_N/l_{sf} leads to :

$$\Delta\mu(z) = \beta e \overset{*}{\rho_F} J\,(z - z_A) \tag{19}$$

$$F(z) = \overset{*}{\rho_F} J = (\rho\uparrow + \rho\downarrow)\, J/4, \qquad J\pm = J/2, \tag{20}$$

in a magnetic layer and

$$\Delta\mu\,(z) = \beta e \overset{*}{\rho_N} J t_F \tag{21}$$

$$F(z) = \overset{*}{\rho_N} J, \qquad J\pm = J/2 \tag{22}$$

in a non-magnetic layer. The amplitude of the oscillations of $\Delta\mu$ is smaller than the maximum value of $\Delta\mu$ at isolated interfaces (Ref[7-9]) by a factor $\approx t_F/l_{sf}$. This is due to the interference between opposite spin accumulations at successive interfaces. It can also be shown that the amplitude of the variation of the current with z are reduced by a factor $\approx t_F t_N/l_{sf}^2$, so that, in Eqs(20,22) limited to first order, the two currents have the constant value J/2 (the assumption of two constant current equal to J/2 is at the basis of the model without spin flip of Zhang and Levy [10] and turns out to be justified in the limit $t \ll l_{sf}$). In the same way, the non linear part of F(z) is of the second order in t_F/l_{sf}, in agreement with the constant value appearing in Eqs(20,22). When one neglects the second order and takes the constant values of F(z) in Eq(20,22), the potential drop accross M bilayers is

$$\Delta V = M\left[(\rho\uparrow + \rho\downarrow)t_F/2 + 2\overset{*}{\rho_N} t_N\right] J/2 = M\left[2\overset{*}{\rho_F} t_F + 2\overset{*}{\rho_N} t_N\right] J/2 \tag{23}$$

as if there was a resistance $M[2\rho_N^* t_N + 2\rho_F^* t_N]$ in each channel.

Eq (23) is for only bulk scattering ($r_b^* = 0$) and in the case of AP ordering.

Extension to the general case (bulk *and* interface scattering) and to both types of ordering (still in the limit $t \ll l_{sf}$) is straighforward. The best way to summarize the results is by Fig.2 which shows that resistances $2\rho_F^* t_F$, r_b^* and $2\rho_N^* t_N$ are in series in each channel for AP ordering. For P ordering, resistances $2\rho_F^* [1-(+)\beta]t_F$, $2r_b^* [1-(+)\gamma]$, $2\rho_N^* t_N$ are in series in the majority (minority) channel.

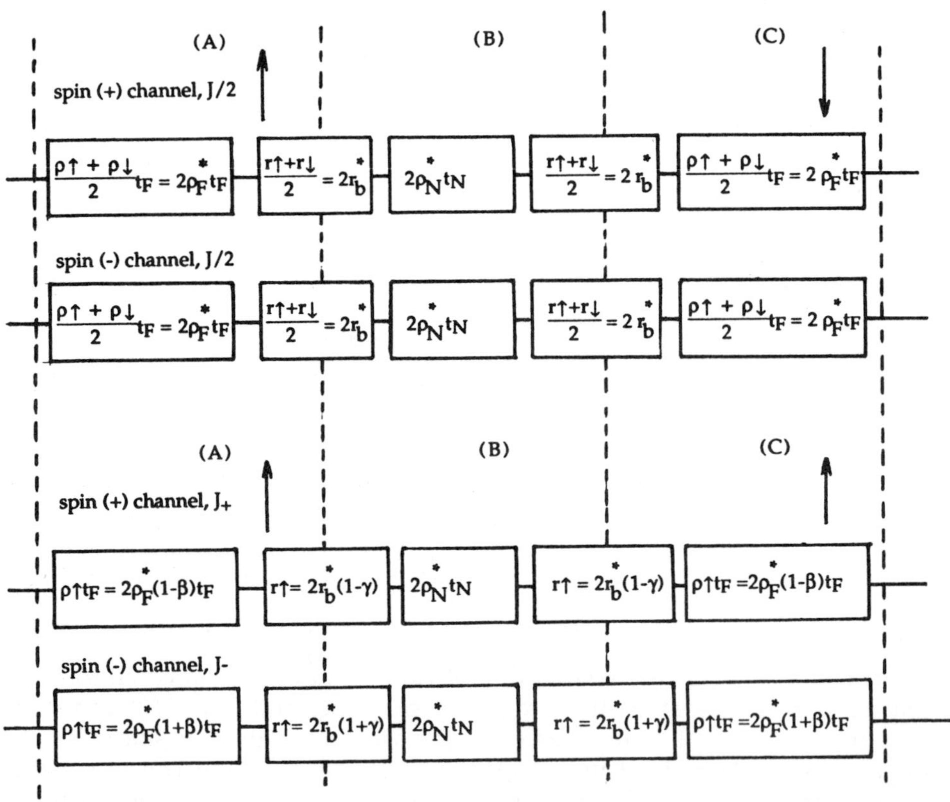

Figure 2 . Equivalent resistance array that can be used to calculate the potential drop accros a multilayer. The potential drop arises only from the pseudo-electric field term F ($\Delta\mu$ is periodic and does not contribute to the potential drop over long distances). The schematics are for the limit $(t_N, t_F) \ll l_{sf}$ in the general case with both bulk and interface spin dependent scattering. The upper schematic is for an antiferromagnetic arrangement and the lower schematic is for a ferromagnetic one. $\rho\uparrow$ and $\rho\downarrow$ are the resistivities due to spin dependent bulk scattering. $r\uparrow$ and $r\downarrow$ are the resistances due to spin dependent interface scattering.

PRACTICAL EXPRESSIONS

The resistor scheme of Fig.2 leads straighforwardly to the type of expression used at Michigan State University [5,6] for the interpretation of the total resistance of Ag/Co and Cu/Co multilayers per unit cross-section.

$$R^{(AP)} = M\left[\rho_F^* t_F + \rho_N^* t_N + 2 r_b^*\right] \tag{24}$$

$$\frac{1}{R^{(P)}} = \frac{1}{M}\left[\frac{1}{2\rho_F^*(1-\beta)t_F + 2\rho_N^* t_N + 4r_b^*(1-\gamma)} + \frac{1}{2\rho_F^*(1+\beta)t_F + 2\rho_N^* t_N + 4r_b^*(1+\gamma)}\right] \tag{25}$$

$$R^{(P)} = R^{(AP)} - \frac{\left[\beta\rho_F^* \frac{t_F}{t_F + t_N} L + 2\gamma r_b^* M\right]^2}{R_{(AP)}} \tag{26}$$

$$\sqrt{(R^{(AP)} - R^{(P)})R^{(AP)}} = \beta\frac{t_F}{t_F + t_N}\rho_F^* L + 2\gamma r_b^* M \tag{27}$$

where $L = M(t_F+t_N)$ is the total thickness of the multilayer. These expressions are meaningful in the range of validity of our calculation i.e $t \ll l_{sf}$ or $M \gg L/l_{sf}$, with expected values of l_{sf} at low temperature above 10^3Å [10]. As in the experiments of Ref [3-6], we suppose that the total L thickness is fixed and we will describe the variation of the magnetoresistance as a function of the number of bilayers M (for $t_F = t_N$ or for t_F fixed and t_N varying). In Fig.3 we show the variation expected from Eq (24,25) for R^{AP} and R^P as a function of M for $\beta = 0$ (Fig3(a)) and $\gamma = 0$ (Fig.3(b)) in the case $t_F = t_N = L/2M$.

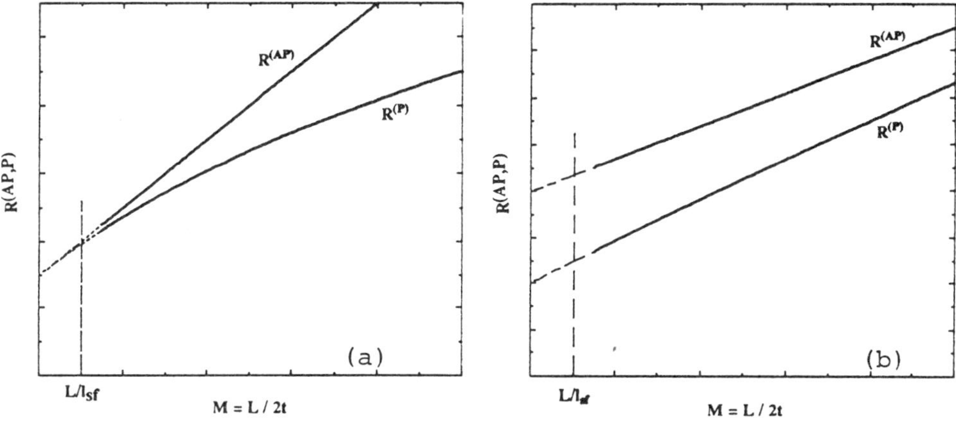

Figure 3 . Predicted variation of the CPP resistance of a multilayer as a function of the number of bilayers M for a fixed total thickness L and the same individual thicknesses for the ferromagnetic and normal layers, $t_F = t_N = t = L/2M$. (a) is for a multilayer with surface spin dependent scattering only (β=0) and (b) with bulk spin dependent scattering only (γ=0). The variations are derived from Eq(24-26). The plots are meaningfull in the validity range of these expressions, i.e $t \ll l_{sf}$ or $M \gg L/l_{sf}$.

For interface spin dependent scattering (Fig.3(a)), $(R^{(AP)} - R^{(P)})$ is an increasing function of M, in agreement with the experimental results on Ag/Co and Co/Cu [3-6]. The resistance $R^{(AP)}$ is a linear function of M. The departure of $R^{(P)}$ from $R^{(AP)}$ starts as M^2 and than, for $M \geq M_c \approx L(\rho_F^* + \rho_N^*)/2r_b^*$, $R^{(P)}$ becomes also linear in M with a smaller slope than $R^{(AP)}$. By increasing ρ_N^* (addition of impurities in Ag), Pratt et all [5,6] have been able to shift M_c and make the crossover from quadratic to linear appear clearly in the experimental range of M for Ag/Co multilayers.

For bulk spin dependent scattering only (Fig.3(b)), $(R^{(AP)}-R^{(P)})$ decreases as M increases. This might be observed for systems with predominant spin dependent scattering within the magnetic layers. As it has been suggested that this occurs for permalloy layers, CPP-MR measurements on, for example, permalloy/Cu multilayers should be of interest.

Some other measurements by Pratt et al [3-6] have been performed on samples with a fixed value for t_F while t_N is varying with M. The predicted variation of $R^{(AP)}$ and $R^{(P)}$ could also be plotted for these conditions. It would appear that, in this case, $(R^{(AP)}-R^{(P)})$ is always an increasing function of M.

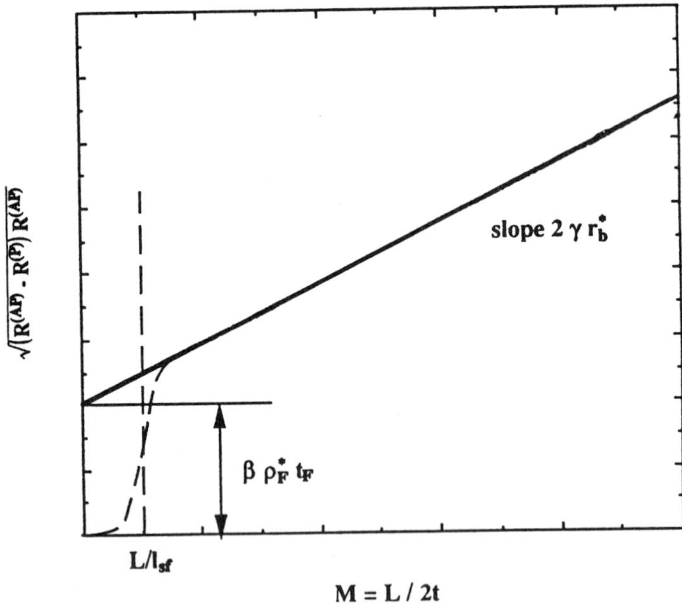

Figure 4. $\sqrt{(r_{AF}-r_F)r_{AF}}$ is plotted as a function of the number of bilayers M for a fixed total thickness L and the same individual thickness for the ferromagnetic and normal layers, $t_F=t_N=t=L/2M$. The solid line is the linear variation expected from Eq.(27) for the limit $t \ll l_{sf}$ or $M \gg l_{sf}$. For $M < L/l_{sf}$. Eq(27) is no longer valid and $\sqrt{(r_{AF}-r_{FM})r_{AF}}$ is expected to drop as $\exp(-t/2l_{sf})$, see dashed line.

Eq (27) presents the interesting property that its right hand side is made of additive contributions from bulk and interface spin dependent scatterings. The plot of $\sqrt{(R^{(AP)}-R^{(P)})R^{(AP)}}$ versus M for L fixed is thus of great interest to identify where the spin scattering occurs, as this has been done at Michigan State University [5,6] and is also shown schematically in Fig. 4 for the case $t_F=t_N=t$. After independent measurements of ρ_F^* and r_b^*, the intercept with the vertical axis gives β , while the slope is characteristic of γ. This type of plot applied to Ag/Co leads to $\beta=0.47$ ($\rho\downarrow/\rho\uparrow=3.4$ in the Co layers) and $\gamma=0.87$ ($r\downarrow/r\uparrow = 12$ at the Ag/Co interfaces)[5,6]. It turns out that the contribution to the CPP-MR from interface spin dependent scattering is predominant at usual thicknesses. We however point out that the relative magnitude of the interface and bulk contributions in Eq(27) depends on the layer thickness (via M) and, for thicknesses of a few hundred of angstroms, the bulk contribution is expected to exceed the interface one.

In the range $M<L/l_{sf}$ where the condition $t \ll l_{sf}$ is not fullfilled, the MR is expected from our general expressions [1,2] to be smaller than predicted from Eq (24-27). More precisely, $(R^{(AP)}-R^{(P)})$ is expected to decrease as $\exp(-t_N/2l_{sf})$ in the limit $t_N \gg l_{sf}$. This deviation from a linear variation is represented by a dotted line in Fig.4. As l_{sf} is expected to be definitely above 10^3Å in Ag or Cu layers but can be strongly reduced by introducing impurities with strong spin-orbit scattering, Pt for example, it should be possible to reduce l_{sf} to a reasonable value and to make the exponential decrease appear in the experimental thickness range.

CONCLUSION

The main difference between the CPP and CIP problems is due to the existence of spin accumulation effects in the CPP case and to the appearance of the spin diffusion length l_{sf} as unique damping length of the current inhomogeneities. The spin diffusion length is fixed by the spin-orbit scattering in the low temperature limit and is expected to be relatively long - above 10^3Å - in multilayers composed of 3d and noble metals (from ESR data [11]).

Our calculation shows that, in the usual conditions where the MFP is much shorter than the spin diffusion length, a Boltzmann equation treatment reduces to the equations of the macroscopic approach introduced by Johnson and Silsbee [7] or van Son et al [9]. We have adapted this microscopic approach to the case of multilayers. Our calculation, because it treats the interplay between spin accumulation at successive interfaces, leads to expressions of the MR definitely different from those obtained by Johnson who added effects calculated for isolated interfaces.

The expressions of the CPP-MR in the limit $t_F,t_N \ll l_{sf}$ are particularly simple and we have justified the analysis of the experimental results already proposed by Pratt et al [5,6]. The typical behavior expected for a multilayer of fixed total thickness as a function of the number of bilayers is shown in Figs.3

and 4. We emphasize that the appearance of additive contributions from bulk and interface spin dependent scattering in Eq(27) is of great interest to separate these contributions (Fig.4).

REFERENCES

1. T. Valet and A. Fert, submitted for publication.
2. T. Valet and A. Fert, Symposium on Magnetic Thin Films, Surfaces and Multilayers, Lyon, (1992).
3. W.P. Pratt Jr., S.F. Lee, J.M. Slaughter, R. Loloee, P.A.Schroeder and J. Bass, Phys. Rev. Lett. 66:3060 (1991).
4. S.F. Lee, W.P. Pratt Jr., R. Loloee, P.A. Schroeder and J. Bass, Phys. Rev. B 46:548 (1992).
5. W.P. Pratt, private communication ; S.F. Lee, W.P. Pratt Jr, R. Loloee, P.A. Schroeder and J. Bass, to appear in J. Mag. Mag. Mat.
6. P.A. Schroeder, J. Bass, P. Holody, S.F. Lee, R. Loloee, W.P. Pratt, Q. Yang, NATO workshop on Magnetism in Systems of Reduced Dimensions, Cargese (France), 1992.
7. M. Johnson and R.H. Silsbee, Phys. Rev. B 35:4959 (1987).
8. M. Johnson, Phys. Rev. Lett. 67:3594 (1991).
9. P.C. van Son, H. van Kempen, and P. Wyder, Phys. Rev. Lett. 58:2271 (1987).
10. S. Zhang and P.M. levy, J. Appl. Phys. 69:4786, (1991) and article of P.M. Levy in the Proceedings.
11. P. Monod and S. Schultz, J. Phys. (Paris) 43:393 (1982).

THE INFLUENCE OF MAGNETISM AND STRUCTURE ON TRANSPORT PROPERTIES AND INTERLAYER COUPLING OF SYSTEMS IN REDUCED DIMENSIONS

Peter M. Levy, Horacio E. Camblong, Zhu-Pei Shi and Shufeng Zhang

Department of Physics
New York University
New York, NY 10003

INTRODUCTION

Recent attention has been focused on magnetic multilayered structures and thin films containing magnetic granules which display a new magnetoresistive (MR) effect. This effect, first found in Fe/Cr multilayered structures, has now been observed in at least a dozen different combinations of magnetic and non-magnetic metallic layers. In addition, two groups have recently observed the effect for thin films containing magnetic particles, e.g., copper with cobalt precipitates.

In this paper, we review the models which have been proposed to explain the MR effect in these layered materials. Some focus on the spin-dependent scattering, in the bulk and at interfaces, of conduction electrons; others accent the spin-dependent potentials in the different layers. The consequences of these models have been worked out by using either the semiclassical Boltzmann equation approach or the Kubo formalism, which requires one to find the electron propagators in the presence of spin-dependent scatterers and potentials. We indicate the approximations made in these different approaches and what limitations they impose on the applicability of the solutions to real systems.

Next, we review our current understanding of the interlayer coupling in magnetic multilayered structures as gleaned from : 1) total energy calculations, 2) perturbative methods, and 3) analytic solutions of a "toy model". We indicate differences in these approaches and in the predictions based on them. In particular, we indicate areas where experimental data can be used to vindicate some of these ideas.

MAGNETORESISTANCE OF MULTILAYERED STRUCTURES AND MAGNETIC GRANULAR FILMS

We start by reviewing the models put forward to understand the giant magnetoresistance in these materials and the methods used to evaluate them.

Magnetism and Structure in Systems of Reduced Dimension
Edited by R.F.C. Farrow *et al.*, Plenum Press, New York, 1993

I. Spin-Dependent Scattering versus Potentials

The Hamiltonian that describes conduction electrons in metallic multilayers is

$$H_M = \frac{p^2}{2m} + V_{\text{pot},M}(z,\sigma) + \sum_i V_{\text{scatt},M}(z - z_i, \sigma) \,, \tag{1}$$

where V_{pot} represents the spin-dependent potentials of the electrons in the different layers, V_{scatt} is the scattering at impurity sites within the bulk of the layers and at the interfaces, and M denotes the magnetic configurations of the layers, e.g., ferro or antiferromagnetic.

Until now attention has been focused on the role of spin-dependent scattering in producing large magnetoresistance in multilayered systems. In past treatments the scattering has been evaluated by using plane wave states [1]. It is time to use the states appropriate to multilayered structures, i.e., to use the eigenstates of

$$H_M^0 = \frac{p^2}{2m} + V_{\text{pot},M}(z,\sigma) \,. \tag{2}$$

This leads to energy eigenvalues of the form

$$\varepsilon_M(\mathbf{k},\sigma) = \varepsilon_{||}(k_{||}) + \varepsilon_M(k_z,\sigma) \,, \tag{3}$$

where $\varepsilon_{||}(k_{||}) = \hbar^2 k_{||}^2/(2m)$ and $k_{||}^2 = (k_x^2 + k_y^2)$, and to eigenstates

$$\phi_M(\mathbf{k},\sigma;\mathbf{r}) = e^{i(k_x x + k_y y)}\, \phi_M(k_z,\sigma;z) \,. \tag{4}$$

1. Unpolarized (s-electron) Conduction Band. If one focuses on an unpolarized band of electrons, the MR effect comes from the spin dependence of the scattering potential. There has been a fair amount of controversy about the relative importance of bulk scattering from within the layers and that coming from the interfaces between the layers. While there seems to be a consensus that interfacial scattering is very strong in systems such as Co/Cu and Fe/Cr, it would be imprudent to entirely overlook that coming from the bulk. This is particularly true for layered structures where the thickness of the magnetic layer t_{m} is very large compared to that of the nonmagnetic layer t_{nm}. For such structures, bulk scattering plays an important role; although it may be weak, it is an *extensive* quantity and increases with size. Interfacial scattering is *intensive*; when expressed as an extensive quantity the scattering per unit length diminishes as the thickness of a layer t_{m} or t_{nm} increases. Therefore, for $t_{\text{m}} \gg t_{\text{nm}}$ and for $t_{\text{m}} + t_{\text{nm}}$ large, bulk scattering, albeit small, should not be neglected.

2. Strength of Scattering. The strength of the spin-dependent scattering in transition metals has been extensively studied by Fert and Campbell [2] From the theoretical work of Friedel and Anderson [3], this depends on the formation of a local impurity moment in the host metal. The experimental surveys and analyses made by Fert and Campbell are a good reference for the strength of the spin-dependent scattering in the *bulk* of the layers; also, Maekawa and his group [4] made a study of how the magnetoresistance varies across the 3d transition metal series based on a phase shift analysis of the scattering. However, the scattering at interfaces is poorly understood, and more work is warranted. From our fits [5] to data on Fe/Cr and Co/Cu, we find that the spin-dependent scattering is quite a bit stronger at interfaces than in the bulk; this has also been remarked by others.

3. Polarized (d-electron) bands. If one focuses one's attention on the conduction coming from the d electrons, which also give rise to the magnetism in the multilayers, then one has to admit polarized conduction electron bands. In multilayered structures this alters the scattering mechanism, as pointed out by Mary Beth Stearns [6], and as can be seen when one describes majority spin conduction electrons from one magnetic layer going into another magnetic layer with its moment at an angle with respect to the first (say antiparallel), i.e., a majority spin electron fitting into a minority band of states. This idea has its antecedents in some work done on magnetic domain scattering [7] and on tunneling of spin-polarized electrons across insulating layers [8].

4. Spin-Dependent Potentials. In the Stoner description of conduction electrons in ferromagnetic layers, one needs to consider different potentials for majority an minority spins. Whereas most theoretical analyses of transport in magnetic multilayers until now have not considered these differences, Hood and Falicov [9] recently updated Camley-Barnás by taking into account the different potentials seen in the magnetic and nonmagnetic layers for ferro and antiferromagnetic order. They find that one needs up to 20 parameters to describe the most general case (this is indeed a staggering number), and that potential barriers have large effects on the magnetoresistance.

To extract the role of spin-dependent potentials on the magnetoresistance of transition metal multilayers, we have used a Kronig-Penney potential in the growth direction, and a constant (zero) potential in the plane of the layers, see Eqs. (2)-(4). We are currently determining the magnetoresistance due to the scattering $\{\sum_i V_{\mathrm{scatt},M}(z-z_i,\sigma)\}$ by using these wavefunctions [10]. The interplay of the amplitude modulation in these functions and of the variation of the intensity of the scattering potential produces the spatial variation in the scattering function that we will use to fit experimental data. It is our aim to assess realistically the relative importance of the potential and scatterers (V_{pot} and V_{scatt}) in producing the large magnetoresistance observed in multilayered structures.

5. Realistic Band Structures. Model calculations of the role of spin-polarized conduction electrons have been based on *one effective* (s-d) *band* [9, 10]. The actual band structure for transition metal multilayers is considerably more complicated. Oguchi and Sasaki [11] used a local density band structure calculation to evaluate the Fermi velocities in a Fe/Cr (3 × 3) superlattice for the F and AF configurations. They find the ratio of the velocities v_F to v_{AF} is 3 to 2 parallel to the layers, and 2.4 to 1 for velocities perpendicular to the layers.

In summary, the models for spin-dependent scattering are quite different from those for the potentials. One focuses on the inhomogeneities (from one layer to the next) in the scattering of the conduction electrons, while the other uses the Bloch-like wavefunctions (instead of plane waves) that are appropriate to multilayered geometries.

II. Boltzmann versus Kubo Approach

1. Layer by Layer or the Real Space Approach.

a) *Semiclassical Boltzmann Equation.* The Boltzmann equation is solved in each layer and the solution is matched at the interfaces between layers. The layer solutions are found by the method first given by Fuchs and Sondheimer for conduction in thin films [12], and then extended to multilayers by Carcia and Suna [13]. The extension to *magnetic* multilayers was made by Camley and Barnás [14].

b) *Green's Functions.* Rather than finding the distribution function for the conduction electrons (Boltzmann approach) one can find the conduction electron propagators (Green's functions) and determine the conductivity of a multilayered structure by using the Kubo linear response formalism. This has been successfully carried out in real space by Vedyayev, Dieny, and Ryzhanova [15]; their quantum solution is in reasonable agreement with the semiclassical approach *provided* a cut-off is introduced in the latter to remove the contribution from electrons whose classical trajectories are nearly parallel to the interfaces.

2. Continuum or the Momentum Space Approach. In this approach, one considers electron states that span the entire multilayered structure, and one focuses on the inhomogeneity of the *scattering* due to the layering [1]. Recently, the effect of the Kronig-Penney–like potentials have been incorporated by using the appropriate Bloch wavefunctions [10]. The one-electron propagators (Green's functions) for multilayered geometries are found by extending the method developed for thin films by Tešanovic, Jarić, and Maekawa [16].

3. Quantum Corrections? The *inability* to simultaneously fit the resistance and magnetoresistance of multilayers within the framework of the semiclassical approach [17], has raised concern that quantum corrections may be needed to remedy the situation. Indeed the solution obtained by Zhang *et al.* [1] by using the Kubo formalism does fit both. To determine whether quantum corrections are necessary, Camblong is studying the solutions of the multilayer problem by using the Kubo formalism in real space [18]. His conclusions are that there are *no* quantum corrections for the magnetic multilayered structures studied to date. Quantum interference terms that represent coherent scattering from different impurity sites are negligible, and quantum size effects are not a consideration for structures over 100 Å thick (in the growth direction). *However*, Camblong does find that the way interface scattering is modeled in the layer-by-layer approach, i.e., by transmission and diffuse scattering coefficients is inadequate. This has been recognized by Johnson and Camley [19]: they corrected it by introducing interfacial regions , which are modeled as additional layers. Camblong has proposed a novel remedy, staying with thin interfaces, by introducing *angle-dependent* transmission coefficients in the semiclassical approach.

III. CPP-MR versus CIP-MR

CPP: Current perpendicular to the plane of the layers. CIP: Current in the plane of the layers.

1. Complementarity of Currents and Fields.

a) For CIP the electric field is uniform throughout the multilayered structure; the current varies from one layer to another, as well as within each layer, see for example Fig. 1 in Ref. [1].

b) For CPP the converse applies, i.e., the current is constant throughout, while the electric field varies.

2. One- and Two-Point Conductivity. For CIP only the one-point conductivity function $\sigma(z)$ is needed; for CPP one needs in general a two-point conductivity $\sigma(z, z')$ because the actual electric field in the structure is *inhomogeneous*.

3. Limiting Cases for CPP. In the limit that the mean free path of the conduction electrons is small compared to the repeat distance for the layering, $\lambda_{\text{mfp}} \ll d_{\text{in}}$, the "local limit", the two-point conductivity reduces to a one-point function and we find the electric field $E(z)$ is proportional to $1/\sigma(z)$ [20]. In the limiting case $\lambda_{\text{mfp}} \gg d_{\text{in}}$, the "homogeneous limit", the field and the one-point conductivity are independent of z.

4. Internal Electric Fields for CPP.

$$E_{\text{tot}}(z) = E_{\text{ext}} + E_{\text{int}}(z)$$

The internal field $E_{\text{int}}(z)$ is like a *chemical potential* or *field* (Lagrange multiplier) to maintain constancy of current despite scattering rates that vary from one layer to another, as well as at the interfaces. When there is no mixing of currents coming from majority and minority spins, the individual currents j^s must be constant throughout the multilayered structure. Now it is necessary to have different fields for the spins because their scattering rates are different, i.e., $E^{\uparrow}(z) \neq E^{\downarrow}(z)$. This is called the CPP(1) model and holds as long as the mean distance for spin flip λ_{sf} is large compared to the layering repeat distance d_{in}. When spin mixing occurs, the currents j^s need not be constant, only $j = \sum_s j^s$ is constant, and one has a unique internal field, i.e., $E^{\uparrow}(z) = E^{\downarrow}(z)$, for $s = \uparrow, \downarrow$. In this case, $\lambda_{\text{sf}} \lesssim d_{\text{in}}$, and we call this CPP(2). The CPP conductivities for the two models are quite different [21].

5. CPP(1)-MR. There are some unusual predictions for the CPP-MR based on the CPP(1) model and using the relation $E(z) \propto 1/\sigma(z)$, which is correct in the local limit, and trivially satisfied in the opposite (homogeneous) limit [20]:

(a) The magnetoresistance is *independent* of the spatial distribution of the scatterers and the magnetic configuration $\mathbf{M}(z)$; rather, the CPP(1)-MR depends only on the average magnetization $\overline{\mathbf{M}} = (\int_0^L dz\, \mathbf{M}(z))/L$, and on the average scattering in the multilayered structure $\Delta = (\int_0^L dz\, \Delta(z))/L$. In addition, the CPP(1)-MR does not decay on the length scale of the mean free path λ_{mfp}; only the spin-flip mean free path λ_{sf} is the relevant length scale. These results are very different from the CIP-MR where the spatial distribution of scatterers and the magnetization are important, and where the MR decays on the length scale of λ_{mfp}. Due to the invariance of the CPP(1)-MR with respect to $M(z)$, it is the same for a multilayer whose magnetic layer moments are randomly oriented, as for one with antiferromagnetic ordered layers; both have $\overline{\mathbf{M}} = 0$.

b) When the thickness of the magnetic and nonmagnetic layers, t_{m} and t_{nm} respectively, are kept the same, the CPP(1)-MR reaches a finite asymptote, see Fig. 1, which comes from the spin-dependent bulk scattering per unit length. This is held *constant* because as we increase t_{nm} we also increase t_{m} to maintain the equality of the two. We note from Fig. 1 that the CPP-MR with $t_{\text{m}} = t_{\text{nm}}$ provides a sensitive test for the presence of spin-dependent bulk scattering. If it is not present, the CPP-MR goes to zero as t_{nm} increases in much the same way as the CIP-MR. The curve for the CPP-MR is predicated on CPP(1), i.e., the assumption that $t_{\text{m}} + t_{\text{nm}} = d_{\text{in}} \ll \lambda_{\text{sf}}$.

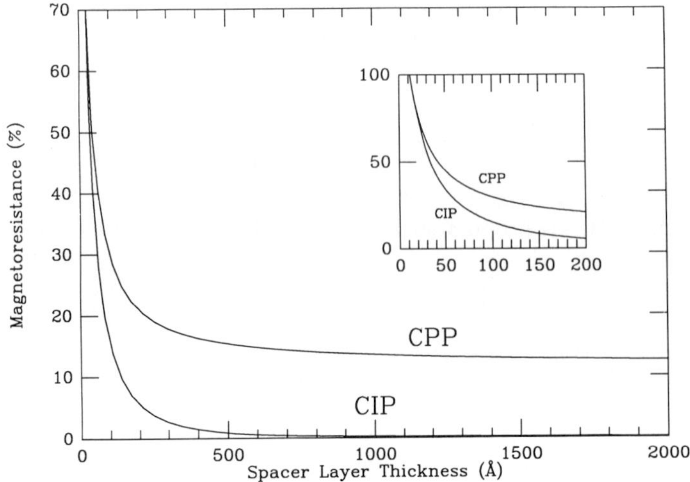

Figure 1. The parameters used for these curves are (see Refs. [1] and [5], for the meaning of the parameters) $\lambda_m = 40$ Å, $\lambda_{nm} = 200$ Å, $\lambda_s = (t_m + t_{nm})/0.6$; $p_s = 0.5$ and $p_b = 0.2$. These are the mean free paths for the magnetic and nonmagnetic layers, and interface, and the ratio of spin-dependent to spin-independent scattering potentials for the interface and the bulk.

Once $t_m + t_{nm} = d_{in} \gtrsim \lambda_{sf}$, one goes over to CPP(2); as $t_m + t_{nm} = d_{in} \gg \lambda_{mfp}$, one is in the local limit, and as mentioned earlier the MR goes to zero. In other words, the CPP-MR maintains its plateau in Fig. 1, only as long as $d_{in} \ll \lambda_{sf}$.

Although the electron mean free path was crucial to understanding the dependence of the CIP-MR on t_{nm} and on t_m, this is not the case for CPP. Here the only relevant length scale is the spin flip distance λ_{sf}. As it is typically an order of magnitude larger than λ_{mfp}, it does not seem to play a role in the interpretation of the CPP-MR results in Co/Ag and Co/Cu [22].

6. Magnetic Granular Films. Thin films of copper or silver containing granules of cobalt have been found to exhibit quite large magnetoresistance, comparable to those in the multilayered structures [23]. One of us (Zhang) has found that the MR in this random alloy is much closer to to CPP(1)-MR than to CIP-MR [24].

INTERLAYER COUPLING

Here, we turn our attention to the second topic we discuss: models and calculations of the coupling between magnetic layers in transition-metal multilayered structures. We do not discuss rare-earth multilayers, as these have received some attention in the past [25].

I. Non-Perturbative Methods

These are total energy calculations which consider a unit cell of the superlattice, evaluate the band structure, and determine the energy of various spin configurations.

The configurations that can be considered are limited by the size of the unit cell used for the calculation, e.g., at the present time one cannot consider incommensurate spin configurations as in chromium; one is limited to commensurate antiferromagnetic and ferromagnetic order. We distinguish between two types of total energy calculations:

1. "First-principle calculations". These consider all the relevant atomic levels to find the electronic band structures of the superlattice. They are similar to calculations of the magnetic order in the elemental transition metals [26], and in the magnetic alloys [27]. These total energy calculations have been done for Fe/Cr [28], Co/Cu and Fe/Cu [29], and Fe/Ru [30]. What emerges is the feeling that really long machine times are needed to calculate the electronic band structure to obtain realistic magnetic moment distributions in superlattices.

2. Tight-binding superlattice band structure calculations. These differ from first-principle calculations in that one uses a *restricted* set of basis functions, and the time for these calculations is reduced considerably. Therefore, much larger unit cells can be considered with a reduced basis; this allows one to probe much more complicated moment configurations. Tight binding calculations of the magnetic configuration in superlattices have been limited to the magnetic d orbitals. Calculations done by Hasegawa [31] have, among other things, corroborated the initial work on Fe/Cr [28]. Edwards and Mathon [32] have looked at Co/Ru, Co/Cr, and Fe/Cr. To date their methods of confining the majority spin electrons to the nonmagnetic spacer layers is more appropriate to cobalt than iron, and they are currently taking into account the effects of "partial electron confinement" on the interlayer coupling. The group in Strasbourg led by Gautier [33] has considered large unit cell superlattices of the type A/B where A= Cr, Mn, V, Pd, Ru; and B = Fe, Co, Ni. They solve for the electron propagators (Green's functions) entering their band structure calculations in real space rather than in momentum space, and find that this reduces their iteration time considerably. This permits them to obtain highly convergent results, and they have extensive results on the spin configuration of relatively thick spacer layers (up to 30 monolayers). In addition, they have been doing very interesting studies of the effects of: 1) roughness and 2) the number of magnetic planes (monolayers) in the magnetic layers, on the interlayer coupling and spin configurations.

What are the ultimate limitations of the first-principle calculations? Is machine time the only impediment, or is the local spin density functional approximation a source of concern? Will this approximation be able to pick up the small energy differences between the different spin configurations that yield the interlayer coupling?

II. Perturbative Methods

Total energy calculations are indubitably needed to obtain the strength of the interlayer coupling, however perturbative methods provide insight into the origin of this coupling. There are two genre of perturbative methods used to determine the interlayer coupling.

1. Extension of Interaction between Two Spins to Sheets of Spins. [34] (a) Within this group there are calculations which focus on the Fermi surface. Bruno and Chappert [35] have extended the analysis of Roth, Zeiger, and Kaplan on the RKKY coupling between spins for nonspherical Fermi surfaces to the RKKY coupling between two sheets of spins. They focus on the *topology* of the Fermi surfaces, e.g., nesting features, , to successfully predict oscillations observed in the interlayer coupling in copper-based systems. Amongst others [36, 37], they have considered the effects of reducing momentum space into the first Brillouin zone; this is known as the aliasing effect. It recognizes that the coupling is probed at lattice sites of periodic structures. For sheets of spins, they find a new feature: multiperiodicity that comes from the discreteness of the distribution of spins on lattice sites in the interfacial planes perpendicular to the direction of the coupling (direction of layer growth). In addition, they have considered experiments on the effects on coupling of different types of disorder (roughness) at interfaces. In this context, Herman and Schrieffer [38] have also considered roughness that stems from magnetic ions *not* occupying periodic lattice sites; they find that this *undoes* the aliasing effect if one uses plane wave states. In addition, they find that, when one goes beyond the plane wave approximation and uses Bloch states, there exists another source of long period oscillations, which they call the "Bloch modulation" effect.

(b) If one does not limit oneself to the coupling coming from the Fermi surface, we have found [39], as have Lacroix and Gavigan [40], an additional coupling mechanism. If the interaction of the local moments of the magnetic layers with the conduction electrons of the spacer layers is modeled by the s-d mixing interaction, there are virtual charge excitation processes (from states *away* from the Fermi surface) that produce antiferromagnetic interlayer coupling. This coupling, due to the mixing interaction, was first studied by Caroli [41] and Falicov [42]. The concept that interfacial roughness acts to filter out rapid oscillations of the coupling was introduced by Wang *et al.* [39]; they showed how the coupling through the induced spin density wave in chromium is removed by the roughness of the same wavelength as the spin density wave.

c) How can one distinguish what part of the coupling comes from spin polarization of the spacer layer, i.e., the RKKY term, from the additional term, known as superexchange [42]? The recent results from the NIST group [43] for a chromium wedge on an iron whisker are crucial to determining this. As we have recently shown [44], the RKKY coupling is not able to reproduce the interlayer coupling observed in Fe/Cr multilayered structures; it is necessary to include the superexchange coupling that comes from the virtual excitation processes that are not tied to the Fermi surface. These additional processes were included in the original calculations of Wang *et al.* for the interlayer coupling in Fe/Cr multilayers. At this time, it is an open question whether the RKKY coupling is sufficient to understand the coupling through noble metal spacer layers; what is clear is that for spacer layers made up of transition metals, e.g., chromium, there is additional coupling.

2. Method of Spin Currents. This is a novel method introduced by Slonczewski [45] for determining the coupling between two magnetic layers. While it was suggested in the context of coupling via an insulating barrier, the method is applicable to metallic spacer layers. Hathaway and Cullen [46] have calculated the interlayer coupling by determining the torque acting on a polarized spin current when it enters a layer magnetized along a direction that is at an angle with respect to the axis of polarization.

III. Analytic Results

Finally, we want to discuss an exact solution that allows one to interpret the validity of the coupling found by perturbative methods.

Deaven, Rokhsar and Johnson [36] consider a one dimensional magnetic multilayer tight binding model with one orbital per site and hopping and exchange interaction between sites. They solve for the spin configurations by iterating their solutions until they are self-consistent. The energy difference between the ferro and antiferromagnetic configurations yields the interlayer coupling. In their model, this coupling comes from the hybridization (mixing) of magnetic and nonmagnetic states at interfaces. They compare their solution with RKKY theory [47] and conclude that perturbation theory is sufficient to understand the *qualitative* features of the fully interacting model system. The perturbative approach is able to predict the correct asymptotic period and decay envelope of the coupling. Also, their solution allows them to show that the largest contribution to the coupling comes from the magnetic layer at the interface; inner planes (layers) are less important. This result lends support to the coupling found by the methods described in Sec. II; in these calculations, only the magnetic layer adjacent to the nonmagnetic one, i.e., the interfacial layer, has been considered.

ACKNOWLEDGEMENTS

This work was supported by the Office of Naval Research Grant No. N00014-91-J-1695, the New York University Research Challenge Fund and the New York University Technology Transfer Fund.

References

[1] S. Zhang, P. M. Levy, and A. Fert, Phys. Rev. B **45**, 8689 (1992); P. M. Levy, S. Zhang, and A. Fert, Phys. Rev. Lett. **65**, 1643 (1990).

[2] A. Fert and I. A. Campbell, J. Phys. F **6**, 849 (1976); I. A. Campbell and A. Fert, in Ferromagnetic Materials, edited by E. P. Wohlfarth (North-Holland, Amsterdan, 1982), Vol. 3, p.769.

[3] P. W. Anderson, Phys. Rev. **124**, 41 (1961); E. Daniel and J. Friedel, in *Proceedings of the Ninth International Conference on Low Temperature Physics*, edited by J. Daunt, P. Edward, F. Milford, and M. Yaqub (Plenum, New York, 1965), p. 933.

[4] J. Inoue, A. Oguri, and S. Maekawa, J. Phys. Soc. Jpn. **60**, 376 (1991); J. Inoue and S. Maekawa, Prog. Theor. Phys. Suppl. **106**, 187(1991); J. Inoue, H. Itoh, and S. Maekawa, J. Phys. Soc. Jpn. **61**, No. 4 (1992).

[5] S. Zhang and P. M. Levy, *Magnetic Surfaces, Thin Films and Multilayers*, edited by S. S. Parkin *et al.*, Mat. Res. Soc. Symp. Proc. **231**, 255 (1992).

[6] M. B. Stearns, J. Magn. Magn. Mater. **104–107**, 1745 (1992).

[7] G. G. Cabrera and L. M. Falicov, Phys. Status Solidi (b) **61**, 539 (1974), and **62**, 217 (1974); L. Berger, J. Appl. Phys. **49**, 2156 (1978).

[8] M. B. Stearns, J. Magn. Magn. Mater. **5**, 167 (1977); J. Slonczewski, Phys. Rev. B **39**, 6995 (1989).

[9] R. Q. Hood and L. M. Falicov, to appear.

[10] S. Zhang *et al.*, to appear.

[11] T. Oguchi and T. Sasaki, *Research Report on Metallic Artificial Superlattices* (Japan) p. 382, March 1992.

[12] K. Fuchs, Proc. Phil. Camb. Soc. **34**, 100 (1938); E. H. Sondheimer, Adv. Phys. **1**, 1 (1952); J. M. Ziman, *Electrons and Phonons* (Oxford University Press, London, 1972), pp. 451–469.

[13] P. F. Carcia and A. Suna, J. Appl. Phys. **54**, 2000 (1983).

[14] R. E. Camley and J. Barnás, Phys. Rev. Lett. **63**, 664 (1989); A. Barthélémy and A. Fert, Phys. Rev. B **43**, 13124 (1991); B. Dieny, Europhys. Lett. **17**, 261 (1992).

[15] A. Vedyayev, B. Dieny, and N. Ryzhanova, Europhys. Lett. **19**, 329 (1992).

[16] Z. Tešanovic, M. V. Jarić and S. Maekawa, Phys. Rev. Lett. **57**, 2760 (1986).

[17] See, for example, Barthélémy and Fert in Ref. [14], and Zhang *et al.* in Ref. [1].

[18] H. E. Camblong and P. M. Levy, to appear.

[19] B. L. Johnson and R. E. Camley, Phys. Rev. B **44**, 9997 (1991).

[20] S. Zhang and P. M. Levy, J. Appl. Phys. **69**, 4786 (1991).

[21] H. E. Camblong, S. Zhang, and P. M. Levy, to appear.

[22] W. P. Pratt, Jr., S.-F. Lee, J. M. Slaughter, R. Loloee, P. A. Schroeder, and J. Bass, Phys. Rev. Lett. **66**, 3060 (1991); (and to be published).

[23] A. E. Berkowitz, S. Zhang *et al.*, Phys. Rev. Lett. **68**, 3745 (1992); J. Q. Xiao, J. S. Jiang, and C. L. Chien *ibid.* **68**, 3749 (1992).

[24] S. Zhang, to appear.

[25] See, for example, Y. Yafet, J. Appl. Phys. **61**, 1058 (1987); Y. Yafet *et al.*, J. Appl. Phys. **63**, 3453 (1988); M. B. Salamon, R. S. Beach, J. A. Borchers, R. W. Erwin, C. P. Flynn, A. Methny, J. J. Rhyne, and F. Tsui, J. Magn. Magn. Mater. **104–107**, 1729 (1992).

[26] V. L. Moruzzi and P. M. Marcus, Phys. Rev. B **45**, 2934 (1992), and references therein.

[27] A. R. Williams, J. Kübler, and C. D. Gelatt, Jr., Phys. Rev. B **19**, 6094 (1979); J. Kübler, J. Magn. Magn. Mater. **20**, 277 (1980); Phys. Lett. **81A**, 81 (1981); J. Kübler, A. R. Williams, and C. B. Sommers, Phys. Rev. B **28** 1745 (1983).

[28] K. Ounadjela, C. B. Sommers and A. Fert, see for example J. Appl. Phys. **67**, 5914 (1990).

[29] F. Herman, J. Sticht and M. van Schilfgaarde, *Magnetic Surfaces, Thin Films and Multilayers*, edited by S. S. P. Parkin *et al.*, Mat. Res. Soc. Symp. Proc. **231**, 195 (1992).

[30] D. Knab and C. Koenig J. Magn. Magn. Mater. **93**, 398 (1991).

[31] H. Hasegawa, Phys. Rev. B **43**, 10803 (1991); J. Phys. Cond. Matter **4**, 169 (1992).

[32] D. M. Edwards, J. Mathon, R. B. Muniz and M. S. Phan, J. Phys. Condens. Matter **3**, 4941 (1991); Phys. Rev. Lett. **67**, 493; **67**, 1476(E) (1991).

[33] D. Stoeffler, K. Ounadjela, and F. Gautier, J. Magn. Magn. Mater. **93**, 386 (1991); D. Stoeffler and F. Gautier, J. Magn. Magn. Mater. **104–107**, 1819 (1992).

[34] Y. Yafet, Phys. Rev. B **36**, 3948 (1987).

[35] P. Bruno and C. Chappert, Phys. Rev. Lett. **67**, 1602 (1991); **67**, 2592(E) (1991); Phys. Rev. B **46**, to appear.

[36] D. M. Deaven, D. S. Rokhsar and M. Johnson, Phys. Rev. B **44**, 5977 (1991).

[37] R. Coehoorn, Phys. Rev. B **44**, 9331 (1991); C. Chappert and J. P. Renard, Europhys. Lett. **15**, 553 (1991).

[38] F. Herman and J. R. Schrieffer, Phys. Rev. B **46**, to appear.

[39] Y. Wang, P. M. Levy and J. L. Fry, Phys. Rev. Lett. **65**, 2732 (1990).

[40] C. Lacroix and J. P. Gavigan, J. Magn. Magn. Mater. **93**, 413 (1991).

[41] B. Caroli, J. Phys. Chem. Solids **28**, 1427 (1967).

[42] C. E. T. Gonçalves da Silva and L. M. Falicov, J. Phys. C **5**, 63 (1972).

[43] J. Unguris, R. J. Celotta, and D. T. Pierce, Phys. Rev. Lett. **67**, 140 (1991); Phys. Rev. Lett. **69**, to appear.

[44] Z.-P. Shi, P. M. Levy and J. L. Fry, to appear.

[45] J. Slonczewski, see Ref. [8].

[46] K. B. Hathaway and J. R. Cullen, J. Magn. Magn. Mater. **104–107**, 1840 (1992).

[47] They introduced aliasing to discuss the period of the oscillations, as did Coehoorn, Chappert and Renard, see Refs. [36] and [37].

MAGNETIC AND MAGNETOTRANSPORT PROPERTIES OF EPITAXIAL (100)Fe/Pd SUPERLATTICES

J. R. Childress[1,2], A. Schuhl[1], J.-M. George[2], O. Durand[1], P. Galtier[1], V. Cros[1], K. Ounadjela[3], R. Kergoat[2] and A. Fert[2]

[1] Laboratoire Central de Recherches
THOMSON-CSF, Domaine de Corbeville
91494 Orsay, France

[2] Laboratoire de Physique des Solides
Université Paris-Sud
91405 Orsay Cédex, France

[3] ICPMS-GEMME
4 rue Blaise Pascal
67070 Strasbourg, France

INTRODUCTION

Indirect exchange-coupling of ferromagnetic ultrathin films has been observed in various multilayer systems, with a number of different elements used for the interlayer (non-ferromagnetic) material. In particular, antiferromagnetic (AFM) coupling *between* the ferromagnetic (FM) layers has been found with both non-magnetic (Cu, Ag, Au) and antiferromagnetic (Cr, Mn) spacers.[1-5] Of special interest for applications is the existence of giant magnetoresistance (MR) effects in these antiferromagnetically-coupled multilayer systems.[6]

To further explore the occurence and mechanism of this AFM coupling in different systems, the use of Palladium as an interlayer is a particularly interesting case. Although itself non-magnetic, Pd displays a very high paramagnetic susceptibility due to a large Stoner enhancement factor.[7] Also, Pd is known for its strong tendency to order ferromagnetically, through polarization from magnetic impurities[7], or possibly through band structure modifications resulting from an expansion of the lattice spacing.[8] Recently, Celinski and Heinrich have studied coupling effects between Fe films in (100)Fe/Pd/Fe trilayers, and

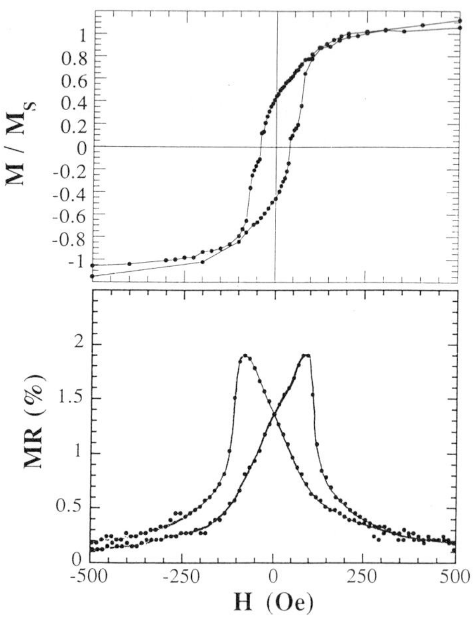

Figure 1. Magnetic hysteresis loop at T=15K (top), and corresponding magnetoresistance curve at T=4.2K (bottom) for a {Fe(20Å)/Pd(30Å)}x10 multilayer grown on a 200Å-thick Pd buffer layer.

found evidence for an oscillating coupling with at least one region (for $t_{Pd}\approx 14$ ML) of antiferromagnetic coupling[3]. The remarkable feature of this result was the combination of an extremely low coupling strength (on the order of 30 Oe) and an abrupt switching behavior (within a few Oe). If associated with spin-valve magnetoresistive effects, this high sensitivity to magnetic fields could be advantageous for low-field device applications. Therefore, we have attempted to confirm the observation of AFM coupling in this system by fabricating *multilayer* (100)Fe/Pd structures, and measuring their magnetotransport properties.

GROWTH AND STRUCTURE

In a previous paper, we have reported on the growth, magnetic properties and magnetotransport properties of (100)Fe/Pd superlattices grown on 200Å-thick Pd buffer layers[9]. The Pd buffer, grown at a substrate temperature of 200°C and at a rate of 4Å/min, was intended to provide a flat starting surface for the growth of the multilayer. Indeed, RHEED patterns for the completed buffer layer displayed sharp streaks typical of a flat surface, although some faceting was clearly visible at the early stages of growth. Best superlattice growth on top of this buffer was then obtained at room temperature. RHEED

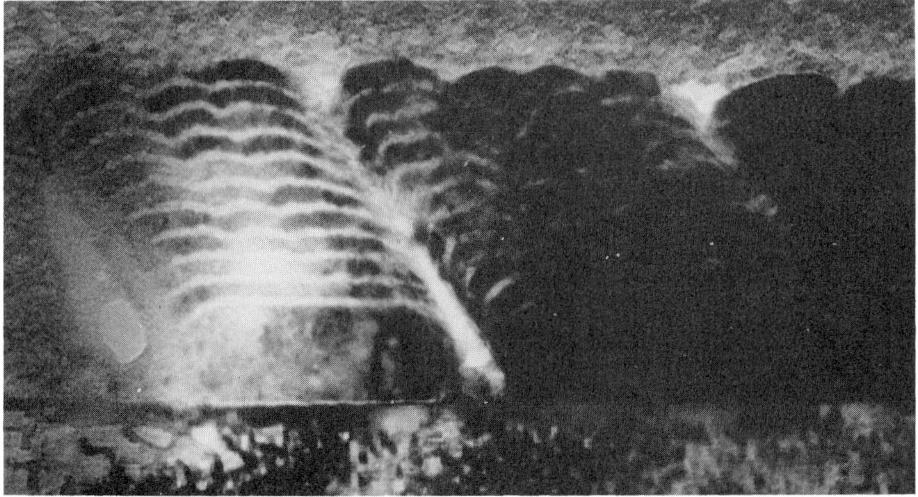

Figure 2. Bright-field cross-sectional TEM Image of a (100)Fe/Pd superlattice grown on a 200Å-thick Pd buffer layer, observed along the MgO (100) direction. The scale marker corresponds to 200 Å. The defects begin at the base of the Pd buffer layer and propagate at an angle with respect to the substrate.

patterns for the layers suggested roughening of the surface (broad streaks with some spottiness), but still indicated monocrystalline growth. Interestingly, these first Fe/Pd structures indeed showed evidence of AFM coupling between the Fe layers (i.e., low remanence at zero field) for Pd thicknesses close to those reported by Celinski and Heinrich. Additionally, MR effects of up to 3% were found[9]. However, as shown in Fig.1, the large MR saturation fields ($H_S \approx$ 500-1000 Oe) and coercive fields ($H_C \approx$ 50-150 Oe) observed were not consistent with the expected weak AFM coupling.

As shown in Fig.2, subsequent observations of the layers by transmission electron microscopy (TEM) revealed a defected, columnar structure, beginning with the buffer layer. Hence, it is possible that the MR effects measured may have been due simply to microstructural disorder, i.e., to a quasi-random distribution of spins at grain boundaries during magnetic reversals, rather than to a true switching of the Fe layers between AFM and FM configurations.

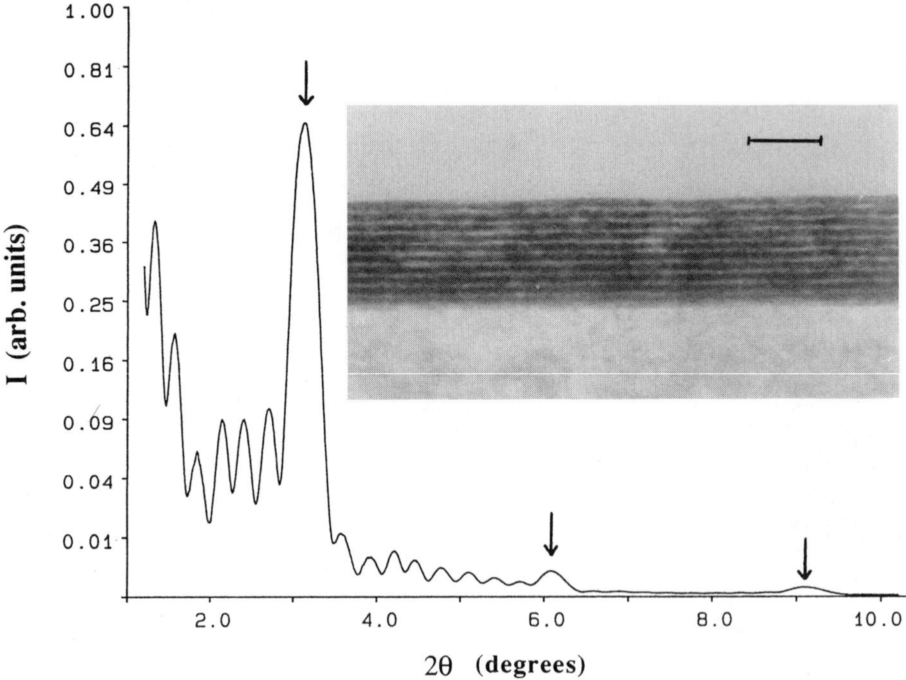

Figure 3. Low-angle x-ray scattering of a {Fe(17Å)/Pd(12Å)}x10 superlattice grown directly onto a (100)MgO substrate at T=50°C. The arrows indicate the Bragg peaks. Inset: Bright-field cross-sectional TEM image of the same multilayer, showing flat and continuous Fe (light bands) and Pd (dark bands) layers. The scale marker corresponds to 200Å.

This example also illustrates the limitation of structural analyses by RHEED, which characterizes the surface only on scales of a few hundred Å's, and may not reveal important features on larger scales. To improve the structural characteristics, and because the observed defects appeared to originate in the Pd buffer layer, the growth of the Fe/Pd superlattices was then attempted *directly* onto the (100)MgO substrate, without the use of any buffer layer. Results on these buffer-free samples are reported in this paper.

The growth of the superlattices was performed on a Riber MBE system, with a pressure during growth in the low 10^{-9} Torr range. The 1cm x 1cm (100)MgO substrates were annealed in UHV for 30 minutes at 450°C immediately prior to deposition. Samples with Fe layer thicknesses ≈ 20 Å and Pd layer thicknesses in the range 10-40 Å were grown directly on top of the MgO, without any kind of buffer, at either T=50°C or room temperature. Some samples were grown in a wedge geometry, with a variable thickness of Pd obtained by tilting the sample manipulator away from the atom beam during Pd deposition. Structural information was obtained *in situ* by reflection high-energy electron diffraction (RHEED). The deposition rates were 4 Å/min for Fe and 2 - 8 Å/min for Pd. Characterization of the completed structures was performed *ex-situ* by x-ray diffraction and TEM.

Fe (bcc structure, a_{Fe}=2.87 Å) grows epitaxially onto the fcc structure of MgO (a_{MgO}=4.20 Å) with the usual 45° rotation of the (001) axis in the plane, so the the <110> Fe azimuth is aligned with the <100> Pd azimuth. The lattice mismatch is then about 4% (a_{Fe} x $\sqrt{2}$ = 4.06 Å). Similarly, fcc Pd (a_{Pd}=3.89 Å), grows on top of Fe with a 45° rotation and also 4% mismatch. RHEED patterns for the first layer of Fe grown onto the MgO substrate show broad streaks characteristic of a nearly-2-dimensional structure with some roughening. As growth progresses, the RHEED streak become progressively narrower, indicating improved flatness for the successive layers throughout the growth of the 10 repeat periods of the multilayer structure. Most importantly, no faceting of the Pd surface is seen, as opposed to what was found for the structures grown on a 200Å-thick Pd buffer.

Evidence for a high-quality multilayered structure is obtained by x-ray diffraction in the θ-2θ geometry (Fig.3), with low-angles spectra showing strong Kiessig fringes and up to 4th-order Bragg peaks. These spectra are much superior to those obtained from multilayers grown on Pd buffers[9]. Direct imaging of the multilayer structure has also been obtained by TEM (Fig.3 inset). The improvement in structural quality over the structures grown on Pd buffers is spectacular, as no grain boundaries can be seen over the entire observation area (lateral distances of several thousand Å's). All the layers are clearly found to be continuous over these distances, with a surface roughness on the order of 5Å. This low roughness has also been confirmed by atomic force microscopy of the top surface of the multilayers. Also, growth at T=50°C resulted in lower roughness compared to growth at room temperature.

MAGNETIC PROPERTIES

The magnetic properties of the multilayers were measured at 15K by superconducting quantum interference device (SQUID) magnetometry and at 300K by magnetic field gradient (MFG) magnetometry. The dramatic improvement of the structural characteristics, compared to the samples grown on Pd buffers, is found to directly result in superior magnetic characteristics. Narrow hysteresis loops (H_C ≈ 10-20 Oe) are obtained at both room temperature and 15 K, suggesting a low density of pinning defects in the 20Å Fe layers.

As expected, the saturation magnetization M_S of the Fe/Pd multilayers, as measured by SQUID in fields of up to 10 kOe, is larger than that expected from the Fe layers alone. Thus it is clear that the Pd near the interfaces must be strongly polarized, either by the proximity of the Fe surface, or by the presence of Fe impurities diffused into the Pd layers. Although fluctuations in this extra magnetization are seen from sample to sample, we find an average $M_S \approx 2150$ emu/cm$^3{}_{Fe}$ at 15 K, which is similar to that already found in sputtered Fe/Pd multilayers.[10] Assuming only an interfacial polarization of the Pd, this corresponds to an enhancement of about 3 μ_B per interfacial atom. However, the number of Pd layers at each

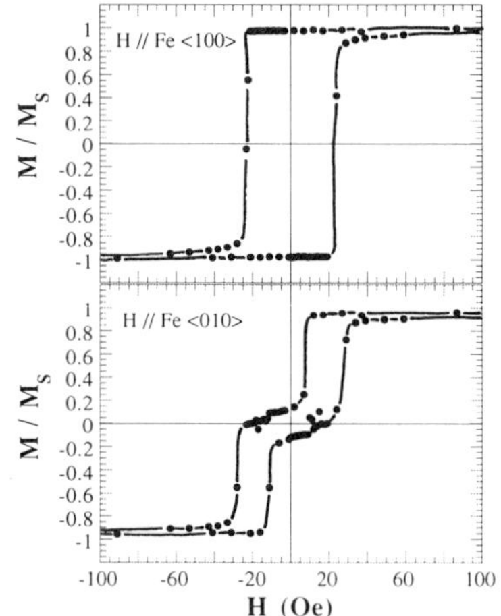

Figure 4. Magnetic hysteresis loops of a {Fe(14Å)/Pd(26Å)}x10 superlattice at T=15 K. Top: the external field is applied in the <100> in-plane easy direction of Fe. Bottom: the external field is applied in the <010> in-plane easy direction of Fe.

interface which participate in this polarization cannot be known at the present time. A theoretical simulation of the Fe/Pd interface[11] yields moments of 0.3 and 0.2 μ_B for the first and second Pd layers away from the interface, respectively, and an enhancement of the Fe magnetization by about 1μ_B. In comparison, the larger total moment induced at the Fe/Pd interface in our case could signal the presence of some intermixing (in particular that due to interface roughness), since a single Fe impurity in Pd can potentially result in a polarization cloud of up to 10 μ_B.

Numerous hysteresis loops have been obtained on samples with various Pd layers thicknesses. Here we report on several important findings which are representative of what is observed in this system. Fig.4 shows hysteresis loops obtained for a multilayer sample with t_{Pd}≈26Å, in the two easy directions (<100> and <010>) of Fe. Surprisingly, two different hysteresis loops are obtained in these nominally equivalent directions. In one direction (closed circles, which we may call the <100>Fe direction) the hysteresis loop is nearly square, with a low coercive field $H_C \approx 20$ Oe and remanence $M_R/M_S \approx 1$. In the other (open squares, <010>Fe), the magnetization reversal occurs in two distinct steps, in a way reminiscent to that found by Celinski and Heinrich. In particular, the field at which the first reversal occurs is also near 30 Oe, and this process also occurs in just a few Oe. Similar anisotropic behaviors are found for all other samples, regardless of the exact thicknesses of Pd, although the actual shapes of the loops vary (often, the non-square loop is simply canted, with low remanent magnetization).

It is tempting to immediately interpret the canted or step loops as representing AFM coupling between successive Fe layers. However, the presence of in-plane uniaxial anisotropy is itself sufficient to explain some of the magnetization behavior. For example, step loops may be explained by assuming that the anisotropy energy is large enough (i.e., larger than the nucleation energy) to cause the Fe moments magnetized in the <010> (hardest) direction to spontaneously flip towards the <100> (easiest) direction when the externally applied field is sufficiently small. In that case, the magnetization (which is measured in the <010> direction) is effectively zero at low fields, and no antiferromagnetic arrangement of successive Fe layers needs to be invoked. Indeed, recent Kerr effect observations of the domain structure in these samples as a function of applied field does support this interpretation of the hysteresis loops. These will be described in a forthcoming paper. As of now, the cause of this anisotropy has not been determined. Possible factors may be the presence of small thickness gradients in the Fe layers, or an effect due to the angle of incidence of the atom beam with the substrate (≈30°). Indeed, preliminary results on samples grown onto rotating MgO substrates indicate that most of the anisotropy is removed. Hence it does not appear to be an intrinsic property of this epitaxial system, as is observed for example in the case of Fe grown onto GaAs.

Still, other observations cannot be explained by the presence of this anisotropy alone. For example, using a multilayer with wedge-shaped Pd layers, hysteresis loops have been obtained as a function of Pd thickness. As shown in Fig.5, we find that their characteristics at T=300 K depends sensitively on the exact Pd thickness. In one easy direction (labeled <100>), there is an oscillation between square and canted hysteresis loops as the Pd thickness is varied, similar to that observed in the case of an oscillating indirect exchange coupling between Fe layers. However, true interlayer exchange coupling should be isotropic. Here, canted loops are *never* found in both easy directions simultaneously, as shown by the square loops obtained in the other (<010>) easy direction.

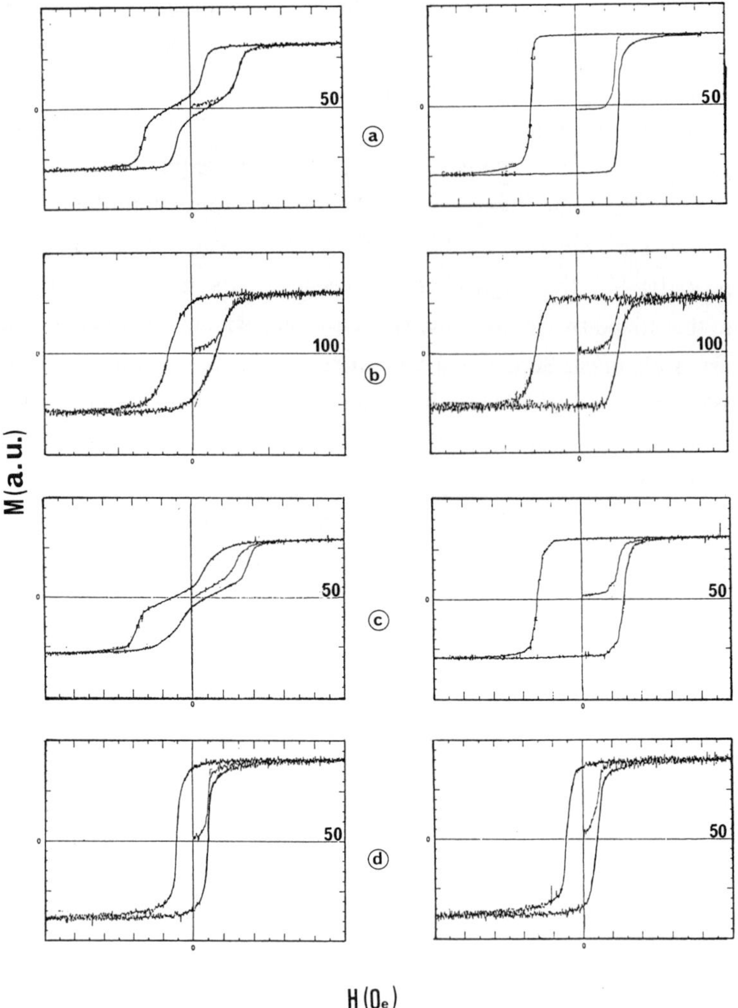

Figure 5. Magnetic hysteresis loops of (100)Fe/Pd superlattices with varying Pd layer thickness, at T=300K. Left side: the external field is applied in one of the two in-plane easy direction of Fe. Right side: the external field is applied in the other in-plane easy direction of Fe. Pd thicknesses are (a) 12.5Å, (b) 14Å, (c) 20.5Å, (d) 25.5Å. Note that the field scale in (b) is twice as large.

Hence no conclusive evidence of antiferromagnetic coupling between the Fe layers has been found in our MBE-grown Fe/Pd superlattices. As mentioned before, we are currently fabricating and analyzing new samples grown on a rotating substrate holder, to minimize any anisotropy effects directly related to the deposition process, and facilitate the detection of possible intrinsic (isotropic) coupling effects.

MAGNETOTRANSPORT PROPERTIES

Finally, the magnetotransport properties of these multilayers have been measured. No significant spin-valve magnetoresistance is found, even in the case of samples displaying strong canted or step hysteresis loops. This supports the conclusion that no AFM coupling of Fe is observed in our Fe/Pd superlattices, although one must keep in mind that the strength of the spin-dependent scattering at the Fe/Pd interface is not known. Still, the sharp switching behavior of the magnetization results in extremely sharp resistive transitions. For example, Fig.6 shows the results obtained for a sample with $t_{Pd} \approx 34$Å. When the field is applied in a <100> direction, the magnetization reversal causes a sudden increase in the resistivity, of about 1% over 4 Oe, which relaxes to its original values when the field reaches about -100 Oe. By contrast, when the field is applied along the <010> direction, no such large jump is observed. Resistivity changes as sharp as 2.7% over 5 Oe have been found in other samples. Such sensitivity begins to be of interest for low-field application. However, considering that the magnitude of the anisotropic magnetoresistance in this system is close to 3%, these effects appear to be due mainly to different anisotropy contributions for different spin orientations attained during magnetic reversals.

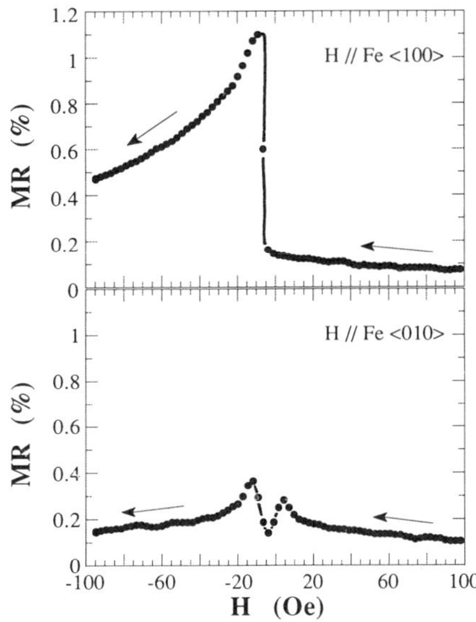

Figure 6. Resistance of a {Fe(17Å)/Pd(34Å)}x10 superlattice as a function of the applied field, at T=4.2 K. Top: the external field is applied in the <100> in-plane easy direction of Fe. Bottom: the external field is applied in the <010> in-plane easy direction of Fe.

CONCLUSIONS

In conclusion, we have found that high-quality epitaxial (100)Fe/Pd superlattices can be grown on (100)MgO substrates without the use of any buffer layer. These structures exhibit two non-equivalent in-plane easy directions of magnetization for the Fe. This causes the hysteresis loops along one easy direction to be strongly canted, resembling those expected for the case of antiferromagnetic exchange coupling between Fe layers. However, no direct evidence for isotropic AFM coupling between Fe layers through the polarized Pd layers has been found. Accordingly, the sharp magnetoresistive effects observed appear to result only from the anisotropic magnetic behavior.

Acknowledgments

This work supported in part by EEC contract ESPRIT3 No. 6146.

REFERENCES

1. D.H. Mosca, F. Petroff, A. Fert, P.A. Schroeder, W.P. Pratt and R. Laloee, J. Mag. Mag. Mat. **94**, L1 (1991).
2. C.A. dos Santos, B. Rodmacq, M. Vaezzadeh and B. George, Appl. Phys. Lett. **59**, 126 (1991).
3. Z. Celinski and B. Heinrich, J. Magn. Magn. Mat. 99, L25 (1991).
4. P. Grünberg, R. Schreiber, Y. Pang, M.B. Brodsky and H. Sowers, Phys. Rev. Lett. **57**, 2442 (1986).
5. S.T. Purcell, M.T. Johnson, N.W.E. Mc Gee, R. Coehoorn and W. Hoving, Phys. Rev. B **45**, 13064 (1992).
6. M.N. Baibich, J.M. Broto, A. Fert, F. Nguyen Van Dau, F. Petroff, P. Etienne, G. Creuzet, A. Friederich and J. Chazelas, Phys. Rev. Lett. **66**, 2472 (1988).
7. G. J. Nieuwenhuys, advances in Physics 24, 515-590 (1975).
8. Z. Celinski, B. Heinrich, J. F. Cohran, W. B. Muir and A. S. Arott, Phys. Rev. Lett. 65, 1156 (1990).
9. A. Schuhl, J.R. Childress, J.-M. George, P. Galtier, O. Durand, A. Barthelemy and A. Fert, J. Mag. Mag. Mat., to be published.
10. F.J.A. den Broeder, H.C. Donkersloot, H.J.G. Draaisma and W.J.M. de Jonge, J. Appl. Phys. **61**, 4317 (1987).
11. S. Blügel, B. Drittler, R. Zeller and P.H. Dederichs, Appl. Phys. A **49**, 457 (1989).

BILINEAR AND BIQUADRATIC EXCHANGE COUPLING IN BCC Fe/Cu/Fe TRILAYERS. EXCHANGE COUPLING IN Fe WHISKER/Cr/Fe(001) STRUCTURES

B. Heinrich, Z. Celinski, J.F. Cochran, and M. From

Physics Department
Simon Fraser University
Burnaby, B.C.
Canada, V5A 1S6

Abstract

Surface Magneto-Optical Effect (SMOKE) studies of the exchange coupling in bcc Fe/Cu/Fe(001) structures and Brillouin Light Scattering (BLS) results on Fe(001) whisker/Cr/Fe(001) structures are presented. It is shown that the interfaces in bcc Fe/Cu/Fe(001) trilayers can be significantly improved by choosing an appropriate growth procedure. The exchange coupling in bcc Fe/Cu/Fe trilayers was studied as a function of the interlayer thickness. The interpretation of magnetization loops for Fe/Cu/Fe trilayers requires the simultaneous presence of bilinear and biquadratic exchange coupling between the magnetic layers. Computer calculations were used to determine the strength of the bilinear and biquadratic exchange couplings. It is shown that the strength of the biquadratic exchange coupling increases with increasing terrace width. The measured values of the bilinear and biquadratic exchange coupling were compared with a model recently proposed by Slonczewski, which treats the exchange coupling in trilayers with imperfect interfaces. Slonczewski's model was used to deconvolute the data to obtain the intrinsic behavior of the bilinear exchange coupling in bcc Fe/Cu/Fe(001) trilayers. It is shown that the exchange coupling unobscured by interface roughness exhibits a strong short wavelength oscillatory behavior which is in agreement with recent first principles band calculations.

The thickness dependence of the exchange coupling through Cr(001) layers has been investigated by means of BLS using films grown on single crystal whisker substrates. A theory of the frequency dependence of the BLS intensity has been worked out for a complex structure consisting of a thin ferromagnetic film separated from a bulk ferromagnetic substrate by a non-magnetic spacer laqyer; this trilayer is covered by a second non-magnetic overlayer. The theory gives a good description of the light scattering data for Cr spacer layer films thicker than 12 ML. Cr spacer layers thinner than 12 ML give rise to thin film BLS peaks whose frequency varies more slowly with applied magnetic field, and whose intensity falls off more rapidly with increasing applied field than is predicted by the model calculations.

Introduction

The study of the exchange interaction between ultrathin ferromagnetic layers separated by ultrathin non-ferromagnetic interlayers has enjoyed a great deal of attention from both theorists and experimentalists. This paper provides a detailed study of the bilinear and biquadratic exchange coupling in bcc Fe/Cu/Fe(001) trilayers and presents our recent measurements of the exchange coupling through Cr(001) using Fe(001) whisker/Cr/Fe(001) samples.

The exchange coupling through non-ferromagnetic interlayers is strongly affected by the interface roughness. The ability to observe short period oscillations depends sensitively on the quality of the interfaces.

The role of interface roughness has been recently treated by Slonczewski[1]. Slonczewski showed that when the exchange interaction changes rapidly with the number of monolayers (ML) and the interface consists of randomly distributed atomic terraces then an additional term $- J_1 \sin^2\theta$ arises in the effective exchange interaction which couples two ferromagnetic layers through a non-magnetic interlayer:

$$E = - J_o \cos\theta - J_1 \sin^2\theta \qquad (1)$$

where θ is the angle between the magnetic moments and J_1 is always positive: it is commonly known as the biquadratic exchange energy. Its presence in ultrathin films was first identified experimentally in the studies of the exchange interaction in the bcc Fe/Cr/Fe and in the fcc Co/Cu/Co(001) systems[2,3].

Our studies of the exchange coupling were carried out using BLS and SMOKE techniques. The intention of this paper is to present our quantitative studies of the exchange coupling in samples which possess interfaces with large atomic terraces. In bcc Fe/Cu/Fe(001) trilayers the first Fe layer was grown at elevated substrate temperatures in order to decrease the number of atomic steps and thereby to increase the area of the individual atomic terraces at the Fe/Cu interface. The Fe(001) whiskers are known to exhibit nearly perfect surfaces[3,4]. The perfectness of Fe(001) whisker templates resulted in magnetic properties which have not been observed in samples prepared on less perfect substrates: e.g. the surface uniaxial anisotropy in a few ML thick Fe(001) using an Fe(001) whisker as a substrate[3] is enhanced significantly compared with that grown on bulk Ag(001) substrates; SEMPA and SMOKE measurements[5,6] revealed the presence of short wavelength oscillations (2ML) in the exchange coupling.

Growth Studies

The epitaxial growth was carried out in a PHI-400 MBE machine equipped for Reflection High Energy Electron Diffraction (RHEED) and Auger electron spectroscopy.

Growth of Fe/Cu/Fe(001) Trilayers

The RHEED studies showed that the Fe growth on Ag(001) at room temperature (RT) proceeds in a quasi layer by layer growth in which the surface roughness is confined to the top two atomic layers[4]. The Fe growths on Fe(001), Cr(001) and Ag(001) templates at room temperature (RT) exhibit a well defined splitting in the RHEED streaks[7]. The shape of the observed streak splitting is characteristic of the intersection of Ewald's sphere with reciprocal rods consisting of alternating segments of hollow cylinders and straight lines. We refer to this structure as the Henzler structure, for it was observed by Henzler and his coworkers[8] in their LEED studies of W(011) grown on W(011), and explained by them to be the result of the formation of clusters of atoms in one atom high islands with a strong correlation of the distances of separation between the island centers. The growth of Fe(001) on Fe(001), Ag(001) and Cr(001) templates (all having a good lateral match to bcc Fe(001)) at RT proceeds in the manner described above.

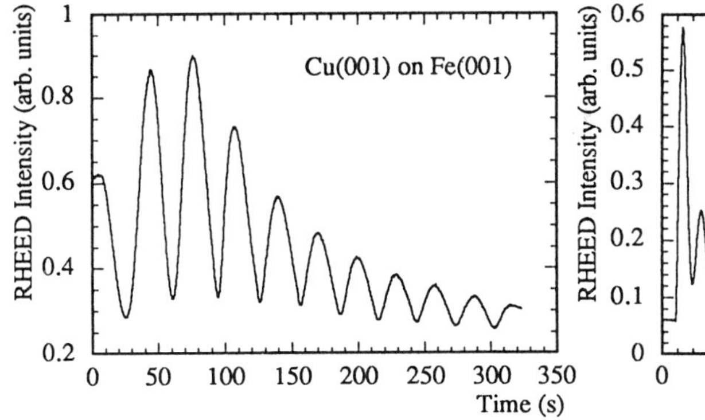

Fig. 1 RHEED intensity oscillations measured at the specular spot during the growth at room temperature of metastable bcc Cu(001) on bcc Fe(001). The electron beam angle of incidence was ~1°.

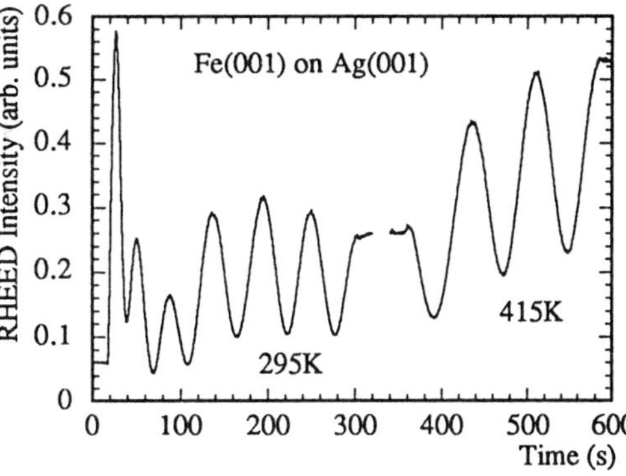

Fig. 2 RHEED intensity oscillations measured at the specular spot during the growth of bcc Fe(001) on a Ag(001) single crystal substrate at 295K and Fe(001) on a Fe(001) template at 415K. The electron beam angle of incidence was ~1°.

The RHEED streak splittings represent a characteristic lateral spacing. It can be shown by a computer simulation that the reciprocal of the RHEED streak splitting is proportional to the average spacing between atomic islands (terraces). At room temperature the nucleation centers of a newly forming atomic Fe(001) layer are separated approximately by 6-12 nm (determined from the RHEED streak splitting). When a Cu layer is grown on Fe surfaces showing the RHEED streak splitting, the splitting disappears abruptly. One Cu(001) atomic layer is sufficient to remove any trace of it.

For Cu grown at room temperature on Fe(001) the RHEED streaks are sharp and the specular spot shows strong RHEED oscillations, see Fig. 1. The growth of bcc Cu on Fe(001) proceeds very likely in the same manner as the Fe growth on Ag(001). The absence of RHEED streak splitting and the sharpness of the RHEED patterns indicate that the average atomic terrace width is significantly larger than that obtained for Fe grown on Ag(001) at RT. The width of the RHEED streaks in the early stages of the Cu growth is nearly the same as that of the Ag(001) substrate and therefore the average separation between Cu(001) atomic terraces should exceed 40 nm.

At a critical thickness of 11-12ML the Cu(001) overlayers start to undergo a weak but noticeable surface reconstruction[9]. A similar situation was observed in bcc Ni(001) overlayers where the critical thickness is much smaller (3-4ML) and where the surface reconstruction is eventually accompanied by an appreciable lattice transformation[10,11]. Reconstructed Cu overlayers which are thicker than 12ML most likely undergo a significant lattice transformation as well. The RHEED intensity oscillations were studied using the specular spot and with the primary electron beam directed 1° with respect to the sample surface. For such small angles the RHEED intensity oscillations are strongly affected by the electron attenuation in the top atomic layer[4]. In this configuration the role of the electron interference is not dominant and the specular spot intensity is more governed by the near surface attenuation. Strong RHEED oscillations are expected to be observed for the growths in which an appreciable surface roughness is confined only to a maximum of two top atomic layers[4,7]. The RHEED intensity maxima occur very likely when the partial filling is confined mostly to the top atomic layer. On the other hand the RHEED intensity minima correspond most likely to the roughest surface during the growth. In this case one can expect the surface roughness to be confined to the top two atomic layers. The amplitude of RHEED intensity oscillations during the growth of Cu, see Fig. 1, gradually decreases with increasing thickness. This trend is even more pronounced as the critical thickness is approached. Therefore it appears that the average atomic terrace width decreases with an increasing Cu thickness.

The surface of the first Fe layer grown at room temperature does not change sufficiently upon annealing to produce any observable changes in the RHEED patterns: Henzler's splitting remains unchanged even at 510 K. The Ag Auger peak intensity measurements indicate that no appreciable interdiffusion between the Ag(001) substrate and a 6ML thick Fe layer occurs for temperatures up to 470 K. However the RHEED patterns change significantly when the growth is carried out at raised temperatures. In our sample preparation procedure the first 5-6ML of Fe were grown at RT to protect the Ag/Fe interface from atomic intermixing. Then the substrate temperature was raised and 3-4 additional atomic layers of Fe were added. The growth at a raised substrate temperature exhibited a behavior which was similar to that observed during the growth of Fe on Fe whisker substrates[4]. The Henzler RHEED streak splitting of Fe on Ag(001) sharpened and decreased with increasing substrate temperature and disappeared completely for temperatures greater than 410 K. The amplitude of the RHEED intensity oscillations increased substantially, ~2-3 fold, see Fig. 2, compared with that observed during the growth at RT. The first period was definitely longer than subsequent periods which were regular and corresponded to 1ML formation. At a substrate temperature of ~420 K the Henzler's streak splitting of Fe disappeared after one additional atomic layer was deposited. The corresponding RHEED patterns were sharp and comparable with the best patterns observed.

In Fe/Cu/Fe trilayers with thin Cu interlayers (1-4ML) deposited on Fe layers which were prepared using raised substrate temperatures the Fe/Cu and Cu/Fe interfaces consisted of large and comparable atomic terraces. With increasing Cu thickness the average terrace width in the Cu/Fe interface decreases and for thicknesses approaching the critical thickness (12ML) it definitely becomes smaller than that at the Fe/Cu interface.

The Growth of Fe(001) whisker/Cr/Fe(001) Structures

Well prepared Fe(001) whiskers represent the best available metallic templates and are characterized by atomic terraces whose dimensions are in excess of 1 μm. The cross-section of a typical Fe(001) whisker was rectangular, approximately .15-.2 mm square, and a typical whisker length lay between 7-15 mm. The whisker surfaces were bounded by {100} planes. The Fe whiskers were prepared by chemical vapor deposition. The Fe whisker surface was cleaned in UHV by the following procedure: The whiskers was first sputtered at room temperature in order to remove any residual surface contamination which was created during the whisker growth. The whisker was then brought to 550°C and sputter cleaned for 90-120 minutes using a 2keV Ar^+ ion beam. The sputtering at elevated temperatures is necessary to remove the subsurface carbon contamination. The whisker was then brought briefly (for appr. 10 min.) to 700°C and then quickly cooled to room temperature. Fe whiskers prepared in this way exhibited nearly perfect RHEED patterns. Only the intersects with Ewald's sphere were observed. The diffraction spots were very sharp showing that the reciprocal lattice rods were very narrow. Typical RHEED streaks were not observed, only very sharp and very short streaks accompanied the main diffraction spots. Kikuchi bands were very sharp but showed very low intensity. All of these diffraction features indicate that the Fe whiskers possessed atomically smooth surfaces with atomic terraces whose dimensions far exceeded the instrumental resolution of our RHEED system.

Cr was grown at an elevated substrate temperatures. The best Cr growth was achieved for substrate temperatures between 350 to 400 °C. The intensity oscillations of the specular spot showed well defined cusps and the oscillations remained unattenuated during the whole growth, see Fig. 3. The presence of well defined cusps is the strongest indication that the growth proceeds in a nearly perfect layer by layer mode[4,12,13]. The RHEED intensity oscillations were studied by directing the RHEED electron beam at an angle which satisfies the second anti-Brag condition (zero intensity for a half filled atomic layer). The Cr RHEED patterns were very similar to those observed for Fe(001) whiskers. In all the Cr growths one could observe somewhat more pronounced streaks. At RHEED intensity maxima the specular spot reaches a maximum intensity and the RHEED streaks are very weak and very short. At RHEED intensity minima the specular diffraction spot disappears entirely (for the 2nd anti-Brag condition[13]) and the RHEED streaks increase their intensity and are visibly broader than those observed at RHEED maxima. At RHEED intensity minima the width and intensity of the RHEED streaks are increased due to the surface roughness caused by a half filled atomic layer. All the above RHEED features can be explained by assuming that the growth proceeds in a layer by layer mode and that the intensity of the electron diffracted beams is strongly affected by the wave intereference between the two top atomic layers[13]. At RHEED intensity maxima one expects the top atomic layer to be completely filled and at RHEED intensity minima the top atomic layer should be half filled. However some growths especially those grown at substrate temperatures below 350 °C showed less perfect RHEED intensity oscillations. In these growths well defined cusps in RHEED intensity oscillations were usually observed only for the first 2-3 oscillations and then the RHEED intensity oscillations became very sinusoidal. In some cases the RHEED oscillations showed a noticeable attenuation with an increasing film thickness. However in all the samples studied the quality of the RHEED patterns of the Cr overlayers grown on Fe whisker substrates far surpassessed those observed during the growth on bulk single crystal substrates prepared by standard metallurgical means.

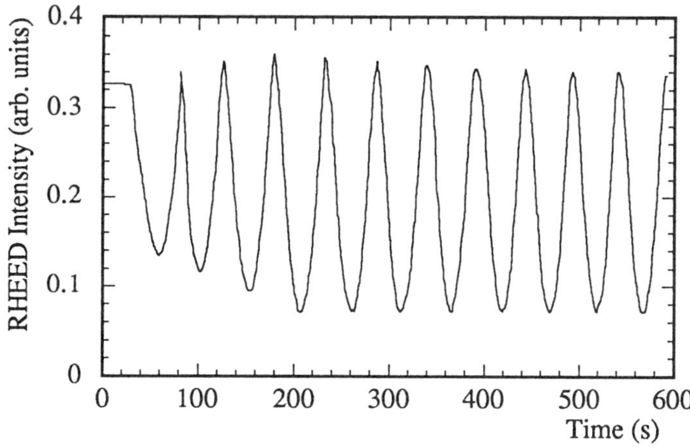

Fig. 3 RHEED intensity oscillations measured at the specular spot during the growth at 350 °C of bcc Cr(001) on bcc Fe(001). The electron beam angle of incidence was 3°, close to the second anti-Bragg condition.

A 6ML Fe ultrathin film was deposited over the Cr layer at substrate temperatures ranging between 30-60 °C. The Fe layer was then covered by a 10ML thick Ag(001) layer. The Ag(001) layer was used to increase the surface uniaxial anisotropy of the Fe(001) layer in order to enhance the difference between the resonance frequencies of the Fe whisker and the Fe(001) ultrathin film. A maximum of 3 samples were prepared at any given growth. Our intention was to determine the exchange coupling in well defined interfaces. The samples were prepared usually by terminating the growth at RHEED intensity maxima, but one sample was grown by terminating the two Cr growths at RHEED intensity minima (13.5 and 14.5 ML on whisker B). Wedged samples were not used intentionally since one is not able to pick accurately the areas (used during the magnetic measurements) corresponding to RHEED intensity maxima and minima respectively.

Magnetic studies

The magnetic studies were carried out using Surface Magneto-Optical Kerr Effect (SMOKE) and Brillouin Light Scattering (BLS). The merits of both techniques and the interpretation of the measured data have been presented in detail in our recent papers[14,15].

bcc Fe/Cu/Fe(001) studies

The studies in the present paper were restricted mostly to samples which maintained an unreconstructed bcc Cu structure (2-12ML thick) and therefore the measured exchange coupling was unobscured by the Cu layer lattice transformation.

Typical SMOKE data for the applied field along one of the in-plane fourfold easy {100} axes are shown in Fig. 4. Two critical fields are clearly visible in Fig. 4. The upper critical field, H_{c1}, corresponds to the field at which the magnetic moments of the individual Fe layers start to deviate from the direction of the dc external field, the second critical field, H_{c2}, corresponds to reaching an antiferromagnetic configuration along the easy axis in which the magnetization in one layer lies along the field direction, and the magnetization in the other layer lies opposite to the field direction (note that we use Fe films of uneven thicknesses). If the magnetization loops are calculated using fourfold anisotropy and only bilinear exchange coupling the main features of the observed loops are preserved, but two obvious differences become apparent. Firstly, the calculated magnetic moment at the critical field H_{c1} shows a well defined jump; a jump in the magnetization is not observed in Fig. 4. Secondly, the observed position of the second critical field H_{c2} is usually at a lower field than that calculated from the value of the exchange coupling determined from H_{c1}. In fact in some samples the second critical field is absent altogether and the total magnetic moment in zero applied magnetic field corresponds to a configuration in which the individual trilayer magnetic moments are nearly oriented along the mutually perpendicular easy magnetic axes ({100}) with the thicker Fe layer oriented along the direction of applied dc field. This observation is not predicted if one uses only the bilinear term. One always needs to decrease the value of the antiferromagnetic exchange coupling with an increasing angle between the magnetic moments in order to bring H_{c2} into agreement with the measured value. This behavior was observed already in our first reported SMOKE measurements on Fe/Cu/Fe samples[9], but it was not until later during the studies of fcc Co/Cu/Cu(001) that we explained[14] the measured magnetization loops using an angular dependent exchange coupling of the form

$$J = J_o - J_1 \cos\theta \tag{2}$$

with positive J_1 lowering the energy for rotation through 90 degrees. Similar behavior was found also by Ruhring[12] et al in Fe/Cr/Fe samples where they introduced the concept of a biquadratic exchange coupling

$$E = - J_1 \sin^2(\theta) \tag{3}$$

Both descriptions are equivalent. Recently Slonczewski identified a possible origin for such behavior. He showed that interface roughness resulting in variations of the interlayer thickness together with short wavelength oscillations (layer by layer) in the exchange coupling resulted in the presence of biquadratic exchange coupling. The strength of the biquadratic exchange coupling is given by[1]

$$J_1 = \frac{2(\Delta J)^2 L}{\pi^3 A} g(f) \sum_{m=1}^{\infty} \left\{ \frac{\coth[\pi(2m-1)\frac{D_1}{L}]}{(2m-1)^3} + \frac{\coth[\pi(2m-1)\frac{D_2}{L}]}{(2m-1)^3} \right\} \tag{4}$$

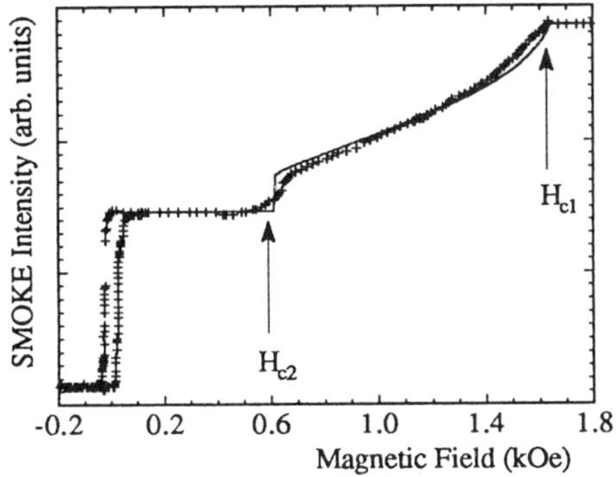

Fig. 4. Typical hysteresis loops measured by SMOKE technique. Sample 9.4Fe/12Cu/16Fe. The applied field lies along the easy magnetic axis {100}. The solid line is a calculated curve using magnetic parameters from Table I.

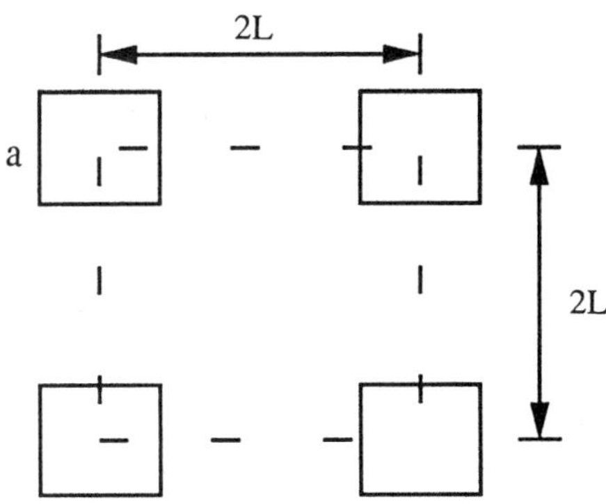

Fig. 5. Schematic diagram of the last layer fractionally covered. Clusters with an average area a^2 are separated by an average distance 2L.

where A is the bulk exchange coupling constant, ΔJ is one half of the change in exchange coupling from odd to even ML (thus it is the slowly varying amplitude of the short period oscillations), D_1, D_2 are the thicknesses of the individual ferromagnetic layers, 2L is the average distance between atomic terraces, see Fig. 5, and g(f) is a function of the partial coverage.

The field dependence of all of our magnetization loops measured by means of SMOKE are better described when biquadratic exchange is included. The biquadratic exchange term removes the jump in the calculated magnetic moment at H_{c1} and moves the second critical field H_{c2} towards lower dc fields. The results of our analyses are summarized in Table I. The values of bilinear and biquadratic exchange couplings obtained by fitting the magnetization loops are listed in columns 2 and 3. The fitting was carried out by assuming that the magnetization process follows the path of minimum energy [16]. For completeness columns 4 and 5 include the analyses in which the measured critical field H_{c1} is interpreted by assuming a bilinear exchange coupling only. Column 4 corresponds to the rotational magnetization process and column 5 corresponds to a magnetization process following the path of minimum energy. The measured trilayers were mostly prepared with the Cu interlayer grown at RT. The substrate temperature during the Cu growth significantly affects the values of J_0 and J_1, compare entries 1 with 2, 4 with 5, and 7 with 8 in Table I. J_1 is increased by growing Cu above RT. The dependence of J_0 and J_1 on Cu layer thicknesses are shown in Fig.6.

In samples grown at elevated temperatures the thickness dependence of the exchange coupling is similar to that observed in Fe/Cu/Fe trilayers which were entirely prepared at RT[17]. However there was a significant difference. The samples with the first Fe layer prepared at ~420 K showed a distinct decrease in magnitude of the antiferromagnetic exchange coupling for an interlayer thickness of 10.5-11ML (-.16 ergs/cm^2). The exchange coupling is stronger at 10ML (-.2 ergs/cm^2) and at 12ML (-.22 ergs/cm^2). All Fe/Cu/Fe trilayers with a Cu interlayer thickness less than 9ML showed only ferromagnetic coupling. Samples with a Cu interlayer thickness equal to or less than 7ML showed only the acoustic FMR peak and therefore possessed strong ferromagnetic coupling.

The rapid variation of the exchange coupling for the Cu interlayers 9-12ML thick, see Fig. 6a, and the presence of the biquadratic exchange coupling in all our measurements, see Fig. 6b, strongly indicates that the exchange coupling in bcc Cu(001) has short-wavelength oscillations. It is therefore reasonable to explore the applicability of the Slonczewski model to our measurements.

Table 1. The results of the exchange coupling as a function of Cu interlayer thickness deduced from SMOKE measurements. All results listed are in units of ergs/cm^2. J_o and J_1 denote the bilinear and biquadratic exchange coupling respectively, see eqs. 1 and 2, and assuming the path of minimum energy using the following parameters: $4\pi M_{eff}$ = 6.08 kG, $2K_1/M_S$ = .31 kOe for a 9.4 ML Fe(001) layer; and $4\pi M_{eff}$ = 15.52 kG, $2K_1/M_S$ = .47 kOe for a 16 ML Fe(001) layer. Bilinear rotation and Bilinear minimum denote exchange couplings which were calculated from H_{c1} based on purely rotational processes or upon magnetization process which follow the path of minimum energy respectively. Easy and hard indicates the easy ({100}) and hard ({110}) in-plane magnetic axes.

Sample and growth temperatures (K)	J_o	J_1	Bilinear rotation	Bilinear minimum
1. 9.4Fe/9Cu/16Fe 420 310 295	-0.059	0.055	-0.169	-0.152
2. 9.4Fe/9Cu/16Fe 420 340 295	-0.015	0.10	-0.215	-0.207
3. 9Fe/10Cu/16Fe 420 295 295	-0.084	0.060	-0.204	-0.195
4. 9.4Fe/11Cu/16Fe 420 330 295	-0.074	0.050	-0.174	-0.159
5. 9.4Fe/12Cu/17Fe 400 295 295	-0.198	0.017	-0.233	-0.220
6. 9.4Fe/12Cu/16Fe 430 325 295	Easy -0.237 Hard -0.221	0.027 0.035	-0.291 -0.283	-0.289 -0.281
7. 9.4Fe/14Cu/16Fe 420 295 295	-0.103	0.012	-0.137	-0.110

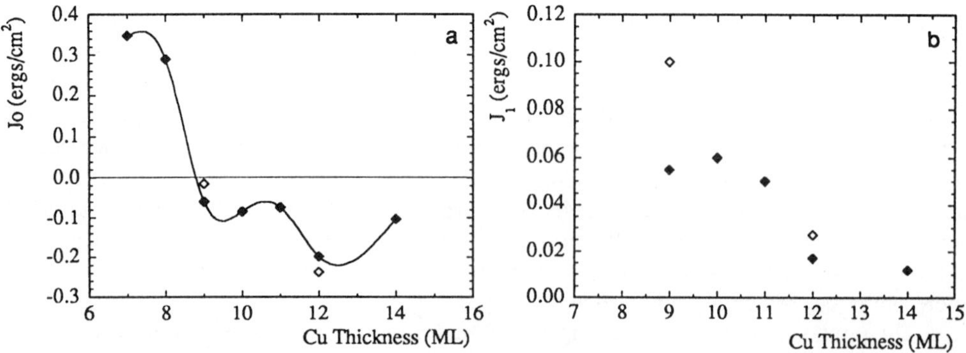

Fig. 6. (a) The bilinear exchange coupling, J_0, vs. Cu interlayer thickness (ML). (b) The biquadratic exchange coupling, J_1, vs. Cu interlayer thickness. For 9 and 12 ML Cu two samples were prepared: the open symbols correspond to samples which were grown at raised Cu substrate temperatures (see Table I)

Most of the Fe layer and Cu interlayer growths were terminated at the RHEED intensity maxima. We assume that their corresponding interfaces can be described by two atomic levels, see also the section on Growth Studies. The Cu interlayer "N ML" thick has its interface formed by a completely filled N^{th} layer but the N+1 layer is partially filled with mesas. Mesas occupy a fractional area $f=a^2/(2L)^2$, where a^2 represents the average area of the mesas and 2L represents the average distance between mesas, see Fig. 5. The growth studies showed that it is the second Cu/Fe interface which has smaller terraces and therefore the above growth parameters describe the Cu/Fe interface. The Slonczewski model includes several parameters. The variations in the exchange coupling are described by a parameter $\Delta J=(J(N+1)-J(N))/2$ where J(N) represents the intrinsic exchange coupling through N atomic layers of Cu which would be measured in the absence of interface roughness. The interface roughness is described by two sets of parameters: (1) The parameters L(N) describe the length scale in which atomic terraces are distributed across the interface, see Fig.5. (2) The parameters f(N) are fractional coverages for each thickness.

The idea is to identify the intrinsic bilinear exchange coupling J(N) from measured values of $J_0(N)$ and $J_1(N)$. In order to do this we need additional information for f(N) or L(N). The parameter f(N) can be estimated from the RHEED intensity oscillations. The RHEED intensity oscillations allow one to determine the overlayer thickness. We have found in all our growth studies that the period of the RHEED oscillations corresponding to 1ML formation (stationary period) is observed only after several initial oscillations. The first few oscillations involve a phase slip which results in a longer period. Therefore the total thickness of a given film in ML can be determined by dividing the total growth time (for constant evaporation rate) by the stationary period. Using this approach one finds, see Fig.1, that the Cu interlayer terminated at the Nth RHEED maximum corresponds to the thickness N+.5 (in ML); and therefore analyses were carried out using f=.5 for all Cu interlayers.

The measured bilinear exchange coupling is given by

$$J_0(N) = J(N) + \{ J(N+1) - J(N) \}f, \tag{5}$$

and the biquadratic exchange coupling is given by formula (4). In order to determine J(N) and L(N) one has to identify one additional parameter. The bilinear exchange coupling strength is slowly varying for Cu interlayer thicknesses of 10-11ML, see Fig. 6a. Therefore we used L for 10ML thick Cu as an additional parameter (L(10)) which allowed us then to determine the values of J(N) and L(N) for all the other samples measured. Using equations (4) and (5) one can determine values for J(10) and J(11). This approach then allows one to propagate the analysis. Equation (5) determines J(N) and equation (4) allows us to evaluate L(N) for a given Cu interlayer. The results of such an analysis are summarized in Table II for L(10)=150 Å. The exchange coupling J(N) shows an oscillatory behavior, see Fig. 7. The parameter L decreases slowly with an increasing Cu thickness. Note that the exchange coupling J(13) is less than J(12) as expected from the measured overall thickness dependence of the exchange coupling. This shows that the choice f=.5 is very realistic.

The exchange coupling, J(N), corresponding to a perfect interface shows strong short wavelength (~2.2 ML) oscillations. The Philips group observed also 2ML oscillations through bcc Cu(001) (d_{Cu}>10ML) in their SMOKE studies using Fe(001) whisker/Cu/Fe(001) samples[18]. The maximum of antiferromagnetic coupling was found at an even number of Cu ML in agreement with our results. Herman[19] et al. have carried out first principles calculations of the exchange coupling through bcc Cu in Fe/Cu/Fe structures. The thickness dependence of the exchange coupling, see Fig. 1 in reference (19), shows a behavior similar to that observed in our measurements, Fig. 6a. This similarity is even more obvious if their calculations are compared with Figs. 7. The first principles calculations by Herman et al show that the exchange coupling crosses from ferromagnetic to antiferromagnetic coupling at d_{Cu}= 8Å, and then a rapid variation in the antiferromagnetic coupling occurs with a separation between antiferromagnetic maxima of 3.4 Å (2.4ML). The antiferromagnetic coupling strength reaches

Table II The exchange interaction and average terrace separation 2L for "structurally perfect" Fe/Cu/Fe trilayers as a function of the Cu interlayer thickness. The pair of J(N) and L(N) were evaluated by using 2L(10)=300Å.

Cu Thickness (ML)	f	J(N) ($\frac{ergs}{cm^2}$)	2L(Å)
7.0	0.50	.59	–
8.0	0.50	.11	–
9.0	0.50	.476	270
10.0	0.50	-.594	300
11.0	0.5	.426	270
12.0	0.50	-.574	210
13.	0.50	.178	–

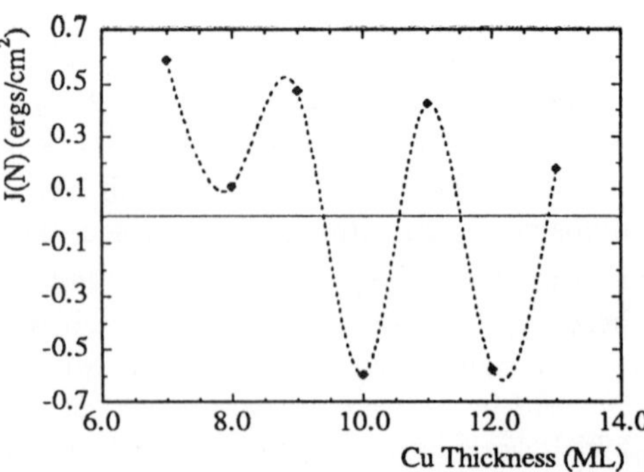

Fig. 7. Values deduced for the thickness dependence of the exchange coupling J(N) which pertain to a smooth interface. These coupling strengths have been calculated from the data using the Slonczewski model[1] assuming f=.5 for all Cu interlayers, and 2L(10)=300 Å (see the text). The solid line is a cubic spline fit to guide the eyes.

maximum values near 8Å and 12Å with a slightly positive ferromagnetic coupling in between (at 10Å). Our values for J(N) show a similar trend. The measured periodicity (2.2ML) of short wavelength oscillations is very close to that obtained from the first principle calculations by Herman[19] (2.4ML). The significant difference between the measured results and the calculations appears to be in the first crossover from ferromagnetic to antiferromagnertic coupling. The measurements show that this first crossover occurs for 9-10ML thick Cu. The first principles calculations show the crossover at approximately 5.5 ML of Cu. We do not intend at this point to discuss this difference.

The strength of the oscillatory exchange coupling J(N), see Fig. 7, is similar to that calculated by Bruno and Chappert[20] for fcc Cu(001) (.4 ergs/cm^2). The first principles calculations for bcc Fe/Cu/Fe(001) by Herman[19] et al. gave significantly larger values for the exchange coupling (25. ergs/cm^2). Deaven[21] et al showed that it is difficult to obtain correct quantitative results for the amplitude of the oscillations but that the period of the oscillations is usually well described by present theoretical methods.

Recent calculations by Edwards[22] showed that the biquadratic exchange coupling can be also generated by an intrinsic mechanism. In model calculations it is predicted that it can be equal to approximately 1/8 (12%) of the bilinear contribution. In our samples the biquadratic exchange coupling is very thickness dependent and can be comparable to the bilinear counterpart. Slonczewski's mechanism is definitely a major contributor to the strength of the biquadratic exchange coupling. However for those samples which exhibit a small biquadratic exchange coupling part of this coupling term could be intrinsic in origin. The analysis of the data would not be changed in an appreciable way. The Slonczewski's contribution would be decreased by the contribution arising from the intrinsic contribution and this would require a decrease of the terrace separation 2L. One can try to estimate the value of the intrinsic biquadratic exchange by choosing a reasonable value for the terrace separation. An acceptable choice of 2L=100 Å for a sample with a 12ML Cu thick interlayer would require a decrease of the Slonczewski contribution by a factor 2 and therefore approximately half of the measured biquadratic exchange coupling for the 12ML Cu sample could be ascribed to the intrinsic contribution (.008 ergs/cm^2). This would suggest an intrinsic biquadratic exchange coupling which is approximately 5% of the bilinear contribution.

bcc Fe(whisker)/Cr/Fe(001) Studies

We have measured the frequency shift of 0.5145 mm light back-scattered from composite specimens consisting of an Fe(001) iron whisker single crystal substrate upon which was grown Cr(001) layers of various thicknesses, followed by the deposition of 6 monolayers (ML) of Fe(001). This sandwich structure was covered by 10 ML of Ag(001) plus 10 ML of Au(001). The gold overlayer protected the underlying layers of metal from becoming oxidized when the specimen was removed from the ultra-high vacuum preparation facility.

Thin film specimens were prepared using 4 different whisker substrates; these have been designated A,B,C, and D. The whisker surface preparation and the thin film depositions have been described in the section on Growth Studies. In most instances two different Cr layer thicknesses were deposited during one growth cycle; in one instance a specimen was prepared in which 3 different Cr film thicknesses were grown on the same whisker (11,12, and 13 ML).

The light scattering spectra were measured using 0.5145 mm single mode incident laser light and a standard 4+2 pass Sandercock interferometer[23] operated in the back-scattering configuration. An angle of incidence of 45° was used for most of the measurements. In some cases spectra were measured for angles of incidence ranging from 2.5° to 70°; frequency shifts with angle of incidence were less than 1 GHz. The incident light intensity at the specimen was approximately 120 mW. No dependence of the frequency of the signals on incident power could be observed for power levels ranging from 50 to 150 mW. The frequency interval between the Rayleigh peaks, the free spectral range (FSR), was divided into 112 equally spaced channels. Most of the data were obtained using FSR= 100 GHz so that each channel corresponded to a frequency interval of 0.9 GHz. Surface mode peak positions (see Fig.8) were found by fitting the data to an empirical linewidth function by means of least squares. The functions used were a combination of a straight line (in order to account for a slowly varying background) and either a Lorentzian or a Gaussian lineshape. Values obtained for a peak frequency were the same for the Lorentzian and the Gaussian fitting functions. It is estimated that the peak frequency corresponding to a surface mode (labeled SM in Fig.8) could be determined with a precision of ±0.1 GHz.

The scattered light signals measured using the iron whisker substrates were found to be very weak compared with signals measured for iron films grown on silver single crystal substrates. The thin film signal measured for the 36 ML Cr spacer layer (see Fig.8), the most intense thin film signal which was observed, was approximately 20-fold less intense than the signal observed for a 9 ML Fe(001) layer grown on a bulk Ag(001) substrate and covered by 7 ML of Ag(001) plus 10 ML of Au(001). For ease of comparison, the data shown in Fig.8 for each magnetic field value have been converted to the number of counts in each channel corresponding to 500 scans for an incident laser power of 100 mW incident on the specimen (approximately 170 mW output at the laser in our case). We used a standard dwell-time of 2 msec per channel so that 500 scans corresponds to a collection time of 1 second per channel; the total time required to collect data for 500 scans was 4.3 minutes. In many instances, particularly for field values greater than ~5 kOe, data were collected using 2000 to 3000 scans.

Results of the BLS measurements

Graphs of scattered light intensity vs. frequency shift for several values of applied magnetic field are shown in Fig.8 for a 36 ML thick (51Å) Cr(001) spacer layer. The applied magnetic field was aligned along a cube axis in the plane of the whisker surface. The three very strong lines at -100, 0, and +100 GHz correspond to the unshifted laser frequency; they are a consequence of a relatively strong background signal caused by imperfections in the specimen surface. Two weak but prominent signals can be seen on the right-hand side of the figure corresponding to light that has been scattered with an increase in frequency. The feature that has been labeled TF corresponds to light that has been up-shifted due to having been scattered from a spin-wave in the 6 ML Fe(001) film that was grown on the 36 ML thick

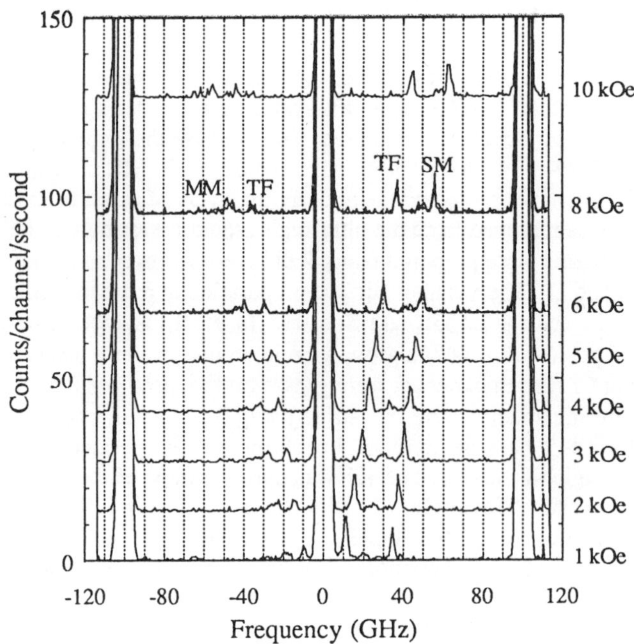

Fig.8. BLS intensity vs frequency shift measured using bulk iron whisker substrate C which supported a 6 ML Fe(001) film separated from the bulk by a Cr(001) layer 36 ML thick. Data for eight different magnetic fields are shown in the figure; the data for different fields have been displaced by an offset proportional to the field strength. The data were obtained using 0.5145 mm laser light incident at 45°, and using an f-2 collection lens in the back-scattering configuration. The data have been normalized to an incident power of 100 mW and a data collection time of 1 second per channel; 224 channels equals 200 GHz. SM- the bulk surface mode; TF- the thin film modes; MM- the bulk spin-wave manifold.

Cr(001) spacer layer. The feature labeled SM corresponds to light that has been up-shifted in frequency as a result of having been scattered by a surface spin-wave in the bulk whisker substrate. Between the SM and TF peaks there is another weaker feature that is due to light scattered by spin-waves near the edge of the bulk spin-wave manifold in the whisker. This feature has a counterpart on the left-hand side of the figure (labeled MM in Fig.8). The bulk spin-wave edge feature is more pronounced on the frequency down-shifted side of the spectrum because there is no surface mode to drain away oscillator strength. The intensity of the thin film peak corresponding to down-shifted light frequencies is weaker than the thin film peak corresponding to up-shifted frequencies. Similar features were observed for all of the specimens which we examined, although the intensities and frequencies of the thin film peaks varied a great deal from one specimen to another. The strength of the surface mode was found to be relatively constant from specimen to specimen and from run to run, although variations of a factor two in intensity were not uncommon due to the difficulty in maintaining the interferometer alignment over a long time interval and the difficulty in establishing the optimum position of the whisker relative to the focal point of the collection lens. The strength of the surface mode peak served as a built-in standard against which the intensities of the thin film peaks could be measured.

The variation of thin film frequencies with Cr film thickness has been plotted in Fig.9 for an applied field of 5 kOe. The thin film frequencies appeared to be reproducible from one specimen to another for Cr interlayer thicknesses greater than 12 ML. See in particular the data points in Fig.9 for 12 and 13 ML thick Cr interlayers for which each point represents a different growth; three different whiskers were used to obtain the four data points shown for

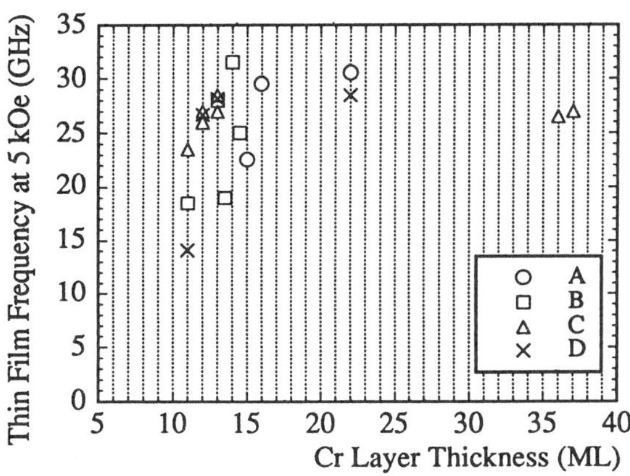

Fig.9. Frequencies of the thin film spin-wave mode corresponding to an applied magnetic field of 5 kOe measured for 6 ML Fe(001) films grown on a Cr(001) spacer layer of variable thickness which was itself grown on a bulk Fe(001) single crystal whisker. Data for four different whiskers are shown in the figure: these substrates are labeled A,B,C, and D. An exchange coupling of 0.10 ergs/cm^2 between the thin film and the substrate would produce a frequency shift of ~2.2 GHz.

the 13 ML thick Cr spacer layer. The frequencies associated with the data points for 13,14,15,and 16 ML appear to oscillate with the two ML period reported by Unguris[5] et al and by Purcell[6] et al. However, the data for Cr thicknesses less than 12 ML do not follow this trend, and the character of the magnetic field dependence of the thin film peak frequencies for 10 and 11 ML thick interlayers was quite different from that observed for the thicker Cr layers (see below).

In one instance the Cr growth was deliberately terminated at a RHEED oscillation intensity minimum so as to produce a rough Cr layer: two different Cr thicknesses were deposited on whisker C and these corresponded to layers $13\frac{1}{2}$ and $14\frac{1}{2}$ ML thick. The resulting thin film frequencies were considerably reduced compared with the thin film frequencies measured using 13 and 14 ML thick Cr layers grown on the <u>same whisker substrate.</u> The 13 and 14 ML growths were terminated on a RHEED intensity maximum, and it must be emphasized that the quality of growth of the Cr layers, and of the 6 ML Fe(001) layers grown on the Cr, was very similar for both the 13,14 ML and the $13\frac{1}{2}$, $14\frac{1}{2}$ ML specimens. We believe that an increase in specimen roughness leads to a decrease in thin film frequency (the exchange coupling between the thin iron film and the substrate becomes more strongly anti-ferromagnetic).

Data obtained for other applied fields are similar to that shown in Fig.9, and in most cases the thin film modes, the bulk manifold edge frequencies, and the bulk surface mode frequency all exhibited a monotonic increase with increasing magnetic field, and a relatively slow decrease in intensity with increasing magnetic field. This standard pattern was broken by the specimens corresponding to a 10 ML spacer layer grown on substrate B, and by the 11 ML spacer layers grown on substrates B, and D. In these three cases the frequency of the thin film mode displayed relatively little dependence upon applied magnetic field, see Fig.10. The intensity of the thin film peaks for these specimens depended very strongly on applied field; the intensities were strong at low fields but fell off at higher fields at a rate which depended upon the particular specimen. For example, the intensity of the thin film peak for 10 ML of Cr(001) grown on substrate B became too small to measure for fields greater than 3 kOe, but thin film peak intensities for the 11 ML spacer layer grown on whisker B could be measured over the

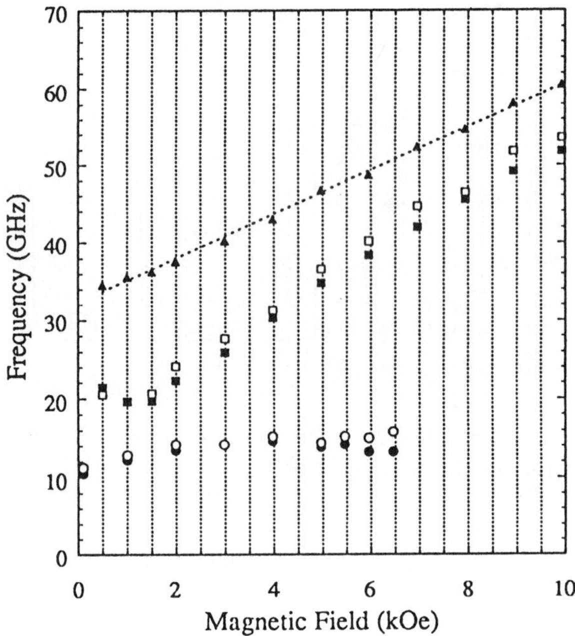

Fig.10. The magnetic field dependence of the BLS frequencies observed for a 6 ML Fe(001) film grown on an 11 ML Cr(001) spacer layer which was grown on whisker substrate D. The data points shown correspond to the bulk surface mode ▲ , the up-shifted bulk spin-wave manifold edge □, the down-shifted bulk manifold edge ■, the up-shifted thin film signal ○ , and the down-shifted thin film signal ● . The intensities of the thin film modes become too small to measure for fields greater than 6.5 kOe.

entire range from 1 to 10 kOe: it should be noted that these two specimens were prepared during the same growth cycle. For the case shown in Fig.10, specimens were prepared on whisker D during the same growth cycle which contained both an 11 ML and a 22 ML Cr(001) spacer layer; data for the 11 ML spacer layer are shown in the figure. The thin film peaks in this case became too weak to measure for fields greater than 6.5 kOe. The intensity of the peak at 6.5 kOe had fallen to 1/5 of its value at 1 kOe; the width in frequency of the thin film peak increased slowly with magnetic field, but the peak at 6.5 kOe was no more than twice as wide as the peak at 1 kOe. It should be noted that the behaviour of the thin film peaks measured for the 22 ML thick Cr layer region on this specimen was very similar to that which was observed for a 22 ML thick Cr interlayer grown on whisker A.

The frequencies associated with the bulk surface mode were found to be relatively constant from specimen to specimen, although variations of ±1 GHz were not uncommon. In one experiment the frequency of the surface mode was measured as a function of position along the length of the whisker; the results are shown in Fig.11. For this experiment, Cr layers 22 ML and 11 ML thick were deposited on whisker D during the same growth cycle. It is quite clear from Fig.11 that the surface mode frequency is sensitive to the Cr layer at the level of ~1 GHz; the transition from the 22 ML to the 11 ML region is clearly apparent. However, it is a most interesting result that the surface mode frequency varied with position within the region which was covered by the nominal 11 ML. This observation is very likely related to the peculiar magnetic field dependence of the thin film frequency that was observed for the 10 and 11 ML thick Cr interlayers.

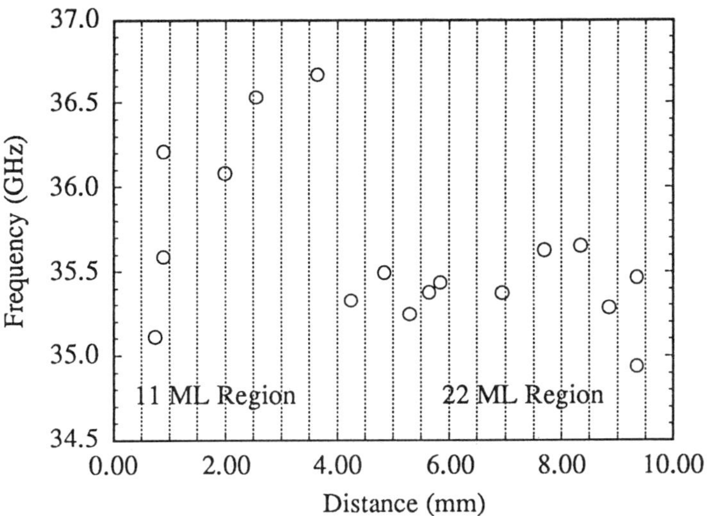

Fig.11. The frequency of the bulk surface mode for a magnetic field of 1 kOe measured as a function of position along whisker D. The region from 0-4 mm carried an 11 ML Cr(001) spacer layer upon which was grown a 6 ML Fe(001) film. The region from 4-10 mm carried a 22 ML Cr(001) spacer layer upon which was grown a 6 ML Fe(001) film.

Comparison with a Simple Model

We have used a theory similar to that described by Camley and Mills[24] to calculate the magnetic field dependence of the intensity of light scattered from a composite sample made up of a magnetic slab, infinitely thick, upon which are deposited a non-magnetic spacer layer of thickness d_s, a thin magnetic film of thickness d, and a non-magnetic cover layer of thickness d_1. The magnetic materials are assumed to be cubic with the surface normal oriented along a cube axis; the cubic axes of the magnetic film and the bulk magnet are assumed to be parallel. The magnetic regions are characterized by a four-fold cubic magnetic anisotropies that may be different for the thin film and the bulk. The thin film is assumed to interact with the bulk magnet via a surface contact exchange interaction of the form

$$E = - J \cos\theta_{12} \quad (6)$$

where θ_{12} is the angle between the thin film magnetization and the magnetization at the surface of the bulk slab. The magnetization in the thin film is assumed to be uniform across the film thickness; this is an excellent approximation for films thinner than ~30 Å because the exchange interaction makes a spatial variation on such a small scale very costly in energy[25-29]. Both the thin film and the bulk have been endowed with a uniaxial magnetic surface energy term of the form

$$E_s = - K_u \cos\theta \quad (7)$$

where θ is the angle between the magnetization and the surface normal. For positive K_u the surface energy corresponds to a torque which is such as to turn the magnetization out of the plane and into the direction of the surface normal. The numerical value of K_u for iron depends upon the metal with which it is in contact. For bulk Fe(001) covered with Ag[30,31] K_u= 0.7 ergs/cm^2 , and for the Fe(001)-Au interface[30,31] K_u= 0.45 ergs/cm^2. Similar values have been reported for ultra-thin iron films[30,31]. The surface energy term, eqn.(7), results in some surface pinning of the bulk spin-wave modes, but this pinning has only a small effect on the frequency associated with the surface mode (SM) and with the spin-wave manifold edge (MM) (~0.6 GHz for a change in surface energy from 0 to 1 $ergs/cm^2$). The effect of the surface energy term is more pronounced in the case of the thin film mode. The surface torque due to the surface energy term effectively acts on all of the spins across the film thickness because the exchange interaction is sufficiently strong to hold all of the spins parallel across a film that is only 6 ML thick. As a result, the thin film behaves as if it possessed an effective magnetization[25-29] given by

$$4\pi M_{eff} = 4\pi M_S - (2K_{UA}/dM_A) \quad (8)$$

where d is the film thickness, M_A is its saturation magnetization, and K_{UA} is the sum of the surface energies for the two film surfaces. As an example, if K_{UA}= 1 erg/cm^2, M_A= 1.7 kOe, and d= 6 ML (8.6 Å for an iron film) the surface energy term in (8) contributes 13.7 kOe to the effective magnetization. The uniaxial surface energy shifts the thin film resonant frequency to a value which is less than the edge of the bulk spin-wave manifold since, approximately,

$$(\omega/\gamma)^2 = H\,(H + 4\pi M_{eff}). \quad (9)$$

A calculation of scattered light intensity vs frequency is shown in Fig.12 for an applied magnetic field of 10 kOe and zero coupling between the film and the bulk. This model calculation uses realistic magnetic parameters for iron: the bulk surface energy parameter has been rather arbitrarily assigned the value 0.5 $ergs/cm^2$ since the characteristic surface energy for the iron-chromium interface is not known; the saturation magnetization for iron has been taken to be 21.2 kOe in order to secure agreement between the observed and calculated surface mode frequencies; and, finally, the surface energy for the film has been chosen to be K_{UA}= 0.73 $ergs/cm^2$ in order that the calculated thin film peak frequency agree with the observed frequency (44 GHz). The model calculation reproduces the features observed in the BLS experiment, compare Figs.8 and 12. The signal strength has been calculated taking into account the collection lens optics[28,32], the finite frequency interval associated with a collection channel (0.89 GHz), and the instrumental broadening associated with the Fabry-Perot interferometer. The intensity calculated for the surface mode peak at 10 kOe is 1850 counts/channel/sec for an incident power of 100 mW. The observed surface mode intensity at 10 kOe (Fig.8) is ~10 counts/channel/sec. However, the calculated intensity must be reduced to take into account optical losses in the interferometer and to take into account a photomultiplier efficiency of ~10%. These two corrections together reduce the calculated intensity by a factor of 200 for our 4+2 pass Fabry-Perot system. Upon applying this correction the calculated surface mode peak intensity becomes 9 counts/channel/sec, a figure which is in reasonable agreement with the experimental observations considering that the calculations were based upon the assumption of perfectly smooth interfaces.
The model calculation indicates that the thin film frequency should be very insensitive to the properties of the bulk substrate. It is, however, very sensitive to the strength of the exchange coupling parameter J (see eqn.(6)) and to the strength of the surface energy term K_{UA} (see eqn.(7)). The variability of the thin mode frequencies for Cr interlayers which are nominally the same would appear to indicate that the surface energy parameter, or the exchange coupling parameter, or both, are sensitive to the details of the Cr interlayer structure or composition; an intermixing of the iron and chromium can not be ruled out. Moreover, the peculiar dependence of the thin film frequency and mode intensity on applied magnetic field observed for the 11 ML Cr interlayer specimens cannot be explained by our simple model. More experiments are required to clarify the factors which influence the exchange coupling through thin Cr films.

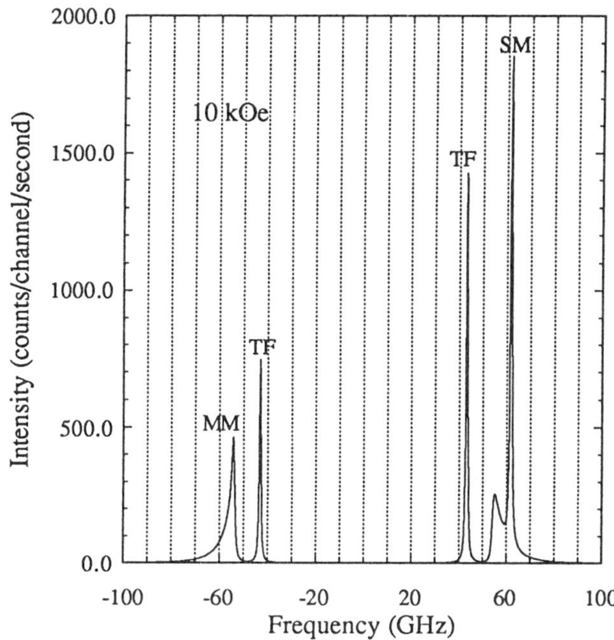

Fig.12. Calculated BLS intensity at 10 kOe as a function of frequency for a bulk iron Fe(001) substrate covered by a 36 ML Cr(001) spacer layer (51.5 Å thick), plus a 6ML Fe(001) iron film (8.6 Å thick) and a 20 ML gold overlayer (28.6 Å). The exchange coupling between the thin iron film and the bulk iron substrate was assumed to be zero. Magnetizations of $4\pi M_S$= 21.2 kOe were used both for the film and the substrate, and their g-factors were set equal to g= 2.09. The Gilbert damping parameters were taken to be 0.7×10^8 Hz and 1.8×10^8 Hz for the bulk and thin film. The bulk exchange stiffness parameter A= 2.0×10^{-6}. The first order magnetocrystalline anisotropy constants used were $K_1 = 4.76 \times 10^5$ ergs/cc for the bulk and $K_1 = 1.14 \times 10^5$ ergs/cc for the thin film. The bulk uniaxial surface energy parameter was taken to be K_{UB}= 0.5 ergs/cm^2; the thin film uniaxial surface energy parameter was taken to be K_{UA}= 0.73 ergs/cm^2 in order to obtain the observed thin film frequencies (TF) of -43 and +44 GHz. The surface mode frequency (SM) is 62 GHz, and the bulk spin-wave manifold frequency (MM) is 54 GHz. The intensity corresponds to an incident laser power of 100 mW, an f-2 collection lens, and data collected for 1 second in channels whose frequency widths are 0.89 GHz. The optical dielectric constants used were: Fe[33]: ε= 0.40+16.41i; Cr[33]: ε= -2.67+19.19i; Au[34]: ε= -3.55+2.93i. The optical coupling parameter used[35] was K= 0.31+0.24i, where $\Delta(4\pi \mathbf{P}) = K(\mathbf{E} \times \frac{\mathbf{M}}{M_S})$.

Conclusions

It has been shown that the average atomic terrace width in Fe/Cu/Fe trilayers can be significantly increased by increasing the substrate temperature during the growth of the first Fe layer and that Cr(001) layers can be grown with a surface perfectness close to that of Fe whiskers.

The exchange coupling in Fe/Cu/Fe trilayers was studied as a function of the Cu interlayer thickness. The studies were carried out using the SMOKE technique. The magnetization loops for Fe/Cu/Fe trilayers can only be well explained by including biquadratic exchange coupling. Values for the bilinear and biquadratic exchange coupling terms were determined from a detailed analysis of the observed magnetization loops. It has been shown that contrary to intuitive expectations the strength of the biquadratic exchange coupling term increases with increasing terrace width.

A quantitative analysis of the measured bilinear and biquadratic exchange coupling contributions was carried out using the Slonczewski model[1] for biquadratic exchange coupling. It has been shown that the exchange coupling J(N) for N perfectly smooth monolayers exhibits an oscillatory behavior. Its amplitude has been estimated. The observed thickness dependence of the exchange coupling has been compared with the recent first principle calculations by Herman[4] et al. The measured period (2.2ML) of short wavelength oscillations is very close to that calculated in the first principle calculations by Herman[4] et al (2.4ML).

The thickness dependence of the exchange coupling through Cr(001) layers has been investigated by means of BLS using films grown on single crystal whisker substrates. A theory of the frequency dependence of the BLS intensity has been worked out for a complex structure consisting of a thin ferromagnetic film separated from a bulk ferromagnetic substrate by a non-magnetic spacer layer; this trilayer is covered by a second non-magnetic overlayer. The theory gives a good description of the light scattering data for Cr spacer layer films thicker than 12 ML. Cr spacer layers thinner than 12 ML give rise to thin film BLS peaks whose frequency varies more slowly with applied magnetic field, and whose intensity falls off more rapidly with increasing applied field than is predicted by the model calculations.

Acknowledgement

The authors would like to thank the Natural Sciences and Engineering Research Council of Canada for grants that supported this work.

References

1. J. C. Slonczewski, Phys. Rev. Lett., **67**, 3172 (1991).
2. M. Rührig, R. Schäfer, A.Hubert, R. Mosler, J.A. Wolf, S. Demokritov, and P. Grünberg, Phys.Status Solidi A **125**, 635 (1991).
3. B. Heinrich, K.B. Urquhart, J.R. Dutcher, J.F. Cochran, A.S. Arrott, D.A. Steigerwald, W.F. Egelhoff, Jr., J.Appl.Phys. **63**,3863,(1988).
4. A.S. Arrott, B. Heinrich and S.T. Purcell, Kinetics of Ordering and Growth at Surface, ed. by M.G. Lagally, Plenum Press, New York, (1990), p.321.
5. J. Unguris, R.J. Cellota, D.T. Pierce, Phys.Rev.Lett., **67**, 140, (1991).
6. S.T. Purcell, W. Folkerts, M.T. Johnson, N.W.E. McGee, K. Jager, J.aan deStegge, W.B. Zeper, W. Hoving, P. Grünberg, Phys.Rev. Lett. **67**, 903, (1991).
7. B. Heinrich, A.S. Arrott, J.F. Cochran, Z. Celinski and K. Myrtle, NATO Advanced Study Institute on the Science and Technology of Nanostructured Magnetic Materials, Greece 1990, eds. G.Hajipanais and G.A. Prinz, Plenum Press, New York, 1991, p.15.
8. P. Hahn, J. Clabes, M. Henzler, J. Appl. Phys. **51**, 2079 (1980).
9. B.Heinrich, Z.Celinski,J.F.Cochran, W.B. Muir, J.Rudd, Q.M. Zhong, A.S.Arrott, K.Myrtle, and J. Kirschner, Phys. Rev. Lett., **64**, 673 (1990).
10. B. Heinrich, S.T. Purcell, J.R. Dutcher, J.F. Cochran, and A.S. Arrott, Phys. Rev. **B 38**, 12879 (1988).
11. D.J. Jiang, N. Alberding, A.J. Seary, B. Heinrich, and E.D. Crozier, Physics B **158**, 662 (1989).
12. S.T. Purcell, A.S. Arrott, B. Heinrich, J.Vac.Sci.Techn. **B6**, 794, (1988).
13. J.M. Van Hove, P.R. Pukite, P.I. Cohen, J.Vac.Sci.Techn. **B3**, 563, (1985).

14. B. Heinrich, J. Kirschner, M. Kowalewski, J.F. Cochran, Z. Celinski, and A.S. Arrott, Phys. Rev. **B 44**, 9348 (1991).
15. J.F. Cochran , J.Rudd, W.B.Muir, B.Heinrich, and Z.Celinski, Phys.Rev.**B42**, 508 (1990).
16. B. Dieny, J.P. Gavigan, and J.P. Rebouillat, J. Phys. C2, 159 (1990); B. Dieny and J.P. Gavigan, J. Phys. **C2**, 187 (1990); Meeting, Anaheim (1991), to be published.
17. Z. Celinski and B. Heinrich, J. Mag. Mag. Mater., **99**, L25 (1991).
18. R. Coehoorn, M.T. Johnson, W. Folkerts, S.T. Purcell, M.A.M. Gijs, A. de Veirman, Y. Suzuki, Magnetism and Structure in Systems of Reduced Dimension, NATO Advanced research Workshop, June, 1992, Cargese, France.
19. F. Herman, J. Sticht, and M. van Schilfgaarde, Proc. Spring MRS, 1991.
20. P. Bruno, C. Chappert, Phys.Rev.Lett. **67**, 1602, (1991).
21. D.M. Deaven, D.S. Rokhsar, and M. Johnson, Phys.Rev. B **44**, 5977 (1991).
22. D.M. Edwards , presented at the NATO Advanced Research Workshop on "Magnetism and Structure in Systems of Reduced Dimension", Cargese, France, June, 1992.
23. J.R.Sandercock. Topics in Applied Physics, Vol.51: Light Scattering in Solids III. Edited by M.Cardona and G.Güntherodt. Springer-Verlag, Berlin, 1982. Chpt.6.
24. R.E.Camley and D.L.Mills, Phys.Rev.B18, 4821 (1978).
25. B.Heinrich, S.T.Purcell, J.R.Dutcher, K.B.Urquhart, J.F.Cochran, and A.S.Arrott, Phys.Rev.**B38**, 12879 (1988).
26. R.P.Erickson and D.L.Mills, Phys.Rev.**B43**, 10715 (1991).
27. D.L.Mills in Ultrathin Magnetic Structures, edited by J.A.C.Bland and B.Heinrich. To be published by Springer-Verlag.
28. J.F.Cochran in Ultrathin Magnetic Structures, edited by J.A.C.Bland and B.Heinrich. To be published by Springer.
29. B.Heinrich in Ultrathin Magnetic Structures, edited by J.A.C.Bland and B.Heinrich. To be published by Springer.
30. B.Heinrich, J.F.Cochran, A.S.Arrott, S.T.Purcell, K.B.Urquhart, J.R.Dutcher, and W.F.Egelhoff,Jr., Appl.Phys.**A49**, 473 (1989).
31. J.F.Cochran, B.Heinrich, A.S.Arrott, K.B.Urquhart, J.R.Dutcher, and S.T.Purcell, J.de Physique, Colloque C8, Supplement au no.12,**49**, C8-1671 (1988).
32. W.Wettling, M.G.Cottam, and J.R.Sandercock, J.Phys.**C8**, 211 (1975).
33. P.B.Johnson and R.W.Christy, Phys.Rev.**B9**, 5056 (1974).
34. P.Joensen, J.C.Irwin, J.F.Cochran, and A.E.Curzon, J.Opt.Soc.Amer.63, 1556 (1973).
35. G.S.Krinchik and V.A.Artem`ev, Zh.Eksp.Teor.Fiz.,**53**,1901 (1967). (English transl:Soviet Physics JETP,**26**, 1080 (1968).)

WHAT RARE EARTHS TEACH ABOUT THIN FILM MAGNETISM

C. P. Flynn and F. Tsui

Physics Department
University of Illinois
Urbana, Ill. USA 61801

INTRODUCTION

Rare earth metals exhibit rich and widely investigated magnetic behavior in the bulk. Comprehensive reviews are offered by Elliott (1972), Coqblin (1977) and Jenson and Mackintosh (1991). The 4f core electrons remain distinct from the conduction states and couple according to Hund's rules to form model magnetic materials, at least in simple compounds. The metallic state entails certain interesting complications. In particular, a portion of the conduction band usually has d symmetry and the ionic moment is somewhat modified. Second, the moments disturb the conduction electrons and this results in the famous 'RKKY' exchange coupling among the moments. There is, in addition, a coupling of the ionic charge distribution to the crystal field, which in the materials of interest here is associated with a hexagonal close packed crystal structure.

These characteristics make rare earths excellent materials with which to pursue model magnetic behavior in metals. The properties are better understood than those of the 3d transition metals, on which this conference is mainly focused, and may often be simulated by relatively simple model Hamiltonians. All the more is this useful in the case of thin film systems where additional complications of geometry, dimensionality, epitaxial deformations, and so on, lead to new and often still poorly understood modifications of bulk behavior. The purpose of the present paper is to provide a brief summary of the insights rare earth systems offer about the specific consequences of thin film geometry for magnetic properties.

Three particular areas of thin film studies are summarized in the text that follows. They are **interlayer coupling**, which includes the RKKY coupling of successive layers in multilayer stacks, **monolayer magnetism** and the result of reduced dimensionality on spontaneous magnetism, and the consequences of **lattice clamping and epitaxial strain,** introduced by the epitaxial constraint, on the resulting thin film magnetism. These three areas of ongoing research effort are discussed consecutively in what follows.

INTERLAYER COUPLING

The RKKY interaction was first recognized for rare earths (see the reviews cited above) and similar conduction electron disturbances in simpler metals were later treated by Friedel and coworkers (Friedel, 1954, Blandin, Daniel and Friedel, 1952). In the latter formulation the RKKY effect is the limiting case of a precise partial wave expansion, when linearized for s waves only, and in the Born approximation to the scattering amplitude. Its oscillatory form then derives directly from the cutoff of state occupancy at the Fermi level. A slightly different description is appropriate in the rare earths. There, the generalized response $\chi(q)$ has near-singular behavior, with peaks along the c axis associated with strong nesting at almost parallel features of the Fermi surface. These peaks lie at values of q that explain the observed oscillatory spin wave phases of the heavy rare earths, such as the helimagnetic phase that occurs at 85K to 180K in bulk Dy. Note particularly that this

Magnetism and Structure in Systems of Reduced Dimension
Edited by R.F.C. Farrow *et al.*, Plenum Press, New York, 1993

characteristic response structure gives the rare earths a propensity for oscillatory interlayer coupling along the c axis, which is largely lacking in the more isotropic response of simple metals with almost spherical Fermi surfaces. Any actual superlattice periodicity that is caused by the response to this oscillatory coupling immediately changes the Fermi surface, thereby complicating the theoretical problem. The description of simple metals has been verified by nuclear resonance experiments that probe the radial disturbance cause by magnetic and nonmagnetic impurity centers, and similar effects have been observed in the liquid state where most structure of the Fermi ball is surely averaged away. For a brief review of this area see Flynn (1972) p742 and following.

Kwo and coworkers (1985) first reported oscillatory interlayer coupling in c axis multilayer systems of Gd alternating with Y, and in work by Yaffet et al (1989) treated the RKKY interaction as its explanation. Concurrent research by Erwin and coworkers (1987) on c axis superlattices in which Dy alternates with Y revealed several important results: (a) the response wavelengths in the Dy and Y layers each agree well with values predicted from the calculated $\chi(q)$; (b) helical Dy spin waves maintain coherence, including chirality, over many superlattice periods; (c) Dy layers couple through Y spacers up to 150A thick, with a coherence length that decreases with the inverse spacing.

By growing superlattices along the a and b axes instead, Tsui and coworkers (1991) have explored the range as a function also of orientation, to find extreme anisotropy. In real space each spin appears to couple in Y at up to 150A in separation along c, but fails to couple comparably at even 25A along b. The consequently ellipsoidal interaction volume is neverthless large enough to explain the cooperative magnetism observed (see Gotaas,et al, 1987) in the form of spiral phases for magnetic impurities at dilutions of order 1% in Y. All these facts are also in semiquantitative agreement with theoretical predictions for the calculated rare earth band structures. In addition, Beach and coworkers (1992) have observed the coupling characteristics through spacer layers made from the metals Lu and Sc, both of which resemble Y in valence structure. The coupling along c falls off from Y to Lu, much as predicted, and the Lu coupling wavelength agrees well with the calculated value; for Sc even the c axis coupling is unobservably weak (Tsui, 1992), and a shorter range is indeed expected from the breadth of its predicted peak in $\chi(q)$.

At least three current ideas warrant mention as deserving clarification. One is how the coupling range through Y spacers can possibly be longer than the mean free path determined by transport measurements (see, eg. the article by Legvold, 1972). A second is how magnetization-induced deformations (see below) can enhance the range of alternating layer magnetization beyond the coherence length of the ferromagnetism itself. Third, is the question of interfacial electronic states, particularly in the presence of limited free paths, and the possible identification of shared band structure near interfaces.

What important lessons for 3d transition metals can one draw from these results of investigations on rare earths? For one thing the RKKY interaction does exist, much as predicted. More specifically, the transmission of chirality from one layer to the next reveals that the coupling is not merely interfacial (see eg Wang, Levy and Fry, 1991), as might too readily be assumed, but instead is distributed some distance through the bulk. This conclusion is also consistent with the known role of oscillatory coupling in cohesion, as discussed for example by Heine and Weaire (1970) The further experimental fact of spin waves in Cr tells us that the 3d metals can have similar traits. Coupling through still simpler metal spacers as observed by Unguris, Celotta and Pierce (1991) may have slightly different roots consistent with the near-isotropy of electronic structure, as mentioned above. The physical processes that might cause this coupling have nevertheless been studied quantitatively since the 1950s, and for effects that lie beyond the linear response relevant to the RKKY approximation. Detailed investigations of the behavior in Cu and Ag host lattices were made in the 1960s (see eg. the review by Flynn, 1972). Current efforts attempt to treat the coupling characteristics of Fe layers through simple metal layers about 100A thick by semiquantitative theory (Edwards et al 1991)

MONOLAYER MAGNETISM

The past three decades have revolutionized our understanding of phase transitions. For a readable review see Stanley (1971). Of particular interest here is the way dimensionality affects the character of magnetic phase transitions. There are two issues described in what follows that pertain to the order observed at surfaces. First, wetting phenomena affect the degree of order that exists in surface layers as the bulk undergoes a transition. Second, a restriction to two dimensions (as measured by the ratio of thickness to coherence length) can itself have a profound influence on phase relationships.

Tsui (1992) has examined the magnetism of monolayer Gd-Y alloys. To enhance the magnetism, many layers were grown, using mica substrates, and with interlayer spacings of Y over 150A thick to eliminate interlayer coupling (see above). The magnetism is reduced mainly to in-plane effects by the internal field. As expected, the alloy Curie temperatures derived from their high temperature paramagnetism point unambiguously to strongly ferromagnetic in-plane coupling among the spins of pure Gd monolayers in Y, and of Gd-Y monolayer alloys down to 10% Gd. Yet no spontaneous ordering occurs. Instead, the monolayers exhibit apparent spin glass behavior with its attendant spin glass freezing and viscous response. Extremely precise scaling is observed in reduced temperature and field, although with unexpected exponents. However, the spin glass behavior itself is just as predicted by the x-y model for two dimensional arrays of spins (Kosterlitz and Thouless, 1978), as is appropriate for the monolayer geometry and the effective reorientation freedom of the spins (a strong anisotropy with easy axes perpendicular to c is observed). An interesting complication is that even in bulk Gd the considerable in-plane (basal) component of the magnetism is always disordered (Cable and Kohler, 1982); any connection with the surface result remains still to be elucidated.

The ordering of Cu_3Au has proved a more profitable system for study of wetting than an analogous first order magnetic phase change, because the order present in the surface layer alone can be measured using RHEED patterns and other analytical tools. This has been demonstrated explicitly by Dura (1991) and Huang (1992) upon growing monolayers of disordered alloy on the ordered crystal by MBE, whereupon the superlattice lines from the chemical order vanish almost completely. Through the first order (discontinuous) transition undergone by the bulk phase, the surface order instead falls to zero continuously, and with a power law exponent of about 0.6 in reduced temperature. This result is much as predicted from mean field theory (Lipowsky and Speth, 1983) which shows that a disordered surface layer penetrates into the bulk, and diverges in thickness as the bulk critical temperature is approached. The (110) and (111) surfaces behave similarly even though the former couples to the crystal chemistry (the alternating Cu and CuAu layers have differing interactions with the surface) while the latter does not.

The direct relevance of these results to the magnetism of Fe monolayers with perpendicular anisotropy may be almost nonexistent. What the results do show, however, is that even in systems which can be modeled by simple Hamiltonians there are subtleties in the degree to which surfaces can exhibit ordered structure. The apparent order bears no simple relationship to **local** interactions alone, but rather responds in addition to global symmetry and extended geometry. It seems more than likely that similar considerations remain relevant for 3d surface magnetism also, although their precise application in the case of itenerant magnetism is just that much more difficult to predict.

LATTICE CLAMPING AND STRAIN

Rarely are epitaxial films able to maintain their own preferred lattice geometries. Instead, the substrate generally stretches the film and also constrains their relative lattice spacing. The former is known as epitaxial strain and the latter is a manifestation of lattice clamping. Although not usually so recognized, these are separate processes that have distinct and frequently important consequences for the magnetism of epitaxial films. For example, a first order phase change in the film cannot be accompanied by its normal dimensional changes because the surface in contact with the substrate is constrained. Just as in a solid-to-liquid transition undergone at constant volume, the result in the epitaxial case at constant in-plane geometry is in general that the epilayer enters a two-phase region of its phase diagram. Of course, the balance between the phases depends on the particular in-plane strain at which the film exists, and in this way the separate effect of epitaxial strain enters the behavior. What follows is a brief review of recent research that pursues these ideas using rare earth metals as model materials.

For the purpose of model experiments that explore the principles involved, Dy is an excellent choice because it exhibits a second order paramagnetic-to-spiral antiferromagnetic transition at about 180 K in the bulk, and a further transition to a ferromagnetic state at about 85 K. Both ordered states are built on strong in-plane magnetism directed along the easy b axes, and with an interlayer coupling derived from the RKKY oscillatory interaction (see above) that dominates the helical antiferromagnetic phase. The in-plane magnetization associated with ferromagnetism causes a large (of order 0.3%) orthorhombic distortion of the lattice. The effects of lattice clamping and strain on first and second order transitions are thus open to investigation.

Erwin and coworkers (1987) find that c axis epitaxy on Y, which stretches Dy by 1.6%, completely suppresses the ferromagnetic phase below 0 K in sufficiently thin Dy layers, although a field-induced metastable magnetized state exists at low temperature. In the single sufficiently-studied case of b axis Dy on Y, Tsui (1992) is able to explain quantitatively the effect of clamping on the detailed energy balance in thin films and superlattices. The observed critical fields for magnetization thus become predictable from the bulk behavior and the observed turn angles, although the latter remain in need of fundamental description. In agreement with earlier results using magnetic templates by Farrow and coworkers (1989), Beach et al (1992) determine for Dy on Lu, which compresses Dy by 1.2%, that the ferromagnetic Curie temperature is raised to about 160 K. The spectacular difference between the Y and Lu induced changes is a consequence of clamping the first order transition at different, opposing strains. In neither case is the Neel (second order) transition temperature affected significantly, which points unambiguously to the role of its magnetostrictive strain in shifting the first order transition.

Beach and coworkers (1992) discover that the clamped ferromagnetic phase of c axis Dy /Lu superlattices distorts both the Dy and its surrounding Lu into 500A domains. This is a new effect in which each domain magnetizes along one of the six equivalent easy directions. The orthorhombic strains in the domains achieve a substantial fraction of the values observed for unconstrained bulk magnetostriction. In this way the spontaneous transition resembles a cooperative Jahn Teller effect. The additional effects of field-induced magnetization remains a poorly understood composite of several processes including two-phase phenomena, domain boundary behavior, and rotation from the easy axes.

In thicker films the magnetoelastic strain is relieved by the plastic processes of dislocation bowing (which changes the number of atomic planes), although in no case has the behavior been sufficiently analyzed. In a detailed study of transition temperatures and critical fields as functions of film thickness for Dy, Tsui (1992) finds that thicknesses of order 1μ are required before bulk-like behavior is restored. Major relaxation changes occur in films over 60A thick. Once again, the occurence of field induced magnetization is a complicated and poorly understood result of several contributing processes.

Are there lessons here to learn about 3d magnetism also? Certainly, epitaxial strain must once again cause a smooth tuning of the magnetic phase diagram with lattice geometry in the thinnest films. Clamping effects, however, are likely to be much weaker than for rare earths because 3d magnetoestriction is generally so much smaller. In films over about 100 A thick, plastic relief must again tend to restore bulk-like behavior much as in rare earth films.

SUMMARY

This paper is itself merely a brief summary of ongoing research in a field with a long and distinguished history, in which many important problems remain still to be resolved. Magnetism is revealed as relatively robust, such that thin films often have properties that approximate those of the bulk. Epitaxial strain and clamping can nevertheless tune magnetic phase diagrams over wide ranges of behavior. Reduced dimensionality causes specific and important changes of the global behavior for given local properties. Electronic mechanisms that couple magnetic layers are reasonably well understood, to a degree that is generally semiquantitative or better. However detailed considerations of the way phase change influences band structure, or of real band structure in multilayers, remain in their infancy.

ACKNOWLEDGEMENTS

The authors are indebted to R W Erwin, M V Klein, J J Rhyne, M B Salamon, R Beach, J Borchers, R Du, J Dura, J C Huang, S Kong and A Matheny for valuable contributions. This research was supported in part by the US National Science Foundation.

REFERENCES

Beach, R.S. 1992, Thesis, University of Illinois, unpublished.
Blandin,A., Daniel,E. and Friedel,J., 1952, *Phil. Mag.* 43:153.
Cable, J.W. and Kohler,W.C., 1982, *J. Appl. Phys.* 53:1904.
Coqblin, B., 1977, "The Electronic Structure of Rare Earth Metals," Academic Press, London.
Dura, J.A., 1991, Thesis, University of Illinois, unpublished.
Elliott, R.J.,1972, "Magnetic Properties of Rare Earth Metals," Plenum, New York.
Edwards, D.M., et. al., 1991, Phys. Rev. Lett. 67:492 and 1476.
Erwin, R.W., et al., 1987, *Phys. Rev* . B35: 6808.
Farrow, R.F.C., et al., 1989, MRS Symp. Proc. 151, MRS, New York.
Flynn, C.P., 1972, "Point Defects and Diffusion," Oxford UP.
Friedel,J., 1954, *Adv. Phys.* 3:446.
Gotaas, J.A. et al., 1987, *J. Appl. Phys.* 61:3415.
Heine,V and Weaire,d., 1970, *in* "Solid State Physics," v 24, Academic Press, New York.
Huang, J.C.A., 1992, Thesis, University of Illinois, unpublished.
Jensen, J. and Mackintosh, A.R., 1991, "Rare Earth Magnetism" Oxford UP.
Kosterlitz, J.M. and Thouless, P.J., 1978, *in* "Low Temperature Physics," D.E.Brewer ed., North Holland, Amsterdam.
Kwo, J., et al., 1985, *in* "Layered Structure Epitaxy and Interfaces," J.H Gibson and L.R.Dawson eds., MRS, New York.
Legvolt.S., 1972, in "Magnetic Properties of Rare Easrth Metals," R.J Elliott ed., Plenum, New York.
Lipowsky, R. and Speth, W., 1983, *Phys. Rev.*, B28:3983.
Stanley, H.E., 1971 "Introduction to Phase Transitions and Critical Phenomena," Oxford UP.
Tsui, F. et al., 1991, *Phys. Rev.* B43:13320.
Tsui, F., 1992, Thesis, University of Illinois, unpublished.
Unguris,J., Celotta,R.J., and Pierce,D.T., 1991, *Phys Rev Lett.* 67:140.
Wang,Y., Levy,P.M. and Fry,J.L, 1990, *Phys. Rev. Lett..* 65:2732.
Yaffet,Y. et al., 1989, *J. Appl. Phys.* 63:3453.

EFFECTS OF STRUCTURE AND CHEMICAL ORDERING ON MAGNETO OPTICAL SPECTRA IN THE $CO_{1-x}PT_x$ ALLOY SYSTEM

D.Weller[1], H.Brändle[2], R.F.C.Farrow[1], R.F. Marks[1] and G.R. Harp[1]

[1] IBM Research Division, Almaden Research Center, San Jose,CA 95120, USA
[2] present address: Balzers AG, FL-9496 Balzers, Fürstentum Liechtenstein

ABSTRACT

The magneto-optical spectra of MBE grown, epitaxial $Co_{1-x}Pt_x$ alloys are analyzed in the range 0.7-5.3eV photon energy and compared to a series of spectra of polycrystalline $Co_{1-x}Pt_x$ alloys, grown by e-beam evaporation. A strong dependence on structural properties and on the effect of chemical ordering is observed. A determination of the off-diagonal conductivity tensor elements confirms the presence of strong Pt 5d band related interband transitions, mainly in the ultraviolet.

INTRODUCTION

Magneto-optical effects have found widespread applications in magnetic thin film research. This includes magnetic hysteresis loop characterization, usually at fixed wavelength and spectroscopic investigations in the ~0.5→5.5eV photon energy regime. These measurements can be carried out under atmospheric conditions and represent a convenient method to access magnetic and electronic thin film parameters. Recent developments in this field include the observation of plasma resonance effects in multilayers [1, 2] and quantum confinement effects in ultrathin Fe films and multilayers [3]. Also efforts have been undertaken to expand the photon energy range further into the ultraviolet energy region in air (up to ~6eV or in vacuum using synchrotron radiation [5, 6].

In this report we focus on polar Kerr spectroscopy in the range 0.7-5.3eV (air, room temperature) and relate Kerr spectra to structural and magnetic properties of $Co_{1-x}Pt_x$ alloy thin films, grown both by conventional electron beam evaporation in high-vacuum (HV) and by MBE techniques in ultra-high-vacuum (UHV). $Co_{1-x}Pt_x$ is an ideal model system to study such a relationship for several reasons:

(i) Co and Pt are mutually miscible and form homogeneous solid solutions over the entire composition range [7, 8].

(ii) Chemically disordered $Co_{1-x}Pt_x$ alloys are ferromagnetically ordered at room temperature in the range $x \lesssim 0.9$.

(iii) Two phases of chemical ordering have been established in the bulk phase diagram [7]: the tetragonal ($L1_0$) Co_1Pt_1 phase and the primitive cubic ($L1_2$) phase. Intermetallic compound formation is expected to occur in relatively wide composition regions around the 1:1 and 1:3 compositions.

(iv) Possibly also an ordered Co_3Pt ($L1_2$) phase exists, as reported by Sanchez et al. [9], who have calculated the equilibrium phase diagram for $Co_{1-x}Pt_x$. However, no conclusive experimental evidence for this phase has been reported and the most recent listing [10] of intermetallic phases excludes Co_3Pt.

Finally, $Co_{1-x}Pt_x$ alloys have attracted renewed attention as magneto-optic recording materials, since they show, besides large magneto-optical effects [11, 12], large perpendicular magnetic anisotropies and suitable hysteresis properties [13-16]. The origin of that anisotropy has been discussed in terms of an inhomogeneous alloy model [15] and the possibility of $CoPt_3$ ordered alloy formation at relatively low temperatures of 200-300°C has been pointed out [17-20].

FILM DEPOSITION AND STRUCTURAL PROPERTIES

Films of ~100nm thickness were prepared using e-beam evaporation from separate Co and Pt sources in a background pressure of (a) 10^{-7}mbar onto fused silica at 200°C and (b) 10^{-10}mbar by MBE growth onto basal-plane sapphire at 600°C. For details about the presently used HV deposition system we refer to previous publications [13-15]. MBE growth was performed in a VG 80-M MBE system using e-beam sources for both Co and Pt [17]. Structural analysis of the films was made using X-ray diffraction (XRD) and transmission electron microscopy (TEM). In the case of MBE films the epitaxial relations were studied in situ with RHEED and LEED and ex-situ additionally with grazing-incidence XRD (GIXRD) and TEM. All film compositions were determined with an absolute accuracy of ~±2at% and a relative accuracy of ~±0.2at% using X-ray fluorescense spectroscopy (XRF).

a) Polycrystalline alloys: All HV grown films were polycrystalline and strongly fcc(111) textured with the [111] axis normal to the film plane. The average grain sizes were in the order of 10-15nm, as e.g deduced from the width of the CoPt(111) diffraction peak in X-ray θ-2θ scans [14]. A systematic expansion of the fcc lattice parameter with increasing Pt content is observed. Fig.1 demonstrates this by showing the dependence of the measured d_{111}-spacing on the film composition. We find rough agreement with Vegard's law for homogeneous solid solutions, which predicts a *linear* expansion of the lattice. We note, however, that a simple linear fit to the data is not possible, indicating deviations from the perfect, homogeneous elastic strain case. This may be due to slight structural inhomogeneities, which in turn may be responsible for the observed strong magnetic anisotropies in these films [15, 18].

b) MBE $CoPt_3$: The epitaxial relations were: $CoPt_3(110)\|Al_2O_3(10.0)$ and $CoPt_3[111]\|Al_2O_3\ [0001]$. The in-plane lattice misfit between film and substrate is -2.2%. The in-plane (fcc) structural coherence length of a 140nm thick film was ~300nm [21] and the same film was reported to exhibit a chemical ordering parameter of S=(0.3±0.1) and a long range chemical coherence length of 8nm [20]. The existence of chemical ordering was confirmed by the presence of (100), (110) and (211) diffraction features in GIXRD and TED patterns. These features were systematically absent in the random alloys and are related to the primitive cubic symmetry. Another 20nm thick film, which was grown under identical conditions to the 140nm film investigated in this study, exhibited an ordering parameter of S=(0.8±0.2) and a chemical in-plane coherence length of 4nm [19].

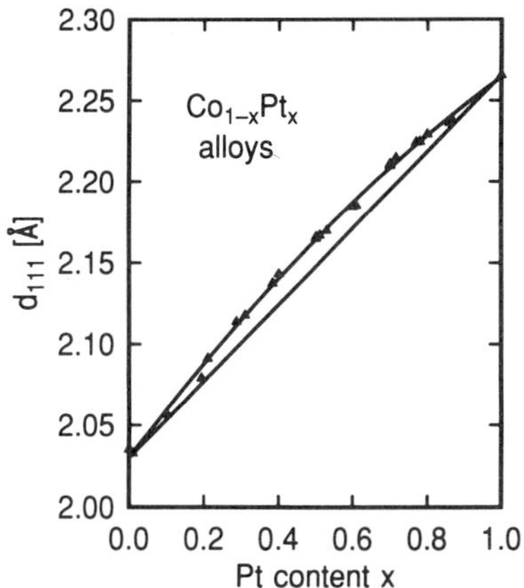

Figure 1. Measured d_{111}-spacings of ~100nm thick HV evaporated, polycrystalline $Co_{1-x}Pt_x$ alloys as function of the Pt content x. The films were grown on fused silica substrates held at ~200°C during deposition.

c) MBE CoPt: A superposition of two epitaxial relations was found: $CoPt(100)\|Al_2O_3(10.0)$ and $CoPt(100)\|Al_2O_3(11.0)$ with $CoPt[111]\|Al_2O_3[00.1]$. However, the intensity of the (100) diffraction compared with the (200) peak was only ~10% which is far less than that (~200%) expected for chemically ordered CoPt. We conclude that the CoPt film is largely disordered.

d) MBE Co_3Pt: The epitaxial relationship was identical to that for $CoPt_3$, except for the lattice mismatch, which is -7.2% in this case. Weak and diffuse (110) and (211) diffractions were observed in both GIXRD and TED. The amount of chemical ordering was limited to $\lesssim$10% of the sample volume and the chemical coherence length was estimated to be $\lesssim$4nm.

Both series of polycrystalline and epitaxial $Co_{1-x}Pt_x$ alloy films showed room temperature magnetizations consistent with those for random $Co_{1-x}Pt_x$ alloys. A considerable reduction of the Curie temperature and therefore the room temperature saturation magnetization, however, is expected if chemical ordering is induced, e.g. by annealing below the disorder-order transition temperatures [22], which are $T_{D-O}^{CoPt_3}$ ~750°C and T_{D-O}^{CoPt} ~825°C according to the CoPt bulk phase diagram [7, 8]. Note that under non-equilibrium thin film growth conditions, the diffusivity can be much higher than in a bulk annealing process, and thermodynamically stable ordered phases may be formed well below $T_{D\text{-}O}$. This explains, why e.g. the present epitaxial $CoPt_3$ film indeed shows some evidence for chemical ordering. Nevertheless, this film still had a high room temperature magnetization, $M_S=(364\pm20)$kA/m, and posesses a magnetic behavior like that of a random rather than an ordered alloy. We attribute this to the observation of a short chemical coherence length of only ~4-8nm, allowing for random coordination of the Co atoms at the boundaries between ordered regions.

Chemical ordering, as evidenced by a reduced Curie temperature and room temperature magnetization was successfully induced by annealing a 140nm thick epitaxial $Co_{27}Pt_{73}$ film at $T_A \simeq 680°$ in UHV ($L1_2$-phase).

In the case of Co_1Pt_1 ($L1_0$-phase) we were able to directly grow an (ordered) alloy with reduced room temperature magnetization. This was achieved by growing a $Co_{\sim 50}Pt_{\sim 50}$ film of 60nm thickness on a 15nm thick Pt seed layer on MgO(100) at ~700°C. Work on these type of films, however, is still in progress and will not be reported here yet.

In the following, we will first discuss magnetic and magneto-optical properties of random $Co_{1-x}Pt_x$ alloys and will then investigate the effect of (at least partial) chemical ordering on MO spectra in the $CoPt_3$ case. Finally the effect of vacuum annealing on magneto-optical spectra of two polycrystalline $Co_{1-x}Pt_x$ alloy compositions will be investigated.

MAGNETIC AND MAGNETO-OPTICAL PROPERTIES

Magneto-optical properties of the present random polycrystalline $Co_{1-x}Pt_x$ alloys ($0<x<0.9$) have been discussed in detail before [12]. Here we present for the first time spectra of epitaxial and UHV annealed $Co_{1-x}Pt_x$ alloy films. We follow the sign convention introduced by Reim and Schoenes [23] and define the complex Kerr rotation as $\tilde{\Phi}_K = \theta_K - i\varepsilon_K$ and the complex refractive index $\tilde{n}$ = n-ik, where θ_K and ε_K denote polar Kerr rotation and ellipticity, respectively and n and k are the refractive and absorption index, respectively. Φ and $\tilde{n}$ are connected to each other through the conductivity tensor $\tilde{\sigma} = \sigma_{1ij} + i\sigma_{2ij}$ according to the well known equation [23]:

$$\tilde{\Phi} = \frac{\tilde{\sigma}_{xy}}{i\tilde{n}\tilde{\sigma}_{xx}} \tag{1}$$

Here $\tilde{\sigma}_{xy}$ and $\tilde{\sigma}_{xx}$ are the complex off-diagonal and diagonal conductivity tensor elements, respectively.

Fig.2 shows the complex Kerr effect spectra for MBE grown Co_3Pt, CoPt and $CoPt_3$ films in comparison with an epitaxial hcp Co film, which was grown on c-axis oriented sapphire, buffered with a thin Pt seed layer. These films are epitaxial but essentially chemically disordered, as described above. The spectral dependences are similar to those of polycrystalline, random $Co_{1-x}Pt_x$ alloys ($x>0.4$) with two dominating features at 4.2eV and 1.4eV. The Kerr rotation spectra for MBE grown Co_3Pt, however, show a striking shift of the UV peak to 4.5-4.6eV. A similar, but less pronounced shift was observed in some polycrystalline (HV evaporated) films with compositions near the Co_3Pt phase [12].

The observed spectral differences are not surprising because a significant change in the bandstructure is expected on going from Co_3Pt to $CoPt_3$. First, the lattice is expanded by about 5% (see Fig.1!) and second the distribution of Co-Co distances is expected to change. For the simple cubic $L1_2$ structure, these would be $a_0^{Co_3Pt}/\sqrt{2} = 0.257nm$ and $a_0^{CoPt_3} = 0.384nm$. Thus, even though the discussed alloys are random, the distribution or average Co-Co distance strongly increases with increasing Pt content.

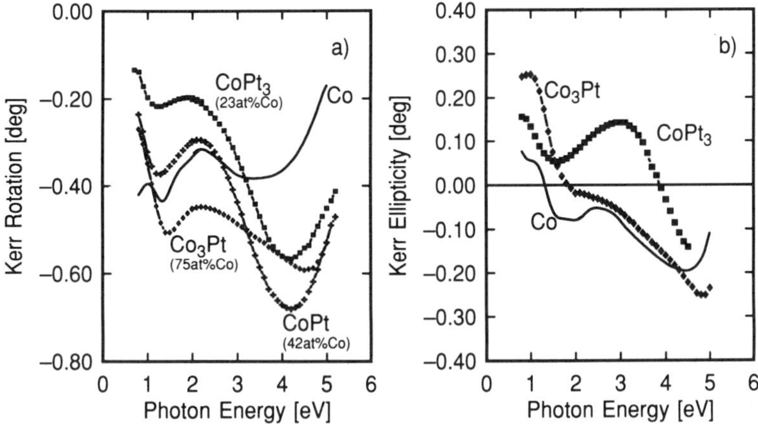

Figure 2. Complex Kerr spectra of MBE grown $Co_{1-x}Pt_x$ alloy films. The measured compositions of these alloys, which were close to those of the chemically ordered CoPt phases, are indicated in parentheses. The growth temperarture was 600°C in all cases, which is well below the disorder-order transition temperatures for CoPt ($L1_0$) and $CoPt_3$ ($L1_2$). A comparison to epitaxial, hcp Co is made.

To compare the magneto-optically active interband transitions and separate out optical constants effects, which are incorporated in $\tilde{n}$ and $\tilde{\sigma}_{xx}$ according to eq.(1), we have calculated the functions $\sigma_{2xy}(\hbar\omega)$ and $\omega\sigma_{2xy}(\hbar\omega)$ [24] from the measured quantities θ_K, ε_K and the optical constants n and k for the present hcp-Co, the Co_3Pt and the $CoPt_3$ film (for a discussion of this procedure see e.g. Erskine et al. [24]). The results are summarized in Fig.3a+b. A plot of $\omega\sigma_{2xy}$ (Fig.3b!) suppresses the contributions of intraband transitions which have a ω^{-1}-dependence for $\hbar\omega > 0.5$-1eV. Peaks in $\omega\sigma_{2xy}$ as shown in Fig.3b are associated with magneto-optically active interband transitions. Without attempting to assign the observed peaks to electronic transitions within the bandstructure of Co and $Co_{1-x}Pt_x$ alloys, we still can make the following experimental statements: In Co a sharp, dominant transition at 1.3eV and a less pronounced shoulder around 4.5eV are observed. Co_3Pt basically shows two transitions with comparable spectral weight, at 1.6eV and at 4.7eV. In $CoPt_3$, the clearly dominating interband transition is located at 4.2eV, and a less pronounced transition appears at 1.4eV. The strong interband transition at 4.2eV is clearly due to the addition of Pt. From the large spectral weight of this transition we conclude, that strong dipole allowed Pt 5d band transitions, e.g. d→p transitions [12] are responsible. On the other hand, the observation of relatively broad features in both σ_{2xy} and $\omega\sigma_{2xy}$ lead us to conclude, that the underlying interband transitions are not strongly localized, which is consistent with a width of the (occupied) Pt5d bands in the order of 6eV, as e.g. observed in X-ray photoemission experiments [11]. In fact, the present experiments indicate that Pt related transitions extend all the way into the infrared region and clearly also contribute to the 1.3eV feature. Otherwise, in a simple dilution model, this peak would be expected to decrease according to the Co dilution with Pt, which it clearly does not.

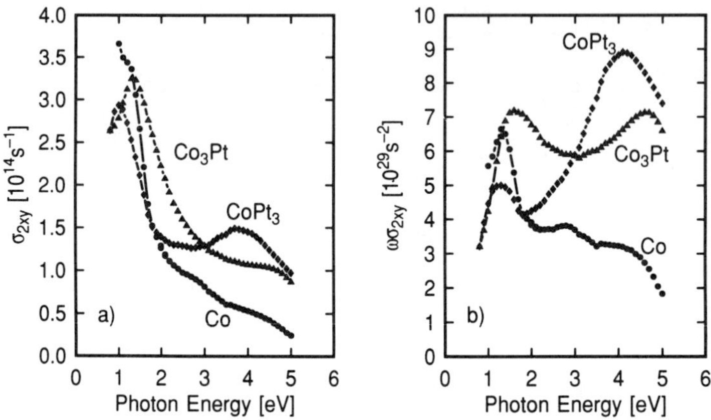

Figure 3. a: Spectral dependence of the imaginary part of the off-diagonal conductivity tensor element σ_{2xy} of the films discussed in Fig.2. b: Same dependence of the function $\omega\sigma_{2xy}$.

Another experimental manifestation of this effect is seen in Fig.4, which shows the composition dependent Kerr rotation at different photon energies. We can restrict the present discussion to the Kerr rotation alone, as the optical constants are smooth functions of hν and do not alter the spectral dependences in the energy region under discussion (0.8-5eV) as the composition is changed (see e.g. ref. [4, 12]). The 4.2eV data have been published before [12] and compared with respective multilayers. Here we focus on the influence of the photon energy. The sharp increase of $|\theta_K|$ in the range $0 \leq x \leq 0.6$ can clearly be attributed to spin-polarized Pt and reflects the strong Pt contributions to the magneto-optical Kerr effect. In the 4.2eV curve, the reduction of $|\theta_K|$ for $x > 0.6$ is readily explained by the drop of the Curie temperature of $Co_{1-x}Pt_x$ alloys, which passes through room temperature at $x \simeq 0.9$. The influence of Pt diminishes as infrared wavelengths are approached, but is still clearly visible at $h\nu = 1.5$eV. At 0.8eV, the lowest energy in the present investigations, an almost linear decrease of $|\theta_K|$ is observed, strongly resembling the measured magnetization behavior. This point is specifically illustrated in Fig.4b, where we have plotted the normalized room temperature saturation magnetization M_S and the normalized 0.8eV Kerr rotation vs the Pt content x. (For further discussion of the magnetization properties see ref. [14,15]). The data were normalized to the maximum values in each series. The fact, that the infrared Kerr rotation follows M_S within an error of $\sim\pm5\%$ is perhaps fortuitous. It implies that at 0.8eV the magneto-optical activity reflects the total magnetic moment in the system, which consists of the composition average of the Co ($1.64\mu_B$) and the Pt ($0.26\mu_B$) moments (numbers taken from polarized neutron scattering experiments of bulk $CoPt_3$ ordered compounds [25]!).

To conclude the presented results so far, we can say that magneto-optical and optical spectroscopy unambiguously demonstrates that Pt plays an important role in the uv Kerr rotation in the $Co_{1-x}Pt_x$ alloy system. It clearly dominates the spectra for Pt-rich compositions at energies around 4eV. The fact, that σ_{2xy} does not always exhibit a distinct peak structure means that the Pt electronic levels involved in the magneto-optical interband transitions are not strongly localized, but rather extend far into the infrared. As a consequence, we expect that the Kerr rotation spectra will strongly depend on the local chemical environment, which can be modified, e.g. by annealing and/or film growth at elevated temperatures. Changes at basically all energies are expected, although the effect of Pt should most strongly be visible in the ultraviolet region.

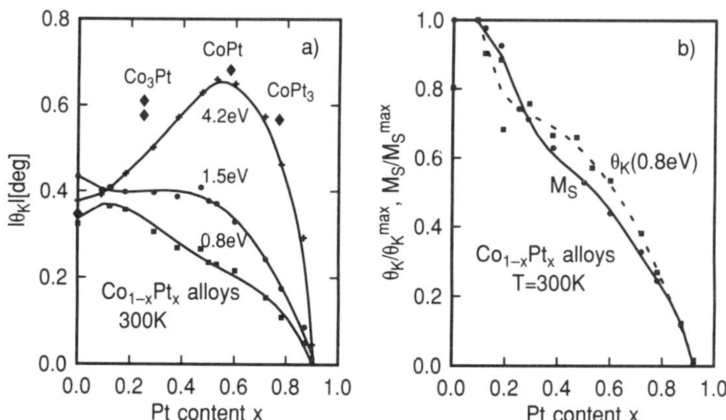

Figure 4. a: Composition dependence of the room temperature Kerr rotation of HV evaporated $Co_{1-x}Pt_x$ alloy films of composition in the range $0\leq x\leq 0.92$ at different photon energies $h\nu = 0.8$, 1,5 and 4.2eV. Data points of MBE grown films are also included.
b: Normalized Kerr rotation at 0.8eV (---) and magnetization M_S at room temperature (—) of the $Co_{1-x}Pt_x$ alloy series in a)

EFFECT OF CHEMICAL ORDERING

Vacuum annealing is used in the following to change structural parameters and to induce changes in the local chemical environment in $Co_{1-x}Pt_x$ alloy films. A $\sim$100°C drop of the Curie temperature upon conversion of the random fcc to the ordered tetragonal (f.c.t.) Co_1Pt_1 phase and a $\sim$200°C drop upon conversion of the random fcc $CoPt_3$ into the chemically ordered fcc $L1_2$ phase have been reported in respective thin film and bulk annealing experiments (see e.g. discussions by Treves et al. [26] and Sanchez et al. [9]). Responsible for this drop in T_C is the change of the Co-Co exchange interaction. Sanchez et al. [9] have also pointed out, that a drastic change of the exchange induced Pt moment is expected.

Table 1. Average nearest neighbor coordination of Co and Pt atoms in random and chemically ordered $Co_{1-x}Pt_x$ phases.

	$Co_{50}Pt_{50}$ random	Co_1Pt_1 ordered $L1_0$	$Co_{25}Pt_{75}$ random	$CoPt_3$ ordered $L1_2$
Co	6Pt/6Co	8Pt/4Co	9Pt/3Co	12Pt/0Co
Pt	6Co/6Pt	8Co/4Pt	3Co/9Pt	4Co/8Pt

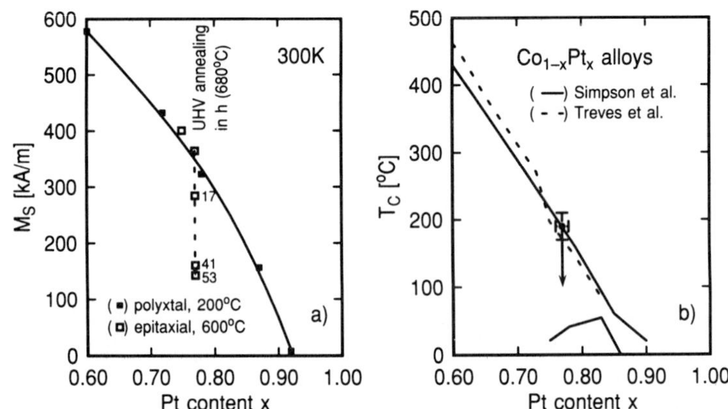

Figure 5. Room temperature magnetization (a) and Curie temperatures (b) of random and (partially) chemically ordered $Co_{1-x}Pt_x$ alloys in the composition range $0.6 \leq x \leq 0.92$. This is the relevant range for MO recording applications, both from the point of view of perpendicular magnetic anisotropy (see [14, 15]) and suitable Curie temperatures. The following literature data are quoted in (b): Simpson et al. (see ref.[22]) and Treves et al. (see ref.[26]).

Before we discuss the effect of annealing on magneto-optical spectra, we summarize in Table 1 the changes in the nearest-neighbor environment of Co and Pt that occur upon chemical ordering. The main effect is, that the average number of Co atoms next to Pt atoms increases, which therefore should enhance the exchange-induced Pt magnetic moment, as pointed out by Sanchez [9]. On the other hand, Co atoms have a lower Co coordination, which leads to the experimentally observed drop in the Curie temperature. This holds for both ordered phases discussed in Table 1.

Fig.5 illustrates these changes for the $CoPt_3$ composition. The figure summarizes room temperature saturation magnetization measurements on random, polycrystalline alloys and MBE grown films at various stages of annealing and shows literature data of the Curie temperatures of random and chemically ordered alloys. The Curie temperature data were taken from the work of Simpson et al.[22] and Treves et al.[26]. We show these data in the range $0.6 \leq x \leq 0.92$, in which perpendicular magnetic anisotropy has been reported [14, 15] and mention that this is the relevant composition range for applications of these alloys as magneto-optic recording materials. Obviously, at 680°C the room temperature magnetization is reduced by about 50% after a total annealing time of 53h in UHV. All measurements (VSM magnetometry and MO spectroscopy) were performed on the same 7x7mm piece of sample, which was put back into the UHV chamber after each annealing/characterization cycle and heat treated again. The quoted annealing times refer to the total time at 680°C and do not include heating and cooling times.

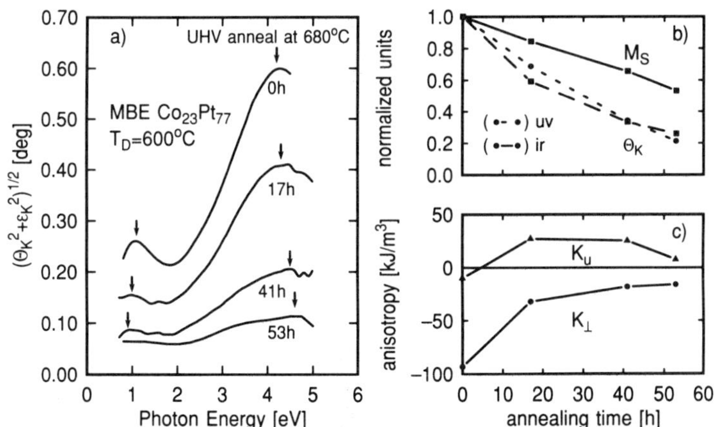

Figure 6. a: Polar Kerr rotation spectra of a 140nm thick, epitaxial $Co_{23}Pt_{77}$ film at different stages of chemical ordering induced by UHV annealing at ~680°C.
b: Normalized room temperature magnetization and Kerr rotations at 0.8 and 4.2eV vs annealing time. The data are relative to the initial values prior to annealing in UHV.
c: Total and intrinsic magnetic anisotropy energy $K_\perp$ and K_u vs annealing time. K_u was determined by adding the demagnetization energy according to: $K_u = K_\perp + 1/2M_S^2$

We now turn to the effect of chemical ordering on magneto-optic spectra of the present 140nm thick, MBE grown $CoPt_3$ film (actual composition: $Co_{0.23}Pt_{0.77}$). As Fig.6a shows, the room temperature Kerr effect indeed drops continuously with increased total annealing time. We have plotted here the absolute Kerr effect $|\tilde{\Phi}| = (\theta_K^2 + \varepsilon_K^2)^{1/2}$ which includes the measured Kerr ellipticity ε_K and mention that plotting the Kerr rotation θ_K alone, doesn't change the following discussion. At first glance, the spectra seem to simply reflect the drop of the magnetization and do not indicate a strong rearrangement between Co and Pt related magneto-optical transitions. We do, however, see some changes, perhaps less dramatic as expected from the above discussion, which are clearly indicative of an electronic structure change. The arrows in Fig.6a mark the experimental peak positions. Both the uv and the ir peaks are found to shift. In the case of the (Pt dominated) uv peak, the peak position moves from 4.2eV to ~4.6eV and in the case of the (Co dominated) ir peak, a shift from 1.1eV down to at least 0.8eV is observed, after 53h total annealing time. There is not much change in the ratio of the uv and ir peak. Both peaks decay down to about 20% of their initial value at about the same rate. However, a much stronger drop of the Kerr rotation compared to the room temperature magnetization is observed. This point is explicitly shown in Fig.6b, where we have plotted the normalized magnetization and Kerr effect values as function of the annealing time. Clearly M_S and $|\tilde{\Phi}|$ do not follow each other, showing again that there is no simple linear relationship between the Kerr effect and the magnetization in this system (see previous discussion of Fig.4!). A full understanding of the present effects can only come from an understanding of the electronic structure changes upon chemical ordering. This has clearly not been achieved yet! We are attempting to solve this problem by first-principles MO calculations and refer to future publications of Sticht et al. [27].

Before turning to two other examples of local chemical environment effects on magneto-optical spectra, we mention that we have also observed changes in the magnetic anisotropy of the present films. This has to be seen within the context of identifying the origin of the perpendicular magnetic anisotropy in polycrystalline, fcc(111) textured $Co_{1-x}Pt_x$ alloys ($0.5 \le x \le 0.9$) (see particularly ref. [13-16]). The present epitaxial $Co_{23}Pt_{77}$ alloy film has an easy in-plane magnetization axis for all annealing conditions, as expressed in Fig.6c, where we have plotted the total and intrinsic magnetic anisotropy energies $K_\perp$ and $K_u = K_\perp + 1/2\mu_0 M_S^2$, respectively, as function of the total annealing time. These measurements were done by torque magnetometry. Interestingly, a positive (perpendicular) anisotropy K_u can be induced and the in-plane tendency is much reduced after 53h of annealing. We do, however, not believe, that this induced anisotropy is intrinsically related to the ordered cubic fcc phase of $CoPt_3$. It rather points to the development of local inhomogeneities, perhaps clusters of Pt rich regions (2at% excess Pt in this film!), which could give rise to the geometrical symmetry break underlying the observation of the present change of the magnetic anisotropy. We note, that polycrystalline films grown at temperatures of 200-300°C on amorphous substrates (fused silica) show 10-20 times larger K_u values (see discussion in ref. [15]). Annealing experiments carried out on such films, destroy the perpendicular magnetic anisotropy, rather than inducing it. This corroborates the above arguments.

Finally, we discuss in Fig.7 the effect of annealing on a UHV grown 125nm thick $Co_{36}Pt_{64}$ film (T_D=300°C) (Fig.7a!) and a HV grown 96nm thick $Co_{25}Pt_{75}$ film (T_D=200°C) (Fig.7b!). These films, which were grown on fused silica substrates, were polycrystalline with strong fcc(111) texture, as revealed by the (002)/(111) peak ratio in X-ray diffraction. Annealing for 5h at 680°C. further improved the (111) texturing and resulted in grain growth, as deduced from a considerable decrease of the width of the CoPt(111) diffraction peak. The Kerr rotation spectra are shown before and after annealing in Fig.7. Without further structural and/or chemical order characterization, we can clearly conclude, that annealing indeed affects the spectral dependence in magneto-optics, most dramatically in the present UHV grown film. While a ~9% increase of the Kerr rotation at 4.5eV is observed, it's value decreases by ~30% at 2eV and stays constant at 3.3eV and 0.8-0.9eV. This, perhaps most conclusively, shows that there is no linear relationship between the Kerr rotation and the magnetization. A single wavelength measurement, e.g. at the most commonly used He-Ne laser wavelength of 633nm (2eV) would have indicated a drop of almost 30%, whereas measurements in the infrared would have shown no change at all, etc.. A similar situation, although less pronounced, is observed in the case of the HV grown film (Fig.7b). Interestingly the Kerr rotation increases at all observed energies. In the infrared a distinct peak evolves at 1.2eV and the uv peak is enhanced by 17%, again clearly indicating the electronic structure change associated with annealing. We have shown these two examples in order to point out that indeed strong spectral changes can be observed upon annealing. For a detailed interpretation of the origin of these effects, we have to refer to a future publication, which addresses the structural and chemical order changes more quantitatively.

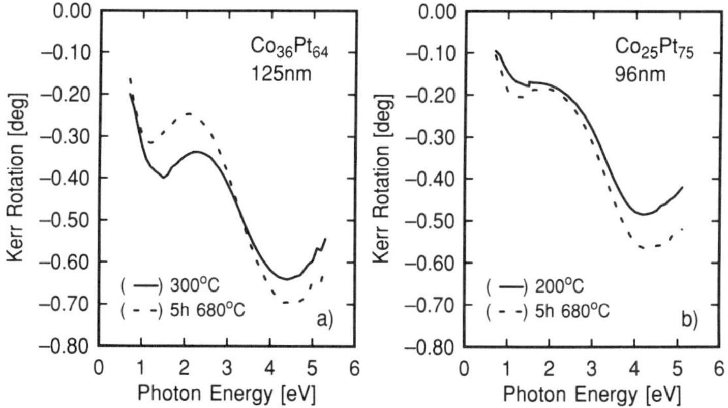

Figure 7. a. Kerr rotation spectra in the range $0.7 \leq h\nu \leq 5.3$eV of a UHV grown, polycrystalline, 125nm thick $Co_{36}Pt_{64}$ film (MBE673) before (—) and after (---) annealing for 5h at 680°C.
b: same for a HV grown, polycrystalline 96nm thick $Co_{25}Pt_{75}$ film (EB143)

Finally we point out that these experimental findings are consistent with recent theoretical work on magneto-optical effects, which shows that there is no simple (or linear) relationship between the exchange splitting parameter and the magneto-optical activity [27-29].

CONCLUSIONS

Chemical ordering in $Co_{1-x}Pt_x$ alloy films has a strong influence on the wavelength dependence of the magneto-optical Kerr effect, making magneto-optical spectroscopy a sensitive tool to probe local chemical environment effects. Pt directly contributes to the Kerr effect through strong 5d-band related interband transition, most likely dipole allowed d→p or p→d transitions, as concluded from the strength and (energetic) width of the observed magneto-optical activity σ_{2xy}. Pt contributes at all energies (0.7-5.3eV), that were investigated in this study, but shows its strongest response in the ultraviolet (~4eV).

ACKNOWLEDGEMENT

The authors are grateful to G. Gorman and R. Swope for numerous XRD and XRF analyses, to J.C. Scott for optical constants measurements, to C.J. Chien, D. Dobbertin and R. Geiss for TEM and TED measurements and to M. Toney for GIXRD studies of a 20nm thick film of epitaxial $CoPt_3$ using synchrotron radiation. Fruitful discussions with J. Sticht on the theoretical aspects of MO spectra and with C. Chappert on the anisotropy mechanism in $Co_{1-x}Pt_x$ alloys are gratefully acknowledged.

REFERENCES

1 T. Katayama, Y. Suzuki, H. Awano, Y. Nishihara and and N. Koshizuka, Phys. Rev. Lett. **60**, 1426 (1988).
2 W. Reim and D. Weller, Appl. Phys. Lett. **53**, 2453-2454 (1988).
3 Y. Suzuki, T. Katayama, S. Yoshida, T. Tanaka, and K. Sato, Phys. Rev. Lett. **68**, 3355 (1992).
4 K. Sato, H. Hongu, H. Ikekame, J. Watanabe, K. Tsuzukiyama, and Y. Togami, M. Fujisawa, T. Fukazawa, Jpn. J. Appl. Phys. **31**, 3603-3607 (1992).
5 Y. Suzuki, T. Katayama, MORIS'92, Tucson (paper Mb-5) (to be published) (1992).
6 N.K. Flevaris, S. Logothetidis, J. Petalas, P. Kielar, M. Nyvlt and V. Parizek, S. Visnovsky, and R. Krishnan, J. Magn. Magn. Mater.(to appear) (1992).
7 T. Massalski, Published by Materials Information Society, Materials Park, Ohio **2**, (1990).
8 M. Hansen, K. Anderko, McGraw Hill, New York (1958).
9 J.M. Sanchez, J.L. Morán-López, C. Leroux, and M.C. Cadeville, J. Phys.: Condens. Matter **1**, 491-496 (1989).
10 Edited by P. Villars and L.D. Calvert, Published by Materials Information Society, Materials Park, Ohio **2**, (1991).
11 D. Weller, W. Reim, Appl. Phys. A **49**, 599-618 (1989).
12 H. Brändle, D. Weller, S.S.P. Parkin, J.C. Scott, C.-J. Lin, IEEE Trans. Mag. **28**, 2967-2969 (1992).
13 C.-J. Lin, G. Gorman, Appl. Phys. Lett. **61**, 1600 (1992).
14 D. Weller, H. Brändle, G. Gorman, C.-J. Lin and H. Notarys, Appl. Phys. Lett. **61**, 2726 (1992).
15 D. Weller, H. Brändle, C. Chappert, Symp. Mag. Ultrathin Films, Multilayers and Surfaces Lyon, Sept. 1992, J. Magn. Magn. Mater. (accepted) (1992).
16 R.F.C. Farrow, R.H. Geiss, G.L. Gorman, G. Harp, R.F. Marks and E.E. Marinero, MORIS'92, Tucson, AZ, December 7-9 (1992).

17 R.F.C. Farrow, C.H. Lee, R.F. Marks, G. Harp, M. Toney and T.A. Rabedeau, D. Weller, H. Brändle, NATO Advanced Research Workshop on Magnetism and Structure in Systems of Reduced Dimensions **Cargese, France, June15-19**, (1992).

18 D. Weller, C. Chappert, H. Brändle, G. Gorman, R.F.C. Farrow and R. Marks, G. Harp, MORIS'92, Tucson (to be published) (1992).

19 M. Toney, unpublished data (1992).

20 T.C. Huang, R. Savoy, R.F.C. Farrow, R.F. Marks, Appl. Phys. Lett. (submitted) (1992).

21 R.F.C. Farrow et al., J. Appl. Phys. (to be published) (1992).

22 A.W. Simpson and R.H. Tredgold, Proc. Phys. Soc. **B67**, 38 (1954).

23 W. Reim, J. Schoenes in:, Handbook on Ferromagnetic Materials **5**, 133-236 (1990).

24 J.L.Erskine, E. A. Stern, Phys. Rev. B **8**, 1239-1255 (1973).

25 F. Menzinger, A. Paoletti, Phys. Rev. B **143**, 365-372 (1966).

26 D. Treves, J.T. Jacobs, E. Sawatzky, J. Appl. Phys. **46**, 2760-2765 (1975).

27 J. Sticht (private communication), (1992).

28 D.K. Misemer, J. Magn. Magn. Mater. **72**, 267 (1988).

29 P.M. Oppeneer, J. Sticht, T. Maurer, J. Kübler, Z. Phys. B - Condensed Matter **88**, 309-315 (1992).

CHEMICAL ORDERING AND MAGNETIC ANISOTROPY IN MBE-GROWN Co/Pt MULTILAYERS *

R.F.C. Farrow, C.H. Lee, R.F. Marks, G.R. Harp, M.F. Toney, T.A. Rabedeau, D. Weller, and H. Brändle

IBM Research Division, Almaden Research Center
650 Harry Road, San Jose, CA 95120-6099

Abstract

In this paper we report the results of a synchrotron X-ray diffraction study of epitaxial, [111]-oriented Co/Pt multilayers which demonstrates the existence of interfacial mixing and extensive chemical ordering in the multilayers. We also report new magnetic data for epitaxial Co/Pt multilayers, grown along the 3 major axes: [111], [110] and [001], which demonstrate that both the interface anisotropy (K_S) and volume anisotropy (K_V) constants depend on the growth axis of the multilayer. This study suggests that chemical interdiffusion and chemical ordering within the interfaces, as well as strain, should be incorporated into a realistic model for the magnetic anisotropy.

1. Introduction

The origin of the perpendicular magnetic anisotropy in Co/Pt multilayers remains unresolved. Such multilayers have potential as optical storage media and recent studies have shown[1-3] that the perpendicular anisotropy is maximized for multilayers with <111> texture along the growth axis. In order to develop a realistic model for the magnetic anisotropy of Co-based multilayers, including Co/Pt, it is necessary to fully characterize them structurally and magnetically. For example, Engel et al[4] recently reported that the orientation dependence of the magnetic anisotropy in Co/Pd multilayers is due to a difference in the volume term (K_V) of the effective anisotropy and that the interface anisotropy term (K_S) is identical for all orientations. Differences in the volume term were attributed to differences in film

* This work was supported in part by the Office of Naval Research (Contract N00014-87-C-0339).

strain for the different growth orientations, a point made earlier by den Broeder et al[5]. However, this model did not consider interdiffusion nor that the d-spacings are different in Co-rich and Pd-rich regions of the multilayer. On the other hand, for epitaxial Co/Pt multilayers, it was recently proposed by Chien et al[6] that the Co/Pt interfaces were not only chemically interdiffused but contained the chemically ordered phase $CoPt_3$.

Since the suggestion of the chemically ordered $CoPt_3$ phase in multilayers was novel and since the bulk phase diagram shows that ordered $CoPt_3$ forms readily, we undertook a search for this phase in [111]-oriented multilayers, grown at various temperatures, using X-ray diffraction. We report the results of these investigations here and find that the multilayers contain significant amounts of ordered $CoPt_3$. We also report the results of magnetic characterization of epitaxial Co/Pt multilayers grown along the 3 major axes, [111], [110] and [001], and covering a wide (2-20Å) range of Co thickness. We find that the interface (K_S) and volume anisotropies (K_V) are both dependent on the growth axis of the superlattice. Our results confirm that chemical interdiffusion and chemical ordering within the interfaces should not be neglected in a realistic model for the magnetic anisotropy and that models based solely on strain of pure Co films may not be realistic.

2. Experimental Techniques

The multilayers described in this paper were prepared in a VG 80-M MBE system (VG Semicon Ltd.) using e-beam sources for both Co and Pt. During growth the background pressure was $\sim 1.10^{-10}$ mbar and Co, Pt growth rates were $\sim$ 0.2Å/s. The beam fluxes were controlled using a Leybold Sentinel III optical system with feedback to the filament emission of the e-gun power supply.

Beam lines X10B and X20C at NSLS (Brookhaven) were used for the X-ray measurements. On X10B, a vertical focussing mirror and horizontally focussing monochromator were used and the energy was 8.019 keV (1.5461Å). While on X20C, the beam was horizontally and vertically focussed with a toroidal mirror and the energy was 10.179 keV (1.218Å). In both cases, a Ge solid state detector was used to eliminate Co fluorescence and the detector resolution was set by 1mrad Soller slits (in-plane) and 10mrad (X10B), 20mrad (X20C) out-of-plane. The sample was immersed in He gas.

We prepared [111], [110] and [001] -oriented multilayers on GaAs substrates as described earlier[1,2]. In addition, [111] oriented superlattices were grown on (0001) sapphire substrates using a 30Å thick seed film of (111) Pt grown onto the sapphire at 600° C. This seed film established the [111] axis of the multilayer since Pt was found to grow epitaxially on sapphire with Pt[111] || Al_2O_3 [0001] and Pt(110) || Al_2O_3 (10$\bar{1}$0). Figure 1 illustrates this epitaxial relationship. The basal plane unit cell of sapphire is shown in Figure 1(a). The atomic arrangement in Pt(111) is shown in Figure 1(b). There is a geometric fit between sapphire and Pt units indicated by the solid lines. Note that in this setting the second neighbour Pt-Pt distances in the (111) plane form a hexagon which is only 0.9% larger than the basal plane unit cell (a = 4.7628Å) of sapphire. This geometric relationship suggests that the Pt atoms occupy the hollow sites in the quasi close -packed oxygen plane of sapphire (0001).

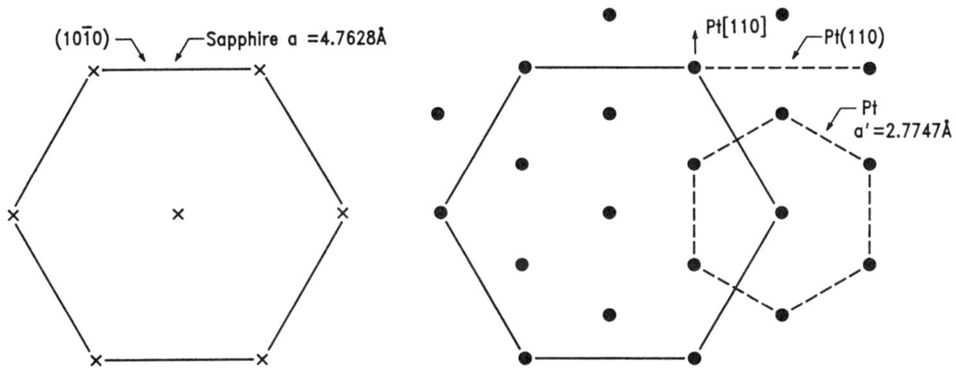

Figure 1. Schematic diagram of the epitaxial relation between Pt(111) and basal-plane sapphire ie sapphire (0001).

Two orientations of Pt (which differ by a rotation of 180° about the [111] axis) were observed. A more complete description of the growth, structural and chemical characterisation of the Pt seed film can be found elsewhere[7] , however, it is worth noting here that the Pt seed film is continuous at 30Å and the Pt [111] axis has a rocking curve width (FWHM) of ~0.5°. The interface between the Pt and sapphire is chemically stable at temperatures up to at least 600°C which permits growth of multilayers at elevated temperatures without the segregation effects observed[8] in the case of Ag-seeded multilayers.

The Co/Pt multilayers were grown at substrate temperatures of 100, 200 and 300°C. A schematic diagram of the sapphire-based multilayers is shown in Figure 2. Magnetic characterization of the multilayers was made by polar Kerr magnetometry, torque magnetometry and vibrating sample magnetometry. The sapphire-based multilayers exhibited perpendicular anisotropy similar to that of [111]-oriented multilayers grown grown on GaAs($\overline{111}$) substrates. Typical hysteresis properties are illustrated by the Kerr loops in Figure 3.

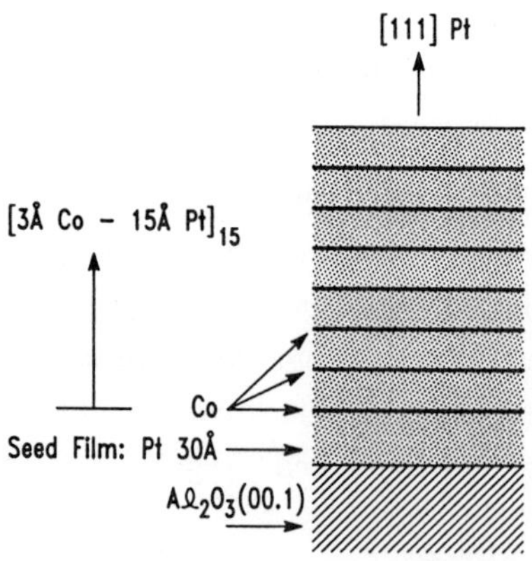

Figure 2. Schematic diagram of Co/Pt multilayer structures grown on sapphire (0001).

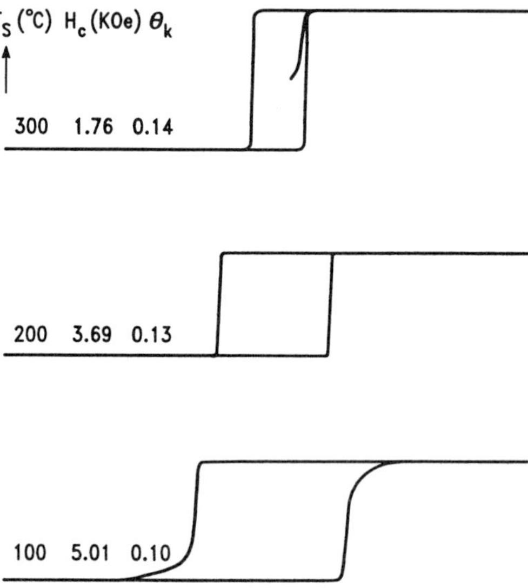

Figure 3. Polar Kerr loops for [111] oriented Co-Pt multilayers grown on basal-plane sapphire at the temperatures indicated. The coercivities and Kerr rotations are indicated. Measurements were at a wavelength of 633 nm.

At all three growth temperatures the loops exhibited perpendicular anisotropy with 100% remanence but the loop squareness was greater and coercivity smaller for the higher growth temperatures. These sapphire-based multilayers were used for the synchrotron X-ray diffraction studies since the seed film was thinner and produced less intense Bragg peaks than the Ag seed film used for growth on GaAs. The [111] axis of the multilayers grown on sapphire had a rocking curve width (FWHM ~2.1°) similar to that for multilayers grown on GaAs at the same temperature

(100°C). At a growth temperature of 300°C the rocking curve width was ~1.4° in both cases.

3. Magnetic Characterization of the Multilayers

A detailed description of magnetic properties of the present multilayer films will be given elsewhere[9]. Here we briefly summarize the results of that study. Magnetization measurements were made for all the multilayers grown on GaAs substrates and Figure 4 shows the saturation magnetization, measured at room temperature. M_S^{eff} refers to the magnetization divided by the Co volume in the multilayer and can be compared with the value for bulk (hcp) Co of 1422 kA/m.

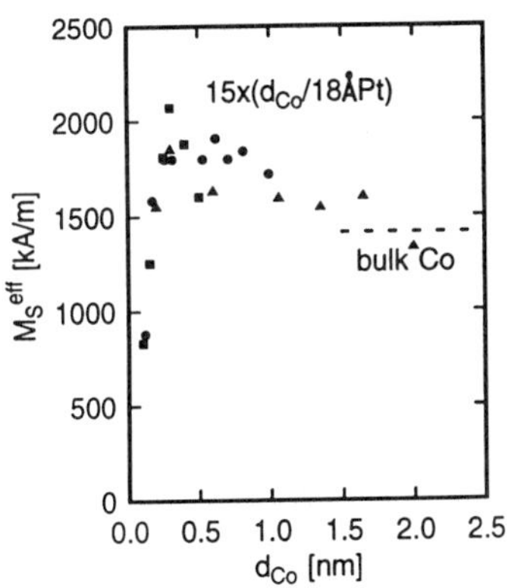

Figure 4. Saturation magnetization data for epitaxial Co-Pt multilayers. ▲ [111] multilayers, • [110] multilayers, and ■ [001] multilayers.

An enhanced magnetization is evident over the thickness range ~ 0.3-1.5 nm. This enhancement can be attributed to polarization of interfacial Pt by the Co. At larger Co thicknesses the magnetization approaches that of bulk Co since the relative number of interfacial atoms is diminished. On the other hand, for Co thicknesses below 0.3 nm the magnetization per Co volume drops off indicating a dilute Co alloy with reduced Curie temperature and/or the formation of a 2D magnetic layer (Co or Co-Pt alloy) which can also lead to a reduction in T_C, due to reduced symmetry.

Both torque magnetometry and VSM measurements confirmed that the [110]-oriented multilayers had a uniaxial in-plane anisotropy. The easy directions were [001] and [00$\bar{1}$], while along the hard directions ([1$\bar{1}$0] and [$\bar{1}$10]) , large

(>10KOe) fields were needed to saturate the magnetization. For these samples the anisotropy energy can be written phenomenologically as:

$$E_A = K_0 + K_{\parallel}^{eff}\sin^2\theta\cos^2\phi + K_{\perp}^{eff}\sin^2\theta$$

where ϕ is the angle between the magnetization and the in-plane easy axis [001], θ the angle between the magnetization and the sample normal [110], $K_{\perp}^{eff}$ and $K_{\parallel}^{eff}$ are the anisotropy constants and K_0 is a constant, independent of θ and ϕ. Torque magnetometry measurements were made for $\phi = 90°$, $\phi = 0°$ and $\theta = 90°$. For the three respective orientations the anisotropy constants $K_{\perp}^{eff}$, $K_{\perp}^{eff} + K_{\parallel}^{eff}$ and $K_{\parallel}^{eff}$ were derived. Figure 5 shows the data. In this figure the line through the upper set of data points is the sum of the fits to the $K_{\parallel}^{eff}$ and $K_{\perp}^{eff}$ data. These fits are shown in the figure. There is good agreement between the $K_{\parallel}^{eff} + K_{\perp}^{eff}$ data points and the sum curve, indicating that the data are self-consistent.

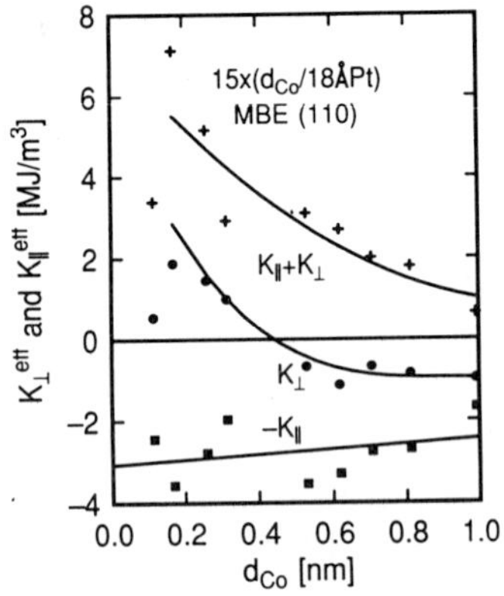

Figure 5. Anisotropy constants $K_{\perp}^{eff}$ and $K_{\parallel}^{eff}$ versus Co thickness in [110] oriented Co-Pt multilayers. (See text).

We now turn our attention back to comparison of the [111], [001] and [110] multilayers. If the effective anisotropy energy is written for all three orientations of the multilayers as:

$$K_{\perp}^{eff}d_{Co} = K_V^{eff}d_{Co} + 2K$$

then plots of $K_{\perp}^{eff}d_{Co}$ versus d_{Co} should give lines of slope K_V^{eff} and intercept $2K_S$. Figure 6 shows the results of such a plot.

The values of K_S and K_V^{eff} are indicated in the figure. Note that each multilayer orientation has a different slope and intercept. The value of K_S for the [111] multilayers, 0.97 mJ/m^2, is larger than that (0.8 mJ/m^2) found in highly textured <111> multilayers grown by evaporation[3] and smaller than that (1.15mJ/m^2) for MBE-grown Co/Pt sandwiches[10]. It is clear that the ansisotropy for the [001] multilayers is driven by a very large negative contribution to K_V^{eff} as has been observed for Co/Pd multilayers by den Broeder et al[5] and Engel et al[4]. This is likely due to the large strain for this orientation[11].

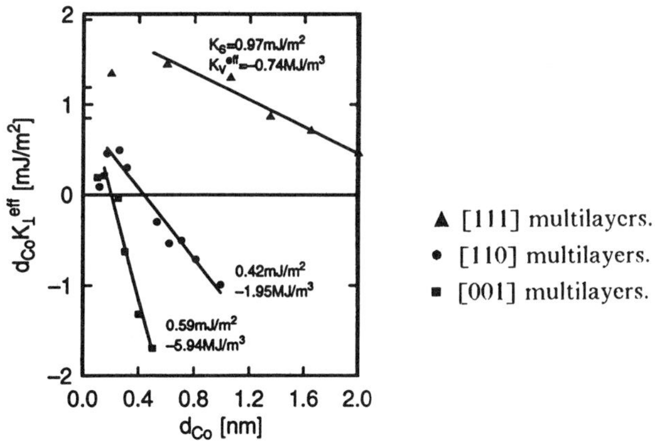

Figure 6. Plot of d_{Co} times $K^{eff}_\perp$ as a function of d_{Co} for epitaxial Co-Pt multilayers.

4. Structural Characterization of the Multilayers

Characterization of the multilayers, using in situ XPD (X-ray photoelectron diffraction)[8] and X-ray diffraction[12] had earlier showed that significant (2-4ML) interfacial interdiffusion was present for all growth orientations. Thus we have carried out a search for the chemically ordered phase $CoPt_3$ using synchrotron X-ray diffraction. A brief account of this work has been published elsewhere[13].

Figure 7 shows a schematic diagram of the ordered and disordered phases of the $CoPt_3$ alloy, together with the atomic arrangements in the (111) plane. In the ordered phase, the Co atoms occupy the corners of the unit cell whilst the Pt atoms occupy the face-centered sites. A comparison of the distribution of atoms in the (111) plane for both cases shows that there is a doubling of the size of the unit cell in the ordered phase. This doubling results in superstructure peaks in the reciprocal lattice which are absent for the disordered alloy. Figure 8 illustrates this. It shows a schematic diagram of the in-plane diffraction pattern ([111] zone-axis) for the multilayer. The circled crosses indicate the fundamental peaks. These are observed whether or not alloy ordering is present. They have Miller indices which are either all odd or all even. The elliptical spots indicate the position of superstructure peaks

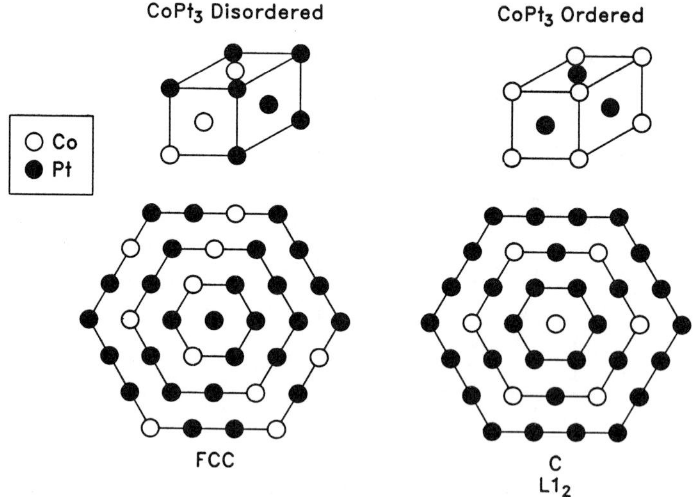

Figure 7. Schematic diagram of random alloy (Co to Pt ratio 1:3) and ordered $CoPt_3$ phase. Note the doubling of real space periodicity for the ordered structure.

which exist only if alloy ordering is present. These have mixed indices. The shape and intensity of these peaks depends on the extent and perfection of alloy ordering. The crosses show the tails from multilayer diffraction rods. This is diffraction that extends along the [111] direction from, for example, the {200} and $\{1\bar{1}1\}$ diffraction peaks. It results from the multilayer character of the sample and is similar to specular diffraction 'rods'.

The in-plane diffraction measurements included radial and phi scans indicated in Figure 8 by the arrows on solid and dashed lines, respectively. Radial scans were made through $(1\bar{1}0)$ and $(2\bar{2}0)$ peaks whilst a phi scan was made around the circle of {110} peaks. These scans were supplemented by out of plane scans through the (110) peak to probe the ordering normal to the plane of the Co-Pt interfaces.

Figures 9 and 10 show the results of the radial scans for the multilayers grown at 100 and 200°C respectively. The (2$\bar{2}$0) peaks are intense and relatively sharp, with ΔQ (FWHM) values of 0.05 and 0.035 Å^{-1} respectively. These correspond to coherence lengths (defined as 0.9 (2π/FWHM)) of 110Å and 160Å, respectively, and can be associated with the in-plane domain size. On the other hand, the (1$\bar{1}$0) scans show much weaker and broader peaks with widths corresponding to coherence

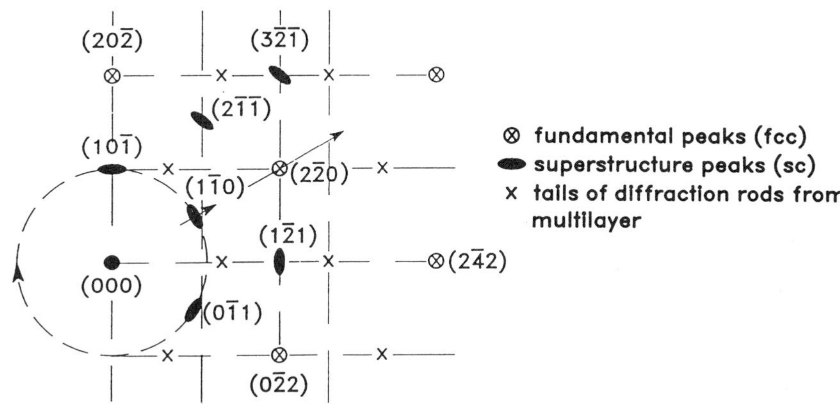

Figure 8. Schematic diagram of in-plane diffraction pattern (<111> zone axis) of Co-Pt multilayer.

lengths of only 30 and 15Å (100°C and 200°C growth, respectively). Thus alloy ordering is present but with a coherence length for chemical ordering of only ~ 10 interatomic spacings in the (111) plane. Out-of-plane scans for the (1$\bar{1}$0) peak showed a ΔQ of ~ 0.5Å^{-1} consistent with a 11Å extent of chemical ordering normal to the interfaces. Phi-scans revealed {1$\bar{1}$0} peaks with 6-fold symmetry as illustrated in Figure 11. This 6-fold symmetry, rather than 3-fold, is expected due to the twinning of the [111]-oriented multilayers. X-ray diffraction measurements were also conducted on [001]-oriented multilayers grown on GaAs to investigate the formation of a chemically ordered alloy for this growth orientation. We observed X-ray scattering at the location expected for diffraction from an ordered alloy (e.g., (100) diffraction peaks). Unfortunately, this location also corresponds to the expected position of X-ray scattering features due to the termination of the GaAs lattice. Despite this ambiguity, we believe that some of the observed intensity is due to

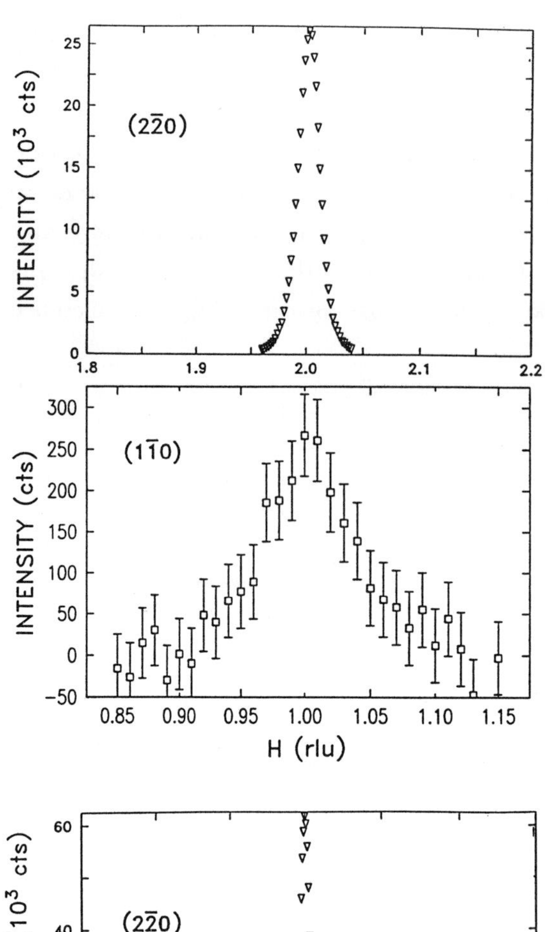

Figure 9. Radial (H$\bar{H}$0) scans: for 100°C multilayer. Note: background subtracted for (1$\bar{1}$0) peak.

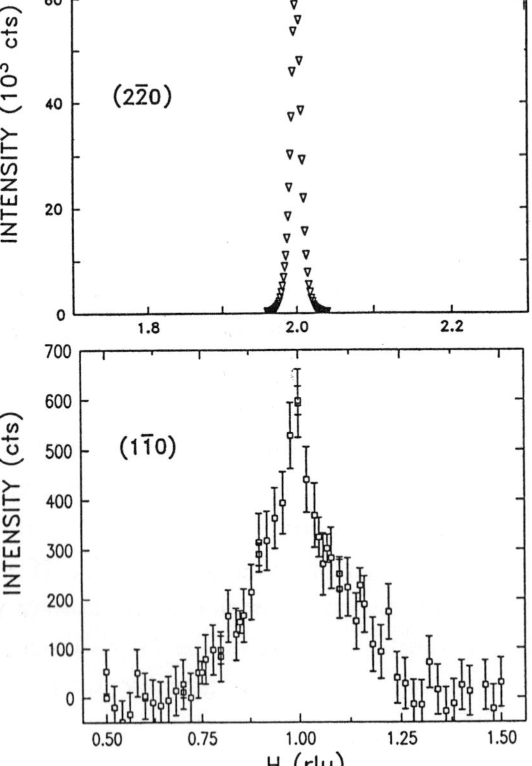

Figure 10. Radial (H$\bar{H}$0) scans: for 200°C multilayer. Note: background subtracted for (1$\bar{1}$0) peak.

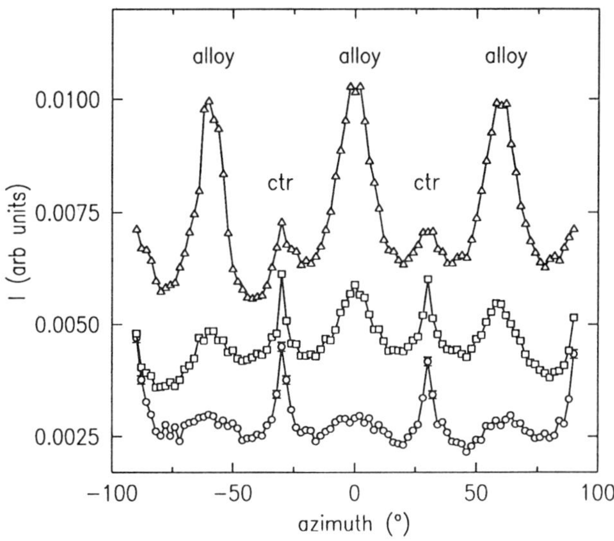

Figure 11. Phi scans through {110} Bragg peaks for 100°C, 200°C and 300°C multilayers. The $CoPt_3$ ordered alloy peaks and peaks due to tails of diffraction from the multilayer rods (crystal-truncation rods- ctr) are indicated. Note that the intensity scales for the data have been offset to show the progressive increase in alloy ordering with film growth temperature: O 100°C, □ 200°C, and △ 300°C.

ordered alloy formation because this feature was quite broad, similar to our observation for ordered alloys in the [111]-oriented multilayers.

Discussion

The diffraction data clearly demonstrate the doubling of real space periodicity characteristic of alloy ordering. However, such doubling can, in principle, arise from one of three ordered phases reported [14] for Co-Pt alloys: $CoPt_3$, CoPt and Co_3Pt. We believe that the first of these (illustrated in Figure 7) is present for the following reasons. First, mixing occurs[8,12] over 3-4 monolayers which favors a Pt-rich environment for Co. Second, we have made parallel studies of epitaxial alloy films grown, by co-evaporation, on sapphire (0001) at temperatures as high as 600°C. The other phases (CoPt and Co_3Pt) do not occur, even at the appropriate compositions and with subsequent annealing for several hours at this temperature. $CoPt_3$, on the other hand, forms[15] spontaneously during growth at temperatures at least as low as 300°C. Finally, since during multilayer growth, Co arrives at a pure Pt surface and has a tendency [16] to prefer Pt neighbors, it is natural that the Co forms the first monolayer of the $CoPt_3$ phase.

Estimates of the amount of $CoPt_3$ formed were made from the ratio of the integrated intensities of the $\{1\bar{1}0\}$ superstructure peaks and the $\{1\bar{1}1\}$ fcc multilayer peaks. For the 100 and 200°C multilayers the fraction of Co in the ordered phase is $20 \pm 10\%$ and $60 \pm 30\%$, respectively (this is the ratio of Co in the ordered alloy to total Co in the multilayer). For the 300°C multilayer, we estimate this fraction is about 90%. These large fractions should be incorporated into models for the magnetic anisotropy.

A consequence of this study is that it raises the question of whether a partially ordered alloy film, of composition close to $CoPt_3$, exhibits perpendicular magnetic anisotropy when it is not constrained within a multilayer. We find that this is indeed the case and that both epitaxial and polycrystalline alloy films can exhibit perpendicular magnetic anisotropy with high coercivity. A description of this work is in progress and will be published elsewhere[17,18].

Conclusions

• Evidence for the existence of the chemically ordered phase $CoPt_3$ within [111] oriented Co-Pt multilayers is presented.

• The extent of this phase is considerable, comprising $60 \pm 30\%$ of all the Co in the 200°C multilayer and $20 \pm 10\%$ in the 100°C multilayer.

• The existence of this phase confirms and extends our previous findings[9-10] of Co-Pt interdiffusion.

• Our results show that interfacial mixing and chemical ordering should be incorporated into a model for the origin of the magnetic anisotropy of Co/Pt multilayers.

• This study raises the question as to whether ordered phases are a general phenomenon in multilayers comprising miscible elements eg. Co-Pd, Fe-Pd, Fe-Pt, Ni-Pt, Co-Ni and many other important systems of magnetic interest.

• This study suggests and has led to the discovery of perpendicular anisotropy, high-coercivity, Co-Pt alloy films for compositions near $CoPt_3$[17,18].

References

1. C.H. Lee, R.F.C. Farrow, C.J. Lin, E.E. Marinero, C.J. Chien, Phys. Rev.B, 42, 11384 (1990).

2. C.H. Lee, R.F.C. Farrow, B.D.Hermsmeier, R.F. Marks, W.R. Bennett, C.J. Lin, E.E. Marinero, P.D. Kirchner, C.J. Chien, J. Magn.Magn. Mat., 93, 592 (1991).

3. C.J. Lin, G.L. Gorman, C.H. Lee, R.F.C. Farrow, E.E. Marinero, H.V. Do, H. Notarys, C.J. Chien, J. Magn. Magn. Mat. 93, 194 (1991).

4. B.N. Engel, C. D. England, R.A. Van Leeuwen, M.H. Wiedmann, C.M. Falco, Phys. Rev. Lett., 67, 1910 (1991).

5. F.J.A. den Broeder, D. Kuiper, H.C. Donkersloot, W. Hoving, Applied Physics, A 49, 507 (1989).

6. C.J. Chien, B.M. Clemens, S.B. Hagstrom, R.F.C. Farrow, C.H. Lee, E.E. Marinero, C.J. Lin, Mat. Res. Soc. Symp. Proc. 231, 465 (1991).

7. R.F.C. Farrow, G.R. Harp, R.F. Marks, T.A. Rabedeau, M.F. Toney, to be submitted to J. Crystal Growth.

8. (a) B.D. Hermsmeier, R.F.C. Farrow, C.H. Lee, E.E. Marinero, C.J. Lin, R.F. Marks, C.J. Chien, J. Appl. Phys. 69, 5646 (1991).

(b) R.F.C. Farrow, B.D. Hermsmeier, C.H. Lee, R.F. Marks, E.E. Marinero, C.J. Lin, C.J. Chien, S.B. Hagstrom, Mat. Res. Soc. Symp. Proc. Vol. 229, 115 (1991).

9. D. Weller, R.F.C. Farrow, C.H. Lee, C. Chappert, R.F. Marks, in preparation.

10. N.W.E. McGee, M.T. Johnson, J.J. de Vries, J. aan de Stegge, Submitted to J Appl.Phys. (1992).

11. T.A. Rabedeau, M.F. Toney, R.F.C. Farrow, R.F. Marks, G.R.Harp, in preparation.

12. X. Yan, T. Egami, E.E. Marinero, R.F.C. Farrow, C.H. Lee, J. Mat. Res. 7, 1309 (1992).

13. M.F. Toney, R.F.C. Farrow, R.F. Marks, G. Harp, T.A. Rabedeau, Mat. Res. Soc. Symp. Proc. Volume 262-F2: "Mechanisms of Epitaxial Growth", 237, (1992).

14. (a) "Binary Alloy Phase Diagrams", 2nd Edition. Editor in Chief, T. Massalski, Volume 2. Published by Materials Information Society, Materials Park, Ohio (1990).

(b) J.M. Sanchez, J.L. Morán-López, C. Leroux, M.C. Cadeville, J. Phys. C: Solid State Phys. 21, L1091 (1988).

15. M.F. Toney, T.A. Rabedeau, R.F.C. Farrow, R.F. Marks, G.R. Harp, in preparation.

16. M. Hansen, "Constitution of Binary Alloys", 2nd Edition, page 493. McGraw Hill (1958).

17. R.F.C. Farrow, R.H. Geiss, G.L. Gorman, G.R. Harp, R.F. Marks, E.E. Marinero, Proceedings of Magneto-Optical Recording International Symposium, December 7-9, 1992. Tucson, AZ.. To appear in a special issue of Journal of the Magnetics Society of Japan.

18. R.F.C. Farrow, D. Weller, R.H. Geiss, G.L. Gorman, G.R. Harp, R.F. Marks, E.E. Marinero, to be submitted to Applied Physics Letters.

EXAFS STUDY OF MAGNETIC METAL OVERLAYERS

D. Chandesris, H. Magnan [1], O. Heckmann and S. Pizzini

LURE, Université de Paris Sud F-91405 ORSAY France
[1] and SRSIM, CEN Saclay F-91191 GIF sur YVETTE France

INTRODUCTION

The knowledge of the crystallography of metastable magnetic films prepared by deposition on suitable substrates is the key for the understanding of their unusual magnetic properties. Surface EXAFS[1] is a very attractive technique to measure it. It gives to a high precision the shape of the first neighbour shell[2], including its possible asymmetry[3] and its thermal broadening[4,5] which is related to the elastic force constant between nearest neighbours (nn) in the film. Moreover, the linear polarization of the synchrotron radiation reveals information about the anisotropy of the crystallographic structure. Moreover EXAFS is a selective method: by measuring the EXAFS oscillations above the K edge of iron (or cobalt) we are sure to be sensitive only to the local order in the iron (or cobalt) film. Then, films of any thickness, coated films and multilayers can be characterized with the same precision. This can be illustrated by the examples of iron on copper and cobalt on copper metastable structures.

EXPERIMENTAL PROCEDURE

The EXAFS experiments were performed at the Laboratoire pour l'Utilisation du Rayonnement Electromagnétique (LURE) on the surface EXAFS set-up using a Si (311) double crystal monochromator installed on the DCI storage ring. The samples Fe/Cu (100) and Co/Cu (100) were prepared in a UHV chamber connected to the X-ray beam line. Iron (or cobalt) is evaporated from a high purity wire heated by electron bombardment in a vacuum better than $5x10^{-10}$ mbar (base pressure is $2x10^{-10}$ mbar). It is deposited at a rate of about 1 ML per min. on a clean Cu (100) surface checked by LEED and Auger spectroscopy. The evaporation rate is calibrated prior to the evaporation with a quartz

Magnetism and Structure in Systems of Reduced Dimension
Edited by R.F.C. Farrow *et al.*, Plenum Press, New York, 1993

microbalance. The thickness of the film is controlled by Auger spectroscopy and by measuring the X-ray absorption edge jump of the sample at the K edge of iron. Using this calibration, we estimate the thickness of the grown film with an accuracy of 10%. The local atomic structure of the Fe/Cu (100) and Co/Cu (100) films is then studied in situ.

The variations of the X-ray absorption coefficient of the samples are measured above the K edge of iron (7112 eV) or cobalt (7709 eV) in the total yield mode. The EXAFS spectra of the Fe-Cu multilayers are recorded ex situ above the K edge of iron and copper. Quantitative analysis of the EXAFS data is done using the general formula giving the EXAFS modulation function, χ, in the case of a linear polarized light and for a K edge[2,5]:

$$\chi(k) = -\sum_i (N_i^*/ kr_i^2) \exp(-2r_i/\lambda) \exp(-2C_{2i} k^2) B_i(k) \sin[2kr_i+\Phi_i(k)] \quad (1)$$

where the sum runs over all the shells i of neighbours situated at the distance r_i of the absorbing atom, $B_i(k)$ their backscattering amplitude, λ is the mean free path of the photoelectron, k its wave number ; $[2kr_i+\Phi_i(k)]$ is the total phase shift taking into account atomic potentials. C_{2i} is the mean square relative displacement between the absorbing atom and the neighbours i (including static and thermal disorder) and N_i^* is the effective number of atoms of the shell situated at the distance r_i:

$$N_i^* = \sum 3 \cos^2\alpha_i^j \quad (2)$$

where α_i^j is the angle between the electric field vector of the X-rays and the direction of the bond between the absorbing atom and the neighbour j of the shell i.

The backscattering amplitude $B_i(k)$ and phase shift $\Phi_i(k)$ amplitude are determined experimentally from reference EXAFS spectra: for cobalt (copper) films, we used bulk cobalt (copper). For iron films, we used the functions of cobalt slightly adjusted to optimize the data of bulk iron (the amplitude function is corrected for a low temperature disorder of 2.7×10^{-3} Å^2 and is found equal to that one calculated by McKale et al.[6]).

Fe ON Cu(100) : RECONSTRUCTION AND THIN FILM REORDERING BY COATING

A lot of work has been devoted to the study of metastable fcc iron films grown on Cu (100): it is shown that these films have magnetic properties strongly dependent on their thickness and their growth temperature[7-11] and a structural study[12] (done by a measure of the angular distribution of the $2p_{3/2}$ core level photoemission intensity) shows that the temperature of the substrate during the growth of the film influences its quality.

Experimental EXAFS spectra recorded at 77 K on two different 3.5 ML thick films grown respectively at 370 K and 300 K are represented in figure 1; it is clear that these films have different structures. The choice of 370 K for the higher growth temperature is governed by the fact that Auger spectroscopy indicates that iron is neither agglomerating on copper at this temperature nor intermixing significantly.

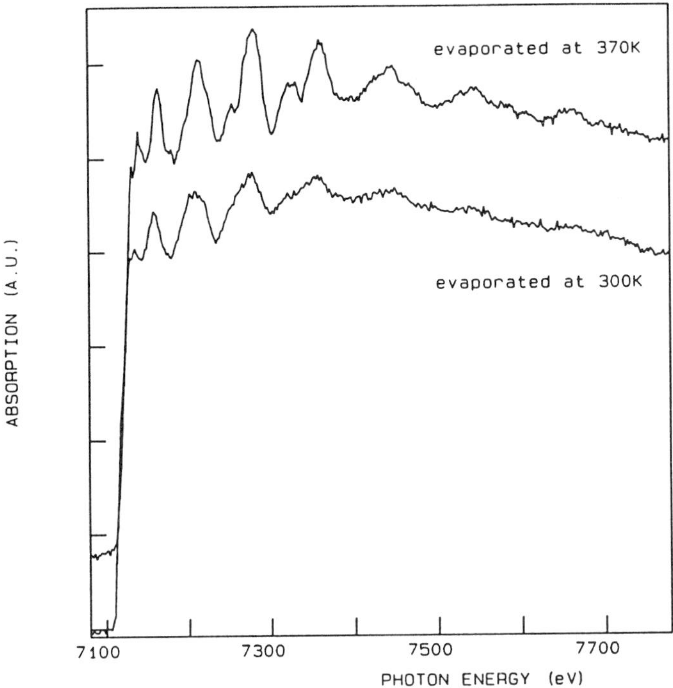

Figure 1. EXAFS spectra recorded at 77 K at the iron K edge for 3.5 ML thick films of iron grown respectively at 370 K and 300 K on Cu (100).

The Fourier transforms of these spectra are shown in figure 2 and compared to that of fcc copper and bcc iron. The 370 K prepared film has a spectrum similar to that of fcc copper with four well defined first shells of neighbours, confirming the existence of a well ordered fcc Fe film. The room temperature prepared one is less ordered: the first peak is weaker and the higher ones are quasi absent.

Due to the two-dimensional character of thin films, we can expect nearest neighbour bonds of different type if they are in-plane bonds (parallel to the interface plane) or out-of-plane bonds. The linear polarization of the synchrotron radiation allows to separate these two contributions since the contribution of each bond is weighted by a factor $\cos^2\alpha$ (equation 2) where α is the angle between the bond and the polarization direction of the X-rays. Most of the samples have been studied both in grazing incidence and in normal incidence to test their crystallographic anisotropy. The first shell radial distribution function (RDF) gaussian parameters are reported in table 1 for samples of different thicknesses and different growth temperatures. The measured coordination numbers are compared to the effective coordination numbers calculated for a perfect epitaxial fcc thin film of the same thickness, $N_{fcc}^* = \sum 3 \cos^2\alpha_j$. In our case, every iron atom has 4 nn in its plane, 4 above (missing for the top layer) and 4 below. When the polarization of the light is perpendicular to the surface

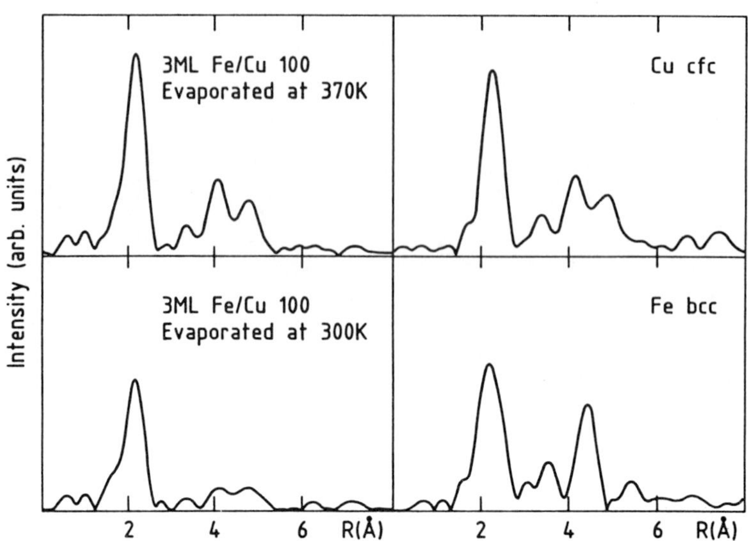

Figure 2. Fourier transforms between $k = 3.5$ Å^{-1} and $k = 11.3$ Å^{-1} for EXAFS spectra recorded at 77 K for 3 ML Fe/Cu (100) grown at 300 K, 3 ML Fe/ Cu (100) grown at 370 K and for bulk bcc iron above the iron K edge; and for bulk fcc copper above the copper K edge.

Table 1. Structural parameter results of least square fits of the first shell of neighbours for different films of Fe/Cu (100). N_{exp} is the experimental coordination number and N_{fcc}* is the effective coordination number for a perfect fcc epitaxial layer of the same thickness. R is the mean nn distance, C_2 the width of the RDF, and $\Delta\sigma^2$ the variation of the Debye-Waller factor between 300 and 77 K.

Film	Thickness (± 10%)	Growth Temp.	Incidence	N_{exp} (± 1)	N_{fcc}*	R (Å) (± 0.03)	C_2 (Å^2) (± 0.5 x10^{-3})	$\Delta\sigma^2$ (Å^2) (± 0.5 x10^{-3})
1	3 ML	300 K	Normal	11	11	2.52	7 x 10^{-3}	3 x 10^{-3}
2	3.5 ML	300 K	Normal	11	11.1	2.52	8 x 10^{-3}	3 x 10^{-3}
			Grazing	10.5	10.5	2.52	11 x 10^{-3}	<0.5 x 10^{-3}
3	8.4 ML	300 K	Normal	11.5	11.6	2.52	3.5 x 10^{-3}	4 x 10^{-3}
			Grazing	11.5	11.4	2.51	3.5 x 10^{-3}	3 x 10^{-3}
4	3.5 ML	370 K	Normal	11	11.1	2.53	2 x 10^{-3}	4 x 10^{-3}
			Grazing	10.5	10.5	2.53	2 x 10^{-3}	4 x 10^{-3}
5	6 ML	570 K (heated)	Normal	12	12	2.52	2 x 10^{-3}	---------

(grazing incidence) only the interlayer bonds contribute to the EXAFS signal but in normal incidence (polarization parallel to the surface) both interlayer bonds and intralayer bonds are contributing with the same weight.

The 3.5 ML thick film prepared at 370 K has the number of nn expected for an fcc structure at (2.53 ± .02) Å with a Gaussian width $C_2 = (2 \pm .5) \times 10^{-3}$ Å^2, the one prepared at room temperature has the same number of nn at (2.52 ± .02) Å but with higher values of C_2. This increase of the first shell Gaussian width is coherent with the loss of intensity in the Fourier transforms of the higher shells (figure 2). A broader RDF can reflect the presence of random static disorder in the layer or the existence of different nn distances which cannot be resolved.

Firstly, we will focus on the samples prepared at room temperature and try to describe their structure. The mean nn distance is always the same in grazing incidence and in normal incidence: this means that the interlayer bonds have the same length than the intralayer ones. Thus the broadening of the first shell RDF in these films is not due to an homogeneous tetragonal deformation of the fcc structure by compression or expansion along the 100 direction. In sample 2 the first shell Gaussian width (C_2) is larger in grazing incidence than in normal incidence. Since in normal incidence the intralayer bonds have the same weight than the interlayer ones, we can evaluate a Gaussian width for the intralayer bonds at $(5 \pm 1) \times 10^{-3}$ Å^2 while it is $(11 \pm .5) \times 10^{-3}$ Å^2 for the interlayer bonds: it is mainly the interlayer bonds which have not a well defined length. What is then the structure of this thin film? It can be described as a fcc distorted structure since the number of nn and the mean interatomic distance are identical to those of the well ordered fcc structure of iron (sample 4). We know also that the distortion influences mainly the interlayer bonds: the distortion is in the direction perpendicular to the interface. However EXAFS is not able to conclude if the interlayer bonds have randomly fluctuating lengths around the mean value or if they have well defined values. By simulating the interlayer RDF by a sum of three well defined shells of neighbours[13] we have shown that at least 50% of the atoms are displaced from the mean distance and that their displacement from the mean distance is at least 0.1 Å. These values allow to exclude another possible structure: flat layers of iron but different interlayer spacings. The fluctuation of 0.1 Å in the nn distances is too important and would be seen in the LEED studies. We can then conclude that the film is highly reconstructed (the displacement of the atoms around the positions they would have in the ordered fcc structure is important) and that the distortion is in the whole film, i.e. it does not concern only the surface atoms.

On similar films, different LEED reconstructions have been observed: a 4x1 and a 5x1[14,15]. Lu et al.[16] have done a precise quantitative LEED study: they cannot reproduce their experimental data assuming flat layers of iron. We show here that the reconstruction induces mainly distortion in the direction perpendicular to the interface. On the other hand MEED and RHEED experiments[14,17] show that at the first stages of deposition iron does not grow layer by layer on Cu(100) at 300 K. The growth before 5 ML is described as a bilayer by bilayer growth by Glatzel et al.[18] or it is characterized by iron agglomeration by Thomassen et al.[14]. The reconstruction that we observe by SEXAFS proves that the stable structure of these thin films is highly distorted compared to a perfect epitaxy and can be related to the fact that even at the first stages of adsorption, iron forms two layers high islands on Cu (100).

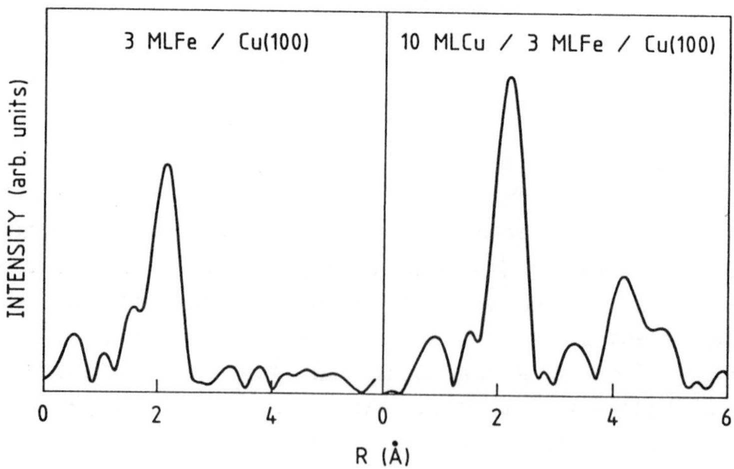

Figure 3. Fourier Transforms between k = 3.5 $Å^{-1}$ and k = 11.3 $Å^{-1}$ for EXAFS spectra recorded at 77 K for 3 ML Fe/Cu (100) grown at 300 K and this sample coated at 300K with 10 ML of Cu.

When we increase the thickness of the film grown at room temperature above 4 ML, we observe a decrease of the first shell width and at 8.5 ML (sample 3) an isotropic fcc structure with a bulk like static disorder and well defined higher shells of neighbours is obtained (for thicknesses higher than 11 ML the films begin to assume a bcc structure). This observation of a bulk-like fcc structure for films thicker than 5 ML is coherent with experiments which show a beautiful layer by layer growth for thicknesses higher than 5 ML. In fact, we observe a real modification of the structure in the whole film: the structure of the 8.5 ML thick film cannot be modelled as the superposition of 3.5 ML with the structure of sample 2 and 5 ML of a well ordered structure: even the first iron layers at the interface with the copper surface are reordered when the thickness of the iron film is increased.

The same modification of the structure of the interface is clearly identified in a Cu-Fe-Cu (100) sandwich structure. A sample identical to sample 2 has been coated at room temperature by 10 ML of copper: the Fourier Transforms of the EXAFS spectra are shown in figure 3. The structure of the Fe film is completely reordered by the upper Cu layer; it is characteristic of a well ordered fcc structure and the distortion has completely disappeared. This is a clear evidence for the importance of long range effects in the stabilization of different structures. Moreover it shows that the structure of a thin film can be completely different if it includes a free surface or if it is coated, and that it can be mistaken to assume that a thin film on a substrate is a good simulation of an interface. Another interesting result that we can deduce from this experiment is that the density of defects is not necessarily cumulative in an epitaxial film, and that a coating can have an ordering effect.

Moreover, one expects unusual elastic properties in these films. Daum et al.[19] indicate the presence of repulsive stress in Fe films prepared at 350 K. Temperature dependent EXAFS measurements give the amplitude of the thermal vibrations along the different bonds which are related to the effective force constants between nearest neighbours[5]. We have compared the filtered first shell EXAFS signals recorded on the same samples at room temperature,

$\chi(k,300K)$, and at 77K, $\chi(k,77K)$. Between these two temperatures we do not observe any structural change for films prepared at room temperature or above. Then we measure the variation of the Debye-Waller factor between these two temperatures, $\Delta\sigma^2 = C_2(300K)$-$C_2(77K)$ by the relation [4,5]:

$$\ln[\chi(k,77K)/\chi(k,300K)] = 2k^2\Delta\sigma^2 \qquad (3)$$

The results are displayed in table 1 last column. The most striking result is for sample 2 in grazing incidence: the amplitude of the thermal vibrations is about zero in the direction perpendicular to the surface. This means an effective interlayer force constant very high and a very anisotropic stress. This stress seems to be released when the isotropic fcc structure is achieved by increasing the thickness of the film (sample 3) or its growth temperature (sample 4). In these films $\Delta\sigma^2$ is the same in normal incidence and in grazing incidence and has an intermediate value between those of bulk fcc copper and bcc iron.

This study shows that, at least, two different structures are obtained when growing Fe on Cu(100). An isotropic well ordered fcc structure (for films of any thickness grown at temperatures higher than 350 K and for films grown at room temperature but thicker than 6 ML) and an anisotropically distorted fcc structure. As shown in table 1, in these two different structures the mean nn distance is the same, 2.52 ± .02 Å, and the number of these neighbours is the same. If we relate our results to the known magnetic properties determined in these films we can say that the films which have been found ferromagnetic with an easy axis perpendicular to the surface in references [7-9] have the anisotropic distorted fcc structure. The films found antiferromagnetic or paramagnetic by Macedo and Keune[10] are fcc. Pappas et al.[9] found an irreversible loss of magnetization on heating their thin films above 350 K; at this temperature they observe an irreversible change of the LEED. We show that these modifications are related to a transition between the distorted structure and the well ordered fcc structure.

A clear result of our work is that the ferromagnetic distorted thin films have the same lattice parameter as the well ordered ones prepared at higher temperature (even for films which have been annealed at 570 K and which have intermixed with copper the lattice parameter is the same (sample 5)). The modifications of the magnetism cannot be explained by changes in the volume of the unit cell as it was often suggested; they seem to be linked to a reconstruction of the entire film. The anisotropy of the elastic force constant that we have observed may play a role in the stabilization of the perpendicular ferromagnetism of these thin films. Moreover our measurements show unambiguously that by increasing the thickness of a solid film or by coating it with a metal, dramatic changes may be observed in the structure of the whole film and epitaxial defects may even be corrected.

Co ON Cu(100) : TETRAGONAL DISTORTION

Thin metastable films of cobalt grown on Cu (100) are also the subject of many magnetic and structural studies (see the papers of C.M. Schneider and co-workers and A. Cebollada and co-workers at this workshop). MEED oscillations combined with in situ STM experiments[20] show that cobalt grows layer by layer on Cu (100) except for the first two

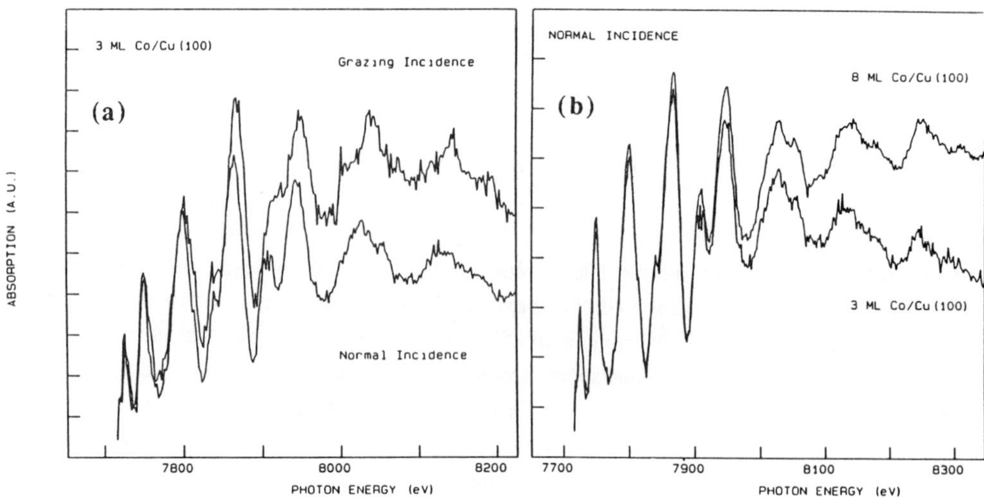

Figure 4. EXAFS spectra recorded at 77 K at the cobalt K edge
a) for 3 ML Co/Cu (100) grown at 300K in normal incidence and in grazing incidence.
b) for 3 ML Co/Cu (100) and 8 ML Co/Cu (100) grown at 300K in normal incidence.

layers. The films are always ferromagnetic, independent of the film thickness, and the remanent magnetization is always parallel to the interface[21], with a strong in-plane anisotropy : the [110] direction is the easy axis[22,23]. The Curie temperature of films evaporated at 450 K is strongly increasing with the film thickness from 130 K for 1.5 ML to 500 K for 2.5 ML[22]. LEED results indicate that cobalt epitaxially grows on Cu (100) with an fcc tetragonal structure[24].

Surface EXAFS with linearly polarized light can here give a precise measure of this metastable structure. The experimental conditions are identical to that ones described for the Fe/Cu(100) system. The X-ray absorption coefficient of thin films of cobalt of thicknesses between 2 and 8 monolayers grown at room temperature have been measured above the K edge of cobalt (7709 eV) both in grazing incidence and in normal incidence. They are reproduced on figure 4. On figure 4a, we directly see the tetragonal distortion of the 3 ML thick film: the frequency of the EXAFS oscillations is higher in normal incidence than in grazing incidence which means a longer nearest neighbour (nn) distance for intralayer bonds. On the other hand, on figure 4b, we clearly see that in normal incidence, the frequency of the EXAFS oscillations is the same for a 3 ML and an 8 ML thick films. The Fourier transforms of the EXAFS spectra of the 3 ML thick films are shown on figure 5: in normal incidence it is characteristic of an fcc well ordered structure, but in grazing incidence there is a loss of intensity and a displacement of the fourth shell. In a perfect fcc structure the fourth neighbours are alined with the first ones and the focusing effect[25] of the first shell is responsible for the anomalous high intensity of this shell. Its modification in grazing incidence must also be related to the structural anisotropy of these films.

The geometry of the Co/Cu (100) is the same one as for the Fe/Cu (100). In grazing incidence, EXAFS is sensitive only to the interlayer bonds (out of plane) and gives a precise measure of the out of plane nn distance. We then deduce the in plane nn distance from the normal incidence EXAFS spectra where the intralayer bonds and the interlayer ones contribute with the same weight. The results for samples of different thicknesses are summarized in table 2: for all the samples the in plane nn distance is 2.55 Å (± 0.01 Å) which is exactly the nn distance in bulk copper, and the out-of plane nn distance is 2.50 Å

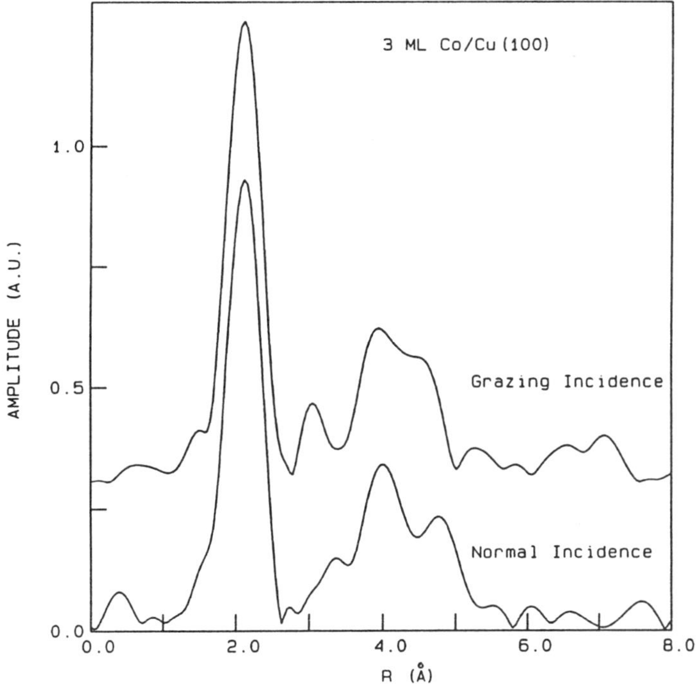

Figure 5. Fourier Transforms between $k = 2.55$ $Å^{-1}$ and $k = 11.64$ $Å^{-1}$ for EXAFS spectra recorded at 77 K in normal incidence and in grazing incidence for 3 ML Co/Cu (100) grown at 300 K.

(± 0.01 Å) which is the nn distance of bulk cobalt. So the tetragonal deformation of the films appears to be an expansion parallel to the interface to adapt exactly the in-plane nn distance of the film to substrate's one, but the out of plane nn distances keep the same value than in bulk cobalt. Then the lattice parameter perpendicular to the interface is contracted by 4% with respect to the parallel one. This tetragonal deformation of 4 % is the same for all the thicknesses up to 8 ML, and conversely to the Fe/Cu (100) system, there is no change of the first shell width when the thickness of the film increases.

Table 2. Structural parameter results of least square fits of the first shell of neighbours for different films of Co/Cu (100). N* is the effective coordination number for a perfect epitaxial film calculated for the in plane bonds and out of plane bonds in normal and grazing incidence. R is the mean nn distance determined by the fits for the in plane bonds and the out of plane bonds.

thickness (± 10%)	Incidence		N*	R (Å) (± 0.01 Å)
2 ML	Normal	in plane:	6	2.536
		out of plane:	4.5	2.51
	Grazing	in plane:	0	
		out of plane:	9	2.51
3 ML	Normal	in plane:	6	2.545
		out of plane:	5	2.51
	Grazing	in plane:	0	
		out of plane:	10	2.51
5 ML	Normal	in plane:	6	2.55
		out of plane:	5.4	2.50
	Grazing	in plane:	0	
		out of plane:	10.8	2.50
8 ML	Normal	in plane:	6	2.55
		out of plane:	5.6	2.50
	Grazing	in plane:	0	
		out of plane:	11.2	2.50

In summary, this EXAFS study shows that we observe about always the same structure when growing cobalt on copper (100) between 2 and 8 monolayers: a tetragonally distorted fcc structure. The cobalt is in perfect epitaxy on the substrate with an in-plane expanded nearest neighbour distance equal to that of bulk copper. This expansion means that there is an evident loss of symmetry: there is no more 111 symmetry axis. The width of the radial distribution function is also independent of the thickness. Even the thinner films are well ordered. The more constant magnetic properties measured in these films (always ferromagnetic and in-plane easy axis) are certainly due to this unique structure. Further temperature and polarization dependent experiments are needed to determine if the elastic properties of these films are also completely identical.

Fe-Cu MULTILAYERS: bcc Cu to fcc Cu

We will now focus on the structure of Fe/Cu multilayers prepared by sputtering.The samples are 60 x (15 Å Fe / t_{Cu} Cu) multilayers with t_{Cu}=6, 13, 24, 42 and 60 Å on a 50 Å Fe buffer deposited on Si(100) substrates at -10°C in a UHV compatible sputtering system[26]. The interest of this multilayer system arises from the oscillatory variation of the interlayer exchange as a function of the copper thickness, with a large magnetoresistance in the thickness ranges where the exchange is antiferromagnetic[27]. The first peak corresponding to the first antiferromagnetic range for Cu thickness around 14 Å is high and sharply defined, but as the Cu thickness increases, the oscillations are less clearly defined and their amplitude is small. Also, in contrast to what is observed in other multilayers and predicted by the theory, the amplitude of the maxima increases slightly as the Cu thickness goes from 25 Å to 50 Å.

These multilayers are characterized ex-situ by X-ray absorption spectroscopy both at the Fe K edge and Cu K edge in order to determine independently the structure of the Fe layers and of the Cu layers.

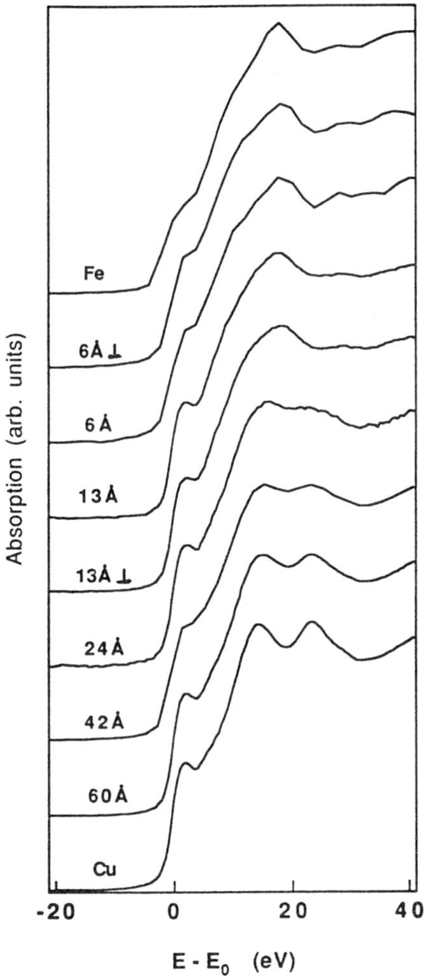

Figure 6. Near edge X ray absorption spectra for 60 x (Fe 15 Å/ Cu t_{Cu}) multilayers at the Cu K edge for t_{Cu} = 6, 13, 24, 42, 60 Å in comparison with bulk fcc Cu at the Cu K edge and bulk bcc Fe at the Fe K edge. Unless specified the spectra were recorded with the polarization of the X ray parallel to the surface, when ⊥ is specified the polarization of the X ray is perpendicular to the surface (grazing incidence).

The results[28] of both near-edge and EXAFS data at the Fe K edge show that the structure of the Fe layers is bcc and largely unaffected by the thickness of the Cu films in the multilayer. On the other hand, the Cu structure is strongly dependent on the Cu film thickness. One can see on figure 6 that for Cu thicknesses of 6 Å and 13 Å the near edge spectra have the typical features of the bcc structure: thin Cu films sandwiched between 15 Å thick Fe layers in the sputtered Fe/Cu multilayers grow with the bcc structure imposed by the Fe substrate. A quantitative analysis shows that the 6 Å thick Cu interlayers have a bcc structure which is isotropic and fairly ordered, as emphasized by the presence of large-distance peaks in the Fourier transform of the EXAFS spectra. The nearest neighbour distance in the Cu bcc structure is about 1% larger than the Fe-Fe distance in bcc Fe, and appears to be the same in the growth plane and in the growth direction (there is no tetragonal distortion). The second shell of atoms is characterized by very high Gaussian width, expressing the presence of a static disorder much larger than in the bulk copper. For 13 Å of Cu (t_{Cu}=13 Å) the bcc structure within the film is highly disordered and beyond this thickness the strained Cu layers tend to relax to the fcc structure typical of the bulk Cu phase. For t_{Cu}=24, 42 and 60 Å the near edge spectra have the typical features of the fcc structure of bulk copper (figure 6). The analysis of the EXAFS spectra of these samples gives a first neighbour shell characteristic of an fcc structure with 12 atoms at 2.55 Å, very close to the Cu-Cu distance in the bulk Cu phase. The Gaussian width associated with this shell decreases as the film thickness increases. The process of relaxation is gradual, and a more and more ordered fcc phase is formed as the Cu film thickness increases.

These results are in agreement with recent data on Fe(001)/Cu(001)/Fe(001) sandwiches prepared by MBE[29], where Cu was found to have the bcc structure of the Fe(001) substrate up to Cu thicknesses of 14-16 Å.

The first peak of antiferromagnetic exchange in these Fe-Cu multilayers is clearly correlated to a copper thickness range in which the structure of Cu is bcc (disordered but without tetragonal distortion). For thicknesses above about 24 Å, the structure of the Cu layers is a more and more ordered fcc. Since the electron mean free path is expected to be longer in ordered structures, the progressive ordering probably accounts for the "anomalous" increase of the magnetoresistance as the Cu thickness goes from 24 Å to 60 Å.

In conclusion, we think that EXAFS and surface EXAFS are techniques which allow a precise structural characterization of thin films and multilayers. In these systems, we have shown that different kind of metastable structures can exist depending on the thickness of the layers, their growth temperature and their coating. These different structures can be connected to the original magnetic properties of these systems.

REFERENCES

1. J. Stöhr, D. Denley, and P. Perfetti, Phys. Rev. B **18**, 4132 (1978);
P. H. Citrin, P. Eisenberger, and R. C. Hewitt, Phys. Rev. Lett. **41**, 309 (1978).
2. D. E. Sayers, E. A. Stern and F. W. Lytle, Phys. Rev. Lett. **27**, 1204 (1971).
3. G. Bunker, Nucl. Instr. Meth. **207**, 437 (1983);
H. Magnan, D. Chandesris, G. Rossi, G. Jezequel, K. Hricovini, and J. Lecante, Phys. Rev. B **40**, 9989 (1989).

4. R. B. Greegor and F. W. Lytle, Phys. Rev. B **20**, 4902 (1979).
5. P. Roubin, D. Chandesris, G. Rossi, J. Lecante, M. C. Desjonquères, and G. Tréglia, Phys. Rev. Lett. **56**, 1272 (1986);
P. Roubin, D. Chandesris, G. Rossi, and J. Lecante, J. Phys. F **18**, 1165 (1988).
6. A. G. McKale, G. S. Knapp, and S. K. Chan, Phys. Rev. B **33**, 841 (1986).
7. D. Pescia, M. Stampanoni, G. L. Bona, A. Vaterlaus, R. F. Willis, and F. Meier, Phys. Rev. Lett. **58**, 2126 (1987).
8. C. Liu, E. R. Moog, and S. D. Bader, Phys. Rev. Lett. Phys. **60**, 2422 (1988).
9. D. P. Pappas, K. P. Kämper, and H. Hopster, Phys. Rev. Lett. **64**, 3179 (1990).
10. W. A. A. Macedo and W. Keune, Phys. Rev. Lett. **61**, 475 (1988).
11. W. Schwarzacher, W. Allison, R.F. Willis, J. Penfold, R. C. Ward, I. Jacob, and W. F. Egelhoff, Solid State Commun. **71**, 563 (1989).
12. D. A. Steigerwald, I. Jacob and W. F. Egelhoff, Jr., Surf. Sci. **202**, 472 (1988).
13. H. Magnan, D. Chandesris, B. Villette, O. Heckmann and J. Lecante, Phys. Rev. Lett. **67**, 859 (1991).
14. J. Thomassen, B. Feldmann and M. Wuttig, Surf. Sci. **264**, 406 (1992).
15. C. Egawa, E. M. Mc Cash, and R. F. Willis, Surf. Sci. **215**, L271 (1989).
16. S. H. Lu, J. Quinn, D. Tian, F. Jona, and P. M. Marcus, Surf. Sci. **209**, 378 (1989).
17. D. A. Steigerwald and W. F. Egelhoff, Jr., Surf. Sci. **192**, L887 (1987).
18. H. Glatzel, Th. Fauster, B. M. U. Sherzer and V. Dose, Surf. Sci. **254**, 58 (1991).
19. W. Daum, C. Stuhlmann, and H. Ibach, Phys. Rev. Lett. **60**, 2741 (1988).
20. C. M. Schneider this workshop.
21. M. Stampanoni, Appl. Phys. A **49**, 449 (1989).
22. C. M. Schneider, P. Bressler, P. Schuster, J. Kirschner, J. J. de Miguel, and R. Miranda, Phys. Rev. Lett. **64**, 1059 (1990).
23. H. P. Oepen, M. Benning, H. Ibach, C. M. Schneider, and J. Kirschner, Journal of Magn. Magn. Mat. **86**, L 137 (1990).
24. A. Clarke, G. Jennings, R. F. Willis, P. J. Rous, and J. B. Pendry, Surf. Sci. **187**, 327 (1987).
25. P. A. Lee and J. B. Pendry, Phys. Rev. B **11**, 2795 (1975).
26. J. M. Slaughter, W. P. Pratt, P.A. Schroeder, Rev. Sci. Instrum. **60**, 127 (1989).
27. F. Petroff, A. Barthelemy, D. H. Mosca, D. K. Lottis, A. Fert, P. A. Schroeder, W. P. Pratt, R. Loloee and S. Lequien, Phys. Rev. B **44**, 5355 (1991).
28. S. Pizzini, F. Baudelet, D. Chandesris, A. Fontaine, H. Magnan, J. M. George, F. Petroff, A. Barthelemy, A. Fert, R. Loloee and P. A. Schroeder, Phys. Rev. B **46** (1992).
29. B. Heinrich, Z. Celinski, J. F. Cochran, W. B. Muir, J. Rudd, Q. M. Zhong, A. S. Arrott, K. Myrtle and J. Kirschner, Phys. Rev. Lett. **64**, 673 (1990).

SPIN-DEPENDENT EMPTY ELECTRONIC STATES AT MAGNETIC SURFACES

Markus Donath

Max-Planck-Institut für Plasmaphysik
EURATOM Association
Boltzmannstr. 2
W-8046 Garching b. München, Germany

INTRODUCTION

What can be learnt about magnetism in systems of reduced dimension from measurements of the electronic structure close to the Fermi level? In a microscopic picture the magnetic moments in $3d$ itinerant ferromagnets result from the different number of electrons with spin magnetic moment parallel and antiparallel to a given direction. The energy levels of these two subsets of electrons, called majority and minority electrons, respectively, are separated from each other by an energy-, momentum-, and possibly temperature-dependent exchange splitting ΔE_{ex}. Studying the electronic states and their energy vs. momentum relation $E(\mathbf{k})$ as a function of the quantum number spin helps one to understand ferromagnetic phenomena on a microscopic level. Since ferromagnetism is a collective order phenomenon, novel properties are expected when the dimension of the system is reduced, which means lower symmetry and smaller coordination number of the atoms. Various experimental techniques are currently employed to investigate these systems.[1–4] Electronic states located with their wave functions at the surface of a three-dimensional ferromagnet or at the interface where two different materials meet serve as a local probe where modified magnetic properties may occur. Exploring the spin dependence of these states is essential to develop a microscopic picture of magnetic phenomena at surfaces, in ultrathin films, and in layered structures, e.g. magnitude of magnetic moments, sign and strength of exchange coupling at surfaces and between layers, and magnetoresistance effects. To date the understanding of magnetic phenomena on an electronic-structural basis is far from being complete. This is a challenge to experimental as well as theoretical approaches[5] to investigating and explaining the spin-dependent electronic structure.

Experimentally, the most direct access to the electronic structure is given by photoemission (PE)[6,7] and inverse photoemission (IPE).[8–11] PE provides information on the states below the Fermi level E_F and above the vacuum level E_V, while IPE probes the empty states above E_F, especially the states between E_F and E_V inaccessible to PE. Performing these techniques in a spin-resolved mode promises detailed results on the exchange splitting ΔE_{ex} of the electronic states.[12,13] The low-energy electrons involved give these techniques a surface sensitivity of a few atomic layers only. Consequently, PE and IPE are well suited to studying the very surface as well as states some layers deep, which are almost bulk-like in metals. Interface states are expected to be accessible, depending on how deep the interface is buried.

Completely empty as well as occupied electronic states do not contribute to the magnetic moment. Therefore, the partly occupied/empty states close to E_F called *magnetic* states are of special interest. By hybridization effects, however, states located some eV above or below E_F also carry 'magnetic' information.

Spin-resolved PE work started about 20 years ago.[13–16] For the detection of the electron spin polarization in experiment one has to pay the price of an intensity loss of typically four orders of magnitude. Comparing PE and IPE for quantum energies in the ultraviolet, relevant for studies of the valence band structure, the cross section for the IPE process is 10^4 times smaller owing to phase space factors. IPE in a spin-resolved mode, however, does not suffer from an additional intensity loss, because it is possible to produce spin-polarized electron beams equivalent in intensity to non-spin-polarized beams. This fact, in principle, makes spin-polarized IPE (SPIPE) comparable to spin-resolved PE in respect of measuring time. In practice, the use of high-intensity synchrotron light makes data recording considerably faster for spin-resolved PE than for SPIPE.

This paper reports on SPIPE results concerning the spin-dependent empty electronic states at magnetic surfaces. Special attention will be given to surface electronic states and changes of the spin-dependent electronic structure caused by adsorption of non-magnetic adatoms.

SPIN-POLARIZED INVERSE PHOTOEMISSION

The technique of SPIPE is now ten years old. Its first application in 1982 was to show the minority character of the empty *d* band in nickel observed on Ni(110).[17] The exchange splitting of empty bands at Fe(110) and Fe(001) and their temperature behavior were studied shortly afterwards.[18–20] The next studies reported in the literature were concerned with chemisorption of O and CO on Ni(110), the results being interpreted as an adsorbate-induced reduction of the surface magnetization.[21,22] A subsequent study on Ni(001) was hindered by ill-defined surface magnetization conditions, which therefore did not show any spin splitting in the *sp* bands.[23] A comprehensive study on Ni(110) exhibited spin-split *sp* bands, surface states, and adsorbate-induced states.[24–26] The temperature dependence of a partly empty, and hence truly magnetic *d* band showed collapsing band behavior for spin-up and spin-down states upon approaching the Curie temperature.[27] Results on the temperature dependence of ΔE_{ex} are essential for testing theoretical models to describe ferromagnetism.[19,28–30] Besides electronic-structural studies, the high spin asymmetry of a direct transition into the unoccupied *d* band at Ni(110) was used as a detector for surface magnetization to study the influence of surface roughness and chemisorption on magnetic hysteresis curves.[31] The first SPIPE work on ultrathin films reported in the literature was concerned with investigating the empty states of Fe and Co overlayers on W(001), showing spin-split states in two-dimensional systems.[32] A reinvestigation of

Ni(001) dealt carefully with the sample magnetization.[33] As a result, spin-split *sp* bands as well as surface states were detected.[34–36] Recently, the exchange splitting of image-potential-induced surface states was the subject of extensive experimental work.[35,37,38] A number of groups is currently building and testing new equipment to perform SPIPE studies.[39,40] This reflects the increasing interest in experimental data of the spin-dependent electronic structure above E_F, complementing information gained by spin-resolved PE on the states below.

Figure 1 shows a schematic diagram for SPIPE. Low-energy spin-polarized electrons usually emitted from a GaAs photocathode[41] impinge on a remanently magnetized sample. They may occupy empty states of energy $E_{i\uparrow,\downarrow}$ above E_V and then undergo radiative transitions to lower-lying empty states of energy $E_{f\uparrow,\downarrow}$. The emitted photons are detected either at a fixed energy in an energy-selective Geiger-Müller counter ($\hbar\omega = 9.6 \pm 0.4$ eV or 9.4 ± 0.2 eV)[42] or at variable energy in a monochromator. For intensity reasons, spin-resolved measurements have hitherto been taken exclusively in the isochromat mode. In this mode spectra are recorded by varying the kinetic energy of the incoming electrons, with the angle of incidence θ, the detection angle of the photons α, and the electron spin polarization as parameters. The experimental setup used in this work was equipped with two Geiger-Müller counters at different angles with respect to the electron beam. This allowed us to study the symmetry character of the states involved in the transitions by recording their emission characteristics. Experimental details are reported elsewhere.[25,26,43]

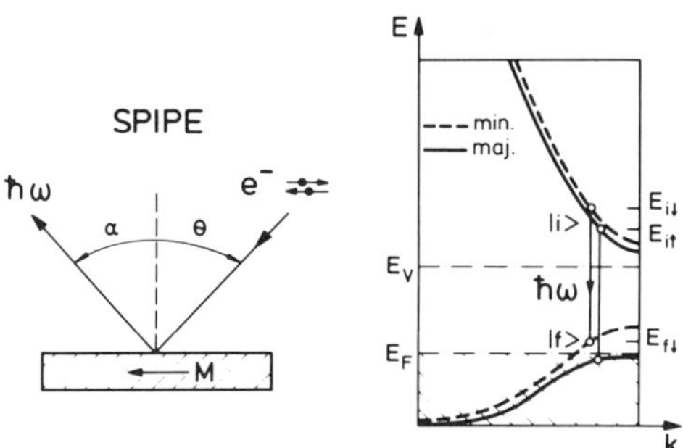

Figure 1. Schematic for spin-polarized inverse photoemission (SPIPE).

Special experimental requirements have to be met to make successful angular- and spin-resolved measurements involving low-energy electrons. The angle between the directions of electron spin polarization and sample magnetization has to be known, which calls for well-defined sample magnetization. In addition, angular-resolved measurements are only possible with samples remanently magnetized in-plane to prevent magnetic stray fields. Samples clamped to a horse-shoe magnet or picture-frame single crystals form a closed magnetic circuit, thereby minimizing stray fields. They can be remanently magnetized by applying a current pulse through a magnetization coil wound around the rear part of the sample. Special attention has to be paid to surfaces with no axis of easy magnetization in the plane of the magnetic circuit[31] or, even more complicated, in the plane of the surface.[33] In any case, the magnetic domain structure in remanence has to be carefully checked, e.g. by Kerr microscopy. In special cases measurements may only be possible at elevated temperatures where the magneto-crystalline anisotropy is sufficiently reduced.[33] Since the surface magnetic domain structure is also influenced by internal or external strain, a way of magnetic sample preparation has to be found for each individual sample to get a defined magnetic domain structure in remanence.

Coming back to Fig. 1, where well-defined sample magnetization is assumed, one sees that the incident spin-polarized electron beam probes either minority or majority bands, depending on its polarization direction. The schematic $E(\mathbf{k})$ diagram shown in the right part of Fig. 1 describes a ferromagnetic material with a partly empty spin-split band, which contributes directly to the magnetic moment of the material. For a given quantum energy defined by the length of the vertical arrow, a transition between minority states is possible, while majority electrons do not find empty final states. Qualitatively, this situation is realized in nickel, where the uppermost d band is partly occupied and responsible for the ferromagnetism. This is confirmed by SPIPE data of a bulk direct transition into empty d states of symmetry Z_2 on the X-(Z)-W line observed on Ni(110) at $\Theta = 25°$ in the $\overline{\Gamma}\overline{X}$ azimuth.[25] The room temperature data are shown in the left panel of Fig. 2. One sees high intensity for minority electrons but almost no intensity for majority electrons. The small majority d-band intensity results from the magnetization being slightly reduced at room temperature ($T/T_C = 0.48$). But not all spectral features close to E_F in data of nickel surfaces reflect transitions into minority d bands. SPIPE data of Ni(111) also displayed in Fig. 2 demonstrate in a convincing way the additional information offered by spin-resolved measurements. The spin-integrated data look almost identical to those obtained for Ni(110). The spin-resolved data, however, exhibit much smaller spin asymmetry leading to a different interpretation of the spectral feature. A detailed study shows that transitions into d states as well as into a surface state are involved.[44] This example demonstrates that spin resolution not only adds information on the spin splitting of electronic states, but also helps to identify them first.

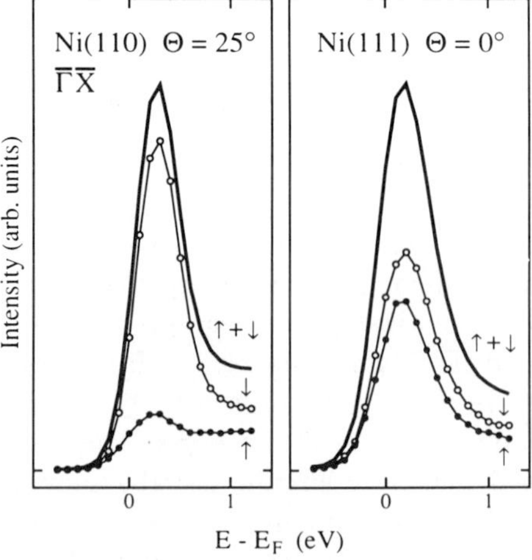

Figure 2. Spin-integrated (↑ + ↓) and spin-resolved (↑, ↓) IPE data ($\hbar\omega = 9.4\,\mathrm{eV}$) of transitions into states just above E_F on Ni(110) and Ni(111). Spin resolution permits the different origins of the spectral features to be identified.

CRYSTAL-INDUCED SURFACE STATES

Concerning optical transitions between states described by bulk band structures the reader is referred to the literature.[17–20,25–27,34–36] This part of the paper focusses on electronic states whose wave functions are concentrated within the topmost layer(s). These states usually appear in gaps of the projected bulk band structure. They are caused by the broken symmetry at the surface and are derived from bulk bands, therefore being called crystal-induced surface states. The magnetic exchange splitting of these states acts as a sensor of surface magnetic properties, which may be quite different from the corresponding bulk properties. Non-spin-resolved PE has detected occupied surface states on nickel as double-peak structures interpreted as energetically separated minority- and majority-spin emissions.[45–47] SPIPE succeeded in detecting empty exchange-split surface states on all three low-index nickel surfaces.[25,26,35,36,44] The size of the observed splittings was found to be of the same order as the splittings of the bulk states they are derived from. These experimental results directly prove that the topmost layers of the samples under investigation are magnetically active. Since PE and IPE probe the electronic states in a **k**-resolved mode, they give detailed information on specific electronic states at selected points of the Brillouin zone. Consequently, quantitative information on **k**-integrated quantities such as slightly reduced or enhanced magnetic moments at the surface cannot be directly deduced from the data.

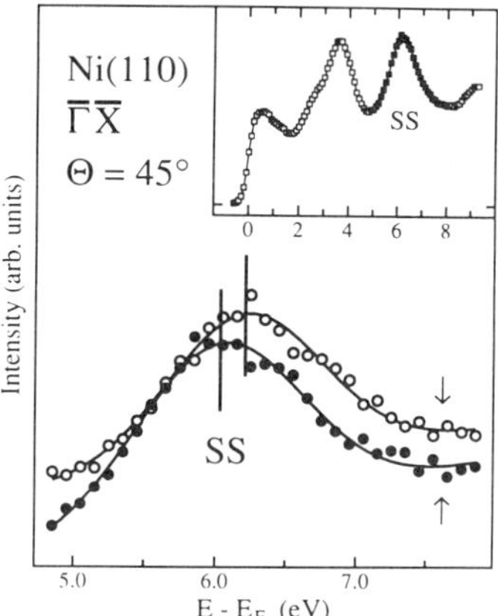

Figure 3. SPIPE data ($\hbar\omega = 9.6\,\mathrm{eV}$) of the crystal-induced surface state on Ni(110) around $\overline{X}$. Inset: Spin-integrated overview spectrum. Data points originating from the surface state emission are emphasized by filled symbols.

Spin-resolved IPE data of the crystal-induced surface state on Ni(110) are displayed in Fig. 3. The surface state appears in a gap around the $\overline{X}$ point of the surface Brillouin zone about 6 eV above E_F. At $\overline{X}$ the gap is confined by the critical points $X_{4'}$ and X_1. The more s-like upper band-gap boundary, which the surface state is derived from, exhibits an exchange splitting $\Delta E_{\rm ex}$ of about 200 meV due to hybridization with the magnetic d bands.[48] The p-like band at the lower edge cannot hybridize with the d bands for symmetry reasons, therefore showing a smaller spin splitting. Experimentally, the surface state is observed to be exchange split by $\Delta E_{ex} = 170 \pm 30$ meV in good agreement with the splitting of the upper band-gap boundary. At least, the result gives no hint of a magnetic moment at the surface significantly different from that in the bulk. A similar situation as described for Ni(110) has been found for Ni(001) at $\overline{X}$ in the $L_{2'}$-L_1 gap.[36]

It should be mentioned that not necessarily both spin counterparts of surface state emission are detected. A spin-resolved PE study on Fe(001) identified minority surface states whose majority counterparts were not observed owing to broadening effects caused by overlap with majority bulk states.[49] In summary, crystal-induced surface states reflect the consequences that the modified symmetry situation at the surface has for the electronic structure. Their exchange splittings serve as a local probe for surface magnetic properties.

IMAGE-POTENTIAL-INDUCED SURFACE STATES

Electrons may also be found in a different kind of surface states whose wave functions are localized a few Å in front of the surface. An electron approaching a conductive surface feels the attractive force of its own image potential, created by the polarization charge it induces at the surface. Provided the reflectivity of the surface is high, i.e. there are no bulk states present to which the electron can couple, the electron may be trapped between the bulk crystal barrier and the image-potential surface barrier, thus giving rise to a Rydberg-like series of bound states: the image-potential-induced surface states.[50–52] The states are pinned to E_V with binding energies of less than 1 eV. As a function of $\mathbf{k}_{||}$ they exhibit free-electron-like dispersion with effective masses m^*/m usually close to one. Unlike crystal-induced surface states, whose wave functions peak predominantly within the topmost atomic layer, the image-potential states are located a few Å outside the surface, therefore only slightly overlapping with

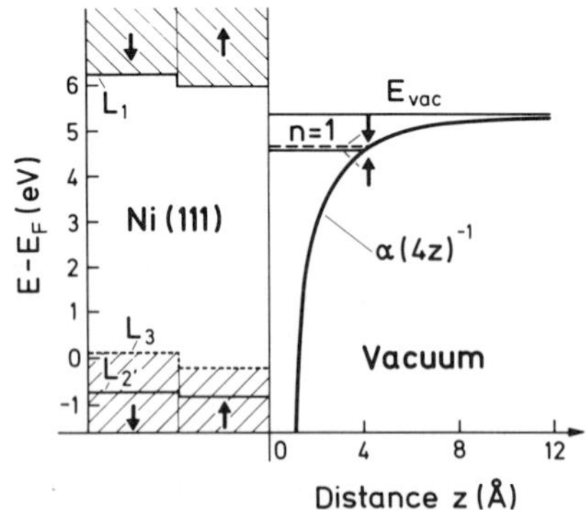

Figure 4. Schematic potential diagram for image-potential surface states on Ni(111), indicating the image-potential barrier outside the crystal and the bulk sp-band gap between $L_{2'}$ and L_1 including the uppermost d band of symmetry L_3.

bulk states. As a consequence, typical linewidths of the states are 20 to 80 meV due to their long lifetime.[53,54] In addition, image states are not expected to show spin splittings of the same size as bulk states or crystal-induced surface states. Calculations within the one-step model of inverse photoemission give for the expected spin splittings values of 100 meV for Fe(110),[55] 13 meV for Ni(001),[36] and 27 meV for Ni(111).[56] The calculations assume the image potential to be spin-independent. The calculated spin splittings are a consequence of the spin-dependent energy positions of the band-gap boundaries, which form the crystal barrier. Since a possible influence of a spin-dependent barrier potential[55,57] has not been taken into account in the calculations, the given numbers are lower limits for the expected spin splittings. Figure 4 gives a schematic potential diagram for Ni(111) demonstrating the spin-dependent height of the crystal barrier due to the spin-split energy levels of the band-gap boundaries resulting in a small difference of the binding energies of the spin-up and spin-down image states.

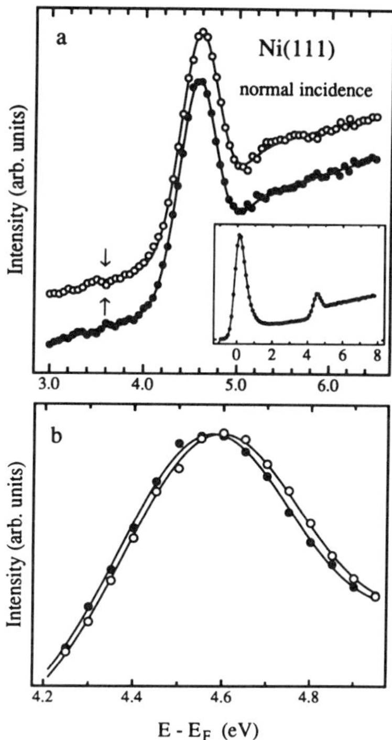

Figure 5. (a) SPIPE data ($\hbar\omega = 9.4\,\text{eV}$) of the image-potential-induced surface state on Ni(111). Inset: Spin-integrated overview spectrum.
(b) Same data on an enlarged energy scale with the spin-dependent background offset suppressed.

Experimentally, the image-potential-induced surface states have been discovered by IPE. A weak spectral feature observed at an energy within a bulk band gap and close to E_V exhibiting free-electron-like $E(\mathbf{k}_{\|})$ dispersion, it was proposed, should be interpreted as an image-potential surface state.[58] Shortly afterwards, $n = 1$ image states were uniquely identified by their pinning to E_V, their photon emission characteristics, their temperature dependence,[59] and their independence on $\mathbf{k}_\perp$.[60] High-resolution two-photon photoemission (2PPE) measurements were able to resolve the first three members of the Rydberg series.[61–63] As a result of the lack of spin resolution, however, 2PPE failed to detect any spin splitting, because the intrinsic linewidths turned out to be larger than the spin splittings. Upper limits for possible spin splittings of the $n = 1$ states were deduced from fitting procedures: 40 meV for Ni(111) (linewidth: 84 meV),[63] 30 meV for Co(0001) (linewidth: 95 meV), and 80 meV for Fe(110) (linewidth: 130 meV).[64]

SPIPE is constrained by its state-of-the-art energy resolution of 300 to 400 meV, yet it permits the two partial spin spectra to be recorded *separately*. Consequently, the detection of spin splittings is not limited by the energy resolution or intrinsic linewidths. On Ni(110), where the image-state emission is weak owing to the lack of a bulk band gap, the adsorption of sulfur produced a well-pronounced spectral feature with a spin splitting of 32±13 meV.[37] A study on clean Ni(001) revealed a hint of nonzero splitting of the image state: ΔE_{ex}=13±13 meV.[35] An image state unambiguously spin-split by 18±3 meV has been detected on Ni(111).[38] The data of this experiment are shown in Fig. 5. The procedures by which the small splitting and its confidence level were deduced from the data are desribed in detail elsewhere.[38] The determined splitting, about one order of magnitude smaller than the splitting of the band-gap boundaries, is in accord with the calculations described above. Consequently, the possible influence of a spin-dependent barrier potential on the splitting of image states seems to be negligible. In summary, information obtained from image states a few Å in front of the surface is of importance in the interaction of electrons at magnetic surfaces and complements information gained from crystal-induced surface states on the first layer and from bulk-like states on the first few layers.

ADSORBATE-INDUCED CHANGES

Adsorption of adatoms different from the surface atoms will certainly change the surface electronic structure. The modification depends on the bonding mechanism and, in general, is not easy to interpret. During recent years special attention has been paid to chalcogen adsorption on transition metals.[65–70] This is partly due to the fact that sulfur is known to be a prototypical adsorbate catalytic poisoner on Ni and Fe, materials often used as catalysts. A further interesting aspect is the modification of the magnetic properties of the ferromagnetic surfaces involved. While for some years experimental studies focussed on adsorbate-induced reduction of the magnetization at the surface,[21,71,72] more recent work has concentrated on specific electronic states. By using spin-polarized Auger electron spectroscopy,[73,74] IPE,[24,25] and PE[75–78] exchange-split adsorbate-induced electronic states have been detected, indicating a nonvanishing magnetic moment even at the non-magnetic adsorbate atoms. This provides a more detailed picture of the interaction of non-magnetic adatoms with ferromagnetic surfaces.

Adsorbate-induced changes of the electronic structure are detected in IPE as energetic shifts and/or (dis)appearance of spectral features whose spin dependence reflects modifications of surface magnetic properties. Usually surface as well as bulk state emissions of the clean surface are quenched by adsorption.[79] Image states are an exception, as one might expect from their quite different origin compared with the 'regular' electronic structure. Image states are shifted according to the work function change because of their pinning to E_V.[59] Their intensity is determined by the crystal reflectivity, which may be changed, either reduced or enhanced,[37] depending on the modified electronic structure at the surface. In many adsorption systems additional spectral features are observed. If the adsorbate forms an ordered overlayer, a new set of reciprocal lattice vectors provides additional elastic scattering channels for the incoming electrons. As a consequence of these surface umklapp processes, the incident electron beam may be diffracted to points in **k**-space where bulk transitions may occur that were not observable on the clean surface. These additional emissions are induced by the adsorbate, but can be interpreted in terms of the bulk band structure.[80–82] Finally, truly adsorbate-induced additional emission from (anti)bonding states may appear, containing essential contributions from adsorbate electronic states. These features in fact contain valuable information on the adsorbate-adsorbate and/or the adsorbate-substrate interaction.

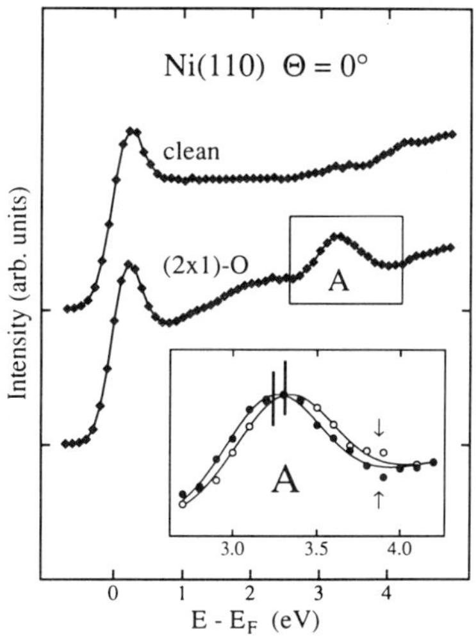

Figure 6. IPE data ($\hbar\omega$ = 9.4 eV) of clean and oxygen-covered Ni(110) for normal electron incidence. Inset: Spin-resolved data of the adsorbate-induced feature A with the spin-dependent background offset suppressed.

As an example, oxygen and sulfur adsorption on Ni(110) is discussed. For a certain oxygen exposure the adsorbate induces a (2x1) reconstruction of the Ni(110) surface whose exact geometry is still the subject of current debate.[83,84] Agreement exists that the oxygen atoms are adsorbed in long-bridge sites along the [001] direction. IPE data for Ni(110)-(2x1)-O are displayed in Fig. 6 in comparison with data of the clean surface. The spectrum for the clean surface only shows one clearly resolved feature close to E_F which can be traced back to indirect, i.e. non-**k**-conserving, transitions into empty d states. Besides almost unchanged d-band emission close to E_F after oxygen adsorption, two adsorbate-induced features appear. Since the broad structure at about 2 eV above E_F can be interpreted in terms of bulk contributions due to adsorbate-induced surface umklapp processes, only structure A centered at about

3.3 eV above E_F may give information on the adsorbate-substrate interaction. This is supported by $E(\mathbf{k}_{||})$ measurements: A shows considerable dispersion in the direction of the Ni rows, but none perpendicular to it.[85] SPIPE results of A shown in the inset of Fig. 6 reveal a magnetic exchange splitting of 80±20 eV, indicating strong ferromagnetic coupling of the adsorbate to the substrate.[24] While oxygen adsorption completely quenches the crystal-induced surface state discussed earlier (Fig. 3), a direct transition into an empty d state shown in Fig. 7 is hardly attenuated at all by the adsorbate. This shows the minor influence of the oxygen on the bulk-like transitions of Ni(110). The observations made by IPE are confirmed by calculations that report a small magnetic moment of about 0.1 Bohr magneton picked up by the oxygen.[86] There is some reduction of the surface magnetic moment, but the oxygen atoms fail to quench it completely.

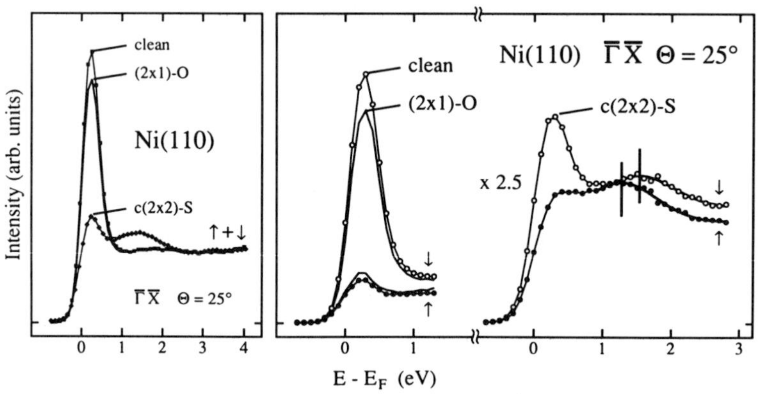

Figure 7. Spin-integrated (left panel) and spin-resolved (right panel) IPE data ($\hbar\omega$ = 9.4 eV) of Ni(110) taken at $\Theta = 25°$ in the $\overline{\Gamma}\overline{X}$ azimuth for three different sample preparations: clean, (2x1)-O, and c(2x2)-S.

Sulfur adsorbs on Ni(110) in the hollow site and forms a c(2x2) overlayer. Figure 7 demonstrates the strong influence sulfur has on the surface electronic structure of Ni(110). The direct transition into empty minority d states is almost completely quenched. The adsorbate-induced feature about 1.2 eV above E_F exhibits a magnetic exchange splitting of about 250 meV, indicating strong hybridization of Ni-3-d and S-3-p states. In fact, measurements for larger $\mathbf{k}_{||}$ show that it crosses the Fermi level, thereby being a truly magnetic band.[25] Target current spectroscopy data for normal electron incidence exhibit a change of the reflectivity around E_V, giving rise to a well-developed image-potential state which is not clearly resolved on the clean surface, because there is no gap of the projected bulk band structure there.[37] This also confirms the much stronger influence of sulfur adatoms on the surface electronic structure compared with oxygen on the same surface. One might argue that the adsorption geometry of sulfur provides more overlap of substrate and adsorbate electron wave functions than in the case of oxygen, where the adatoms sit between the Ni atoms on the exposed Ni rows. Therefore, the difference between (2x1)-O and c(2x2)-S in the electronic structure might be a consequence of the different adsorption geometry rather than a different local bonding. Nevertheless, calculations for O and S on Ni(001), both in the same c(2x2) adsorption geometry, predict a more covalent bonding for S-Ni and a more ionic bonding for O-Ni,[70] resulting in a more reduced surface magnetic moment for S on Ni(001) than for O. This prediction for Ni(001) seems to be also true of Ni(110) on the basis of the discussed IPE data. Calculations for Ni(110)-c(2x2)-S, were they available, would shed more light on this issue. At least, the data confirm the stronger impact of S on the surface electronic structure of Ni(110) compared with O. In summary, spin-resolved measurements of adsorbate-induced changes of the surface electronic structure tell about changes of the magnetic properties and, in some cases, even give new insight into the bonding mechanism.

SUMMARY

In this paper empty spin-dependent electronic states at magnetic surfaces have been discussed on the basis of experimental results obtained by SPIPE. Exchange-split surface states localized within the topmost layer or in front of the surface tell about magnetically active surfaces. Adsorption of non-magnetic adatoms changes the spin-dependent electronic structure, whose investigation provides insight into bonding mechanisms as well as modifications of the magnetic properties. A lot of information on the bulk-surface-vacuum and bulk-adsorbate-vacuum interfaces has been collected during recent years.

Currently, research activity in the field of surface and thin-film magnetism is concentrating on indirect exchange coupling between magnetic layers mediated by non-magnetic spacer layers.[87,88] Magnetic phenomena such as oscillatory coupling behavior as a function of spacer thickness or giant magnetoresistance effects have been detected. Yet, the understanding of these phenomena on an electronic-structural basis is still fragmentary.[89–93] Consequently, detailed information is needed on the spin dependence of interface electronic states in layered systems consisting of ferromagnetic and non-magnetic materials. Spin-resolved PE has already achieved first results on electronic states localized at interfaces.[94–96] It was shown, for example, how a minority surface resonance develops to an interface state with discrete binding energy depending on the number of atomic layers of the overlayer.[95] Non-spin-resolved IPE detected quantum well states as possible mediators of magnetic coupling in superlattices.[97,98] The need for spin-resolved data in this field will certainly influence future activities, promising valuable contributions to a more microscopic understanding of the fascinating phenomena in interface magnetism.

ACKNOWLEDGEMENTS

It is a pleasure to thank V. Dose, K. Ertl, and F. Passek for their enjoyable collaboration and for many stimulating discussions.

REFERENCES

1. L.M. Falicov, D.T. Pierce, S.D. Bader, R. Gronsky, K.B. Hathaway, H.J. Hopster, D.N. Lambeth, S.S.P. Parkin, G. Prinz, M. Salamon, I. K. Schuller, and R.H. Victora, J. Mater. Res. **5**, 1299 (1990).
2. D. Pescia, ed., *Magnetism in Ultrathin Films,* Appl. Phys. A **49 (5+6)** (1989).
3. U. Gradmann, J. Magn. Magn. Mater. **100**, 481 (1991).
4. H.C. Siegmann, J. Phys.: Condens. Matter **4**, 8395 (1992).
5. A.J. Freeman and R. Wu, J. Magn. Magn. Mater. **100**, 497 (1991) and ref. cited.
6. E.W. Plummer and W. Eberhardt, Adv. Chem. Phys. **49**, 533 (1982).
7. F.J. Himpsel, Adv. Phys. **32**, 1 (1983).
8. V. Dose, Surf. Sci. Rep. **5**, 337 (1985).
9. N.V. Smith, Rep. Prog. Phys. **51** 1227 (1988).
10. P.T. Andrews, I.R. Collins, and J.E. Inglesfield, in: *Unoccupied Electronic States*, ed. by J.C. Fuggle and J.E. Inglesfield, Topics in Applied Physics, Vol. **69** (Springer, Berlin, Heidelberg 1992), p. 243.
11. R. Schneider and V. Dose, in: *Unoccupied Electronic States*, ed. by J.C. Fuggle and J.E. Inglesfield, Topics in Applied Physics, Vol. **69** (Springer, Berlin, Heidelberg 1992), p. 277.
12. J. Kirschner: *Polarized Electrons at Surfaces,* Springer Tracts in Modern Physics, Vol. **106** (Springer, Berlin, Heidelberg 1985).
13. R. Feder (ed.): *Polarized Electrons in Surface Physics,* Advanced Series in Surface Science (World Scientific, Singapore 1985).
14. G. Busch, M. Campagna, and H.C. Siegmann, J. Appl. Phys. **41**, 1044 (1970).
15. H.C. Siegmann, F. Meier, M. Erbudak, and M. Landolt, Advances in Electronics and Electron Physics **62**, 1 (1984).
16. E. Kisker, J. Magn. Magn. Mat. **45**, 23 (1984).
17. J. Unguris, A. Seiler, R.J. Celotta, D.T. Pierce, P.D. Johnson, and N.V. Smith, Phys. Rev. Lett. **49**, 1047 (1982).
18. H. Scheidt, M. Glöbl, V. Dose, and J. Kirschner, Phys. Rev. Lett. **51**, 1688 (1983).
19. J. Kirschner, M. Glöbl, V. Dose, and H. Scheidt, Phys. Rev. Lett. **53**, 612 (1984).
20. V. Dose and M. Glöbl, in: *Polarized Electrons in Surface Physics*, Advanced Series in Surface Science, ed. by R. Feder (World Scientific, Singapore 1985), p. 547.
21. A. Seiler, C.S. Feigerle, J.L. Peña, R.J. Celotta, and D.T. Pierce, Phys. Rev. B **32**, 7776 (1985).
22. C.S. Feigerle, A. Seiler, J.L. Peña, R.J. Celotta, and D.T. Pierce, Phys. Rev. Lett. **56**, 2207 (1986).
23. L.E. Klebanoff, R.K. Jones, D.T. Pierce, and R.J. Celotta, Phys. Rev. B **36**, 7849 (1987).
24. G. Schönhense, M. Donath, U. Kolac, and V. Dose, Surf. Sci. **206**, L888 (1988).
25. M. Donath, Appl. Phys. A **49**, 351 (1989).
26. M. Donath, V. Dose, K. Ertl, and U. Kolac, Phys. Rev. B **41**, 5509 (1990).
27. M. Donath and V. Dose, Europhys. Lett. **9**, 821 (1989).
28. W. Nolting, W. Borgieł, V. Dose, and Th. Fauster, Phys. Rev. B **39**, 6962 (1989).
29. W. Borgieł, W. Nolting, and M. Donath, Solid State Commun. **72**, 825 (1989).
30. W. Nolting, J. Braun, G. Borstel, and W. Borgieł, Phys. Scr. **41**, 601 (1990).
31. M. Donath, G. Schönhense, K. Ertl, and V. Dose, Appl. Phys. A **50**, 49 (1990).
32. Q. Cai, R. Avci, and G.J. Lapeyre, Mat. Res. Soc. Symp. Proc. **151**, 65 (1989).
33. K. Starke, K. Ertl, and V. Dose, Phys. Rev. B **46**, 9709 (1992).
34. K. Starke, K. Ertl, M. Donath, and V. Dose, Vacuum **41**, 755 (1990).
35. K. Starke, K. Ertl, and V. Dose, Phys. Rev. B **45**, 6154 (1992).
36. R. Schneider, K. Starke, K. Ertl, M. Donath, V. Dose, J. Braun, M. Graß, and G. Borstel, J. Phys.: Condens. Matter **4**, 4293 (1992).
37. M. Donath and K. Ertl, Surf. Sci. **262**, L49 (1992).
38. F. Passek and M. Donath, Phys. Rev. Lett. **69**, 1101 (1992).
39. W. Grentz, M. Tschudy, B. Reihl, and G. Kaindl, Rev. Sci. Instrum. **61**, 2528 (1990).
40. F. Ciccacci, E. Vescovo, G. Chiaia, S. De Rossi, and M. Tosca, Rev. Sci. Instrum. **63**, 3333 (1992).
41. D.T. Pierce and F. Meier, Phys. Rev. B **13**, 5484 (1976).
42. V. Dose, Th. Fauster, and R. Schneider, Appl. Phys. A **40**, 203 (1986).
43. U. Kolac, M. Donath, K. Ertl, H. Liebl, and V. Dose, Rev. Sci. Instrum. **59**, 1931 (1988).
44. F. Passek, M. Donath, and V. Dose, to be published (1992).
45. E.W. Plummer and W. Eberhardt, Phys. Rev. B **20**, 1444 (1979).
46. W. Eberhardt, E.W. Plummer, K. Horn, and J. Erskine, Phys. Rev. Lett. **45**, 273 (1980).
47. J.L. Erskine, Phys. Rev. Lett. **45**, 1446 (1980).
48. J. Noffke, private communication.
49. N.B. Brookes, A. Clarke, P.D. Johnson, and M. Weinert, Phys. Rev. B **41**, 2643 (1990).
50. P.M. Echenique and J.B. Pendry, J. Phys. C **11**, 2065 (1978).
51. E.G. McRae, Rev. Mod. Phys. **51**, 541 (1979).

52. P.M. Echenique and M.E. Uranga, Surf. Sci. **247**, 125 (1991).
53. P.M. Echenique, F. Flores, and R. Sols, Phys. Rev. Lett. **55**, 2348 (1985).
54. R.W. Schoenlein, J.G. Fujimoto, G.L. Eesley, and T.W. Capehart, Phys. Rev. Lett. **61**, 2596 (1988).
55. G. Borstel and G. Thörner, Surf. Sci. Rep. **8**, 1 (1988).
56. R. Schneider, private communication.
57. R.L. Kautz and B.B. Schwartz, Phys. Rev. B **14**, 2017 (1976).
58. P.D. Johnson and N.V. Smith, Phys. Rev. B **27**, 2527 (1983).
59. V. Dose, W. Altmann, A. Goldmann, U. Kolac, and J. Rogozik, Phys. Rev. Lett. **52**, 1919 (1984).
60. D. Straub and F.J. Himpsel, Phys. Rev. Lett. **52**, 1922 (1984).
61. K. Giesen, F. Hage, F.J. Himpsel, H.J. Riess, and W. Steinmann, Phys. Rev. Lett. **55**, 300 (1985).
62. W. Steinmann, Appl. Phys. A **49**, 365 (1989) and ref. cited.
63. N. Fischer, S. Schuppler, Th. Fauster, and W. Steinmann, Phys. Rev. B **42**, 9717 (1990).
64. R. Fischer, N. Fischer, S. Schuppler, Th. Fauster, and F.J. Himpsel, Phys. Rev. B **46**, 9691 (1992).
65. H. Huang and J. Hermanson, Phys. Rev. B **32**, 6312 (1985).
66. S.R. Chubb, E. Wimmer, and A.J. Freeman, Bull. Am. Phys. Soc. **30**, 599 (1985).
67. S.R. Chubb and W.E. Pickett, Phys. Rev. Lett. **58**, 1248 (1987).
68. S.R. Chubb and W.E. Pickett, Phys. Rev. B **38**, 10227 (1988).
69. S.R. Chubb and W.E. Pickett, Phys. Rev. B **38**, 12700 (1988).
70. C.L. Fu and A.J. Freeman, Phys. Rev. B **40**, 5359 (1989).
71. D.L. Abraham and J. Hopster, Phys. Rev. Lett. **58**, 1352 (1987).
72. H.J. Elmers and U. Gradmann, Surf. Sci. **193**, 94 (1988).
73. R. Allenspach, M. Taborelli, and M. Landolt, Phys. Rev. Lett. **55**, 2599 (1985).
74. B. Sinković, P.D. Johnson, N.B. Brookes, A. Clarke, and N.V. Smith, Phys. Rev. Lett. **62**, 2740 (1989).
75. G. Schönhense, M. Getzlaff, C. Westphal, B. Heidemann, and J. Bansmann, J. Physique C**8**, 1643 (1988).
76. P.D. Johnson, A. Clarke, N.B. Brookes, S.L. Hulbert, B. Sinković, and N.V. Smith, Phys. Rev. Lett. **61**, 2257 (1988).
77. A. Clarke, N.B. Brookes, P.D. Johnson, M. Weinert, B. Sinković, and N.V. Smith, Phys. Rev. B **41**, 9659 (1990).
78. M. Getzlaff, J. Bansmann, C. Westphal, and G. Schönhense, J. Magn. Magn. Mat. **104-107**, 1781 (1992).
79. W. Altmann, M. Donath, V. Dose, and A. Goldmann, Solid State Commun. **53**, 209 (1985).
80. J. Anderson and G.J. Lapeyre, Phys. Rev. Lett. **36**, 376 (1976).
81. D. Westphal and A. Goldmann, Surf. Sci. **126**, 253 (1983).
82. K. Desinger, W. Altmann, and V. Dose, Surf. Sci. **201**, L491 (1988).
83. B. Voigtländer, S. Lehwald, and H. Ibach, Surf. Sci. **225**, 162 (1990).
84. G. Kleinle, J. Wintterlin, G. Ertl, R.J. Behm, F. Jona, and W. Moritz, Surf. Sci. **225**, 171 (1990).
85. K. Desinger, V. Dose, A. Goldmann, W. Jacob, and H. Scheidt, Surf. Sci. **154**, 695 (1985).
86. B. Weimert, J. Noffke, and L. Fritsche, Surf. Sci. **264**, 365 (1992).
87. P. Grünberg, R. Schreiber, Y. Pang, M.B. Brodsky, and H. Sowers, Phys. Rev. Lett. **57**, 2442 (1986) and this volume.
88. S.S.P. Parkin, N. More, and K.P. Roche, Phys. Rev. Lett. **64**, 2304 (1990) and this volume.
89. Y. Wang and P.M. Levy, Phys. Rev. Lett. **65**, 2732 (1990).
90. D.M. Edwards, J. Mathon, R.B. Muniz, and M.S. Phan, Phys. Rev. Lett. **67**, 493 (1991).
91. P. Bruno and C. Chappert, Phys. Rev. Lett. **67**, 1602 (1991).
92. F. Herman and R. Schrieffer, Phys. Rev. B **46**, 5806 (1992).
93. F. Herman, M. van Schilfgaarde, and J. Sticht, Journal of Modern Physics (ICPTM-92 Proceedings), to be published (1992).
94. W. Weber, D.A. Wesner, G. Güntherodt, and U. Linke, Phys. Rev. Lett. **66**, 942 (1991).
95. N.B. Brookes, Y. Chang, and P.D. Johnson, Phys. Rev. Lett. **67**, 354 (1991).
96. O. Rader, C. Carbone, W. Clemens, E. Vescovo, S. Blügel, W. Eberhardt, and W. Gudat, Phys. Rev. B **45**, 13823 (1992).
97. F.J. Himpsel, Phys. Rev. B **44**, 5966 (1991).
98. J.E. Ortega and F.J. Himpsel, Phys. Rev. Lett. **69**, 844 (1992).

EXCHANGE-COUPLING BETWEEN FERROMAGNETS ACROSS NON-METALLIC AMORPHOUS SPACER-LAYERS: Si AND SiO

S. Toscano, B. Briner, and M. Landolt

Laboratorium für Festkörperphysik
ETH-Zürich
CH-8093 Zürich, Switzerland

INTRODUCTION

The discovery of exchange coupling between ferromagnetic films separated by a non-magnetic spacer layer over distances of a few nanometers has provoked a renaissance in the research of magnetism. So far only metals have been studied and long as well as short period oscillations have been observed.[1] The theoretical descriptions all share in common that this phenomenon essentially is a consequence of two properties of the spacer material: the existence and topology of a Fermi surface and the discreteness of the layer thickness. In addition, a certain roughness of the interfaces is made responsible for smearing out short-wavelength oscillations. It remains quite surprising, however, that many studies have revealed similar periods of the coupling for a wide variety of materials and structures. Starting from this situation we set out, in the present investigation, to question the very role of the Fermi surface or, more generally, to investigate the response of the exchange coupling to variations in geometrical structure and electrical conduction properties of the spacer. Thus we extend the investigations to a group of completely different spacer materials: non-metallic and non-crystalline solids. We choose as a first model material amorphous Si,[2] which is a semiconductor, and then amorphous SiO, which is an insulator.

EXPERIMENTAL

To have access to the magnetic properties of the system under investigation we make use of the well characterized technique of Spin Polarized Secondary Electron Emission (SPSEE).[3] The experimental information searched in SPSEE is the spin polarization P_o of the low-energy secondary electrons emitted in vast numbers when a metallic sample is bombarded by a high-energy primary-electron beam. In the present investigation the primary energy is $E_p = 1000$ eV. To obtain the spin information, the emitted secondary electrons are focussed into a beam and are transported by an electron-lens system into the spin analyzer, a 100-keV Mott detector. The polarization is defined as $P = (N\uparrow - N\downarrow)/(N\uparrow + N\downarrow)$, where $N\uparrow(\downarrow)$ represents the number of electrons with spin parallel (antiparallel) to the chosen quantization axis of the detector. P is determined from the backscattering asymmetry arising from spin-orbit coupling of the relativistic electrons scattering at 100 keV from a very thin gold foil. With the present aparatus we are able to measure the two components of the polarization parallel to the surface of the sample.

Magnetism and Structure in Systems of Reduced Dimension
Edited by R.F.C. Farrow *et al.*, Plenum Press, New York, 1993

A detailed characterization of the magnetic properties of a system at a fixed temperature is obtained by measuring the magnetic hysteresis loop. In the case of in-plane magnetization, as in the present one, the applied external field H_{ext} produces stray fields parallel to the sample surface which deflect the low-energy secondary electrons emerging from the sample. As a consequence, the scattering region on the gold foil in the Mott detector varies with H_{ext} and this generates a fictive polarization. In order to eliminate these disturbing stray fields we use as a substrate a ferromagnetic alloy with a very low coercivity. This substrate acts as a "magnetic driver". It is mounted onto a horseshoe electromagnet and exhibits a square hysteresis loop, as shown below on Fig.4 (topmost panel). Ferromagnetic materials, when evaporated on such a substrate, are strongly exchange-coupled to the latter. Therefore they are exposed to a transferred exchange field H_{exch} originating from the substrate which acts as an external magnetic field parallel to the sample surface. In this way we are able to magnetize a very thin polycrystalline Fe film without generating disturbing stray fields. The sample is in thermal contact with a sapphire mounted on a cryostat and can be cooled down to a temperature of 40 K.

The substrate is sputter-cleaned by 3-keV Ne^+ bombardment. All different materials are evaporated by electron bombardment at a substrate temperature of 40 K. The evaporators are mounted on a single flange and can be separated from the rest of the UHV chamber by a gate valve, which allows loading without breaking the vacuum. All films studied in this work are evaporated at a total pressure lower than 5.10^{-10} Torr. With Auger Electron Spectroscopy (AES) we check the cleanliness of the substrate and all films after evaporation. Furthermore, by comparing the relative changes of the Auger intensities of the materials involved, we determine the thickness of the films and the growth mode. The evaporation rate is determined independently by a water-cooled quartz balance.

Fe/Si/Fe TRILAYERS

As already described the Fe/Si/Fe trilayers are evaporated on an amorphous $Fe_{40}Ni_{40}B_{20}$ ribbon kept at 40 K. Typical rates for both Fe and Si are between 1 and 4 Å/min. The interface properties of the system are investigated with AES. If the interface is sharp, i.e. there is no interdiffusion or island formation, then, in a simple model, the variations of the Auger intensities upon evaporation time can be fitted with exponentials. Fig.1 (top) shows the measured Fe and Si Auger-intensities versus evaporation time and the exponential fits. In order to obtain the Si film thickness we use[4] 13 Å for the Inelastic Mean Free Path (IMFP) of electrons in Si at the energy E = 650 eV, corresponding to the Fe LMM-Auger emission. The IMFP in Si at E = 90 eV, corresponding to the Si LMM emission, is a free parameter and comes out to 11.5 Å, in good agreement with the universal curve.[4] The slow initial decrease of the Fe Auger-line and the corresponding slow initial increase of the Si one clearly reveal the presence of a continuous interface when Si is evaporated on Fe at 40 K. The exponentials start to fit the variations of the Fe and Si Auger-intensities at a nominal Si thickness of 13 Å. Correspondingly, at this thickness the Si evaporation rate abruptly increases and remains constant up to at least 35 Å. The structural picture emerging is that of pure Fe and pure Si separated by an intermixing region with a well defined thickness of 13 Å. In other words, the initial interdiffusion of Si in Fe reduces the effective thickness of the pure Si layer by an amount of 13 Å. We note that "pure" in this context means that eventual impurities with concentrations below 1 % cannot be detected by AES.

The same structural analysis has been conducted for Fe evaporated on Si at T = 40 K, shown in Fig.2. Here the exponential curves fit the experiment from the very beginning of the evaporation, and the evaporation rate remains constant. From this we conclude that the Fe/Si interface is sharp, i.e. no interdiffusion occurs. This is consistent with the very low solubility of Fe in Si.[5] For the thickness calibration we use the IMFP in Fe at E = 650 eV of the Fe LMM Auger-line from Ref.4, and the one at E = 90 eV of the Si LMM Auger-line as a free parameter. We note that in the case of Fe we obtain 7 Å, and not 11.5 Å as in the case of Si, which reflects the expected different scattering behaviour of electrons moving in metals or in semiconductors.

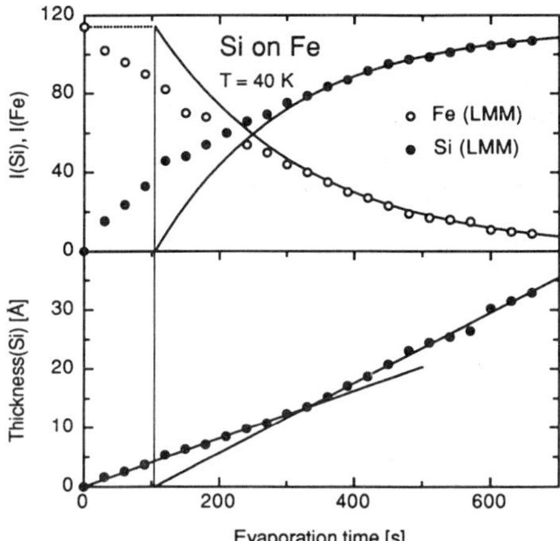

Figure 1. Auger-electron intensities (top) and film thickness (bottom) versus evaporation time of amorphous Si on Fe.

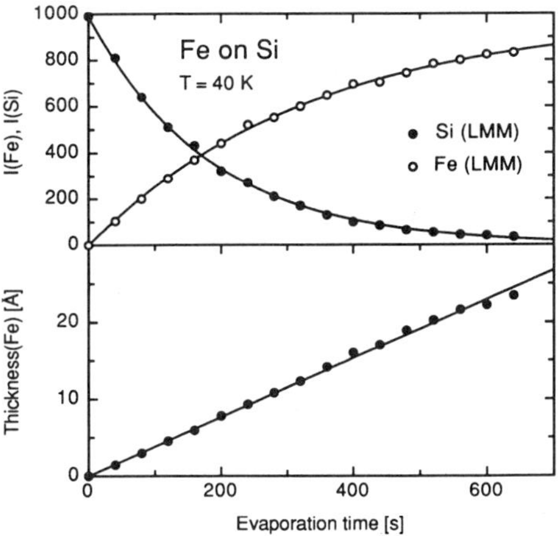

Figure 2. Auger-electron intensities (top) and film thickness (bottom) versus evaporation time of Fe on amorphous Si.

Finally, let us emphasize that when the temperature rises the intermixing region grows. The trilayer structure modifies irreversibly and silicide formation occurs. We have verified this by AES and by observing irreversible changes in the magnetic properties of the samples. Silicide formation up to a spacer thickness of 20 Å has been reported recently in sputtered Fe/Si superlattices fabricated at room temperature by Fullerton et al.,[6] well in line with observations by Dufour et al.[7]

When evaporated at low temperatures Si is known to grow in an amorphous structure. According to Mott and Davis[8] electrical conduction at low temperatures occurs through variable-range hopping and shows a characteristic temperature dependence, which in three dimensions is described as $\sigma = \sigma_0 \exp-(T_0/T)^{1/4}$, where $\sigma_0 = e^2N(E_F)R^2\nu_{ph}$ and $T_0 \approx 16\alpha^3/kN(E_F)$. $R = (0.62)(\alpha N(E_F)kT)^{-1/4}$ is the mean hopping distance, ν_{ph} the electron-phonon interaction constant, and α the inverse decay-length of the localized states. To check the conduction mechanism in the present films we carried out in situ conductivity measurements on films of 1000-3000 Å thickness grown on quartz on the same sample holder under

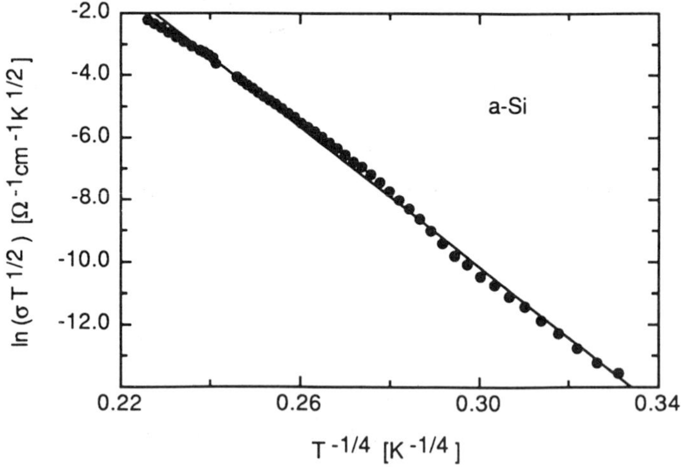

Figure 3 Electrical conductivity σ of 3000 Å-thick amorphous Si versus temperature T in a $\ln(\sigma T^{1/2})$ versus $T^{-1/4}$ plot indicating variable- range hopping.

conditions identical to those of the magnetic experiments. The measured conductivity depicted in Fig.3 exhibits a positive temperature coefficient typical for semiconductors and shows a linear behaviour of $\ln(\sigma T^{1/2})$ vs. $T^{-1/4}$ revealing that conduction in the present films occurs by variable-range hopping. The slope of the fitted straight line leads to a density of states at the Fermi energy of $N(E_F) = 1.5\ 10^{18}$ $eV^{-1}cm^{-3}$ when using $\alpha^{-1} = 7$ Å for the decay length of the localized states.[9] The mean hopping distance for T = 40 K comes out as R = 120 Å which is larger than the Si film thicknesses under consideration. However, we wish to emphasize that conduction does neither occur via delocalized states above the mobility edge nor by n- or p-doping which both would lead to a linear lns vs 1/T behaviour, which is not observed.

We have characterized the magnetic properties of the present system at 40 K by measuring magnetic hysteresis loops. We have carried out measurements first on the $Fe_{40}Ni_{40}B_{20}$ substrate and consecutively on each film after deposition as illustrated below.

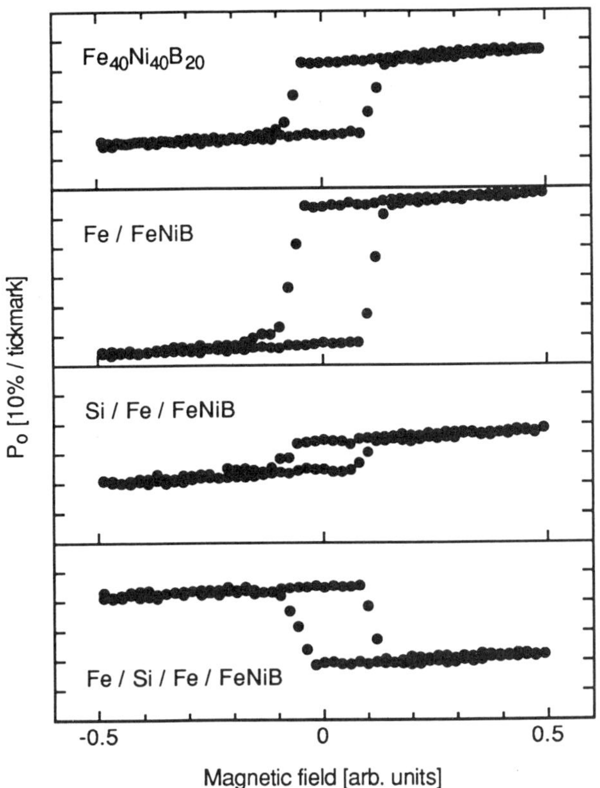

Figure 4. Spin polarization of secondary electrons versus external magnetic field of the $Fe_{40}Ni_{40}B_{20}$ substrate and after deposition of each overlayer of an Fe/ amorphous Si/ Fe sandwich, respectively, at 40 K. The nominal thickness of the amorphous-Si spacer is 18.5 Å.

The first two panels of Fig.4 show the hysteresis loops of the substrate and of a 15 Å-thick polycrystalline Fe film evaporated on it. The Fe film is strongly exchange-coupled to the substrate and therefore exhibits the same coercivity H_c. Thus the hysteresis loop measured on a 15 Å-thick Fe film evaporated on the substrate is exactly the same as the one of the substrate but with increased spin polarization as expected.

On Fe we evaporate a Si film of variable thickness. A corresponding hysteresis loop is shown in Fig.4, third panel, for a nominal Si thickness of 18.5 Å. Here we observe a strongly reduced spin polarization which reflects the admixing of unpolarized electrons from the Si to the polarized electrons from the underlying Fe film. Finally we complete the sandwich structure with a 15 Å-thick Fe film. We call this outermost layer "Fe overlayer".

The interaction between the two Fe films separated by a Si spacer is determined from the hysteresis loop measured on the Fe overlayer (Fig.4, fourth panel) with respect to the magnetization of the Fe film adjacent to the substrate (Fig.4, second panel). The similarity of H_c of the two hystereses establishes the exchange origin of the Fe-overlayer magnetization. In particular, Fig.4 shows a case where the coupling between the two Fe films is antiferromagnetic. In fact, the polarization of the Fe film adjacent to the substrate and that of the Fe overlayer have opposite signs, which indicates antiparallel orientation of the respective magnetizations. The same sign, on the other hand, is expected and found for ferromagnetic coupling. This interpretation of the data is straightforward since SPSEE is surface sensitive enough to only reveal the magnetization of the outermost layer.

We have measured many sets of hysteresis loops as described above for different Si spacer-layer thicknesses up to 30 Å. As a result, Fig.5 shows the spontaneous polarization at remanence P_r of the Fe overlayer versus Si film thickness.

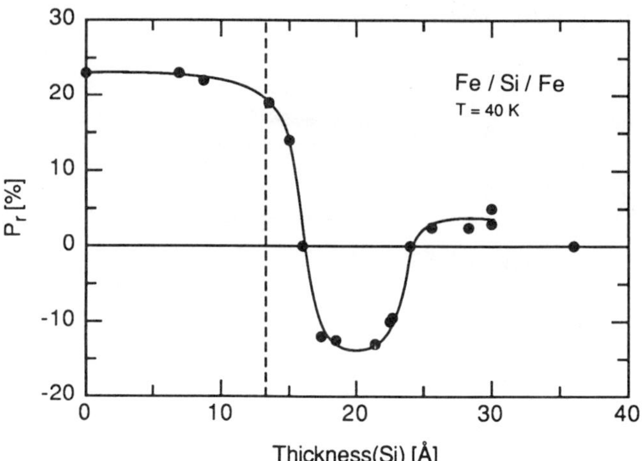

Figure 5. Spin polarization of secondary electrons at remanence of the top Fe layer of the Fe/ amorphous Si/ Fe sandwiches deposited on an $Fe_{40}Ni_{40}B_{20}$ magnetic driver versus Si film thickness.

The exchange coupling is initially ferromagnetic and then goes from ferromagnetic to antiferromagnetic and back to ferromagnetic. The apparent oscillation period is found to be 16 Å. We encounter again a long period oscillation in surprising, maybe fortuitous, similarity with previous measurements on metallic spacers.

We claim that the initial interdiffusion between Fe and Si by no means affects the above results. We observe with the corresponding hysteresis loops that the thin interface layer is strongly ferromagnetically exchange-coupled to the Fe substrate and thus only reduces the effective thickness of the pure Si spacer layer. To visualize this we have drawn a zero-offset line in Fig.5 at 13 Å, which corresponds to the thickness of the interface layer obtained from the Auger analysis on Fig.1. From this we infer that the first transition from ferro- to antiferromagnetic coupling occurs at a Si thickness of 3 Å.

Finally we wish to comment on the decrease of the spin polarization at remanence P_r with increasing Si thickness, depicted in Fig.5. The relationship between P_r and the magnetic exchange-coupling constant J is not trivial. We know that P_r is proportional to the Fe-overlayer magnetization M_{Fe}, which is a function of the external magnetic field H and the temperature T: $P_r \sim M_{Fe}(H,T)$. Following Ref.10 we assume that the magnetic exchange-coupling corresponds to an exchange field H_{exch} transferred from the Fe layer adjacent to the substrate to the one on the top of the trilayer. Then H is the sum of H_{magnet} produced by the horseshoe magnet and H_{exch}, $H = H_{magnet} + H_{exch}$. P_r is measured at vanishing H_{magnet}, which leads to the relation $P_r \sim M_{Fe}(H_{exch},T)$. A quantitative determination of H_{exch} and thus of the coupling strength requires detailed knowledge of $M_{Fe}(H,T)$, which currently is not available. The data reported on Fig.5 are measured at a constant temperature of T = 40 K. From this we infer that the decrease of P_r reflects an incomplete alignment of the atomic spins in the Fe overlayer and reveals the decrease of H_{exch} with increasing thickness of the spacer layer.

Fe/SiO/Fe TRILAYERS

In the case of a metallic spacer, the magnetic exchange-coupling is believed to be a manifestation of the RKKY interaction. The electrons producing this interaction are the ones which are responsible for electrical transport. In amorphous Si, the latter occurs via localized electrons near the Fermi level. The next step is then to investigate the role of such electrons in the coupling mechanism. Along this line we choose SiO as a further spacer material.

The technique of growing SiO thin films is well established: the usual method is thermal evaporation starting from SiO powder.[11-13] On the other hand, the picture of SiO emerging from the literature is not free of controversies, mostly because of the strong sensitivity of the different SiO properties to the preparation conditions. Nevertheless, it has been shown that SiO is not a simple macroscopic mixture of Si and SiO_2, but has its own structure. As regards the electronic properties, we base our description mainly on a comprehensive study by W.J. Ching.[14] The transition from SiO_2 to Si is characterized by the progressive disappearing of the O states and the corresponding progressive filling of Si states in the SiO_2 band gap. A peculiarity of the DOS of SiO is the presence of a "pseudogap", i.e. an energy region with a small number of electron states. Its width is comparable to the SiO_2 band gap of ≈ 9 eV. Various experiments strongly suggest the existence of a band gap similar to the one of amorphous Si.[11,14,15] As in the case of amorphous Si, disorder generates localized states at the band edges and in the gap. However, we expect that oxygen saturates a certain number of dangling bonds, reducing in this way the number of electron states at the Fermi level. Therefore the modification occuring when going from amorphous Si to amorphous SiO essentially is given by a reduction of the number of electron states near the Fermi energy and in the upper part of the valence band. In contrast, little change is expected in the magnitude of the band gap.

The trilayer is evaporated by electron bombardment on a Co-based amorphous substrate at T = 40 K. The evaporation rate lies between 1 and 5 Å/min. The thickness of the two Fe films sandwiching the spacer is 15 Å. The growth mode of SiO on Fe and of Fe on SiO is investigated with AES. Results are presented in Figs.6 and 7.

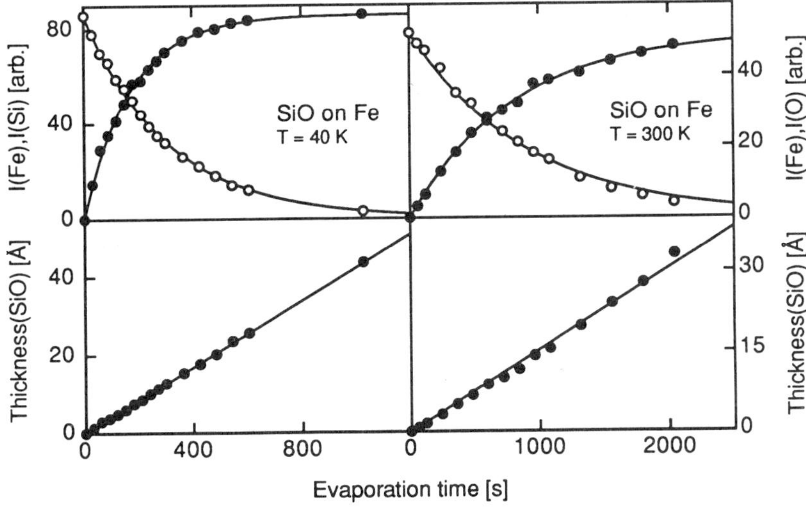

Figure 6. Auger-electron intensities (top) and film thicknesses (bottom) versus evaporation time of amorphous SiO on Fe at T = 40 K and T = 300 K. The upper panel on the left shows the Si(LMM) (dots) and the Fe(LMM) (circles) Auger-lines, that on the right shows the O(KLL) (dots) and the Fe(LMM) (circles) Auger-lines.

Fig.6 shows the growth of SiO on Fe at 40 K and 300 K. In both cases the exponential curves fit well the data points and the growth rate remains constant in time: we conclude that the SiO/Fe interface is sharp at both temperatures, i.e. we exclude interdiffusion or island formation.

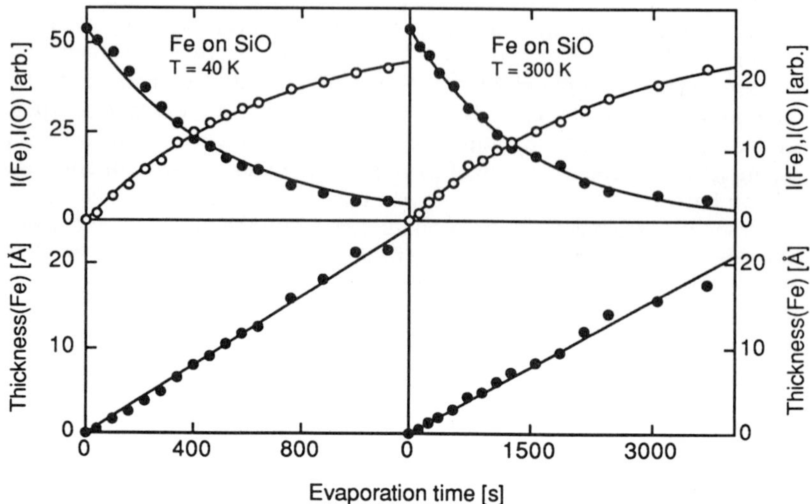

Figure 7. Auger-electron intensities (top) and film thicknesses (bottom) versus evaporation time of Fe on amorphous SiO at T = 40 K and T = 300 K. Both upper panels show the O(KLL) (dots) and the Fe(LMM) (circles) Auger-lines.

The case of Fe evaporated on SiO appears to be more complex, as one can see from Fig.7. At room temperature the fit seems accurate, but at 40 K deviations from the exponentials occur in the early stages of the evaporation, in particular for the O Auger-line. We are tempted to conclude, based on the thermal stability of the system, that the observed deviations arise from an initial island formation rather than from Fe-SiO intermixing. The islands percolate at a nominal Fe thickness between 5 Å and 8 Å. Beyond that the data accomodate on the exponential fit. In summary, the system Fe/SiO/Fe is characterized by sharp interfaces at 40 K and 300 K. No interdiffusion occurs when the temperature is varied between 40 K and 300 K. This is in contrast to the case of Si and enables one to study temperature variations of the exchange coupling..

Amorphous SiO is known to be an insulator or poor semiconductor. We are currently measuring the temperature dependence of the conductivity σ of SiO films sandwiched between Fe along a path perpendicular to the SiO surface. This experimental geometry, which is extremely difficult to realize, is necessary because of the small magnitude of σ. Preliminary results clearly demonstrate that metallic pinholes do not exist for SiO film-thicknesses down to 5 Å. We obseve very low conductivity of the order of $\sigma = 10^{-9}\Omega^{-1}cm^{-1}$ and a positive temperature coefficient reminiscent of a poor semiconductor. An extensive study of the electrical transport across ultrathin SiO films between metals is currently under way in our laboratory.[16]

We have investigated the magnetic exchange-coupling in Fe/SiO/Fe trilayers for various spacer thicknesses and temperatures. The thermal analysis, which can contribute significantly to the understanding of the origin of the coupling, is made possible by the stability of the sharp SiO/Fe and Fe/SiO interfaces which was discussed above. We note that, on the other hand, this thermal analysis cannot be performed in the case of amorphous Si because of interdiffusion of Si in Fe and consequent silicide formation.

Information on the magnetic exchange-coupling constant J again is obtained by comparing the hysteresis loop of the Fe film adjacent to the substrate with that of the Fe film on top of the trilayer. The existence of an exchange interaction is revealed by the equality of the coercive fields. The sign of J is then unequivocally determined by that of the remanent polarization P_r

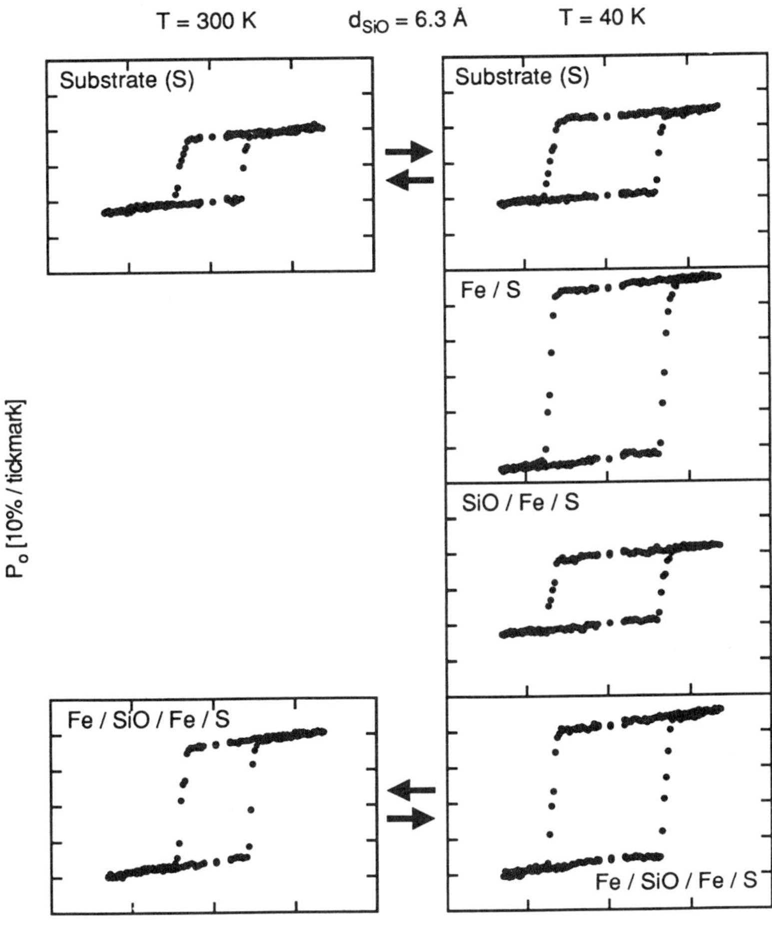

Figure 8. Right: spin polarization of secondary electrons versus external magnetic field of the Co-based substrate and after deposition of each overlayer of an Fe/ amorphous SiO/ Fe sandwich, respectively, at 40 K. Left: spin polarization of secondary electrons versus external magnetic field of the Co-based substrate an after completion of an Fe/ amorphous SiO/ Fe sandwich at 300 K. The thickness of the SiO layer is 6.3 Å.

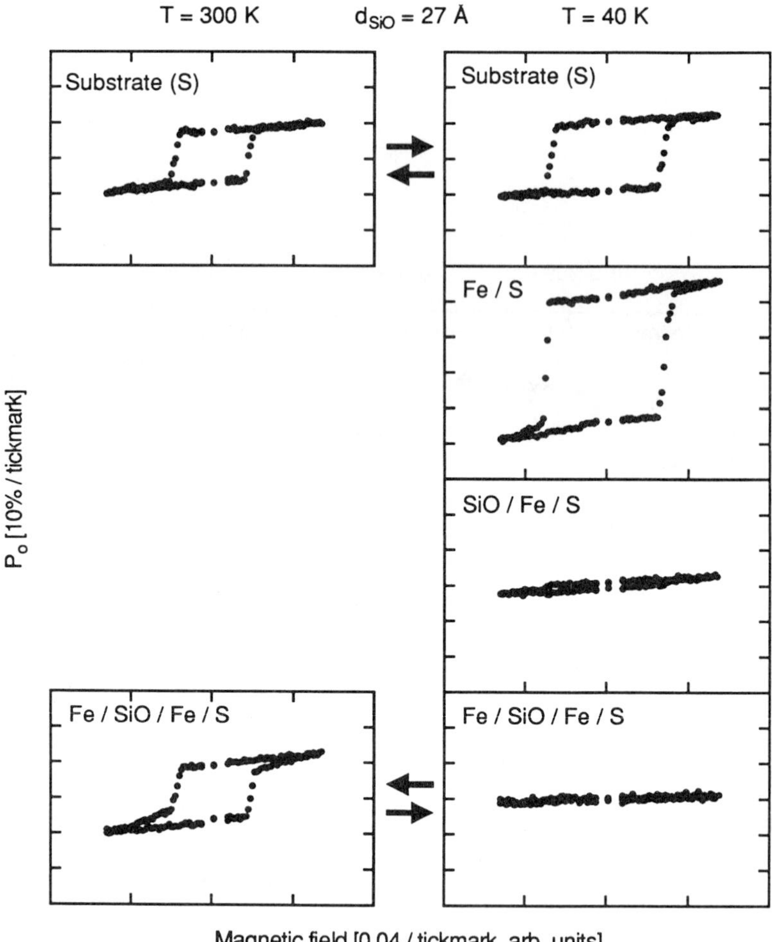

Figure 9. Right: spin polarization of secondary electrons versus external magnetic field of the Co-based substrate and after deposition of each overlayer of an Fe/ amorphous SiO/ Fe sandwich, respectively, at 40 K. Left: spin polarization of secondary electrons versus external magnetic field of the Co-based substrate an after completion of an Fe/ amorphous SiO/ Fe sandwich at 300 K. The thickness of the SiO layer is 27 Å.

of the Fe overlayer, and the magnitude of J can be related to P_r. In the present investigation we are also interested in the T-dependence of J, i.e. in $H_{exch}(T)$.

Two sets of hysteresis loops recorded at different stages of the Fe/SiO/Fe trilayer growth for two SiO thicknesses of 6.3 Å and 27 Å are shown in Figs.8 and 9, respectively.

In both cases the substrate is first characterized at 300 K and then cooled to 40 K. We note the pronounced increase of H_c upon cooling. The trilayers are then evaporated at 40 K. We find ferromagnetic exchange-coupling for 6.3 Å and no coupling for 27 Å, respectively, at 40 K. Subsequently the trilayers are heated to 300 K and cooled again to 40 K. In the case of a spacer layer thickness of 6.3 Å, shown in Fig.8, we find a reversible change of the coercive field observed at the topmost Fe layer, which nicely expresses the exchange nature of the apparent ferromagnetic coupling. Truly exciting, indeed, are the findings for a SiO thickness of 27 Å, depicted on Fig.9. There we observe a fully reversible onset of ferromagnetic exchange-coupling when heating the trilayer from 40 K to 300 K. It is important to note that again the coercive field observed at the topmost Fe layer exactly corresponds to the one of the substrate at the same temperature, expressing again the exchange character of the coupling.

The results on the magnetic exchange-coupling of the Fe/SiO/Fe trilayers are summarized on Fig. 10. It shows the polarization at remanence P_r of the topmost Fe layer versus SiO film thickness for 40 K and 300 K.

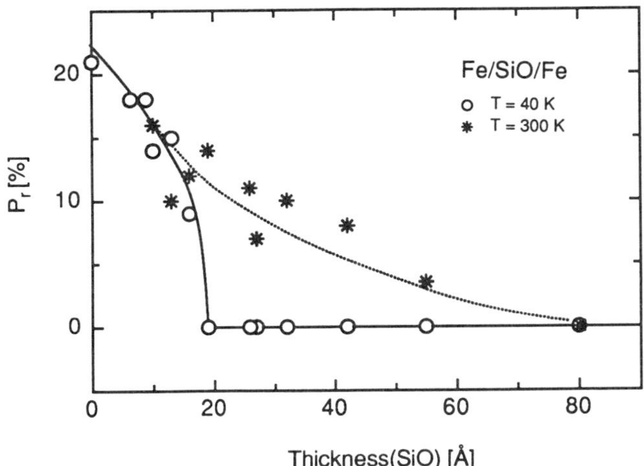

Figure 10. Spin polarization of secondary electrons at remanence of the top Fe layer of the Fe/ amorphous SiO/ Fe sandwiches deposited at 40 K on a Co-based substrate and measured at 40 K and 300 K, respectively, versus SiO film thickness. The two lines are to guide the eye.

We conclude from Fig.10 that ferromagnetic exchange-coupling exists between two Fe films separated by SiO. The coupling does not show any oscillations in the investigated thickness range of the spacer. Furthermore, the interaction is found to be temperature dependent, as shown by the different maximum spacer thicknesses up to which the Fe overlayer is magnetically coupled to the substrate at 40 K and 300 K. This phenomenon is not connected to structural changes of the trilayer: it is fully reversible upon heating and cooling the sample. As already mentioned, no interdiffusion occurs when the temperature is varied between 40 K and 300 K.

DISCUSSION

The trilayers Fe/Si/Fe and Fe/SiO/Fe exhibit magnetic exchange-coupling. For Si the interaction goes from ferromagnetic to antiferromagnetic and finally back to ferromagnetic with increasing spacer thickness with an apparent period of approximately 16 Å. For SiO no oscillations occur; the exchange coupling remains ferromagnetic. The absolute value of J cannot be extracted from the present observations because the detailed knowledge of the M(H,T) curve of the Fe overlayer is missing at present.

In metallic superlattices J reaches values of the order of 10^{-3} J/m^2. In those systems several investigations have shown the persistence of the coupling up to 50-80 Å and the presence of many oscillations even at low temperatures. These findings,[1] together with the apparent temperature dependence of J in the case of Fe/SiO/Fe, which is absent for metallic spacers, give us the feeling of a stronger interaction of a different kind for metallic spacer layers in comparison to non-metallic ones.

The term magnetic exchange-coupling expresses the "communication" between magnetic ions separated by non-magnetic materials. This interaction, mediated by electrons, is direct or indirect depending on the origin of the electrons involved and relies on the electronic structure of the spacer material. Fe/Si/Fe and Fe/SiO/Fe represent a metal-semiconductor-metal junction and a metal-insulator-metal junction, respectively. For these systems four kinds of “communication channels” between the two Fe layers are conceivable:

(a) Via the Boltzmann gas formed by electrons which are thermally excited to the energies of the spacer conduction band (Boltzmann interaction).
(b) Via valence electrons of the spacer which are virtually excited to the conduction band (superexchange interaction).
(c) Via Fe electrons which tunnel through the spacer.
(d) Via localized electrons present in the band gap of the spacer layer (analogous to RKKY-interaction).

Every channel leads to a term in the total free energy which determines the final magnetic state of the trilayers. An evaluation of the weight of the different contributions requires an analytical or numerical analysis based on quantitative details of the electronic structures which are not presently available. Instead, Toscano[17] has performed a qualitative analysis of the Boltzman and superexchange interactions based on calculations by Baltensperger and de Graaf[18] and by Bloembergen and Rowland,[19] respectively. He finds a surprisingly consistent description of the present observations with a weighted superposition of Boltzmann and superexchange interacitons. The model is still qualitative, however, and the importance of direct or indirect tunnelling, which seems to be considerable in electric transport, remains to be quanitfied.

Thus the present contribution still is open for a theoretical discussion. Further experiments with spacer materials with variable state density at the Fermi level reaching from insulators all the way to the metallic state are of great interest and will be undertaken in our laboratory. There potential is to study the various kinds of exchange interactions separately by virtue of atomic engineering.

ACKNOWLEDGEMENTS

We thank H.C. Siegmann for continuous support and many fruitful conversations and W Baltensperger, J. Helman, R. Monnier, H. von Känel and N. Onda for many helpful discussions. We are indepted to K. Brunner for expert technical assistance. Financial support by the Schweizerischer Nationalfonds is gratefully acknowledged.

REFERENCES

1. For a review see the present proceedings and references therein.
2. S. Toscano, B. Briner, H. Hopster and M. Landolt, J. Magn. Magn. Mater. **114**, L6 (1992).
3. M. Landolt, Appl. Phys. **A41**, 83.(1986).
4. M.P. Seah, and W.A. Dench, Surf. and Interf. Anal. **1**, 2 (1979).
5. S.M. Sze, "Physics of Semiconductor Devices", John Wiley, New York (1981).
6. E.E. Fullerton, J.E. Mattson, S.R. Lee, C.H. Sowers, Y.Y. Huang, G. Felcher, S. Bader, and F.T. Parker, J. Magn. Magn. Mater., in press (1992).
7. C. Dufour, A. Bruson, G. Marchal, B. George, and Ph. Mangin, J. Magn. Magn. Mater. **93**, 545 (1991).
8. N.F. Mott and E.A. Davis, "Electronic Processes in Non-Crystalline Materials", Claredon Press, Oxford (1979).
9. Y. Xu, A. Matsuda, and M.R. Beasley, Phys. Rev. **B42**, 1492 (1990).
10. M. Donath, D. Scholl, H.C. Siegmann, and E. Kay, Phys. Rev. **B43**, 3164 (1991).
11. H.R. Philipp, J. Phys. Chem. Solids **32**, 1935 (1971).
12. T. Scimeca, Solid State Commun. **77**, 817 (1991).
13. J.A. Roger, C.H.S. Dupuy, and S.J. Fonash, J. Appl. Phys. **46**, 3102 (1975).
14. W.Y. Ching, Phys. Rev. **B26**, 6610 (1982).
15. K. Hübner, Phys.Status Solidi **A61**, 665 (1980).
16. B. Briner and M. Landolt, to be published.
17. S. Toscano, Dissertation Nr 9915, Eidgenössische Technische Hochschule, Zürich (1992).
18. W. Baltensperger, and A.M. de Graaf, Helv. Phys. Acta **33**, 881 (1960).
19. N. Bloembergen, and T.J. Rowland, Phys. Rev. **97**, 1679 (1955).

INVESTIGATION OF MAGNETIC MICROSTRUCTURE IN FILMS AND MULTILAYERS BY TRANSMISSION ELECTRON MICROSCOPY

J. N. Chapman and D. M. Donnet

Department of Physics and Astronomy
University of Glasgow
Glasgow G12 8QQ
United Kingdom

INTRODUCTION

Thin magnetic alloy films and multilayers frequently display properties markedly different from those found in bulk magnetic materials. Furthermore, many of the desirable properties they do display depend crucially on the way they are grown; in other words, these properties are induced rather than being intrinsic to the material itself. Thus optimisation of their properties is only likely to be achieved if the magnetic microstructure can be determined and its dependence on the physical microstructure investigated.

We therefore have need for a technique which allows the spatial distribution of magnetisation within the specimen to be measured with as high a resolution as possible. The requirement for high resolution follows from the fact that many of the materials of interest are markedly inhomogeneous and an averaging of properties over many structural "units" is unlikely to lead to a clear understanding of fundamental mechanisms. We are concerned not only with the magnetisation distribution of the material in its as-grown state but on how it is affected by application of magnetic fields and changes in temperature. These factors are usually the ones which determine the application potential of the material. When interesting or unexpected magnetic features are observed in a particular region it is necessary to investigate the corresponding local physical properties of the same region. Only by these means is the understanding required likely to be realised.

Electron microscopy has the necessary potential to meet all these requirements but the standard high performance transmission electron microscope (TEM) is well suited only to providing a detailed description of the physical microstructure. The principal problem as far as revealing the magnetic domain structure is concerned is that the specimen is immersed in the high magnetic field of the objective lens and this field is sufficiently high to destroy, or substantially modify, the domain structures of interest. Simply turning off the objective lens and using only existing lenses more distant from the specimen does allow magnetic structures to be imaged but with, at best, modest resolution. In the case of a TEM/STEM instrument equipped with a thermionic electron source, the resolution in either a fixed-beam (CTEM) or a scanning (STEM) mode is rarely better than 50nm and usually substantially worse. Access to both scanning and fixed-beam modes is desirable as different magnetic imaging modes are advantageously used in different studies. Thus, the Fresnel and Foucault modes, which are most efficiently practised in a CTEM (Chapman, 1984) provide information on the domain geometry and the approximate direction of magnetisation in each domain whilst their simplicity of implementation makes them very well suited for use in in-situ dynamic experiments. However, whilst under favourable conditions they can be used to give some information on, for example, the domain wall structure, in general they are unsuitable for deriving quantitative information on the detailed spatial distribution of induction throughout the specimen. For this purpose the differential phase contrast (DPC) imaging mode, which can only be realised in a STEM, is most suitable (Chapman et al., 1987).

In the following section we give a brief desciption of DPC imaging and describe how it can be optimised for the study of magnetic multilayer films. Thereafter, in the third section its use is illustrated with reference to the study of marks written thermomagnetically in evaporated Co/Pt mulilayers.

DIFFERENTIAL PHASE CONTRAST LORENTZ MICROSCOPY

In DPC Lorentz microscopy images are formed by mapping the small deflection that a focussed probe of electrons experiences as it passes through a thin magnetic film. This is shown schematically in figure 1. In all cases magnetic contrast can only arise from components of induction perpendicular to the direction of electron travel. Consequently, if thin films supporting perpendicular magnetisation are under investigation it is usually necessary to tilt the specimen as indicated in figure 1. The deflections are measured using a segmented detector of the kind shown in figure 2. Originally a four segment (quadrant) detector was used but substantial advantages accrue through use of an eight segment device in instances where the magnetic signal is substantially smaller than that of non-magnetic origin. In the case of many alloy films and multilayers the most serious source of non-magnetic contrast is from the crystallites of which the film itself is composed. Indeed, for magnetic multilayers the difficulties are particularly acute because not only is the amount of magnetic material present small (leading to deflections of the electron beam as low as 1μrad) but the intervening non-magnetic layers, and any underlayers, can be the source of a large unwanted signal. Using difference signals from opposite segments of the outer annular detector (A-C, B-D), we have shown that the unwanted non-magnetic signal can be substantially reduced with respect to the magnetic signal of interest (Chapman et al., 1990). This is achieved by using the microscope post-specimen lenses to ensure that the angle subtended by the probe only slightly exceeds that of the inner (solid) quadrant detector. Under these conditions rapidly varying signals, such as those from the crystallite boundaries, are strongly filtered whilst the signal-to-noise ratio of the more slowly varying magnetic signal is substantially enhanced in this *modified* DPC imaging mode.

When MDPC imaging is undertaken in our extended VG HB5, a resolution of approximately 10nm is routinely achievable for investigations of thin alloy films in which the magnetisation lies substantially in plane. This figure corresponds to the minimum probe diameter that can be achieved in the standard specimen plane whilst the specimen is maintained in magnetic field free space. Recently Ploessl et al. (1992) have taken advantage of this high performance to study quantitatively details of the complex two-dimensional magnetisation distributions found in permalloy films. MDPC imaging not only allows accurate maps of the spatial variation of induction to be obtained but also provides detailed information on how the domain wall width can vary, dependent on the nature of the wall. Thus, the width of a wall supporting alternating cross-ties and vortices has been shown to be markedly reduced at a vortex and to suffer a mild reduction in width in the vicinity of a

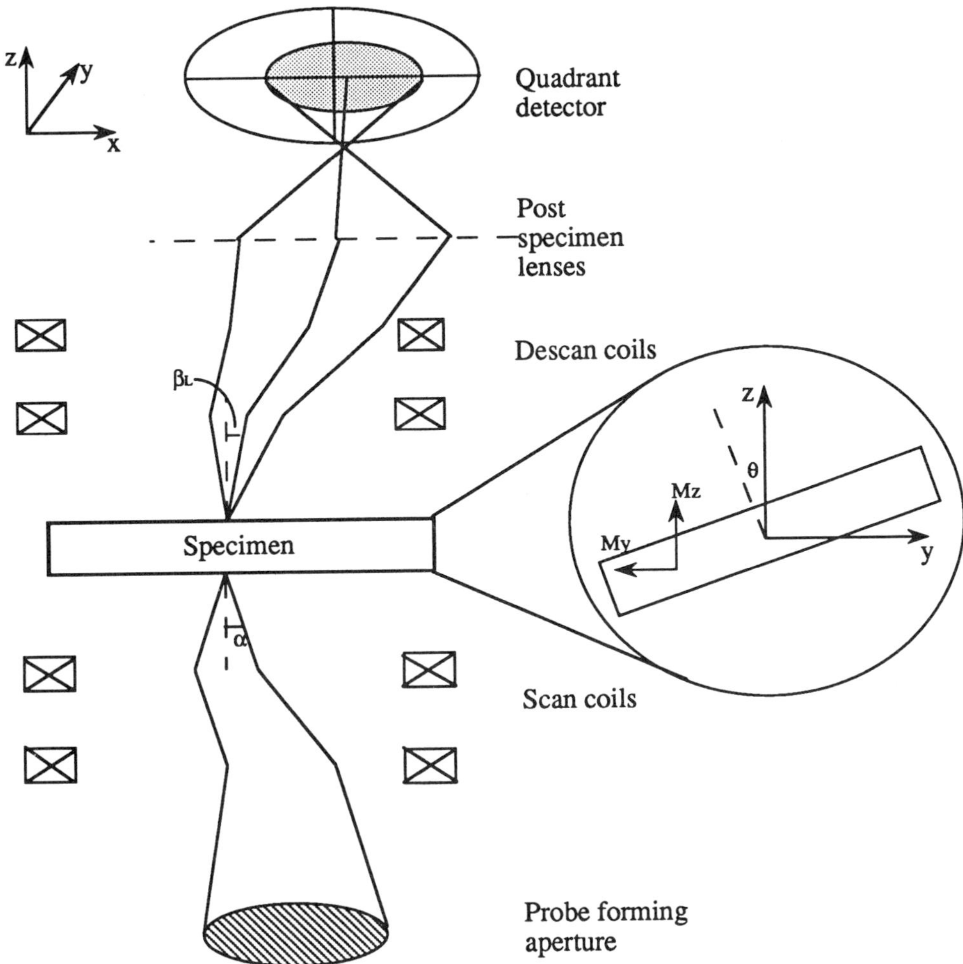

Figure 1. Schematic showing how contrast arises from magnetic specimens in DPC imaging.

cross-tie. These findings are in accord with theoretical modelling (Nakatani et al., 1989) and have significant implications for the wall energy.

WRITTEN MARKS IN Co/Pt MULTILAYER FILMS

Co/Pt multilayers are of great interest as magneto-optic recording media (Greidanus et al., 1989; Zeper et al., 1989). They are typically 20nm thick and contain 10-15 bilayers each of which comprises 2 monolayers of Co and 4-8 monolayers of Pt. Thus the total thickness of Co present is only ≈5nm and, as the easy direction of magnetisation is perpendicular to the plane of the film, it is necessary to tilt the specimen through ≈20° to produce a significant component of induction orthogonal to the electron beam. Even then, Lorentz deflection angles are small (<2μrad.) and we have found that the use of probe angles substantially less than 1mrad is advantageous if high contrast images are to be obtained (Chapman et al., 1992). Under these conditions the best resolution that can be achieved is limited by diffraction to a value of ≈20nm.

To understand the micromagnetic properties of these and other multilayers it is desirable to study their natural domain structure and how this changes under the influence of a field.

Recently we have observed changes directly in the microscope as films were subjected to isothermal remanence and dc-demagnetisation remanence cycles. From this work, insight into the nucleation of domains and their subsequent annihilation as saturation is approached was obtained and a value for the domain wall energy has been determined (Donnet et al., 1992). Here, however, we are concerned with the form of the domains (marks) written thermomagnetically in the film using a focused laser beam and, in particular, how this varies as a function of laser power and bias field.

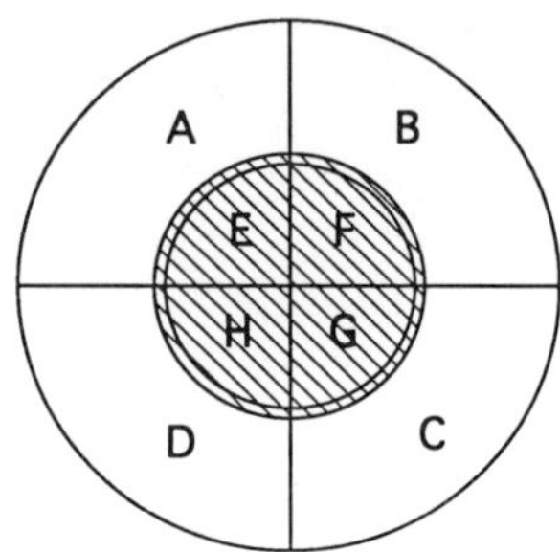

Figure 2. The detector used for (modified) DPC imaging. The shaded area denotes the size of the bright field disc under typical operating conditions.

Figure 3 shows a pair of MDPC images containing a number of written marks. The composition of the film was 1.0nmPt + 15*(0.35nmCo + 1.08nmPt) and it was deposited by electron beam evaporation directly onto a specially prepared silicon nitride membrane suitable for TEM investigations. The room temperature coercivity of the film was 139kA/m. Marks were written using an argon laser operating at 488nm in a modified Kerr microscope (Greidanus et al., 1989). Laser powers were in the range 4 to 20mW whilst bias fields were varied between -10 and 50kA/m. Several points concerning the images are worthy of note. Firstly, the elliptical shape of the marks is a result of the film tilt, the tilt axis being approximately 30° from the vertical in figure 3. Secondly, the contrast in the two MDPC images is quite different and depends on the relation between the tilt axis and the direction of induction mapped. Thus, figure 3a, which maps induction components perpendicular to the tilt axis, shows contrast arising from both the magnetisation within the mark and the stray fields to which it gives rise. In distinction to this, contrast in figure 3b arises from stray field only as there is no component of magnetisation parallel to the tilt axis in a film whose easy axis is orthogonal to the film plane. Also apparent in figure 3 is the increase in the contrast of the marks from top left to bottom right of the images. This is as a result of increasing bias field towards the bottom right-hand corner and to understand what is happening here more detailed images are required.

To better see the effect of varying bias field and laser power, 256x256 pixel 8-bit digital images were recorded from individual marks of the kind shown in figure 3. Each image took approximately 50s to collect and the direction of induction mapped is the vertical direction in the figure, the tilt axis being horizontal. Figure 4 shows a montage of two sequences. In the upper two rows the laser power was held constant at 10mW whilst the bias field was varied in steps of 10kA/m; conversely, in the lower rows the power was varied whilst the bias remained constant at 25kA/m. Immediately apparent is the increase in mark size with laser

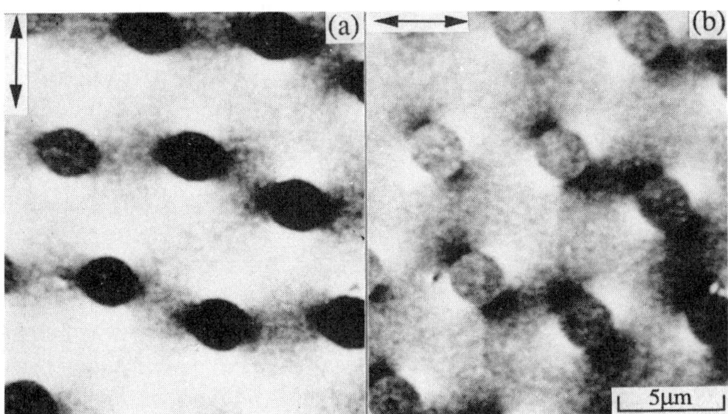

Figure 3. MDPC images of marks written with different bias fields. The arrows denote the directions of induction mapped.

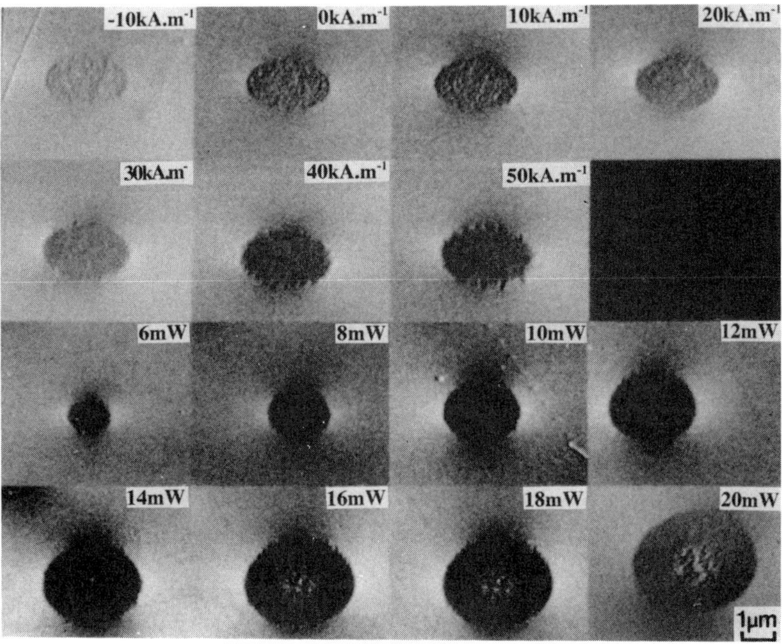

Figure 4. Images of individual marks written with different bias fields and laser powers.

power and its relative insensitivity to bias field. Figure 5 shows the form these variations take. Another notable feature in the images is the appearance and growth of additional contrast towards the centres of the marks when laser powers in excess of 12mW are used. This structure is not entirely magnetic in origin as can be verified by recording bright field images (insensitive to magnetic induction) from the same region of specimen. Indeed, what these images show is that the high laser power has induced grain growth so that at the mark centre the grain size is now ≈50nm, an increase of about a factor of 5 over the grain size elsewhere in the film. Such changes in the physical microstructure are irreversible and as such totally undesirable in an information storage medium designed for multiple over-writing. All further discussion pertains to marks written with lower powers (as were those in the upper two rows of figure 4) where no apparent changes to the physical microstructure can be observed.

Examination of the upper half of figure 4 shows that the contrast in the MDPC images, which is directly related to the magnetic moment of the mark, increases steadily with bias field. Furthermore the reason for low contrast in the marks written with negative or low bias field can be seen to be the presence of small reverse domains within those marks. That marks can be seen at all under the former conditions is a consequence of the high demagnetising fields in the otherwise uniformly magnetised multilayers. Thus, if uniformly magnetised marks are to be obtained, a necessity for a high carrier-to-noise ratio in recording applications, substantial bias fields are required. However, also required for optimum results is that the shape of the written mark be regular. Unfortunately, as figure 5 shows, the disappearance of reverse domains from within the domain at high bias fields is accompanied by an increase in the irregularity of the domain boundary.

Figure 6 shows a trio of images recorded simultaneously using various combinations of the signals available from the eight segment detector which exploit the high spatial resolution capability of the STEM to the full. Figures 6a and b are MDPC images of the kind seen in figure 3 and are formed using signal combinations (A-C) and (B-D) respectively whilst figure 6c is the incoherent bright field image (E+F+G+H). The pixel separation is 5.5nm, corresponding to slight over-sampling. At this magnification the crystallite structure is very apparent in figure 6c whilst, as required, the MDPC images are substantially insensitive to its existence. What is clear from figures 6a and b is the extent of the irregularity of the mark which was written with a laser power of 10mW and a bias field of 25kA/m. Excursions from the expected circular shape on a scale of 200nm are observed. Such irregularities cannot be seen easily using imaging techniques with lower spatial resolution, nor indeed are they so clearly seen in the lower magnification MDPC images shown earlier in this paper as a result of their coarser sampling interval. Despite the irregularity of the boundary shape, figure 6a confirms that the local transition between upward and downward directed magnetisation remains abrupt. Examination of traces perpendicular to the local boundary direction confirms that the transition width is less than 20nm. In other words the domain wall width is below the resolution limit of our experimental study.

DISCUSSION AND CONCLUSIONS

In the preceding sections we have shown that MDPC Lorentz microscopy is well suited to the detailed study of magnetic thin films. Even when the total thickness of magnetic material is very small, high contrast, high resolution images can be obtained and a range of qualitative and quantitative micromagnetic information extracted. Here, we have illustrated the use of the technique in a study of written marks in a Co/Pt multilayer specimen. Whilst such marks have been studied previously by other modes of Lorentz microscopy (Greidanus et al., 1989; Suits et al., 1988) earlier studies have provided less information on the detailed form of the marks. Of particular interest here is the increase in irregularity that is

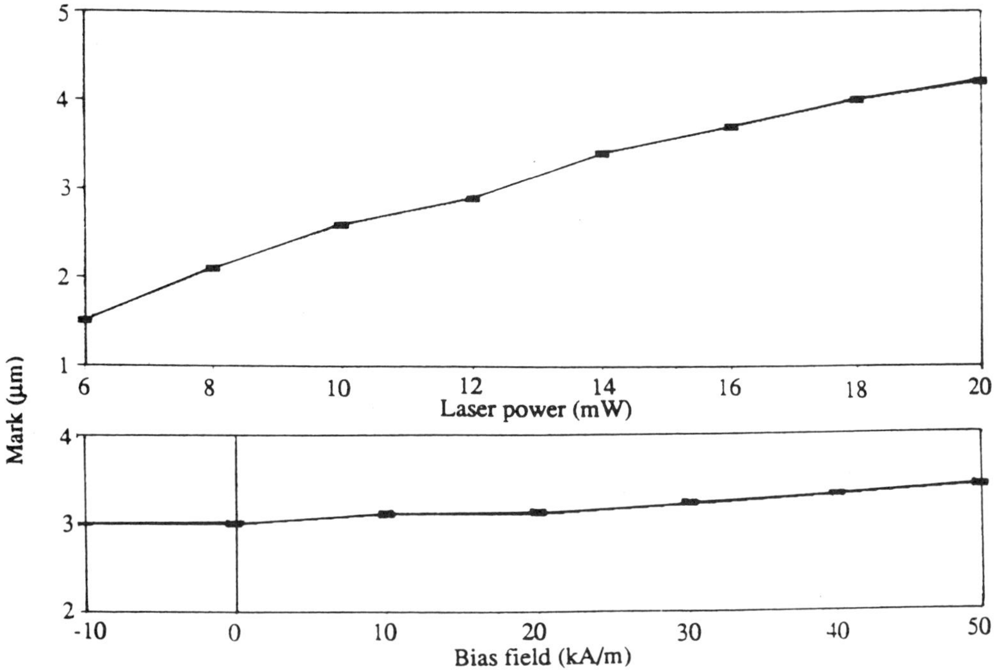

Figure 5. Variation of mark size with laser power at constant bias field and with bias field at constant laser power.

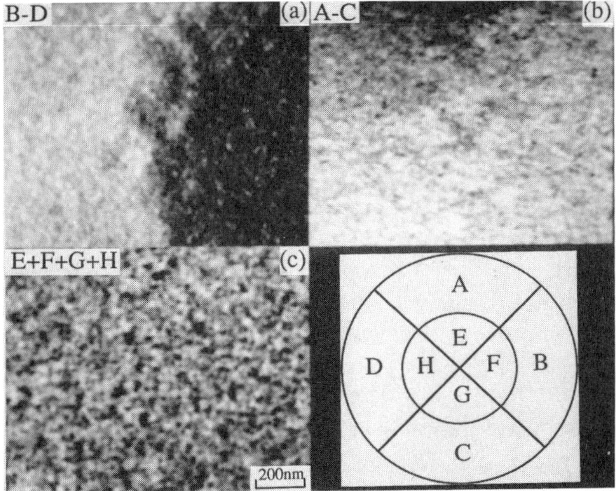

Figure 6. Images of part of a mark formed using the signal combinations shown.

observed with increasing bias field which suggests that optimum recording conditions may be those in which the bias is just high enough to ensure that small reverse domains are eliminated from the interior of the mark. That such irregularities in shape exist is presumably a consequence of local variations in coercivity as a result of microscopic pinning sites. Thus as the local area of film cools following heating by the laser, the force on the domain wall as a result of the conflicting demands of the external field, the demagnetising energy and the wall energy fails to overcome the local coercive field and distortions are introduced into the mark shape. This is in accord with models used by Hansen (1987) and van Kesteren et al. (1991). We should note that these considerations become of increasing importance as the size of the marks decrease as will be the case if shorter wavelength lasers are introduced.

Whilst the brief discussion and the examples cited above show that electron microscopy already provides much information on magnetic systems of reduced dimensionality, we believe that with further instrumental development is desirable. At Glasgow University we will have an instrument optimised for the study of magnetic material in place early in 1993. The new magnetic microscope will be based on a Philips CM20 FEG TEM/STEM and will offer a number of features not readily available on other instruments. Foremost among these are: (i) there will be sufficient room around the specimen for a small "micromagnetic laboratory" to allow a wider range of in-situ experiments to be undertaken, (ii) CTEM, STEM and SEM imaging modes will all be accessible with state-of-the-art resolution for a specimen situated in magnetic field free space and (iii) the instrument will have an analytical capability so that structural and/or compositional information can be obtained from regions where magnetic features of interest are observed. Using the facilities available on the new instrument, we believe that the power of electron-based observation techniques will be significantly extended.

ACKNOWLEDGEMENTS

We would like to thank Drs. W B Zeper and H W van Kesteren for provision of the Co/Pt multilayers and for many useful discussions. We would also like to thank the SERC, Philips Research Laboratories and CAMST for financial support for this work.

REFERENCES

Chapman, J.N., 1984, The investigation of magnetic domain structures in thin foils by electron microscopy, J. Phys. D: Appl. Phys., 17:623.

Chapman, J.N., McVitie, S. and McFadyen, I.R., 1987, Magnetic structure determination by scanning transmission electron microscopy, *in*: "Scanning Electron Microscopy Supplement 1," J. Kirschner, K. Murata and J.A. Venables, eds., AMF O'Hare, Chicago p221.

Chapman, J.N., McFadyen, I.R. and McVitie, S., 1990, Modified differential phase contrast Lorentz microscopy for improved imaging of magnetic structures, IEEE Trans. Mag., 26:1506.

Chapman, J.N., Ploessl, R. and Donnet, D.M., 1992, Differential phase contrast imaging of magnetic materials, Ultramicroscopy. - in press.

Donnet, D.M., Chapman, J.N., van Kesteren H.W. and Zeper, W.B., 1992, Investigation of domain structures in Co/Pt multilayers by modified differential phase contrast microscopy, J. Magn. Magn. Mat. - in press.

Greidanus, F.J.A.M., Zeper, W.B., Jacobs, B.A.J., Spruit, J.H.M. and Carcia, P.F., 1989, Magneto-optical recording in Co/Pt multilayers, Jap. J. Appl. Phys., 28 Supplement 3:37.

Hansen, P., 1987, Thermomagnetic switching in amorphous rare-earth transition metal alloys, J. Appl. Phys., 62:216.

Nakatani, Y., Uesaka Y.and Hayashi, N., 1989, Direct solution of the Landau-Lifshitz-Gilbert equation for micromagnetics, J. Appl. Phys. 66:2485.
Ploessl, R., Chapman, J.N., Thomson, A.M., Zweck, J. and Hoffmann, H., 1992, Investigation of the micromagnetic structure of cross-tie walls in permalloy, submitted to; J. Appl. Phys.
Suits, J.C., Rugar, D. and Lin, C.J., 1988, Thermomagnetic writing in Tb-Fe: modelling and comparison with experiment, J. Appl. Phys., 64:252.
van Kesteren, H.W., den Boef, A.J., Zeper, W.B., Spruit, J.H.M., Jacobs, B.A.J. and Carcia, P.F., 1991, Scanning force microscopy on Co/Pt magneto-optical disks, J. Appl. Phys., 70:2413.
Zeper, W.B., Greidanus, F.J.A.M., Carcia, P.F. and Fincher, C.R., 1989, Perpendicular magnetic anisotropy and magneto-optical Kerr effect of vapor deposited Co/Pt layered structures, J. Appl. Phys. 65:4971.

MAGNETORESISTANCE OF SPIN-VALVE SANDWICHES AND MULTILAYERS (EXPERIMENTS AND THEORIES)

B. Dieny[1], V.S. Speriosu[2], J.P Nozières[3], B.A Gurney[2], A.Vedyayev[4], and N.Ryzhanova[4]

[1]CENG, DRFMC/SP2M/MP, 85X, 38041 Grenoble Cedex, France
[2]IBM Almaden Research Division, 650 Harry Road, San Jose, California 95120-6099
[3]Laboratoire Louis Néel, CNRS, BP 166, 38042 Grenoble Cedex 9 France
[4]Department of Physics, Moscow Lomonosov University, Moscow 119899, Russia

INTRODUCTION

Electrical transport properties of multilayers comprising ferromagnetic layers separated by non-magnetic metallic layers have been studied experimentally by many groups(1). Very large or even giant magnetoresistance (MR) has been obtained in these structures (spin-valve multilayers) in which the relative orientation of the magnetizations in successive ferromagnetic layers can be changed either thanks to the existence of an antiferromagnetic coupling through the spacer layer (as in Fe/Cr(2), Co/Ru(3), Co/Cu(4,5)...) or because of different pinning forces acting on the magnetization of the ferromagnetic layers (different coercivities(6,7) or exchange anisotropy(8-11)). Qualitatively, it is now widely admitted that this particular MR effect results from a coherent interplay between the successive ferromagnetic layers of spin-dependent scattering (SDS) phenomena occuring at the interfaces and/or in the bulk of the ferromagnetic layers.

In this paper, we present experimental and theoretical results on the transport properties of spin-valve sandwiches and multilayers. The experiments have been carried out on series of sandwiches comprising substrate/F1/NM/F2/antiferromagnet, prepared by sputtering. F1 and F2 are ferromagnetic metals, NM is a non-ferromagnetic metal, the magnetization of F2 is constrained by exchange anisotropy (e.g NiFe/FeMn). We first give an overview of the materials that we have investigated and the order of magnitude of the spin-valve MR observed in each case. We next demonstrate that the absolute change of sheet conductance ΔG (sheet conductance per period) between parallel and antiparallel alignment of the magnetizations in the successive ferromagnetic layers is the most appropriate measurable physical quantity for the description and comparison of spin-valve magnetoresistance[12]. We then report a quantitative study of the relative role of interfacial versus bulk spin-dependent scattering in three series of spin-valve sandwiches comprising $Ni_{80}Fe_{20}$, Co or Fe layers with Cu spacers. This study consists of measuring and fitting with Camley and Barnas'theory[13,14], the variation of the sheet conductance and magnetoresistance versus the thickness of the ferromagnetic layers. We demonstrate that the spin-dependent scattering (SDS) is mainly bulk for NiFe/Cu, a mixture of interfacial and bulk SDS for Co/Cu and mainly interfacial for Fe/Cu[15]. We also show that our conclusion on SDS scattering in Co/Cu is corroborated by recent experiments[15,16] in which thin Co layers are introduced at NiFe/Cu interfaces.

On the theoretical side, we have quantitatively interpreted our results within two types of theories both based on the existence of SDS phenomena occuring in the ferromagnetic layers. The first approach is similar to Camley and Barnas' theory[13,14]. Besides the fact that we have been able to fit all our experimental data at low temperature, we also obtained some new and unexpected results[17]. The second approach developed for bulk SDS (the case of NiFe) is a quantum mechanical theory based on the Kubo formalism adapted to finite systems[18]. There are some differences between this theory and Levy et al in the way the SDS is treated (coherent potential approximation) and the appropriate Green functions are derived. An analytical formula for the conductance and magnetoresistance of an infinite multilayer has been obtained. A comparison of the classical and quantum theories has been undertaken in the case of bulk spin-dependent scattering. For the same set of parameters, the values of magnetoresistance calculated from these two types of theories on a broad range of thicknesses of the ferromagnetic and non-magnetic layers are in surprisingly good agreement (they differ by less than 1% in relative value). This result supports the use of classical theories (Fuchs-Sondheimer) to interpret spin-valve MR data even when the thicknesses of the layers are much smaller then the mean-free paths of conduction electrons.

SPIN-VALVE SANDWICHES AND SPIN-VALVE MAGNETORESISTANCE

Spin-valve sandwiches comprise two ferromagnetic layers (F1 and F2) separated by a non-ferromagnetic metal (NM). The magnetization of F2 is constrained by coupling to an adjacent antiferromagnetic layer (by exchange anisotropy). Our samples were prepared in a conventional dc magnetron sputtering system with four targets and a base pressure of 10^{-7} Torr. More details of the sample preparation can be found in references 8-11.

Figures 1a and 1b, respectively, show room-temperature hysteresis loops and magnetoresistance of a sample with structure Si/2∗(Ta 50Å/NiFe 60Å/Cu 22Å/NiFe40Å/FeMn 70Å)/Ta 50Å. On an expanded scale, Fig 1c shows the MR response corresponding to field swept between ± 50Oe. Here we have deposited two repetitions of the spin-valve structure. However we have found that in this type of structure the relative MR amplitude $\Delta R/R$ is independent of the number of periods. This result is not surprising because of the presence of the highly resistive Ta and FeMn layers within each period.The hysteresis of this system consists of two loops. The first one, centered at 6 Oe with a coercive field less than 1 Oe corresponds to the reversal of the two free NiFe layers which have nearly identical properties. The second loop with a coercivity of 100 Oe and an exchange shift of 420 Oe, corresponds to the reversal of the magnetizations of the NiFe layers coupled to FeMn. When the field is swept between ±1kOe, the magnetizations of adjacent NiFe layers change from parallel below 4 Oe and above ≅600 Oe to antiparallel between 8 Oe and ≅250 Oe.

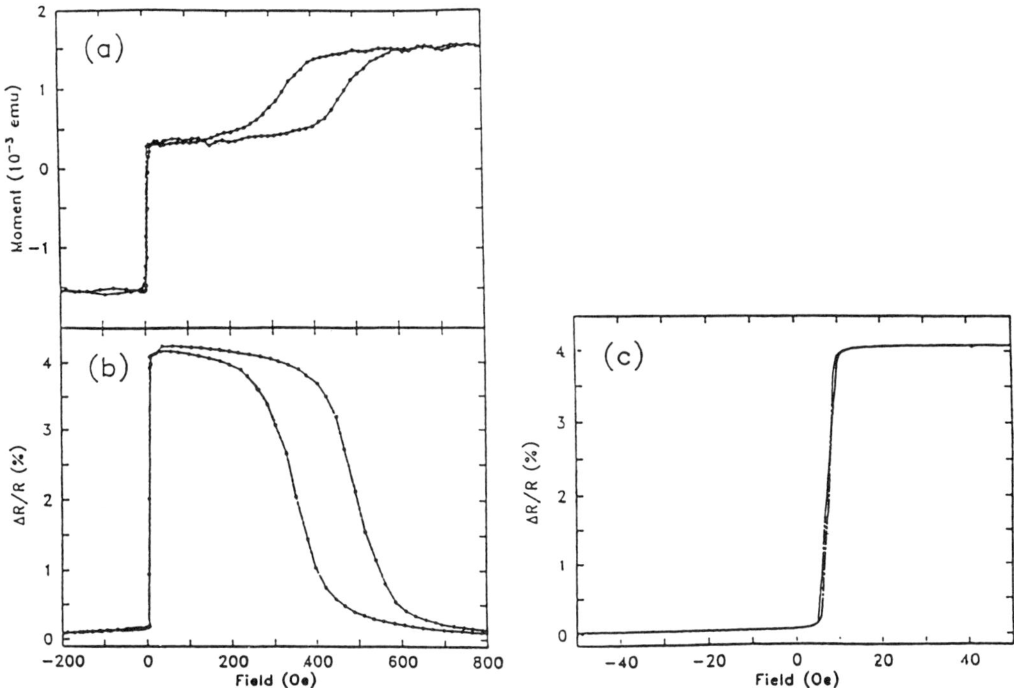

Fig.1.(a) Room temperature hysteresis loop and (b) magnetoresistance for a sample with structure Si/2x(50Å Ta/60Å NiFe/22Å Cu/40Å NiFe/70Å FeMn)/50Å Ta. (a,b) The field was swept between ±1kOe alond the easy axis. (c) MR response for field swept between ±50 Oe.

The 6-Oe shift of the nominally free NiFe layers (see Fig 1c) indicates the presence of a slight ferromagnetic coupling through the Cu interlayer. In this paper we focus on the magnetoresistive properties of these spin-valve sandwiches and not on the interlayer coupling through the non-ferromagnetic spacer. However we point out that in another study[19] on Co 30Å/Cu t_{Cu}/Co 30Å/FeMn 150Å we have observed oscillations in the shift of the hysteresis loop of the unpinned Co layer versus t_{Cu} indicating oscillations in the strength of the coupling between the Co layers through the Cu spacer. These oscillations are similar to those observed in (Co/Cu) multilayers[4,5]. In general spin-valve sandwiches allow the determination of the coupling energy in both cases of ferromagnetic and antiferromagnetic coupling simply from magnetization or magnetoresistance measurements.

The magnetoresistance curve of Fig.1b exhibits a very sharp rise corresponding to the switching of the free layers, followed by a gradual decrease to zero as the pinned layers are reversed. This change of resistance related to the relative alignment of the magnetizations in the successive ferromagnetic layers is now widely named *spin-valve magnetoresistance*[1,10]. It has the same physical origin in the present sandwiches as in Fe/Cr[2,3], Co/Cu[4,5], Ag/Ni[20]...multilayers.

An interesting feature of these spin-valve sandwiches is that we do not rely on the existence of an antiferromagnetic coupling between the ferromagnetic layers to change the relative orientation of the magnetizations of adjacent ferromagnetic layers. Thanks to the asymmetric pinning forces acting on the magnetizations of the ferromagnetic layers it is always possible to switch the alignment of the magnetizations from fully parallel to fully antiparallel. The only requirement is that the thickness of the spacer layer must be large enough to provide a sufficient decoupling of the ferromagnetic layers (typically 15Å of non-magnetic material is enough to considerably reduce the ferromagnetic coupling between ferromagnetic layers through pinholes). It is therefore possible to probe the spin-valve magnetoresistance of almost any system of the form F/NM/NiFe/FeMn where F is a ferromagnetic metal, NM a non ferromagnetic metal and derive information on the spin-dependent scattering properties of the ferromagnetic metal F and of the F/NM interface. The results that we obtained on a wide variety of materials are summarized in the table below .

The largest spin-valve MR values have been obtained with Co and NiFe layers separated by a noble metal (Cu, Ag or Au). With Ag, we observed that for samples deposited at room temperature a minimum thickness of 50Å was required to obtain a good decoupling of the ferromagnetic layers. On the contrary, Rodmacq et al[20,21] observed that very good crystallographic growth of Ag/Ni multilayers could be obtained using the same sputtering techniques by lowering the substrate temperature down to that of liquid nitrogen. This underlines the crucial role of substrate temperature on the growth of Ag on Ni or NiFe.

With rare-earth ferromagnets, no spin-valve MR has been observed. This is not really surprising since spin-dependent scattering properties are expected to be weak in these systems.

Table 1

F1	NM	F2/FeMn	largest ΔR/R at room T	Minimum t_{NM} for decoupling
Co	Cu	Co	9.5%	$t_{Cu} \geq 16Å$
$Ni_{80}Fe_{20}$	Cu	$Ni_{80}Fe_{20}$	5%	$t_{Cu} \geq 18Å$
Ni	Cu	Ni	2.5%	$t_{Cu} \geq 20Å$
Fe	Cu	$Ni_{80}Fe_{20}$	2.5%	$t_{Cu} \geq 18Å$
Gd	Cu	$Ni_{80}Fe_{20}$	0 % at 77K	$t_{Cu} \geq 20Å$
Nd	Cu	$Ni_{80}Fe_{20}$	0 % at 77K	$t_{Cu} \geq 20Å$
Co	Cu	$Ni_{80}Fe_{20}$	6.5%	$t_{Cu} \geq 16Å$
Co	Ag	$Ni_{80}Fe_{20}$	1.5%	$t_{Ag} \geq 50Å$
$Ni_{80}Fe_{20}$	Ag	$Ni_{80}Fe_{20}$	1.2%	$t_{Ag} \geq 50Å$
Co	Au	$Ni_{80}Fe_{20}$	4.5%	$t_{Au} \geq 10Å$
$Ni_{80}Fe_{20}$	Pt	$Ni_{80}Fe_{20}$	0.3%	$t_{Pt} \geq 20Å$
$Ni_{80}Fe_{20}$	Pd	$Ni_{80}Fe_{20}$	0.2%	$t_{Pd} \geq 20Å$
$Ni_{80}Fe_{20}$	Al	$Ni_{80}Fe_{20}$	0 %	$t_{Al} \geq 20Å$
$Ni_{80}Fe_{20}$	V	$Ni_{80}Fe_{20}$	0 %	$t_V \geq 16$ Å
$Ni_{80}Fe_{20}$	Cr	$Ni_{80}Fe_{20}$	0 %	$t_{Cr} \geq 16$ Å
$Ni_{80}Fe_{20}$	Nb	$Ni_{80}Fe_{20}$	0 %	$t_{Nb} \geq 20Å$
$Ni_{80}Fe_{20}$	Ru	$Ni_{80}Fe_{20}$	0 %	$t_{Ru} \geq 15$ Å
$Ni_{80}Fe_{20}$	Ta	$Ni_{80}Fe_{20}$	0 %	$t_{Ta} \geq 15Å$
$Ni_{80}Fe_{20}$	W	$Ni_{80}Fe_{20}$	0 %	$t_W \geq 15$ Å

Indeed the split narrow 4f bands characteristic of the magnetism of these rare earths are usually far away from the Fermi level. Therefore they do not cause a significant difference in density of states for spin ↑ and spin ↓ electrons at the Fermi level and consequently no spin-dependent scattering. Other rare-earths such as Ce for which one of the 4f sub-band is close to ε_F may lead to interesting effects.

Pt and Pd give measurable but very small MR. This has also been found by A.Schuhl et al in Fe/Pd multilayers with antiferromagnetic coupling (see this book, growth and magnetic properties of (100) Fe/Pd multilayers).

Several other non-ferromagnetic transition metals that we tried (Al, V, Cr, Nb, Ru, Ta, W) did not give any measurable MR. We think that the absence of spin-valve MR in these structures is associated with the impossibility for conduction electrons to travel back and forth between the ferromagnetic layers through the spacer or to keep their polarization. This may be due to a too high resistivity of the spacer layer (too short mean-free path compared to the minimum thickness of the spacer layer) (V, Cr, Nb, Ru, Ta, W) , a poor crystallographic quality of the interfaces or of the spacer, or to interdiffusion (this is the case for Al which drastically interdiffuses into NiFe). The formation of paramagnetic layers at the F/NM interfaces (as in Ni/Cu[11]) can also significantly reduce the MR by inducing strong scattering, including spin-flip scattering of incoming or outgoing conduction electrons.

ABSOLUTE CHANGE OF SHEET CONDUCTANCE: FUNDAMENTAL MEASURE OF THE SPIN-VALVE MR

A wide variety of quantities are used in the literature to quantify the longitudinal spin-valve MR: absolute change of resistance ΔR or relative change of resistance normalized by $R_{parallel}$ or $R_{antiparallel}$ ($\Delta R/R_{parallel}$ or $\Delta R/R_{antiparallel}$) or similar quantities expressed in terms of resistivity where the resistivity is a macroscopic average quantity defined by ρ=Rt where R is the sheet resistance and t the total thickness. The use of these various quantities does not allow an easy comparison of all results published in the literature. They are of course related to each other by the following relations:

$$\frac{\Delta G}{G_{parallel}} = \frac{\Delta R}{R_{antiparallel}} \qquad \frac{\Delta G}{G_{antiparallel}} = \frac{\Delta R}{R_{parallel}}$$

$$\Delta R = \frac{\Delta G}{G_{parallel}\, G_{antiparallel}} \qquad \Delta G = \frac{\Delta R}{R_{parallel}\, R_{antiparallel}}$$

Here we want to show that the measurements of both the sheet conductance and magnetoresistance are necessary for a comparison of the transport properties of various samples and that the absolute change of sheet conductance ΔG is the most relevant measurable parameter for quantifying the spin-valve MR from a fundamental point of view. First we give a theoretical argument : At a microscopic scale, the application of an electric field **E** to the system induces a perturbation $g_\sigma(z,v)$ in the Fermi-Dirac equilibrium distribution $f^o_\sigma(z,v)$ of conduction electrons (σ refers to the spin of the electrons): $f_\sigma(z,v) = f^o_\sigma(z,v) + g_\sigma(z,v)$. The change in this perturbation $g_\sigma(z,v)$ with the applied magnetic field provides the most direct description of the magnetoresistance of these structures. However it is a complex function of several variables (position and velocity) which is not accessible experimentally. The current is related to this perturbation by :

$$J = GE = \sum_{\sigma=\uparrow\downarrow} e \int v_x \, d^3v \int g_\sigma(z,v_z)\, dz$$ and the magnetoresistance by

$$G_{parallel}-G_{antiparallel} = \sum_{\sigma=\uparrow\downarrow} \frac{e}{E} \int v_x \, d^3v \int (g_{\sigma\ parallel}(v_z,z)-g_{\sigma\ antiparallel}(v_z,z))\, dz.$$ This relation shows that in this in-plane geometry, there is a *linear relationship* between the macroscopic measurable quantity $\Delta G= G_{parallel}-G_{antiparallel}$ and the change in the distribution $g_{\sigma\ parallel}(v_z,z) - g_{\sigma\ antiparallel}(v_z,z)$ at the microscopic scale. Such a linear relationship does not exist with $\Delta R/R$ or ΔR so that the interpretation of these quantities is not as direct as with ΔG.

To illustrate the implication of this point, as an example we compared the thermal variation of the resistance and magnetoresistance of three samples of composition Si/Co 80Å/Cu 25Å/NiFe 50Å/FeMn 90Å/ Ta 50Å/Cu t_{Cu}/Ta 50Å. These samples have exactly the same magnetic inner part, only the thickness t_{Cu} in the capping layer is varied (t_{Cu}=0Å, 100Å, 300Å). Figures 2a,b,c,d respectively show the thermal variation of the resistance and magnetoresistance expressed as $\Delta R/R$, ΔR or ΔG for these three samples. *$\Delta G(T)$ is found to be exactly the same in shape and amplitude along the series demonstrating the merit of this quantity in characterizing the intrinsic magnetoresistance of these structures. In contrast $\Delta R/R(T)$ and $\Delta R(T)$ are influenced by both the magnetoresistance and the overall conductance of the structures.*

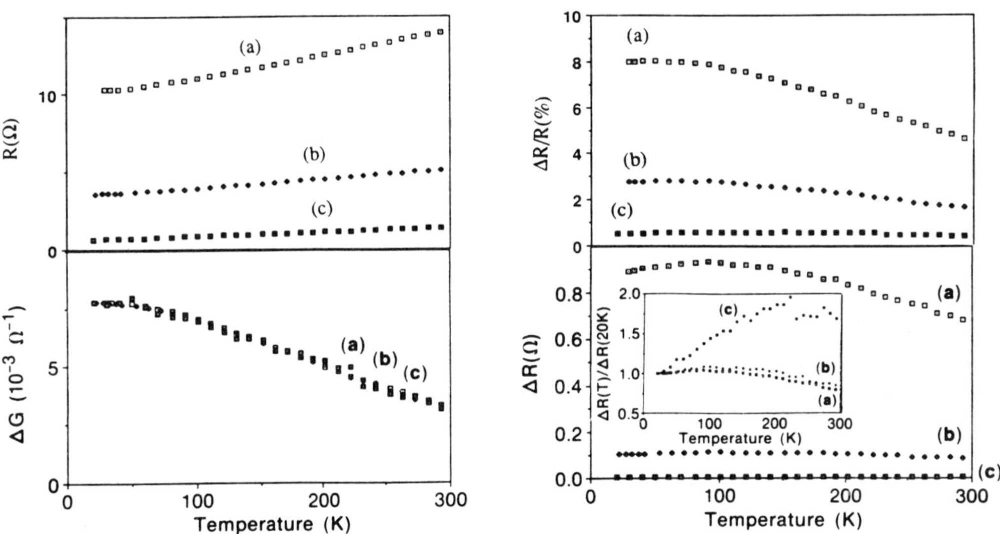

Fig.2 . Comparison of the thermal variation of R, ΔG, $\Delta R/R$ and ΔR for three spin-valve structures of composition Si/Co 80Å/Cu 25Å/FeMn 90Å/Ta 50Å/Cu t_{Cu}/Ta 50Å with (a) t_{Cu}=0Å, (b) t_{Cu}=100Å, (c) t_{Cu}= 300Å.

RELATIVE ROLE OF BULK AND INTERFACIAL SPIN-DEPENDENT SCATTERING IN SPIN-VALVES COMPRISING NiFe/Cu, Co/Cu AND Fe/Cu

We carried out a study of the relative role of bulk and interfacial spin-dependent scattering in three series of spin-valve sandwiches comprising NiFe/Cu, Co/Cu and Fe/Cu. We deposited three series of samples of composition Si/F t_FÅ/Cu 22Å/NiFe 50Å/FeMn 90Å where F= NiFe, Co or Fe and where only the thickness of the nominally free ferromagnetic layer F was varied. For each sample, we measured both the conductance G and magnetoresistance ΔG at 1.5 K.

We then fitted the variation of $G(t_F)$ and $\Delta G(t_F)$ within the Camley and Barnas theory taking into account both bulk spin-dependent scattering (SDS) within the ferromagnetic layer (the bulk SDS is described by two mean-free paths $\lambda_{\uparrow F} \neq \lambda_{\downarrow F}$) and interfacial SDS (described by spin-dependent transmission coefficients through the F/Cu interfaces $T_{\uparrow F/Cu} \neq T_{\downarrow F/Cu}$). As discussed in the next part of this paper (comparison of classical and quantum theories of spin-valve MR), this type of classical theory (extended Fuchs-Sondheimer theory) seems to account well for the spin-valve MR of these structures even when the thicknesses of the layers are much smaller than the mean-free paths of conduction electrons. Quantitatively the fits of $G(t_F)$ and $\Delta G(t_F)$ are made with four fitting parameters : $\lambda_{\uparrow F}$, $\lambda_{\downarrow F}$, $T_{\uparrow F/Cu}$, $T_{\downarrow F/Cu}$. We also take into account an anisotropy in the mean-free paths introduced by the scattering at grain-boudaries (see details in reference 24). The slope of $G(t_F)$ mainly determines the sum $(\lambda_{\uparrow F} + \lambda_{\downarrow F})$.

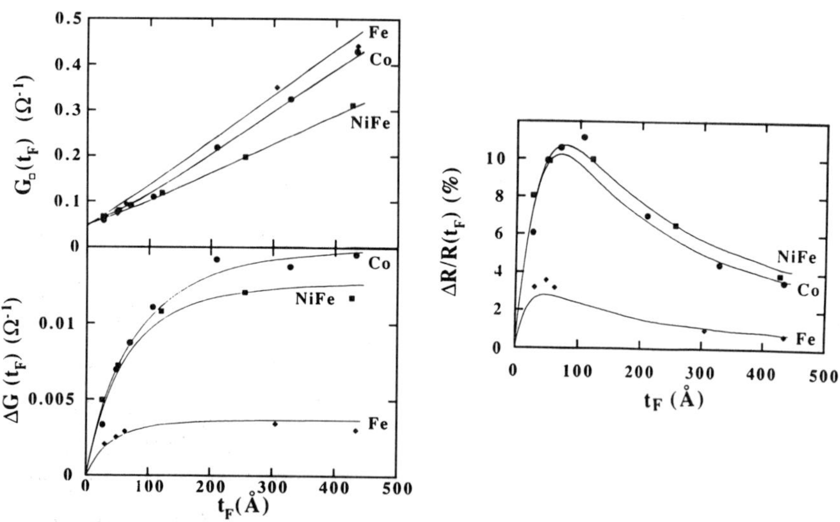

Fig.3 . Variation of the sheet conductance and magnetoresistance (expressed as ΔG or ΔR/R) versus the thickness t_F of the ferromagnetic layer F for three series of spin-valve sandwiches of composition Si/F t_FÅ/Cu 22Å/NiFe 50Å/FeMn 90Å where F=NiFe, Co or Fe. Points=experiments, lines=fits.

The characteristic length of the increase of ΔG versus t_F is related to the longest of the two mean-free paths $\lambda\uparrow_F$ or $\lambda\downarrow_F$. The amplitude of the saturation of $\Delta G(t_F)$ leads to the other parameters. The results are summarized in figure 3 and in the table below. Although at room temperature the MR is larger with Co than with NiFe, it turns out that at low temperature they are quite comparable. In contrast Fe/Cu gives much lower MR as also observed in multilayers[22]. Note that for Fe the curve $\Delta G(t_F)$ saturates much faster than for NiFe or Co. Similarly on $\Delta R/R(t_F)$ the position of the maximum occurs for Fe at lower t_F thickness than for the two other elements. These features are characteristic of the very small role of bulk spin-dependent scattering in Fe/Cu compared to the other elements.

The parameters found for NiFe, Co and Fe at 1.5K are summarized in the table below .

	NiFe	Co	Fe
$\rho_F(\mu\Omega cm)$	15.4	10.7	10.5
$\lambda\uparrow_F$(Å)	110	140	70
$\lambda\downarrow_F$(Å)	10	10	70
$T\uparrow_{F/Cu}$	1	1	1
$T\downarrow_{F/Cu}$	1	0.2	0.6

Table 2. Resistivity, mean-free path for spin ↑ and spin ↓ electrons in the ferromagnetic metal F, coherent transmission coefficient for spin ↑ and ↓ electrons at the F/Cu interfaces at 1.5 K.

In conclusion we found that the spin-dependent scattering in F/Cu is mainly bulk in NiFe, a mixture of bulk and interfacial in Co/Cu and mainly interfacial in Fe/Cu but with a much weaker spin-dependent character than at the Co/Cu interfaces. We suggest that the differences in the spin-dependent scattering properties of these various elements are due to differences in the density of state of d-electron holes at the Fermi level.

We show next that our conclusion on the SDS in NiFe and Co is corroborated by recent experiments[15,16] in which a thin layer of Co is introduced at the NiFe/Cu interfaces or moved away from the interface into the NiFe layers. Two series of samples have been prepared with the composition : Si/Ta 50Å/ NiFe (60-x)Å/Co xÅ/Cu 25Å/Co xÅ/NiFe (25-x)Å/FeMn 100Å/Ta 50Å where $0Å \leq x \leq 12Å$ and Si/Ta 50Å/ NiFe (55-y)Å/Co 5Å/NiFe yÅ/Cu 25Å/NiFe yÅ/Co 5Å/NiFe (20-y)Å/FeMn 100Å/Ta 50Å where $0Å \leq y \leq 10Å$. The variation of $\Delta R/R$ versus x or y is plotted on figures 4a and 4b. $\Delta R/R$ increases dramatically (by a factor 2) as a thin layer of Co (0 to 4 Å thick) is introduced at the NiFe/Cu interfaces. This feature corroborates the existence of a very large interfacial spin-dependent scattering at Co/Cu interfaces. Quantitatively the length of 2.8Å found by empirically fitting the MR curve with an exponential variation (see figure) characterizes the thickness of Co required to establish the electronic properties of the Co/Cu interfaces. A second interesting feature of this variation versus x is the plateau or even slight increase in $\Delta R/R$ above 4Å of Co. This increase can only be explained by considering the bulk SDS contribution within the Co layer.

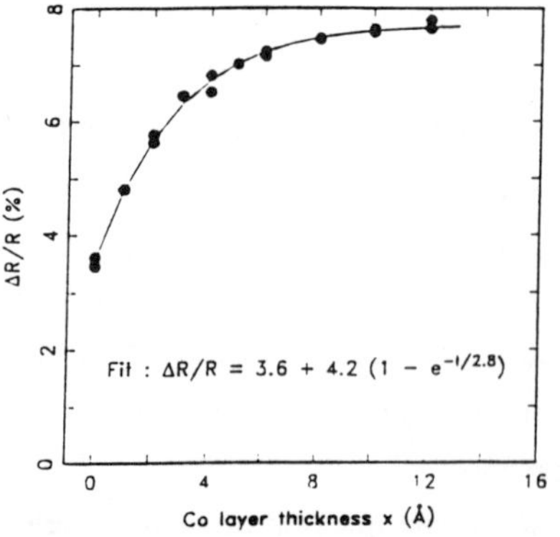

Fig. 4A. Magnetoresistance at room temperature versus thickness of the thin Co layer introduced at the NiFe/Cu interface (see text above). Points = experiments, line = fit.

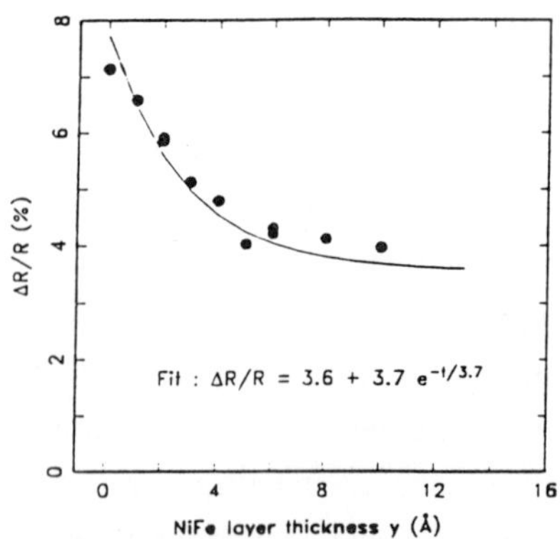

Fig. 4B. Magnetoresistance at room temperature versus position of the Co layer (5Å thick) within the NiFe layers (see text above). Points = experiments, line = fit.

As shown on the calculated curves below (using Camley and Barnas' theory), if the bulk SDS in the thin Co layer were not significant, ΔR/R(x) should decrease once the Co/Cu interface is formed. Since this decrease is not observed, *this experiment confirms the existence of both interfacial and bulk SDS in Co/Cu.* Regarding the second variation (ΔR/R versus y) the MR decreases rapidly towards the value of the corresponding NiFe/Cu/NiFe spin-valve without Co as the Co layer is moved backwards within the NiFe layer. This decrease is due to the disappearance of the spin-filtering Co/Cu interface. The characteristic length of this decrease (3.7Å) is slightly larger than in the former experiment which probably indicates a greater roughness of the NiFe/Cu interfaces than of the Co/Cu interfaces. Again it can be shown (see figure below) that if the bulk SDS in the Co nanolayer were not significant, ΔR/R(y) should go through a minimum once the Co/Cu interface has disappeared (around 4Å of NiFe). Since this minimum is not observed *this again corroborates the existence of significant bulk SDS in Co.*

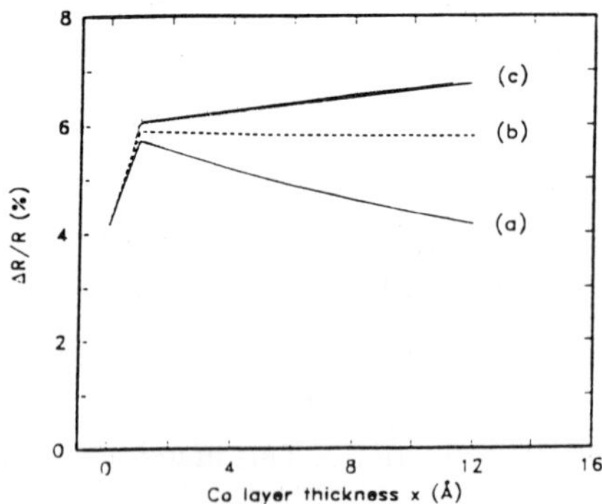

Fig. 5A. Calculated magnetoresistance in the conditions of the experiment of Fig. 4A assuming fixed interfacial SDS ($T_\downarrow$ = 0.2, $T_\uparrow$ = 1) and varying the bulk SDS properties in Co: (a)$\lambda_\downarrow$ = 75Å, $\lambda_\uparrow$ = 75Å; (b)$\lambda_\downarrow$ = 25Å, $\lambda_\uparrow$ = 125Å; (c)$\lambda_\downarrow$ = 7Å, $\lambda_\uparrow$ = 140Å.

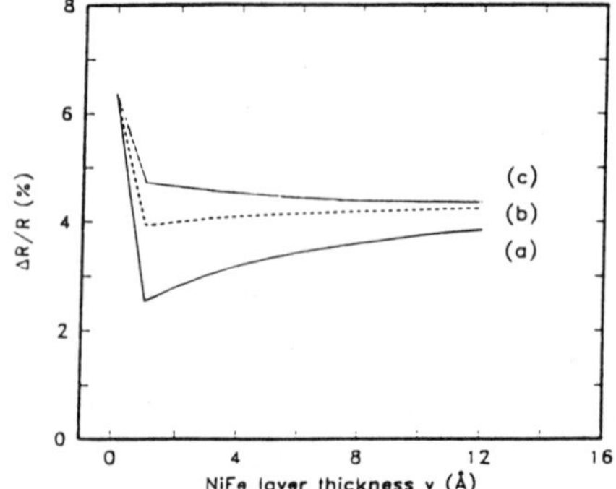

Fig. 5B. Calculated magnetoresistance in the conditions of the experiment of Fig. 4B assuming fixed interfacial SDS ($T_\downarrow$ = 0.2, $T_\uparrow$ = 1) and varying the bulk SDS properties in Co: (a)$\lambda_\downarrow$ = 75Å, $\lambda_\uparrow$ = 75Å; (b)$\lambda_\downarrow$ = 25Å, $\lambda_\uparrow$ = 125Å; (c)$\lambda_\downarrow$ = 7Å, $\lambda_\uparrow$ = 140Å.

The perpendicular transport experiments (CPP) in Co/Cu multilayers (see papers by Schroeder et al and Fert et al in this book) also lead to the same conclusion.

We showed above that each ferromagnetic transition metal and each ferromagnetic metal/non-magnetic metal interface has its own spin-dependent scattering properties. Next we point out that the thicknesses of the layers also influence the relative role of bulk and interfacial SDS. To illustrate this point, we carried out a calculation of the spin-valve MR of an infinite multilayer with the parameters that we found for Co/Cu based spin-valves (coexistence of bulk and interfacial SDS). We made three successive hypotheses :

(i) (B+I): The two contributions (bulk+interfacial SDS) are considered simultaneously in the calculation of the spin-valve MR of this ∞(Co t_{Co}/Cu 10Å/Co t_{Co}/Cu 10Å) multilayer.

$\lambda\uparrow_{Co}$(Å) =140Å $\qquad$ $\lambda\downarrow_{Co}$(Å) =10Å $\qquad$ $T\uparrow_{Co/Cu}=1$ $\qquad$ $T\downarrow_{Co/Cu}=0.2$

(ii) (B): Only the bulk SDS contribution is kept while the interfacial SDS is switched off : $\lambda\uparrow_{Co}$(Å) =140Å $\qquad$ $\lambda\downarrow_{Co}$(Å) =10Å $\qquad$ $T\uparrow_{Co/Cu}=1$ $\qquad$ $T\downarrow_{Co/Cu}=1$

(iii) (I) : Only the interfacial SDS contribution is kept while the bulk SDS is switched off: $\lambda\uparrow_{Co}$(Å) =75Å $\qquad$ $\lambda\downarrow_{Co}$(Å) =75Å $\qquad$ $T\uparrow_{Co/Cu}=1$ $\qquad$ $T\downarrow_{Co/Cu}=0.2$

(The sum $\lambda\uparrow_{Co}+\lambda\downarrow_{Co}$ is kept constant to insure the same resistivity of the Co layer). We then compared the MR in the real situation where the two contributions (B+I) are considered together with the MR according to hypotheses (B) and (I).

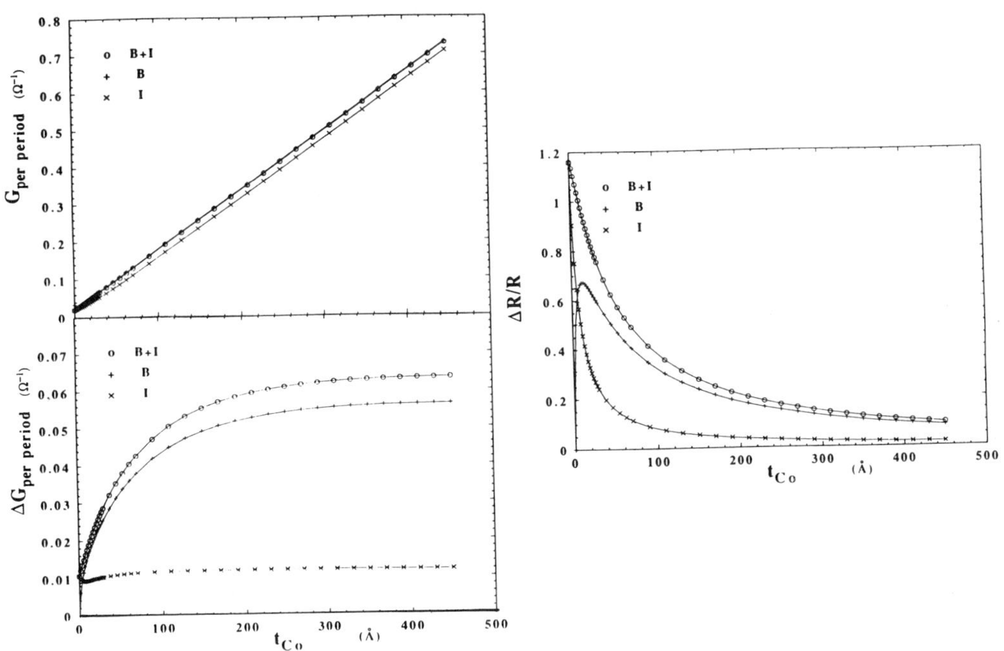

Fig.6 . Conductance per period and magnetoresistance for ∞(Co t_{Co}/Cu 10Å/Co t_{Co}/Cu 10Å) versus t_{Co} for three different hypotheses : bulk+interfacial SDS (B+I), bulk only SDS (B), interfacial only SDS (I) (see text above).

The variation of ΔG versus t_{Co} shows quite different behaviors in the two cases of bulk-only (B) or interfacial-only (I) SDS. ΔG is almost independent of t_{Co} in case (I) while it exhibits a slow saturation (at a rate characterized by $\lambda_{\uparrow Co}$) in the case (B). In $\Delta R/R$ a maximum at finite thickness (15Å) occurs in case (B) while no maximum is observed in case (I) (except in real samples at very low Co thickness when the Co/Cu interfaces disappear).Surprisingly, although within Camley and Barnas'theory no linearity of ΔG with the density of scattering centers can be expected it turns out that $(\Delta G)_{(B+I)}$ is slightly smaller but very close to the sum $(\Delta G)_{(B)}+(\Delta G)_{(I)}$ which means that *in the present case of infinite multilayers the two contributions (bulk and interfacial) are almost additive.* Furthermore it appears very clearly that the relative contribution of bulk and interfacial SDS depends on the thickness of the ferromagnetic layers. *At low Co thickness the interfacial contribution is favoured whereas the bulk contribution plays a more important role at large Co thickness.*

COMPARISON OF CLASSICAL AND QUANTUM THEORIES OF SPIN-VALVE MAGNETORESISTANCE

We carried out a quantitative comparison of two existing theories of giant magnetoresistance in spin-valve multilayers. The basic physical picture is the same in these various theories namely that the MR results form a coherent interplay of the spin-dependent scattering mechanism occuring at the interfaces or in the bulk of the successive ferromagnetic layers.

The first theory (classical), initiated by Camley and Barnas[13,14], is an extension of the Fuchs-Sondheimer theory, which takes into account spin-dependent mean free paths in the ferromagnetic layers and/or spin-dependent transmission coefficients at the ferro/non-magnetic interfaces. This type of theory has been further developed by different groups[23-26] and is now widely used (as we did above) to interpret quantitatively experimental spin-valve MR data obtained on sandwiches or multilayers.

The second theory (Vedyaev, Dieny, Ryzhanova)[18] is a quantum theory. As in the classical theory, the electrons are supposed to form a free electron gas subjected to spin-dependent scattering potentials. Using the Kubo formalism, it allows an exact calculation of the Green functions of the system in the real space in the case of an infinite multilayer with bulk spin-dependent scattering. The theory leads to a formula for the conductance as an integral over momentum which contains non-oscillating and quickly oscillating terms with frequency Dk_F (D is the thickness of the chemical period). These latter terms are due to the effect of quantization of momentum k_Z and are of the order of $(Dk_F)^{-1}$ .We can therefore neglect them in the quasiclassical limit. The final exact expression for the conductance per period in a multilayer ∞(M_1 aÅ/M_2 bÅ/M_3 cÅ/M_2 bÅ) in which M_1, M_2, M_3 are three metals of thicknesses a, b and c characterized by the mean free paths l_1,l_2,l_3 is given by :

$$G\,(a,b,c,l_1,l_2,l_3)= \frac{3}{2}\,\sigma_0{}^{a}\int_0^1 du\,(1-u^2)\,\{a + 2b\frac{l_2}{l_1} + c\frac{l_3}{l_1}$$

$$-\frac{l_1 u}{E_{abc}}\,[\frac{(l_1-l_2)}{l_1}E_aE_{bc} + \frac{(l_2-l_3)}{l_1}B\,E_aE_c$$

$$+\frac{l_2}{l_1}(\frac{l_2-l_1}{l_1})\,E_a\,(E_{bc} - B\,E_c) + \frac{l_2}{l_1}(\frac{l_2-l_3}{l_1})\,E_c\,(E_{ab}-B\,E_a)$$

$$+\frac{l_3}{l_1}(\frac{l_3-l_2}{l_1})\,E_c\,E_{ab} + \frac{l_3}{l_1}(\frac{l_2-l_1}{l_1})\,B\,E_a\,E_c\,]\}$$

$$\text{with } B=\exp\text{-}(\,2\,b\,k_F\,|\mathrm{Im}(u^2+\frac{i}{l_2k_F})^{1/2}|)$$

$$E_a=\{1-\exp\text{-}(\,2\,a\,k_F\,|\mathrm{Im}(u^2+\frac{i}{l_1k_F})^{1/2}|)\}$$

$$E_b=\{1-\exp\text{-}(\,4\,b\,k_F\,|\mathrm{Im}(u^2+\frac{i}{l_2k_F})^{1/2}|)\}$$

$$E_c=\{1-\exp\text{-}(\,2\,c\,k_F\,|\mathrm{Im}(u^2+\frac{i}{l_3k_F})^{1/2}|)\}$$

$$E_{ab}=\{1-\exp\text{-}(\,2\,a\,k_F\,|\mathrm{Im}(u^2+\frac{i}{l_1\,k_F})^{1/2}|)\,\exp\text{-}(\,4\,b\,k_F\,|\mathrm{Im}(u^2+\frac{i}{l_2k_F})^{1/2}|)\}$$

$$E_{bc}=\{1-\exp\text{-}(\,2\,c\,k_F\,|\mathrm{Im}(u^2+\frac{i}{l_3\,k_F})^{1/2}|)\,\exp\text{-}(\,4\,b\,k_F\,|\mathrm{Im}(u^2+\frac{i}{l_2k_F})^{1/2}|)\}$$

$$E_{abc}=\{1-\exp\text{-}(\,2\,a\,k_F\,|\mathrm{Im}(u^2+\frac{i}{l_1\,k_F})^{1/2}|)\,*$$

$$\exp\text{-}(\,4\,b\,k_F\,|\mathrm{Im}(u^2+\frac{i}{l_2k_F})^{1/2}|)\,\exp\text{-}(\,2\,c\,k_F\,|\mathrm{Im}(u^2+\frac{i}{l_3\,k_F})^{1/2}|)\}$$

In this expression, $\sigma_0{}^{a}$ represents the conductivity of the considered species of electron in material M1. In the case of a magnetic multilayer comprising ferromagnetic layers (F characterized by $\lambda_{\uparrow F}$, $\lambda_{\downarrow F}$) alternating with non-magnetic layers (NM characterized by λ_{NM}), the conductances in parallel and antiparallel alignment of the magnetizations in the successive ferromagnetic layers are given by

$G_{parallel}\,(a,b,c) = G(a,b,c,\,\lambda_{\uparrow F}\,,\,\lambda_{NM}\,,\,\lambda_{\uparrow F}) + G(a,b,c,\,\lambda_{\downarrow F},\,\lambda_{NM}\,,\,\lambda_{\downarrow F})$ and
$G_{antiparallel}\,(a,b,c) = G(a,b,c,\,\lambda_{\uparrow F},\,\lambda_{NM}\,,\,\lambda_{\downarrow F}) + G(a,b,c,\,\lambda_{\downarrow F},\,\lambda_{NM},\,\lambda_{\uparrow F})$
(parallel conductance of the two species of conduction electrons).

For our quantitative comparison, we chose the case of ∞(NiFe t_{NiFe}/Cu t_{Cu}/NiFe t_{NiFe}/Cu t_{Cu}) multilayers in which we believe bulk spin-dependent scattering is predominant. The following set of parameters has been used: $\lambda_{\uparrow NiFe}$ = 114Å , $\lambda_{\downarrow NiFe}$ = 12Å, λ_{Cu} = 205Å. We then calculated the magnetoresistance (expressed as ΔG or $\Delta R/R$) versus thickness of the ferromagnetic layers and non-magnetic layers. The results are plotted on the figures below .

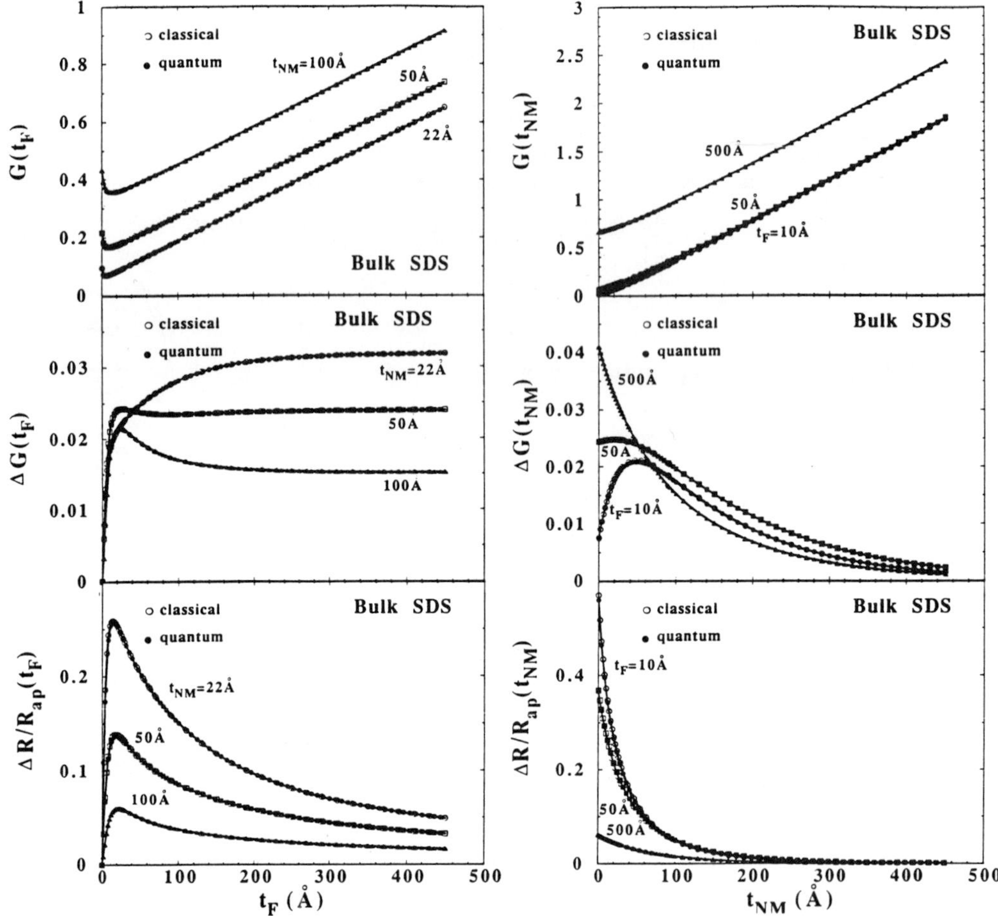

Fig. 7A. Comparison of the conductance per period and magnetoresistance for ∞(F t_F/NM t_{NM}/F t_F/NM t_{NM}) versus t_F calculated from the classical and quantum theories. $\lambda_{\downarrow F}$ = 12Å, $\lambda_{\uparrow F}$ = 114Å, $\lambda_{\downarrow NM}$ = 205Å.

Fig. 7B. Comparison of the conductance per period and magnetoresistance for ∞(F t_F/NM t_{NM}/F t_F/NM t_{NM}) versus t_{NM} calculated from the classical and quantum theories. $\lambda_{\downarrow F}$ = 12Å, $\lambda_{\uparrow F}$ = 114Å, $\lambda_{\downarrow NM}$ = 205Å.

Clearly the agreement between these two very different theories is surprisingly good in the present case of bulk SDS (difference smaller than 1% in relative value) even when the thicknesses of the layers are much smaller than the mean-free paths. A similar comparison in the case of interfacial SDS is in progress. In this later case larger differences seem to exist between the two theories but with a still good agreement (within 10 to 15%). Futhermore in the case of bulk SDS, some unexpected features of the conductance and magnetoresistance found with the classical theory are also present in the quantum theory :

-The conductance of a multilayer n.(ferro/non-mag/ferro/non-mag) goes through a minimum when the thickness of the ferromagnetic layers is increased. This effect results from a confinement of conduction electrons in the non magnetic layer (actually in the layer of lower resistivity) when the thickness of the ferromagnetic layer t_F (of higher resistivity) becomes comparable to the mean-free path λ_F.

-In the present case of bulk spin-dependent scattering, $\Delta G_{per\ period}$ has a marked maximum versus the thickness of the non magnetic layer t_{NM} when the thickness of the ferromagnetic layer t_F is smaller than 50Å. Note that this unexpected feature does not appear in $\Delta R/R$.

-$\Delta G_{per\ period}$ also has a maximum versus t_F for large t_{NM} thickness.

CONCLUSION

Thanks to the use of exchange anisotropy, spin-valve structures of the form F/NM/F/antiferromagnet offer the possibility to test the magnetotransport properties of a wide variety of ferromagnetic (F) and non-ferromagnetic (NM) metals. Moreover in these structures, the two configurations of parallel and antiparallel alignment of the magnetizations in the two ferromagnetic layers are perfectly well defined with single domain magnetization. Therefore this allows one to push forward the quantitative interpretation of the data without any correction due to partial misalignment of the magnetizations. We showed that classical theories (Camley and Barnas) give almost the same quantitative results as the more sophisticated quantum mechanical theory so that the classical approach can be used with confidence to fit experimental data. With this type of extended Fuchs-Sondheimer theory we interpreted the variation of magnetoresistance versus thicknesses of ferromagnetic and non-magnetic layers and discussed the relative role of bulk and interfacial spin-dependent scattering in these structures. We also underlined that both the measurements of the sheet conductance and magnetoresistance are required for a comparison of transport properties between various samples and that ΔG provides a much more direct comparison from a fundamental point of view than other quantities such as $\Delta R/R$ or ΔR.

REFERENCES

1. see for instance proceedings of the international workshop on "spin-valve multilayered structures", Madrid Sept 9-12 (1991) or proceedings of the NATO advanced research workshop on "structure and magnetic properties of systems in reduced dimension", Cargèse, France, June 15-20th 1992, to be published in the NATO series, Plenum Press.
2. M.N.Baibich, J.M.Broto, A.Fert, F.Nguyen Van Dau, F.Petroff, P.Etienne, G.Creuzet, A.Friederich and J.Chazelas, Phys.Rev.Lett., 61 (1988) 2472.

3. S.S.P Parkin, N.More and K.P.Roche, Phys.Rev.Lett, 64 (1990) 2304.
4. D.H.Mosca, F.Petroff, A.Fert, P.A.Schroeder, W.P.Pratt and R.Laloe, J.Magn.Magn.Mater.,94 (1991) L-1.
5. S.S.P.Parkin, R.Bhadra and K.P.Roche, Phys.Rev.Lett, 66 (1991) 2152.
6. C.Dupas, P.Beauvillain, C.Chappert, J.P.Renard, F.Triqui, P.Veillet, E.Velu, and D.Renard, J.Appl.Phys., 67 (1990) 3061.
7. T.Shinjo and H.Yamamoto, J.Phys.Soc.Jpn., 59 (1990) 3061.
8. B. Dieny, V.S. Speriosu, S.S.P. Parkin, B.A. Gurney, D.R. Wilhoit and D. Mauri, Phys. Rev. B 43, 1297 (1991).
9. B. Dieny, V.S. Speriosu, B.A. Gurney, S.S.P. Parkin, D.R. Wilhoit, K.P. Roche, S. Metin, D.T. Peterson and S. Nadimi, J. Mag. Mag. Mat. 93, 101 (1991).
10. B. Dieny, V.S. Speriosu, S. Metin, S.S.P. Parkin, B.A. Gurney, P. Baumgart and D.R. Wilhoit, J. Appl. Phys. 69, 4774 (1991).
11. B. Dieny, P. Humbert, V.S. Speriosu, S. Metin, B.A. Gurney, P. Baumgart, H. Lefakis, Phys. Rev. B.45, 806 (1992).
12. B.Dieny, J.P.Nozières, V.S.Speriosu, B.A.Gurney, D.Wilhoit, to appear in Appl.Phys.Lett. (1992).
13. R.E. Camley and J. Barnas, Phys. Rev. Lett. 63, 664 (1989).
14. J. Barnas, A. Fuss, R.E. Camley, P. Grünberg and W. Zinn, Phys. Rev. B 42, 8110 (1990).
15. B.Dieny, V.S.Speriosu, J.P.Nozieres and B.A.Gurney, to appear in J.Magn.Magn.Mater (1992).
16. S.S.P.Parkin, R.F.Marks, R.F.C.Farrow, G.Harp, this book.
17. B.Dieny, to appear in Journ.phys: Condens.Mater (1992).
18. A. Vedyayev, B.Dieny, N. Ryzhanova, Europh.Lett, 19 (4) (1992) 329.
19. V.S.Speriosu, B.Dieny, P.Humbert, B.A.Gurney, H.Lefakis, Phys.Rev.B (rapid Comm.) 44 (1991) 5358.
20. C.A. dos Santos, B.Rodmacq, M.Vaezzadeh and B.George, Appl.Phys.Lett.59 (1991),126.
21. B.Rodmacq, Journ.Appl.Phys. 70 (8) (1991) 4194.
22. R.Coehoorn, M.T.Johnson, W.Folkerts, S.T.Purcell, M.A.J.Gijs, A.de Vierman, this book.
23. A.Barthélémy, A.Fert, Phys.Rev.B, 43, 13124 (1991).
24. B. Dieny, Europhysics Letters 17, 261 (1992).
25. D.M. Edwards, J. Mathon, and R.B. Muniz, proceedings of the ICMFS meeting, Glasgow, July 1991 and workshop on "spin-valve multilayered structures", Madrid, September 1991.
26. L.M. Falicov, R.Q. Hood, proceedings of the ε-MRS meeting, Sept. 7-10, 1992, Lyon (France).

OSCILLATORY INTERLAYER EXCHANGE INTERACTION AND MAGNETORESISTANCE IN Co/Cu AND Fe/Cu SYSTEMS

R. Coehoorn, M.T. Johnson, W. Folkerts, S.T. Purcell[1], N.W.E. McGee, A. De Veirman and P.J.H. Bloemen[2]

Philips Research Laboratories
P.O. Box 80000
5600 JA Eindhoven
The Netherlands

1. INTRODUCTION

The oscillatory exchange coupling between ferromagnetic layers across non-ferromagnetic interlayers, which has been observed now for a large number of systems, is an intriguing phenomenon. Refs. 1-5 give a selection of papers on pioneering work. The occurence of oscillations was almost immediately after its discovery related to the RKKY oscillatory exchange interactions between magnetic impurities in, for example, noble metal host lattices. The RKKY model[6] predicts that the Heisenberg exchange coupling constant J(z) for a system of two magnetic impurities in a free electron gas with the Fermi-wavevector k_F depends on their mutual distance z as

$$J(z) \propto \frac{\cos 2k_F z}{z^3}, \qquad (1)$$

for $k_F z >> 1$. The periods $\lambda = \pi/k_F$ are quite short, e.g. 2.31 Å for Cu (if treated as a free electron metal). Experimental evidence for such oscillatory exchange interactions comes from the spin-glass behaviour of dilute alloy systems[7]. A related phenomenon is the oscillatory induced spin density around a magnetic impurity in a non-magnetic host, which has been observed for example around transition metal impurities in Cu from NMR- Knight shift measurements[8,9].

The RKKY model predicts that the period of the exchange interaction reflects the Fermi surface dimensions. To what extent can the oscillatory coupling across nearly free electron metals in *layered* systems be described in terms of an RKKY- like model, and would it be possible to 'measure' the Fermi-surface dimensions in the

[1]Present address: Dept. de Physique de Materiaux, Univ. Claude Bernard, Lyon 1, F-69622 Villeurbánne, France.
[2]Affiliation: Eindhoven University of Technology, 5600 MB Eindhoven, The Netherlands.

Magnetism and Structure in Systems of Reduced Dimension
Edited by R.F.C. Farrow *et al.*, Plenum Press, New York, 1993

growth direction by simply measuring the period of the oscillatory coupling? The simplest theoretical treatment of this problem was presented by Yafet [10] and by Hellmann and Baltensperger [11]. They showed that, within the RKKY model, the exchange coupling constant J(z) for a system of two planar continuous ferromagnetic layers at a distance L, with infinitesimal thickness, embedded in a free electron gas with wavevector k_F, is given by

$$J(L) \propto \frac{\sin 2k_F L}{L^2}, \tag{2}$$

for $k_F L >> 1$. In view of the predicted short period of the oscillations, just as for the impurity case, it was initially a surprise that long periods (8-22 Å) were observed in Fe/Cr, Co/Cr, Co/Ru [1] and Co/Cu [2,3] multilayers. However, it was quickly realized that, even in the case of free-electron metallic interlayers, long periods can be expected due to the *aliasing effect,* whereby rapid RKKY-like oscillations are only sampled at discrete thicknesses of the interlayer [12]. Fig. 1 illustrates this phenomenon. The period depends on the interplanar distance, d, within the interlayer, and thereby on the crystal structure and orientation of the interlayer. Mathematically, the experimentally measured period Λ is given by

$$\Lambda = \frac{2\pi}{\left| 2k_F - n\frac{2\pi}{d} \right|}. \tag{3}$$

The Fermi surface spanning vector $2k_F$ is reduced by an integral number (n) times the reciprocal lattice vector $2\pi/d$ of the one-dimensional lattice which is obtained by projecting the interlayer metal on the z-axis, in order to confine the denominator of eq. (3) to the interval $[-\pi/d\ ;\ \pi/d]$. In actuality, n is mostly 0 or 1. Bruno and Chappert [13] have emphasized that this approach is also valid for nearly free electron Fermi surfaces, extremal points of which, connected by vectors $\vec{q}$ parallel to the growth direction, generate oscillation periods equal to $2\pi/|\vec{q}|$. For a given growth direction a multiplicity of oscillation periods may be encountered, each connected to an extremal $\vec{q}$ vector. It should be mentioned here that the relation between the period of the oscillatory coupling and the Fermi surface dimensions has also been discussed by Edwards and Mathon [14] and by Herman et al. [15]

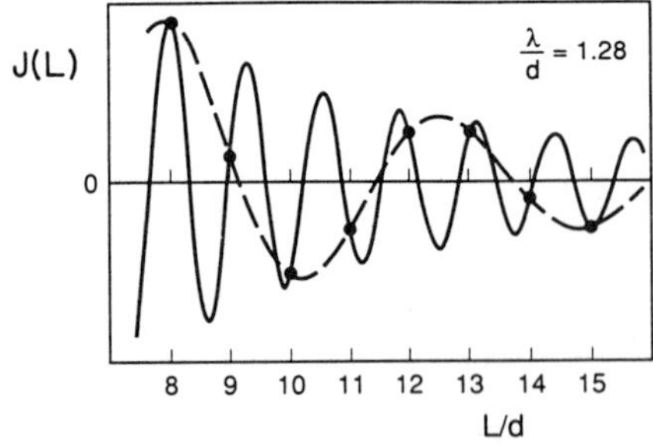

Figure 1. Full curve: the coupling function J(L) for a monovalent fcc(100) metal (arbitrary units), as calculated within the continuum version of the RKKY model. Broken curve: coupling function with the experimentally measured periodicity, for L=Nd (N integer).

The nearly-free-electron noble metals Cu, Ag and Au are the most suitable candidates for a detailed experimental verification of the theoretical approach by Bruno and Chappert. In Sec. 2 of this paper, we give an overview of our experimental studies of the exchange coupling across Cu interlayers in coherent Co/Cu/Co sandwiches with (100), (110) and (111) growth directions, and in coherent Fe/Cu/Fe sandwiches with the (100) growth direction[16,17]. In all cases, the Cu layer was grown in the form of a wedge, but a crucial distinction was its structure, being fcc in the Co/Cu/Co sandwiches but bcc in the Fe/Cu/Fe sample. The samples were grown by Molecular Beam Epitaxy (MBE) on Cu single crystals and Fe single crystalline whisker substrates, respectively. The presentation of the experimental results is preceded by a discussion of the theoretical predictions for these cases. Our results essentially support the proposed relationship between the Fermi-surface dimensions and the periods of the exchange coupling. It should be noted that a similar conclusion has followed from coupling investigations of Fe/Au/Fe (100) [18] and Fe/Ag/Fe (100) [19] samples .

In Sec. 3 we present results of a systematic study of the oscillatory exchange coupling and magnetoresistance in sputtered Co/Cu (100) multilayers, grown on Si(100) substrates, and compare these results with results for sputtered Co/Cu (111) textured systems and (as far as the exchange coupling is concerned) with the results from the MBE-grown systems. Contrary to a recent suggestion by Egelhoff and Kief[20], our results on sputtered and MBE-grown samples indicate that the AF coupling found for sputtered Co/Cu (111)-textured multilayers is of an intrinsic nature, and is not due to a small amount of misaligned crystallites.

2. MBE-GROWN Co/Cu/Co AND Fe/Cu/Fe WEDGE SYSTEMS

2.1. Theoretical Predictions

The relevant $\vec{q}$ vectors, which within the theory of Bruno and Chappert determine the period of the oscillatory exchange interaction across Cu, have been indicated for the fcc (100), (110) and (111) growth directions in figs. 2a-c, respectively. The figures show cross-sections of the Fermi surface, as we obtained from selfconsistent Augmented Spherical Wave (ASW) band structure calculations. Almost identical cross sections, derived from de Haas-van Alphen measurements, were given by Bruno and Chappert in ref. 13. The resulting periods will be given in Table 1, together with the experimental results (discussed in sec. 2.2).

The cross-section of the Fermi surface of hypothetical bcc Cu, which is relevant for the Fe/Cu/Fe (100) system, is shown in fig. 3. The calculated periods are 2.56 and 2.22 monolayers (ML), corresponding to $\vec{q}_1$ and $\vec{q}_2$, respectively.

2.2. Experimental Results

The Co/Cu/Co samples were deposited on Cu single crystal substrates, and the Fe/Cu/Fe sample was prepared by deposition on an Fe whisker. The completed samples were composed as follows:

Cu(100)/Co(60 Å)/Cu wedge (0-40 Å)/Co(60 Å)/Cu(7 Å)/Au(20 Å)
Cu(110)/Co(40 Å)/Cu wedge (0-44 Å)/Co(40 Å)/Cu(7 Å)/Au(20 Å)
Cu(111)/Co(40 Å)/Cu wedge (0-35 Å)/Co(40 Å)/Cu(7 Å)/Au(20 Å)
Fe(100)/Cu wedge (0-40 Å)/Fe(80 Å)/Au(20 Å).

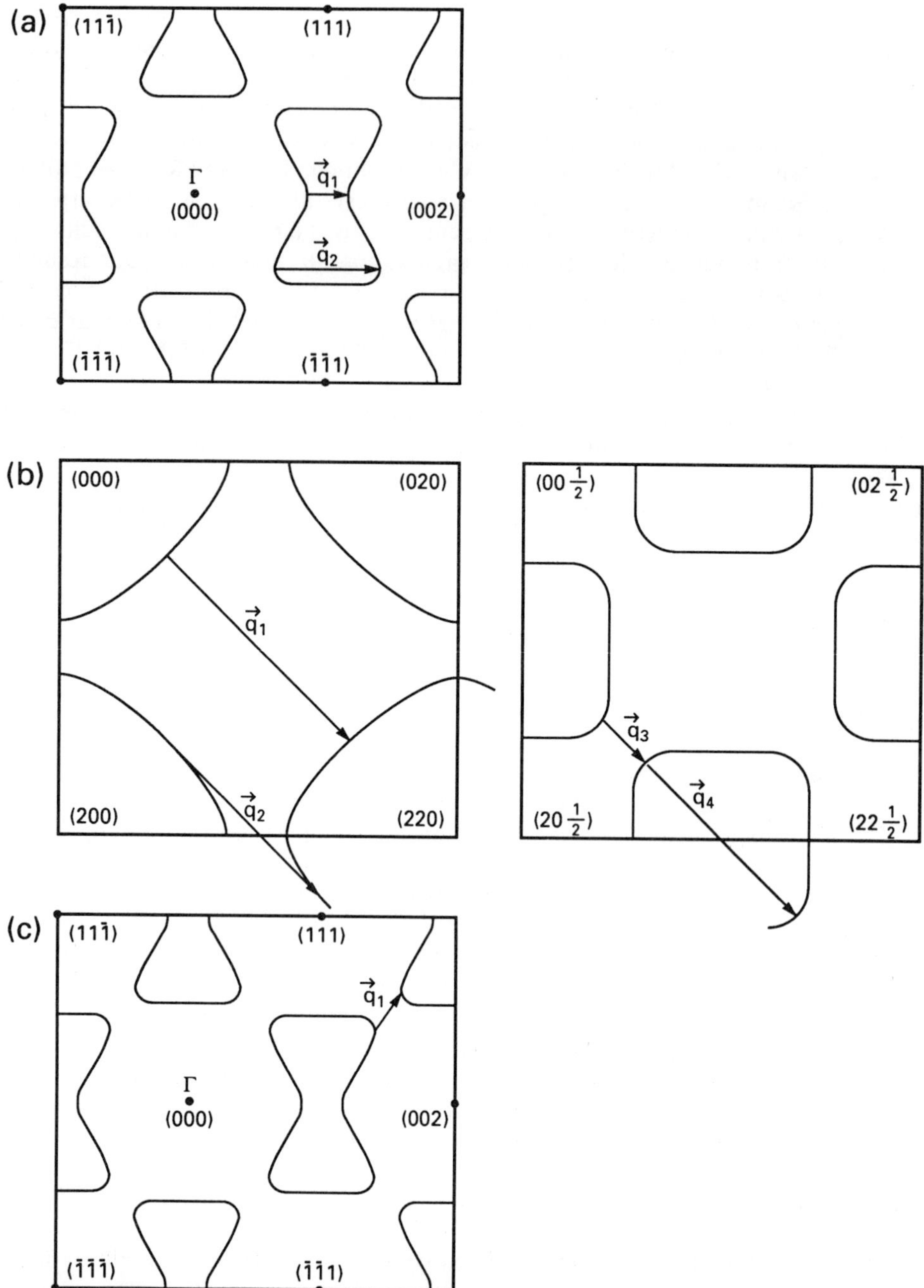

Figure 2. Cross-sections of the Fermi surface of fcc Cu, indicating the q vectors which are predicted to determine the periods of the oscillatory exchange interaction, for (a) (100), (b) (110) and (c) (111) growth directions.

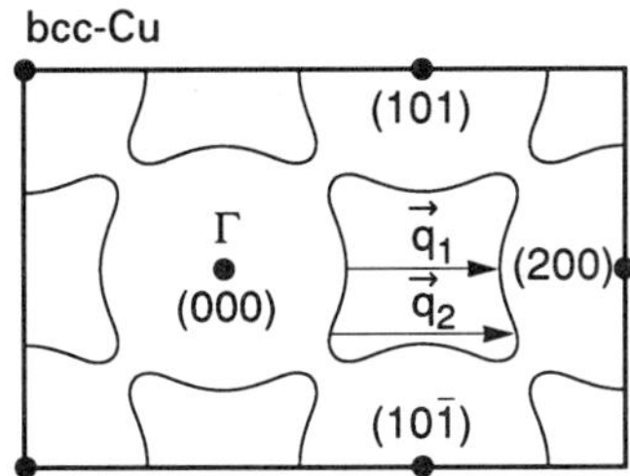

Figure 3. Cross section of the Fermi surface of bcc Cu, indicating the q vectors which are predicted to determine the periods of the oscillatory exchange interaction for the (100) growth direction.

The length of the wedges was typically 1-2 cm. The overlayers were deposited in a multichamber MBE system with a base pressure of 5×10^{-11} mbar. The Cu thicknesses were determined using a quartz monitor, which was calibrated using RHEED oscillations observed during the growth of Cu on Cu(100) and using chemically analyzed reference samples, and subsequently confirmed after deposition using *in situ* Auger Electron Spectroscopy. Substrate temperatures were 50°C during Cu deposition and 20°C during the deposition of the other layers. More details concerning the substrate preparation and deposition are given in refs. 16 and 17. These papers also contain a detailed description of the structure of the samples, as observed from RHEED and LEED studies. It is important to note, however, that in all four samples the Cu wedge shaped interlayers grew coherently, displaying identical in-plane and perpendicular lattice spacing to those of the substrates. In the three Co/Cu samples fcc Cu interlayers with (100), (110) and (111) orientations have been created, whereas in the Fe/Cu sample metastable bcc (100) Cu forms the interlayer.

The antiferromagnetic (AF) coupling strength was determined from the magnetization loops, as measured locally along the length of the wedges using the longitudinal magneto-optical Kerr effect (MOKE). All measurements were carried out at room temperature. Typical examples of such loops are displayed in figs 4(a)-(f) for Co/Cu/Co systems. Figs 4(b), (c) and (f) display cases of AF coupling.

In the following discussion, we assume that the interlayer exchange coupling energy E (per m^2) between two layers with uniform magnetizations M_1 and M_2 is given by the relation

$$E = -J \frac{\vec{M}_1 \cdot \vec{M}_2}{|\vec{M}_1||\vec{M}_2|}. \tag{4}$$

The form of the magnetization loops is determined by the interplay of the AF interlayer exchange coupling, the field energy and the in-plane magnetocrystalline anisotropy[21,22]. For the (100) and (110) systems, the in-plane anisotropy of the Co layers is rather large. Upon the application of a magnetic field along the easy axis, the magnetization first rotates gently, up to the flip field H_{flip}, where the magnetization switches to the saturation magnetization, M_{sat}. The AF coupling constant is given approximately by the relation[22]

$$J \simeq -\mu_o t M_{sat} H_{flip}, \tag{5}$$

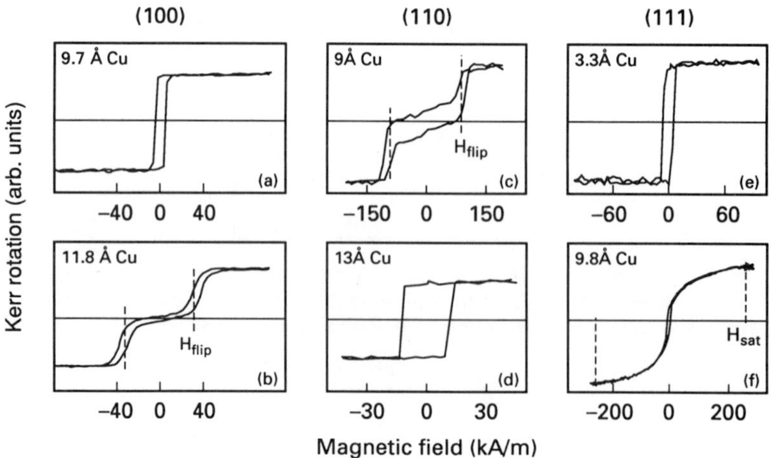

Figure 4. MOKE hysteresis loops for MBE-grown (100), (110) and (111) Co/Cu/Co.

where t is the thickness of a Co layer. For the (100) and (110) grown systems we have found that $H_{flip} << H_K$ (where H_K is the anisotropy field), which implies that eq. (5) is well satisfied. On the other hand, for the (111) systems the in-plane anisotropy of the Co layers is negligible. The magnetization of the AF-coupled layers then increases gradually in a magnetic field, until saturation is reached at the saturation field H_{sat}. The coupling constant is obtained from the relation[22]

$$J \simeq -\frac{1}{2}\mu_o t M_{sat} H_{sat}. \tag{6}$$

Fig. 5 shows the variation of H_{flip} and H_{sat} with the Cu thickness for the Co/Cu systems. The (100) sample displayed 5 peaks in the AF coupling. A fit to the experimental data in terms of a superposition of a long period (Λ_1) and a short period (Λ_2), with a quadratic overall decay of J with increasing Cu thickness, and with the phases, the relative amplitudes and the periods regarded as free parameters, led to the coupling curve which is given in the inset. The periods are $\Lambda_1 = 8.0 \pm 0.5$ ML (14 ± 1 Å), and $\Lambda_2 = 2.6 \pm 0.05$ ML (4.6 ± 0.1 Å). As shown in Table 1, the period of the short oscillation is in excellent agreement with the theoretically predicted value (as derived from $|\vec{q}_2|$ in fig. 2(a)), whereas there is a small difference for Λ_1. The (110) sample displayed only one oscillatory period, whose value of 12.5 Å agrees well with the theoretical period following from the neck diameter of the Fermi surface of Cu (vector $\vec{q}_3$ in fig. 2(b)). The other three (shorter) periods were not observed, not even in a second sample with a shallower wedge.

For the (111) sample, very strong AF coupling, with J larger than for the (100) and (110) samples (see Table 1), was found around $t_{Cu} = 8.5$ Å. It should be noted that even at this Cu thickness a non-zero Kerr signal and a steplike switch were found near zero field. Repeat samples displayed the same phenomenon. We attribute this to the presence of ferromagnetically coupled regions[17]. We obtained only a weak indication of the possible presence of a second AF peak around 20 Å . In order to

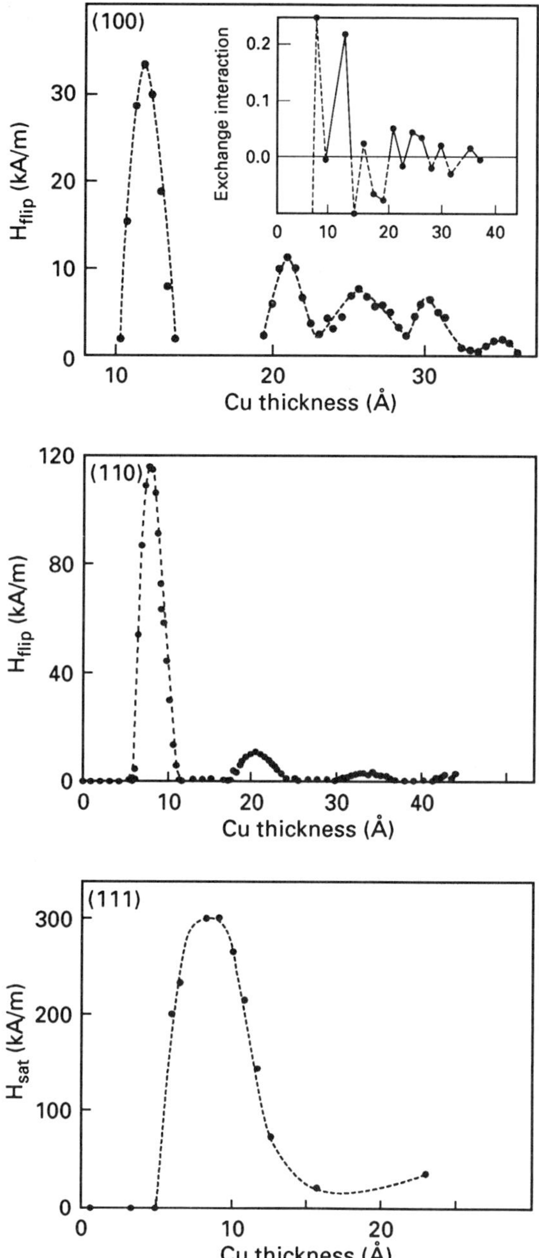

Figure 5. Variation, as a function of Cu thickness, of the flip field H_{flip} for Co/Cu/Co (100) and (110) samples, and of the saturation field H_{sat} for Co/Cu/Co (111) samples.

Table 1. Measured (MBE) and theoretical coupling data for the Co/Cu/Co system. The measured position (t_1), the FWHM (Δt_1) and the coupling constant (J) are indicated for the first AF peak. Periods which have been predicted, but are not observed, are given between parentheses.

	t_1 Å	Δt_1 Å	-J mJ/m^2	Λ (Å) Measured	Λ (Å) Predicted (dHvA)	Λ (Å) Predicted (ASW)
(100)	12	2.5	0.4	14, 4.6	10.6, 4.6	11.6, 4.6
(110)	8.5	2.9	0.7	12.5	12.2 (2.7, 3.2, 4.2)	11.2 (2.7, 3.3, 4.2)
(111)	8.5	5.5	1.1	-	9.4	9.0

enhance the chances to observe a weak second peak, if present, two additional samples were made. In the first sample the Co layers were replaced by Co/Ni multilayers with a perpendicular preferential magnetic anisotropy.[23] These multilayers were terminated, at the interface with the Cu wedge, with a 4 Å Co layer. The perpendicular anisotropy leads to a large energy of walls between domains in which the magnetization is parallel and antiparallel with respect to a uniformly magnetized neighboring layer, favoring perfectly ferromagnetic or antiferromagnetic alignments. A perfectly antiferromagnetic alignment was observed indeed in the first AF peak, but no second peak was found for t_{Cu} below 40 Å . A second sample, with only 5 Å thick Co layers, which also displayed perpendicular anisotropy, was made in order to increase the coupling fields. However, in spite of the finding of AF coupling fields in the first AF peak above the available external fields (estimated maximum coupling field 2400 kA/m), this sample, too, did not reveal a second AF peak.[24] With this reservation our results, combined with the results presented in sec. 3 on sputtered Co/Cu multilayer systems, provide support for the view that the strong AF coupling in (111)-textured Co/Cu multilayers is of an intrinsic nature. This issue had become a matter of dispute (see Egelhoff and Kief[20]). AF coupling in (111)-oriented epitaxial Co/Cu systems has also recently been found by several other groups.[25]

A LEED study of the Fe/Cu wedge/Fe (100) sample revealed that the Cu grows bcc up to (at least) 20 ML[16]. The Fe overlayer also grew in the bcc structure. Fig. 6a shows two typical examples of MOKE hysteresis loops for this sample. The lower figure, for 12 ML Cu, displays an example of AF coupling. The hysteresis loops for AF coupling are slightly more complicated than for the Co/Cu/Co samples discussed above, because of the situation of unequal magnetic layer thicknesses. Coming from high fields, the Fe overlayer starts to rotate at a critical field H_2, and its magnetization becomes oriented antiparallel to the whisker below the critical field H_1. At zero field, a reversal of the magnetization of the entire system occurs. H_1 and H_2 are much smaller than the in-plane anisotropy field of the whisker and to a good approximation the AF coupling constant is given by the relation[22]

$$J \simeq -\frac{1}{2}\mu_o t M_{sat}(H_1 + H_2), \qquad (7)$$

where t is the thickness of the Fe overlayer.

The variation of H_1 and H_2 with the Cu thickness is shown in fig. 6b. Above 10 ML, the coupling becomes AF, and oscillates with a period of 2 ML Cu. The coupling constant at the highest AF peak (at 12 ML Cu) is $J \simeq -0.1 mJ/m^2$, which is much weaker than for the Co/Cu/Co samples. The detection of the 2 ML oscillations further emphasises that extremely high-quality interfaces can be obtained by growth

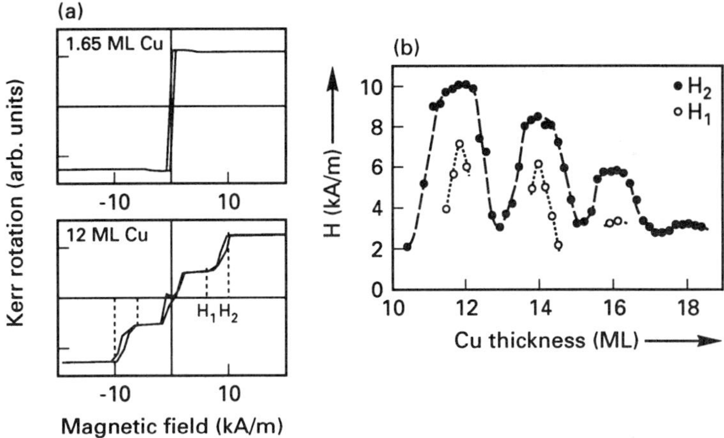

Figure 6.
MBE grown (100) Fe/Cu/Fe: (a) MOKE hysteresis loops (field applied along the whisker); (b) Variation with the Cu thickness of the critical fields H_1 and H_2.

on whisker substrates, as had been demonstrated earlier for Fe/Cr/Fe[26] and Fe/Mn/Fe[27] systems. F (or weak AF) coupling was found up to 10 ML. We note that Heinrich et al. have recently observed an AF peak at 10 ML and a minimum in the F coupling at 8 ML for Fe/Cu/Fe (bcc,(100)) samples grown on Ag[28]. As stated in sec. 2.1, theory predicts the occurence of two oscillations, with periods of 2.56 and 2.22 ML. A series of 4 oscillations of 2 ML, followed by a phase slip, would follow from a period of 2.22 ML, and we have indeed observed 4 AF peaks. However, the thickness interval which was studied is too narrow to give a complete assessment of possible contributions of longer period oscillations (such as the predicted oscillation with $\Lambda = 2.56$ ML).

3. SPUTTERED Co/Cu MULTILAYERS WITH (100) AND (111) TEXTURE

The first observations of an oscillatory exchange interaction across Cu layers were reported by Mosca et al.[3] and by Parkin et al.[2] for sputter-deposited Co/Cu multilayers. These multilayer samples were grown on a Si (100) substrate, covered with an Fe buffer layer, and displayed a giant magnetoresistance (MR) for Cu thicknesses where the coupling was AF. Record values of the MR ratio, defined as $(R_{max} - R_\infty)/R_\infty$, where R_{max} is the highest resistance and R_∞ is the lowest resistance (at high fields), were reported by Parkin et al.[29] For Co(8 Å) /Cu(8.3 Å) multilayers they found an MR ratio of 65 % at room temperature and 115 % at 4.2 K. These samples show a weak (111) preferential orientation. The fact that initially very carefully MBE-grown Co/Cu(111) systems did not show AF coupling led Egelhoff and Kief[20] to the hypothesis that the AF coupling of the sputtered (111)-textured samples was not of an intrinsic nature, but was due to the presence of a small amount of (100)-oriented crystallites.

In this section, we present results of a comparative study of sputtered Co/Cu(100) and (111)-textured multilayers. From our results, three arguments follow which prove the intrinsic nature of the AF coupling in sputtered (111) systems: (i) The first AF peak positions for MBE-grown and sputtered (111) systems agree well, and are different to that of the (100) systems, (ii) No detectable fraction of (100) oriented crystallites is observed by XRD and electron diffraction in strongly AF-coupled sputtered (111) samples, and (iii) Sputtered (111) samples show a *stronger* coupling strength in the first AF peak than (100) MBE-grown and sputtered systems, whereas the coupling strength due to a minority (100) component is expected to be *weaker* than the intrinsic (100) coupling strength.

Results on the preparation and MR for (111) systems have been reported already by other groups[3,29], so our emphasis will be on the XRD and TEM results. Results for sputtered (100) systems have been reported for $t_{Cu} > 18$ Å[30], but not for the region around the first strong AF peak. It is important to note that a comparative study of these two orientations is made possible by the fact that we have used identical deposition conditions. We also note that our study on sputtered (100) Co/Cu systems is part of a more comprehensive study on the oscillatory exchange interaction and MR in Fe-Co-Ni/Cu (fcc,(100)) multilayers[31].

The Co/Cu multilayer samples were prepared by DC-magnetoron sputtering, in a system with a base pressure of 5×10^{-7} torr, using an Ar pressure of 7×10^{-3} torr. The substrate was held at room temperature during deposition. The rates were 2 Å/s (Co) and 3 Å/s (Cu). Three series of samples were deposited:

Si(100)/200 Å Cu/ 25 × (16 Å Co/ t_{Cu} Å Cu)/50 Å Cu (t_{Cu} = 7-50 Å)
Si(100)/200 Å Cu/ n × (16 Å Co/ 10.5 Å Cu)/50 Å Cu (n = 2-50)
Si(100)/50 Å Fe/ 25 × (8 Å Co/ t_{Cu} Å Cu)/20 Å Fe (t_{Cu} = 6-12 Å).

The sample dimensions were 4x12 mm^2. The substrates were cleaned by an HF dip *(ex situ)* and by a glow discharge *(in situ)*. The first two series resulted in strongly oriented (100) textured multilayers (see also ref. 32). Fig. 7a shows a high-angle Cu – Kα X-ray diffraction pattern, for a sample with 50 repetitions and with $t_{Cu} = 10.5$ Å, which displays the main (200) Bragg peak and a number of satellite peaks. The full width at half maximum (FWHM) of the (200) rocking curves varied from 5.6°(2θ) for n = 2 to 2.1° for n = 50. From a fit of the low-angle XRD pattern (not shown) we derived a roughness parameter of the Co/Cu interfaces of 5 Å . The third series, with the Fe buffer layer, resulted in a weak (111) texture. Fig. 7b shows a high angle diffraction pattern for a sample with $t_{Cu} = 9$ Å. Note that only the (111) peak is visible, and no (200) peak (at about 51°). The FWHM of the (111) rocking curve was about 30°, and low angle XRD revealed a Co/Cu interface roughness of about 5 Å .

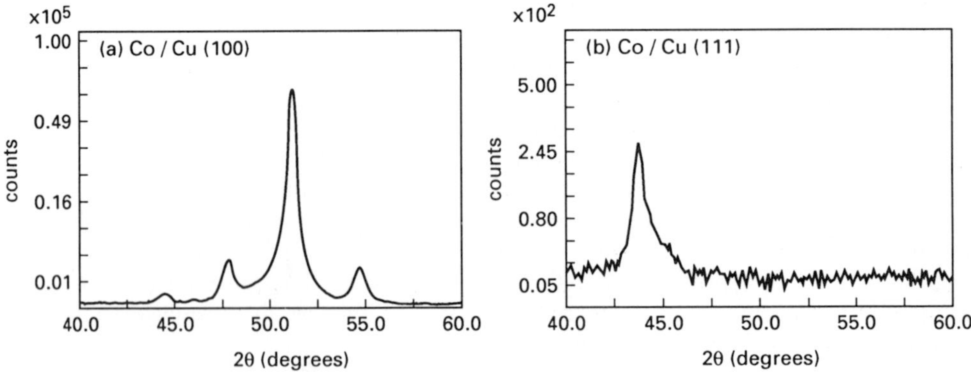

Figure 7. Cu – K$_\alpha$ X-ray diffraction patterns for (a) a 50 × (16 Å Co/10.5 Å Cu) multilayer, and (b) a 25 × (8 Å Co/9 Å Cu) multilayer. Details concerning the buffer and coverage layers are given in the text. The intensity has been plotted on a squareroot scale.

Fig. 8 shows a cross-sectional TEM image of a (100) Co/Cu multilayer, taken along the [0$\bar{1}$1] direction of the superlattice. The photograph reveals rather flat layers, and the occurrence of twinning. The upper diffraction pattern, of the superlattice only, further proves the highly oriented character of the system. The epitaxial relationship with the substrate follows from the lower diffraction pattern, which was taken including the substrate. The Si spots are encircled.

Fig. 9 shows a cross-sectional TEM image of a (111)-textured Co (10 Å)/Cu (9 Å) multilayer (n = 25), deposited on an Fe buffer layer. The sample showed a strong AF coupling. As in the (100) case, the photograph reveals rather flat layers. The non-perfect alignment of the [111] axes of the crystallites is revealed by the diffraction pattern (inset), which, however, shows no visible fraction of [100] oriented crystallites.

Fig. 10a shows the variation of the MR for (100) systems, as measured at 5 K. Maxima in the MR ratio were found at about 10 and 24 Å Cu thickness. It should be noted that the actual MR ratios varied between nominally equal series of samples (see figure), grown in different batches. We have observed from VSM measurements that these differences are related to the remanence in the first peak, that is to the degree of antiparallel alignment. For the sample in series 3 the remanence at the first peak was about half the saturation magnetization, and it was higher for series 1 and

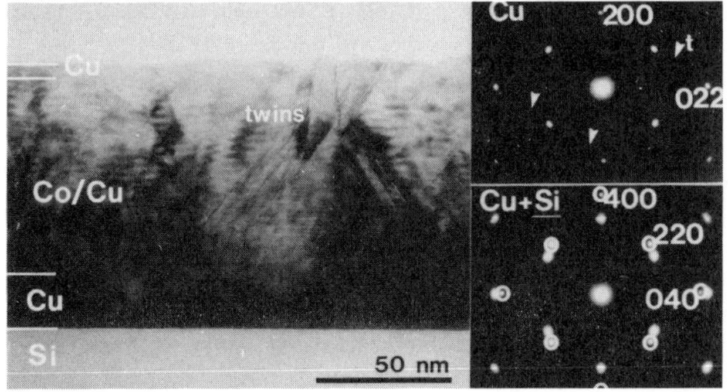

Figure 8. Left: Cross-sectional TEM image of a 25 × (16ÅCo/10.5ÅCu) multilayer. Upper and lower diffraction pattern: multilayer without and including the Si substrate, respectively. Twin defects and twin reflections (t) are indicated.

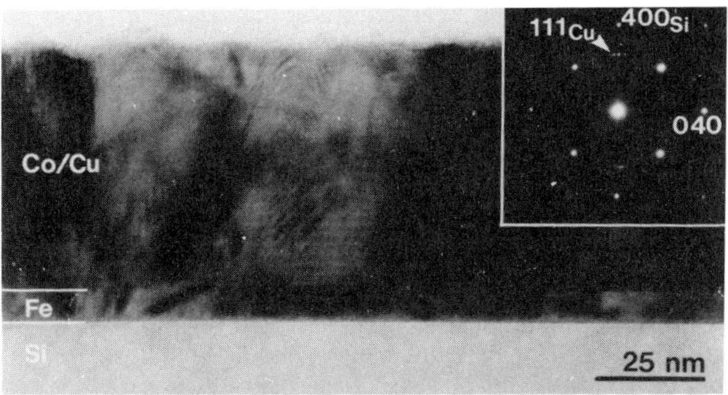

Figure 9. Cross-sectional TEM image of a 25 × (10ÅCo/9ÅCu) multilayer. Inset: diffraction pattern including the substrate.

2. We do not know the precise origin of these variations, but suspect that the small width of the first AF peak makes the remanence very sensitive to interface roughness. In the case of non-ideal interfaces, the occurrence of a significant biquadratic term in the interlayer exchange coupling, resulting in non-collinear coupling[33], is quite likely. We note that the MR ratio of the sputtered (100) samples is, of course, depressed by the shunting due to the Cu buffer and coverage layers. A study of the MR ratio as a function of the number of repetitions, from n=2 to n=50, leads to an extrapolated value ($n \rightarrow \infty$) of $\Delta R/R = 73$ % (5 K, 1st peak) and 44 % (room temperature, RT), for samples in series 3.[34]

The inset in fig. 10a shows the magnetoresistance curve for $t_{Cu} = 10.5$ Å (at the first peak), measured at 5 K. From the saturation field we derive coupling constants $-J = 0.5$ and 0.3 mJ/m² at 5 K and at RT, respectively, neglecting the small in-plane anisotropy and possible higher-order coupling terms. The RT value is only slightly lower than the value of 0.4 mJ/m² found for the MBE-grown sample.

Fig. 10b shows the variation of the MR for (111) systems, measured at 5 K. In agreement with the earlier results[2,3], we find a peak at $t_{Cu} = 9$ Å, with a giant MR ratio. The coupling constants (-J), as derived from the saturation field, are 0.7 and 0.45 mJ/m² at 5 K and RT, respectively. The latter value is close to the RT values of 0.3 and 0.5 mJ/m² reported in refs. 2 and 3, but much smaller than the MBE value of 1.1 mJ/m² . This may be taken as an indication of the better interfacial quality of the MBE sample, as compared to the sputtered samples. It should be noted that the first AF peak in the (111) samples occurs at a slightly lower Cu-thickness than in (100) samples, just as in the case of MBE-samples (fig. 5). This, together with the relative coupling strength strength for (111) and (100) systems, provides further proof that the AF coupling in (111) sputtered samples is not due to a small fraction of [100] – oriented crystallites.

4. CONCLUDING REMARKS

We have shown that the oscillatory exchange interaction across Cu layers shows a variation with the Cu thickness which depends on the orientation and the crystal

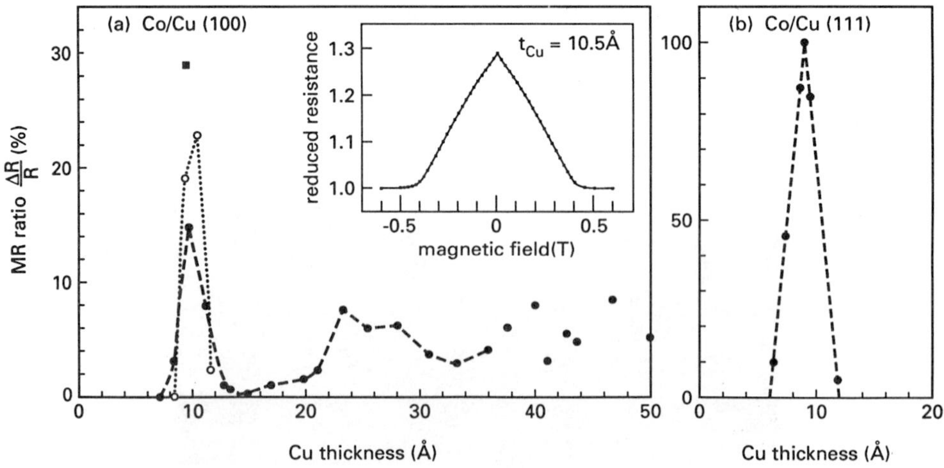

Fig. 10. MR ratio for (a) (100) and (b) (111) textured sputtered Co/Cu multilayers. Closed dots, open dots and squares in fig. (a) denote series 1, 2 and 3, respectively (see text).

structure of the Cu layers. In agreement with the predictions by Bruno and Chappert we have observed multiple periods for Co/Cu/Co (100). However, for the (110) samples, for which a superposition of 4 periods was predicted, only one period was observed. The measured short period for the fcc(100) system is in excellent agreement with theory, but the measured long period for the (100) systems is somewhat greater than predicted, and this is also the case for the (110) direction if a comparison with the ASW results is made (see table 1). It would be of interest to extend these studies to larger Cu thicknesses, in order to investigate whether asymptotic behaviour (phase and period independent of the Cu thickness) has already effectively been reached at the first AF peak.

From a comparsion of the AF peak positions and coupling strengths in MBE-grown and sputtered (100) and (111) samples, and from a structural study of the sputtered samples, we have obtained support for the view that the strong AF coupling in (111) textured Co/Cu multilayers around $t_{Cu} = 9$ Å is of an intrinsic nature. A remaining issue, however, is the exchange coupling at larger Cu thicknesses. The sputtered multilayer samples display AF coupling at $t_{Cu} \simeq 20$ Å[2,3], but our studies on several types of MBE-grown (111) wedge samples have not yielded conclusive proof of intrinsic AF coupling above $t_{Cu} = 15$ Å.

A recent new development is the study of the antiferromagnetic *and* ferromagnetic coupling strength using double-wedge samples with perpendicular directions of the slope of the wedges. Using such a sample, Bloemen et al.[35] have succeeded in proving the existence of a double oscillation in the ferromagnetic coupling strength across Cu (100) between 13 and 20 Å Cu thickness. This behaviour had been predicted from our analysis of the observed antiferromagnetic coupling strength in terms of a superposition of two oscillatory periods, as shown in the inset of the Co/Cu/Co (100) graph in fig. 5.

REFERENCES

1. S.S.P. Parkin, N. More and K.P. Roche, Phys. Rev. Lett. 64: 2304 (1990).
2. S.S.P. Parkin, R. Bhadra and K.P. Roche, Phys. Rev. Lett. 66: 2152 (1991).
3. D.H. Mosca, F. Petroff, A. Fert, P.A. Schroeder, W.P. Pratt, Jr. and R. Loloee, J. Magn. Magn. Mater. 94: L1 (1991).
4. S.S.P. Parkin, Phys. Rev. Lett. 67: 3598 (1991).
5. W.R. Bennett, W. Schwarzacher and W.F. Egelhoff, Jr., Phys. Rev. Lett. 65: 3169 (1990).
6. C. Kittel, in "Quantum Theory of Solids", Wiley, New York (1963).
7. J.A. Mydosh and G.J. Nieuwenhuys, in "Ferromagnetic Materials", ed. E.P. Wohlfarth, North-Holland, Amsterdam (1980), p. 71.
8. D.M. Follstaedt, D. Abbas, T.S. Stakelon and C.P. Slichter, Phys. Rev. B 14: 47 (1976).
9. J.D. Cohen and C.P. Slichter, Phys. Rev. B 22: 45 (1980).
10. Y. Yafet, Phys. Rev. B 36: 3948 (1987).
11. W. Baltensperger and J.S. Helman, Appl. Phys. Lett. 57: 2954 (1990).
12. R. Coehoorn, Phys. Rev. B 44: 9331 (1991); C. Chappert and J.P. Renard, Europhys. Lett. 15: 553 (1991); D.M. Deaven, D.S. Rokhsar and M. Johnson, Phys. Rev. B 44: 5977 (1991).
13. P. Bruno and C. Chappert, Phys. Rev. Lett. 67: 1602 (1991).
14. D.M. Edwards, J. Mathon, R.B. Muniz and M.S. Phan, Phys. Rev. Lett. 67: 493 (1991).

15. F. Herman, J. Sticht and M. van Schilfgaarde, J. Appl. Phys. 69: 4783 (1991).
16. M.T. Johnson, S.T. Purcell, N.W.E. McGee, R. Coehoorn, J. aan de Stegge and W. Hoving, Phys. Rev. Lett. 68: 2688 (1992).
17. M.T. Johnson, R. Coehoorn, J.J. de Vries, N.W.E. McGee, J. aan de Stegge and P.J.H. Bloemen, Phys. Rev. Lett. 69: 969 (1992).
18. A. Fuss, S. Democritov, P. Grünberg and W. Zinn, J. Magn. Magn. Mater. 103: L221 (1992).
19. D.T. Pierce, APS Conference, Indianapolis, March 1992.
20. W.F. Egelhoff, Jr. and M.T. Kief, Phys. Rev. B 45: 7795 (1992).
21. B. Dieny, J.P. Gavigan and J.P. Rebouillat, J. Phys.: Condens. Matter 2: 159 (1990); B. Dieny and J.P. Gavigan, J. Phys.: Condens. Matter 2: 187 (1990).
22. W. Folkerts, J. Magn. Magn. Mater. 94: 302 (1991); W. Folkerts and S.T. Purcell, J. Magn. Magn. Mater. 111: 306 (1992).
23. P.J.H. Bloemen, M.T. Johnson, J. aan de Stegge and W.J.M. de Jonge, J. Magn. Magn. Mater. 116 (1992) (to be published).
24. P.J.H. Bloemen (unpublished).
25. D. Greig et al., J. Magn. Magn. Mater. 110: L239 (1992); J.P. Renard, proceedings of this conference; R. Clarke, proceedings of this conference; J. Kohlhepp, S. Cordes, H.J. Elmers and U. Gradmann, J. Magn. Magn. Mater. 111: L231 (1992).
26. S.T. Purcell, W. Folkerts, M.T. Johnson, N.W.E. McGee, K. Jager, J. aan de Stegge, W.B. Zeper, W. Hoving and P. Grünberg , Phys. Rev. Lett. 67: 903 (1991); J. Unguris, R.J. Celotta and D.T. Pierce, Phys. Rev. Lett. 67: 140 (1991); S. Democritov, J.A. Wolf, P. Grünberg and W. Zinn, in "Magnetic Thin Films, Multilayers and Surfaces", ed. S.S.P. Parkin, MRS Proceedings Series, 231 (to be published).
27. S.T. Purcell, M.T. Johnson, N.W.E. McGee, R. Coehoorn and W. Hoving, Phys. Rev. B 45: 13064 (1992).
28. B. Heinrich, this proceedings.
29. S.S.P. Parkin, Z.G. Li and D.J. Smith, Appl. Phys. Lett. 58: 2710 (1991).
30. F. Giron et al., J. Magn. and Magn. Mater. 104-107: 1887 (1992).
31. R. Coehoorn, A. De Veirman and J.P.W.B. Duchateau, J. Magn. and Magn. Mater. (to be published).
32. F. Giron, P. Boher, K. le Dang and P. Veillet, J. Magn. Magn. Mater. (to be published).
33. J.C. Slonczewski, Phys. Rev. Lett. 67: 3172 (1991).
34. R. Coehoorn (unpublished results).
35. P.J.H. Bloemen, R. van Dalen, W.J.M. de Jonge, M.T. Johnson and J. aan de Stegge (to be published).

NMR STUDIES IN LAYERED MAGNETIC STRUCTURES

H. A. M. de Gronckel and W. J. M. de Jonge

Eindhoven University of Technology
Department of Physics
5600 MB Eindhoven, The Netherlands

INTRODUCTION

One of the major problems in the rapidly developing field of research on artificially layered magnetic materials is the characterization of the structural perfection of the multilayered films, which includes the topology (or sharpness) of the interfaces, the lattice parameters, coherency and strains. At the same time, according to current understanding, these parameters are of vital importance for the physical behavior, which includes anisotropy[1], anomalous magnetoresistance[2] and interlayer coupling[3]. For interfaces and individual layers embedded in the multilayer structure such information is difficult to obtain. In principle, however, techniques such as nuclear magnetic resonance (NMR) and Mössbauer spectroscopy provide the possibility to probe the multilayers and the interfaces on a nanoscopic scale. One of the attractive aspects of NMR is that the nucleus of almost every element can serve as a local structural probe.

In this contribution we will introduce briefly the major principles behind the application of NMR in the research on magnetic metallic multilayers and thin films. A more extensive review can be found elsewhere[4,5].

NUCLEAR INTERACTIONS WITH THE SURROUNDING

The Effective Magnetic Field

A nucleus interacts with its environment in various ways. The Hamiltonian includes magnetic as well as electrostatic terms. In the context of this paper we will only focus on the magnetic interaction. Magnetically the nuclear levels are split by the hyperfine interaction with the unpaired core and valence electrons, the dipolar interactions with atomic magnetic moments of other atoms in the solid and the applied field.

The hyperfine interactions can be represented by an effective field on the nuclear spin $\boldsymbol{I}$

$$\boldsymbol{B}'_{\text{hf}} = \frac{1}{\gamma\hbar}\boldsymbol{A}\cdot\langle\boldsymbol{S}\rangle \tag{1}$$

Magnetism and Structure in Systems of Reduced Dimension
Edited by R.F.C. Farrow *et al.*, Plenum Press, New York, 1993

where $\boldsymbol{S}$ represents the total spin of the atom, γ is the nuclear gyromagnetic ratio and $\boldsymbol{A}$ is a tensor of second rank describing the strength of the interaction. In magnetically ordered materials $\langle \boldsymbol{S} \rangle$ deviates significantly from zero and consequently the hyperfine field $\boldsymbol{B}'_{hf}$ may attain values between 1 T and 100 T. For paramagnetic systems $\langle \boldsymbol{S} \rangle \approx 0$ and thus the hyperfine interaction has a much reduced effect.

The principal sources contributing to the magnetic hyperfine interaction (and thus $\boldsymbol{B}'_{hf}$) are:

-The Fermi contact interaction[6], resulting from the *s* electron density at the nucleus,

-The dipolar interaction between the nuclear magnetic moment and the spin density of the *d* shell,

-The interaction between the nuclear magnetic moment and the unquenched part of the orbital angular momentum of the valence electrons.

For magnetic atoms the isotropic contact interaction is the dominant contribution. The associated effective field $\boldsymbol{B}_{cont}$ is proportional to the spin density "at" the nucleus. Core as well as valence electrons, referred to as the conduction electron (CE) contribution, contribute to this term because of a slight unbalance brought about by the polarization of the *d* shell. In the context of this paper we would like to note that the valence contribution also contains a contribution arising from the hybridization of the valence *s* wave functions with the spin polarized *d* wave functions of the neighboring atoms. This induces a weak valence *s* polarization leading to a, so called, "transferred" contribution to $\boldsymbol{B}_{cont}$. The sign and magnitude of the "transferred" polarization depends essentially on the magnetic moment of the neighboring atoms thus making the contact field sensitive to the local environment (see, e. g., Ref. 7, 8).

Apart from the hyperfine field discussed above, the nucleus also experiences the dipolar field arising from the dipole-dipole interaction between the nuclear magnetic moment and the atomic magnetic moments of all the other atoms in the solid. If we also take into account the effect of an external applied magnetic field $\boldsymbol{B}_{appl}$, the total effective field "sensed" by the nucleus results in a nuclear energy level Zeeman splitting ΔE of magnitude

$$\Delta E = hf = \hbar\gamma\, | \boldsymbol{B}'_{hf} + \boldsymbol{B}_{dip} + \boldsymbol{B}_{appl} | \tag{2}$$

where f is the NMR resonance frequency and the small effects due to Pauli spin paramagnetism, etc., have been neglected.

Local Atomic Environment and Magnetic Hyperfine Field

The magnetic field at the nucleus is sensitive to the local environment. In this section we will discuss the effect of (1) the local structure and symmetry, (2) the presence of "foreign" atoms, i. e. interfaces, and (3) strain.

Local Structure and Symmetry. The sensitivity of $\boldsymbol{B}'_{hf}$ to the local structure and symmetry arises from two effects: the sensitivity of the conduction electron polarization to the local environment and the dependence of the orbital and (atomic) dipolar part of $\boldsymbol{B}'_{hf}$ on the local symmetry. The former only affects the magnitude of $\boldsymbol{B}'_{hf}$ since it contributes to $\boldsymbol{B}'_{hf}$ through the isotropic contact interaction. The latter mainly determines the anisotropy of $\boldsymbol{B}'_{hf}$: Its specific angular dependence will be characteristic of the local symmetry. These features enable one to discriminate uniquely between, for instance, fcc, bcc or stacking faults in the pure phase through the magnitude of $\boldsymbol{B}'_{hf}$[9,10] and between fcc and hcp or fcc and tetragonal deformed fcc by its anisotropy.

The Presence of "Foreign" Atoms. The presence of a "foreign" atom, or impurity, alters the hyperfine field of its neighboring atoms[7,8]. First, the core polarization and valence *s* moment contribution to the hyperfine field may change since substitution of a host atom by

Table 1. Experimentally observed hyperfine field of Co in some diluted alloys. $B_{hf,i}$ denotes the hyperfine field of a Co atom with *i* "foreign" element atoms as nearest neighbor. The data refer to liquid helium temperatures.

element	$B_{hf,0}$ (T)	$B_{hf,1}$ (T)	$B_{hf,2}$ (T)	$B_{hf,3}$ (T)	Ref.
Ti	-21.6	-17.5	-13.9		11
Cr	-21.6	-17.6	-14.0		11
Ni	-21.6	-20.9	-20.0		11
	-21.62	-20.89	-20.14	-19.4	12
Cu	-21.6	-19.7	-17.9		13
	-21.62	-19.8	-17.9	-16.2	14

a "foreign" atom can alter the magnetic moment of the neighbor atoms. The magnitude and range of this moment disturbance depends very strongly on the (combination of) elements. In the second place, the "foreign" atom alters the hyperfine field of its neighbors because it brings about a change in the "transferred" polarization contribution. Although the range of the perturbation can be rather extended, in most experiments only the nearest neighbors to a "foreign" atom can be distinctly observed. If the nearest neighbor shell contains more than one "foreign" atom the magnitude of B'_{hf} changes by an amount about proportional to the number of "foreign" atoms, as is evidenced by experiments on diluted alloys (cf. Table 1).

The sensitivity of B'_{hf} for foreign atoms in the nearest neighbor shell enables one to discriminate between atoms at the interfaces in a multilayer and atoms well inside the constituting layers. Moreover, the dependence of the magnitude of B'_{hf} on the number of "foreign" atoms in the nearest neighbor shell makes that the hyperfine field distribution, i. e. the intensity distribution over the corresponding satellites, reflects the characteristics of the topology of the interfaces.

Strain. Since the magnetic hyperfine field is largely the result of a delicate (un)balance between the spatial distribution of the spin up and spin down electrons it will also be sensitive to distortions of this spatial distribution, i. e., (volume) strain. In the case of a magnetically ordered material under pressure this leads, first of all, to a decrease of the net magnetic moment. This implies a decrease in the exchange polarization forces and thus the core po-

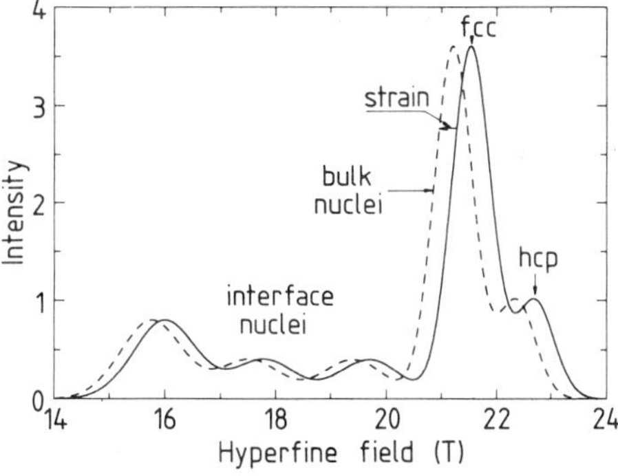

Figure 1. Illustration of some features important for the NMR on magnetic multilayers. The figure demonstrates that atoms at the interface have a different hyperfine field than "bulk" atoms in the inner layers, that for the inner atoms the crystallographic phase determines the magnitude of the hyperfine field and that strain results in an overall shift of the spectrum.

larization at the nucleus decreases under pressure. The net magnetic moment also determines the valence polarization at the nucleus, albeit in a more complicated way, so one might expect it to decrease too. However, the 4*s* electron density at the nucleus increases under pressure due to the (slight) squeezing of the valence electron wave functions, thereby more than compensating the effect of the reduced magnetic moment. As a result, the valence contribution to the hyperfine field increases under pressure and the net change in hyperfine field will be the result of the opposing changes in the core and valence spin densities[15]. In Co and Ni the hyperfine field decreases (becomes more negative) under pressure, while in Fe it increases. Essential is that the relative change in B'_{hf} is proportional to the relative volume change.

To summarize, the effects of structure, foreign atoms and strain on the hyperfine field spectrum are schematically shown in the hypothetical spectrum in Fig. 1.

EXPERIMENTAL ASPECTS

Basically, a NMR experiment is an investigation of the response of the nuclear magnetization to an applied oscillating magnetic field. The response is significant only when the frequency of the oscillation is close to the transition frequency defined in Eq. (2). Resonance can be obtained by sweeping the frequency at constant B_{appl} or sweeping the field at constant frequency. For the broad spectra frequently encountered in ferromagnetic materials spin echo methods are preferable[16,17]. They also offer the possibility of measuring directly the nuclear spin-lattice and spin-spin relaxation times. For more details about the spin echo method and the advantages of pulsed NMR the reader is referred to the literature[16,17,18]. Here only some selected experimental aspects of NMR in magnetic multilayers will be described.

Enhancement

In ferromagnetic materials the quantum resonant transitions of the nuclear spins are not induced by direct action of the external variable RF magnetic field B_1 (with a frequency $f \sim$ Larmor frequency) but by the variable part of the internal local field induced by the field B_1, i. e., the dynamic (transverse) component of B_{loc} (= $B_{hf} + B_{dip}$)[19,20]. The resulting enhancement factor η can be of the order of 10 (for single domain materials).

Not only the RF driving field is enhanced but also the nuclear induction signal: Through the hyperfine interaction the precessing nuclear magnetization induces a coherent precession of the (entire) electronic magnetization. The total transverse magnetization is $(1-\eta)\langle I \rangle_\perp$ and therefore the induced signal in the receiver coil is also enhanced by $|1-\eta|$.

In multidomain ferromagnets the response of the macroscopic magnetization to weak external fields involves domain wall displacements rather than "domain" rotations. Consequently, the local magnetization inside a domain wall experiences a particularly strong rotation resulting in very large enhancement factors[19,21]. Typical values for η for a nucleus inside a domain wall are $\eta \sim 10^2$-10^4.

It is clear that the enormous enhancement of the wall resonance is of great advantage. Unfortunately, the continuous variation of the direction of the magnetization within the domain walls, the possible distribution of magnetic anisotropy as well as the distribution of enhancement factors within and between the domain walls complicates the interpretation of the data (see, e. g., Ref. 16,21,22,23). It is evident that these problems do not occur in the single domain regime, i. e., with applied magnetic fields larger than the saturation field. The drawback is a large reduction of the enhancement factor which necessitates the use of larger samples. Correction of the NMR spectrum for the variation in η is necessary but relatively simple.

Table 2. Nuclear properties (data taken from Ref. 25) and (relative) spin echo NMR sensitivity for the 3*d*-elements. *A* denotes the mass number, *I* is the nuclear spin, γ is the nuclear gyromagnetic ratio and *m* is the nuclear magnetic moment (in multiples of the nuclear magneton μ_N). The relative sensitivity is calculated on basis of Eqs. (3) and (4) and refers to the magnitude of the spin echo signal in a magnetic field of 5 T in the (electronic) non magnetic state and is normalized to unity for (non magnetic) Co. The zero field sensitivity is calculated using the experimental values for the hyperfine field in the ferromagnetic state assuming a (bulk) enhancement factor $\eta \cong 1000$. The columns "enriched" and "nat. abund." refer to the sensitivity for a 100% isotopically enriched sample and a sample with natural abundance of the isotopes respectively. As a reference, at a temperature of 1.4 K about 1000 ML (monolayers) on 1 cm^2 are needed to acquire a signal to noise ratio of unity for non magnetic Co in a magnetic field of 5 T; Ferromagnetic Co in zero field requires about 20,000 times less material, i. e., 0.06 ML on 1 cm^2.

Element	A	I	m (μ_N)	$\gamma/2\pi$ (Mhz/T)	abundance (%)	relative sensitivity enriched	relative sensitivity nat. abund.	zero field sensitivity enriched	zero field sensitivity nat. abund.
Sc	45	7/2	4.75	10.343	100	1.09	1.09		
Ti	47	5/2	0.66	2.400	7.3	0.0076	0.00055		
	49	7/2	1.18	2.4005	5.5	0.0136	0.00075		
V	50	6	3.34	4.2450	0.24	0.201	0.00048		
	51	7/2	5.14	11.19	99.76	1.38	1.38		
Cr	53	3/2	-0.47	2.4065	9.55	0.0033	0.00031		
Mn	55	5/2	3.44	10.501	100	0.633	0.633		
Fe	57	1/2	0.09	1.3758	2.19	0.000122	$2.7 \cdot 10^{-6}$	5.61	0.123
Co	59	7/2	4.62	10.054	100	1	1	$1.86 \cdot 10^4$	$1.86 \cdot 10^4$
Ni	61	3/2	-0.75	3.8047	1.2	0.0129	0.00016	2.89	0.347
Cu	63	3/2	2.23	11.285	69.1	0.337	0.233		
	65	3/2	2.39	12.089	30.9	0.414	0.128		
Zn	67	5/2	0.87	2.663	4.1	0.0103	0.00042		

Feasibility of a NMR Experiment

The rotation of the nuclear magnetization $\boldsymbol{M}$ around the local magnetic field $\boldsymbol{B}_{loc}$ with frequency $2\pi f = \gamma B_{loc}$ induces an oscillatory flux in the receiver coil. Using the principle of reciprocity the EMF E in the coil can be expressed as[24]

$$E = K 2\pi f B_{1,\perp} M_{\perp} V_s \cos(2\pi f t) \tag{3}$$

where $B_{1,\perp}$ is the component perpendicular to B_{loc} of the RF magnetic field B_1 that would be produced, at the sample position, by a unit current in the receiver coil, K (≤ 1) is an "inhomogeneity factor" accounting for the inhomogeneity of B_1 within the sample volume, $M_{\perp}$ is the component of $\boldsymbol{M}$ perpendicular to $\boldsymbol{B}_{loc}$, V_s is the volume of the sample, and phase has been neglected. The magnitude of $M_{\perp}$ is given by[18]

$$M_{\perp} = (1-\eta) M_0 \sin\alpha \sin^2\left(\frac{\alpha}{2}\right) \exp\left(\frac{-2\tau}{T_2}\right) \tag{4}$$

where $\alpha = (1-\eta)\gamma B_{1,\perp} t_w$ is the turning angle of the applied RF pulses, t_w is the duration of the pulses, τ is the separation between the pulses, η is the enhancement factor and T_2 is the spin-spin relaxation time. $M_0 = N\gamma^2\hbar^2 I(I+1)B_{loc}/3k_B T$ is the total magnetic moment generated by N nuclei per unit volume in thermal equilibrium at a temperature T, where I is the nuclear spin angular moment, in accordance with Curie's law.

It follows that the EMF induced in the coil is proportional to the square of the Larmor frequency f and the total number NV_s of resonating nuclei, and inversely proportional to the temperature T. Thus high local fields (ferromagnetic materials) and low temperatures ($T \cong$ 1.4 K) are preferable.

Table 2 compares nuclear properties and the spin echo NMR "sensitivity", for the 3*d*-metals. As a reference we note that in a magnetic field of 5 T and at a temperature of 1.4 K non-magnetic Co requires about 1000 ML (monolayers) on 1 cm^2 for a signal to noise ratio of unity. For ferromagnetic Co in a field of 5 T and at 1.4 K the amount of material necessary for a signal to noise ratio of unity is reduced by a factor of approximately 100 compared to non-magnetic Co, i. e., 10 ML on 1 cm^2, while in zero field about 20,000 times less material is needed.

The spin echo intensity also depends on T_2, the spin-spin relaxation time. T_2 reflects the coupling between the nuclei which also depends on the environment of the nuclei[26]. Because a NMR spectrum of a multilayer essentially consists of lines corresponding to nuclei with different environments T_2 varies throughout the spectrum. When accurate relative intensities of the lines in a spectrum are required, T_2 for each line should be determined and the spin echo intensity should be corrected (cf. Eq. (4)). This is especially important for spectra recorded in low or zero magnetic field.

EXAMPLES OF NMR STUDIES ON MAGNETIC MULTILAYERS

In the following sections we present selected examples of the application of NMR to the research on magnetic multilayers. These examples comprise a systematic study of the *structure* of thin Co films and a systematic study of *interface roughness and topology* and *strain* in Co/Cu and Co/Ni multilayers. Other work is briefly reviewed. Because of sensitivity reasons, most of the NMR work has been done on ferromagnetic Co/X multilayers. However, also examples of NMR on V, Mn, Sb, Cu and Fe nuclei are presented, which show that the possibilities of NMR on magnetic multilayers are certainly not limited to multilayers containing ferromagnetic Co.

Local Structure of the Inner "Bulk" Layers. As we quoted above, the structure of

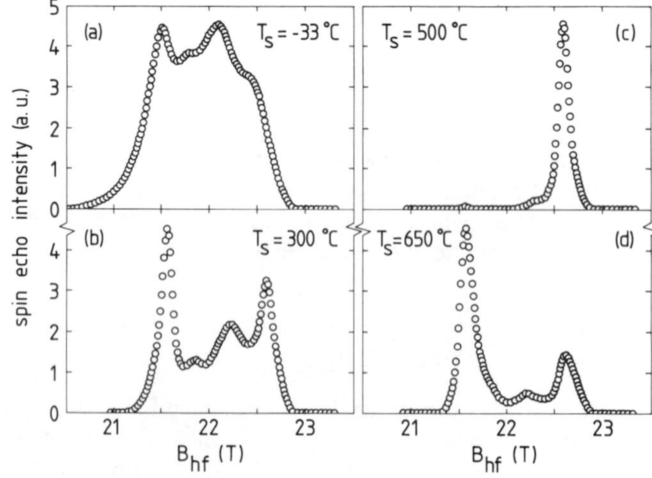

Figure 2. NMR spectra of 1000 Å Co films grown on mica showing the variation in bulk structure as function of the growth temperature T_S. The spectra were recorded at f = 190 Mhz with the magnetic field applied parallel to the film plane and at a temperature of 1.4 K.

the magnetic layers in a multilayer may be of great influence on the actual observed physical properties. The structural properties are strongly determined by the complex interplay between the used growth technique, the preparation conditions, the elements which are grown, etc.. To exemplify the kind of information which can be obtained from NMR we will shortly review some recent results on a particularly basic example: a Co film. The research was undertaken to relate the anisotropy of such a film to the structure.

A typical example of a NMR spectrum of a Co [111] film is shown in Fig. 2b. Four distinct resonance lines can be distinguished. The most intense peak has a resonance field corresponding to that of bulk fcc Co and is therefore assigned to Co atoms in the fcc phase. The signals at the high field side of this fcc line may be attributed to both stacking faults and the hcp phase[9]. On basis of the resonance fields (or, equivalently, B_{hf}) alone a decisive discrimination between these phases cannot be made. However, by recording the spectrum with the field applied perpendicular to the film plane, use can be made of the fact that the hyperfine field of hcp Co is anisotropic (the hyperfine field of hcp Co is approximately 0.86 T lower for the electronic magnetization, i. e., $\boldsymbol{B}_{hf}$, parallel to the [00.1] direction than for $\boldsymbol{B}_{hf}$ perpendicular to this direction[27]) while for stacking faults $\boldsymbol{B}_{hf}$ is almost perfectly isotropic. This thus means that if in the sample hcp Co is present the corresponding resonance line in the spectrum recorded with the field applied perpendicular to the film plane will, compared to the spectrum recorded with the field parallel, show a shift of -0.86 T with respect to the (isotropic) fcc line. This is exactly what can be observed for the line at 22.6 T.

In the way described above, we are able to determine what phases exist within the Co layer and, since the spin echo intensity of a line is proportional to the number of atoms in the corresponding environment, to determine the amount of Co in each phase. This opens the possibility to systematically study the effect of manipulation of preparation conditions on the structure and magnetic anisotropy. Figure 2 contains an example of such a study. This figure compares the NMR spectra for [111] Co films grown on mica substrates at different growth temperatures. As can be clearly seen from the spectra the relative amount of the phases changes with the temperature. At low temperatures the fcc phase and stacking faults dominate. This is in itself remarkable, since for bulk Co below 450 °C hcp is the stable phase. When the growth temperature exceeds 200 °C the number of stacking faults starts to decrease in favor of the pure fcc and hcp phase. At still higher temperatures hcp becomes the stable phase until at 500 °C almost the complete sample consists of hcp Co. When the growth temperature is further increased the percentage hcp phase decreases again in favor of the fcc phase and in accordance with the bulk phase transition[28]. Figure 3 summarizes these

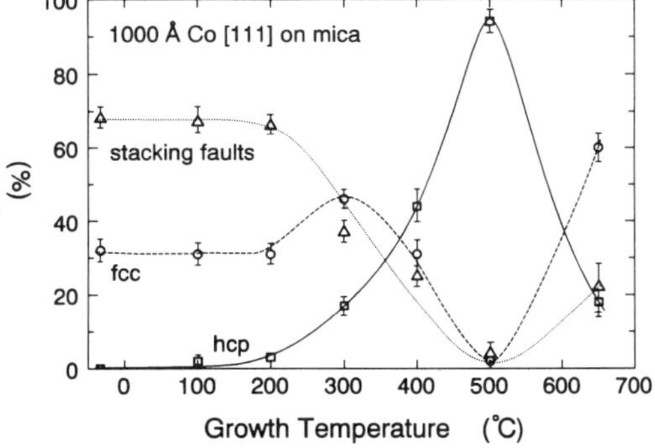

Figure 3. Percentage of the fcc phase (circles), hcp phase (squares) and stacking faults (triangles) in 1000 Å thick [111] Co films grown on mica substrates as function of the growth temperature. The results are derived from the NMR spectra in Fig. 2.

results in a more quantitative form. We should stress that the results shown in Fig. 3 are specific for Co films grown on mica, i. e., the relative amount of the phases and its temperature dependence is also strongly affected by, for instance, the choice of substrate or base layer[29].

The detailed knowledge of the structure of the Co films forms a good starting point for a quantitative interpretation of the observed electronic magnetic anisotropy. A discussion of this topic is, however, out of the scope of this article and therefore the reader is referred to the literature[29].

The analysis described above has also been used to clarify the structure of Co/X multilayers[30,31,32,33]. We will limit ourselves to a few remarks. For [111] Co/Cu multilayers, grown by UHV on oxidized silicon, the Co was shown to be pure fcc below a Co thickness of 20 Å . Above this thickness a small percentage of stacking faults developed[30,31]. In [111] Co/Ni grown by MBE on oxidized silicon the Co was purely fcc for all samples investigated[32]. In Co/Ir, Co/Au and Co/Pd grown by UHV on glass substrates the resonance lines were broadened due to the (small grain) polycrystallinity of the samples. Nevertheless, in all these samples a significant fcc contribution could be observed[30,33,34]. In Co/Au and Co/Pd also hcp contributions were detected[34]. For Co/Au also zero field NMR results have been reported where the Co was found to be purely in the hcp phase[35].

Interesting results were obtained on Co/Fe and Co/Cr multilayers[36]. For these materials it was deduced from the zero field NMR spectra that below a Co thickness of 20 Å (Co/Fe), respectively 10 Å (Co/Cr), Co grows in the bcc phase. For larger thicknesses the fcc and hcp phase are again favoured.

For Co/Mn multilayers was reported from zero field NMR that the structure of Co changed below a Co thickness of 36 Å. A α-Mn texture for the Co was suggested[37].

Not much work has been devoted to the study of local symmetry changes in magnetic multilayers. The study of such effects is focused on the anisotropy. Study of the anisotropy of the main line of a fcc Co/Cu [001] multilayer showed that the hyperfine field is slightly anisotropic and thus the Co layer in this multilayer has no cubic symmetry. This is probably because of tetragonal deformation due to the lattice mismatch between the Co and Cu[38,39]

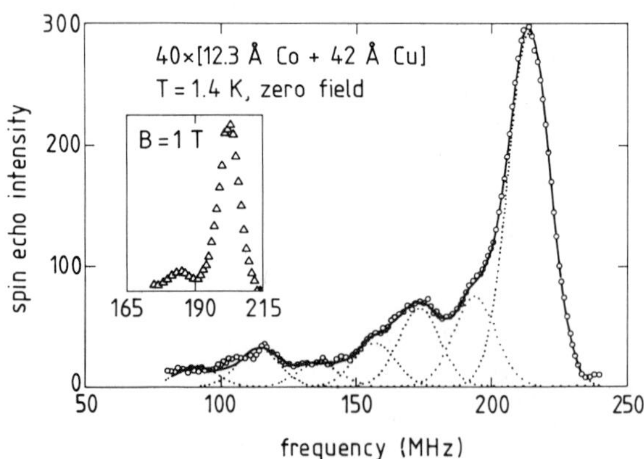

Figure 4. NMR spin echo intensity as function of resonance frequency for a 40× (12 Å Co + 42 Å Cu) [111] multilayer in zero field as recorded. The solid line represents the result of a fit with seven Gaussians. Each Gaussian (denoted by the dotted lines) corresponds to Co atoms in a specific environment. The dependence of the spin echo intensity on the square of the frequency has been incorporated in the fitting program. The inset shows that application of a magnetic field reduces the line broadening due to inhomogeneous magnetization, thereby clearly resolving the first satellite from the main line.

Similar effects have been observed in Co/Ag and Co/V multilayers[40]. Such effects may yield dipolar contributions to the crystal field anisotropy which are generally not taken into consideration.

Interface Roughness and Topology. Fig. 4 shows the spin echo intensity for a 40×(12.3 Å Co + 42 Å Cu) [111] multilayer versus frequency in zero field[31]. Fig. 5 shows the spin echo intensity for a 50×(12.3 Å Co + 42 Å Ni) [111] multilayer as function of the hyperfine field, recorded with the magnetic field ($B_{appl} \geq 1.8$ T) parallel to the film plane[32]. For both materials the main line in the spectrum appears close to the value of fcc surrounded bulk Co (217 MHz, resp. 21.6 T (γ_{Co} = 10.054 MHz/T)). The shift of this line with respect to the bulk value is caused by strain in the Co layer. At the high frequency side there is practically no intensity, indicating that in these samples the amount of hcp Co and the number of stacking faults is very small. Since the intensity ratio between the main line and the lower frequency part of the spectrum increases systematically with t_{Co} as shown in Fig. 6, the part of the spectrum below the main line is assigned to Co atoms at the interfaces where one or more nearest neighbor Co atoms are replaced by Cu, respectively, Ni. This assignment is supported by the observation that the spin-spin relaxation time T_2 for the satellites is typically twice the value measured for the main (bulk) peak, evidencing a different origin of the signals. From the inset of Fig. 6a it appears that for Co/Cu the intensity ratio between the main line and most intense satellite varies as n_{Co}-2, indicating unambiguously that the mixed layer is only one layer thick.

To analyse interface spectra such as shown in Fig. 4 more quantitatively, various contributions should be identified. Since the spectrum in this particular case originates from Co atoms with one or more Cu atoms in their nearest neighbor shell, it should consist of a number of absorption lines shifted with respect to the bulk fcc line by approximately 18 MHz per substituted Cu atom as deduced from experiments in diluted alloys[13] (cf. Table 1). The solid line in Fig. 4 shows that the structure of the spectrum is indeed well fitted by (seven) approximately equally spaced Gaussians (denoted by the dotted lines) with an average spacing of 19 ± 3 MHz.

Although we will not go into detail (see Ref. 4, 31) this result indicates that the various lines can thus be assigned to Co nuclei having 12 Co neighbors ("bulk" atoms) and nuclei having 11 to 6 Co neighbors ("interface atoms"). The existence of Co sites with an environment different from 12 Co neighbors ("bulk" atoms) or 9 Co neighbors (sites in perfectly flat interface layers have 9 Co neighbors for the [111] oriented growth) is obviously related to

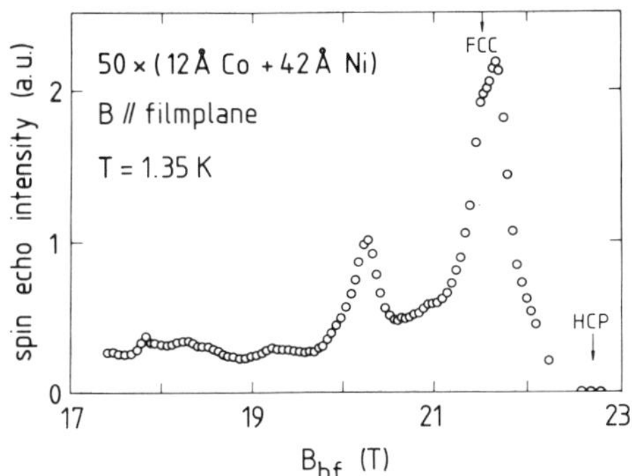

Figure 5. NMR spin echo intensity versus the hyperfine field B_{hf} for a 50×(12 Å Co + 42 Å Ni) [111] multilayer. The spectrum was recorded at 1.35 K with a magnetic field applied parallel to the film plane. The arrows denote the hyperfine field for bulk Co in a fcc and a hcp structure. The spectrum is corrected for enhancement.

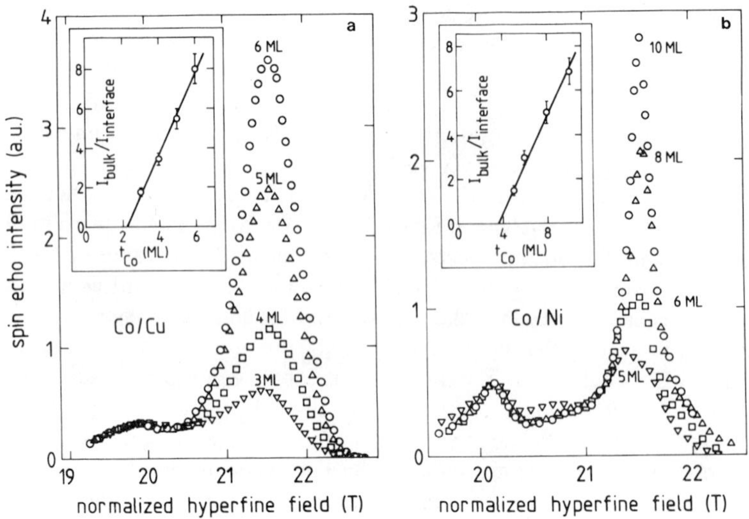

Figure 6. In field spectra of Co_xCu_{21} multilayers (Fig. a, left part of the figure) and of Co_xNi_{42} multilayers (Fig. b, right part of the figure) showing the systematic increase in intensity ratio between the main line and the most intense satellite as function of the Co thickness t_{Co} (expressed in monolayers). For easy comparison the spectra have been normalized to the most intense satellite (in intensity as well as in hyperfine field). The insets show the ratio of intensity of bulk to interface as function of Co thickness.

interface roughness. Assuming a model for the interface topology, one can now compare the statistical occurrence of various surroundings in the model with the experimental intensity of the satellite spectrum. In the present example of mixing in one monolayer rather detailed results could be obtained[31]. In the case of Co/Ni it appears that the intensity ratio between the main line and most intense satellite varies as n_{Co}-4 (see Fig. 6b), indicating that, in contrast to Co/Cu, there are at least two mixed layers at each interface. In such cases the interpretation in terms of a specific interface structure model is much more complicated and hazardous.

Important studies on interface roughness were also reported by Takanashi et al. on Fe/V, Co/Sb and Fe/Mn multilayers[41,42,43,44]. Since this research has already been extensively reviewed elsewhere[5] we will limit ourselves to a few remarks.

Fe/V: The ^{51}V hyperfine field distribution of V atoms at the interface with Fe was recorded in zero field. The spectrum was explained in terms of a diffusion model incorporating atomic short range order. The best fit to the spectra was obtained for an interface mixing of approximately three to five atomic layers, depending on the preparation conditions[41,42].

Co/Sb: The presence of non-magnetic Co at the interface was detected. An analysis of the line shape and comparison with alloy spectra revealed that at the interface compound formation of $CoSb_2$ or $CoSb_3$ had occurred[43].

Fe/Mn: The ^{55}Mn hyperfine field distribution was determined in zero field. The range of the spectrum extended from 20 MHz to about 260 Mhz. No distinct interface lines were observed but merely a continuous distribution of intensity. From the experiments it was concluded that at the interface where Fe was deposited on Mn random mixing occurred over about 15 Å[44].

Other work on interface roughness and topology is fragmentary:

Co/Cu: Part of the interface spectrum has been observed in zero field for Co/Cu [111][14,45] and Co/Cu [001][45]. No modelling of the interface was performed.

Cu/Nb: From the Cu Knight shift the existence of atomic layers with a different electronic structure was established. From the intensity of the signals was concluded that the Nb affected at least four atomic layers of Cu[46].

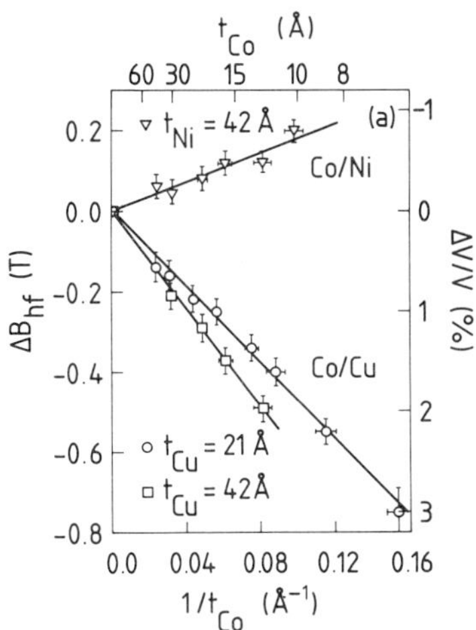

Figure 7. Hyperfine field B_{hf} derived from in-field spectra versus inverse Co thickness $1/t_{Co}$ for [111] Co/Ni and Co/Cu multilayers. The solid lines represent least squares fits of a straight line to the data.

Co/Fe: Signals above the "bulk" Co line were observed and attributed to interface and sub-interface layers[36].

Strain. Application of NMR to study strain in multilayers has sofar been limited to Co/Cu and Co/Ni[31,32] and some earlier data on Co/Pd[33]. In Fig. 7 the shift of the main line in Co/Cu and Co/Ni multilayers relative to the position of the bulk is plotted against $1/t_{Co}$. The "bulk" value B_{hf} = 21.53 T observed for a 1000 Å thick Co layer is used as reference value. It is obvious that in the multilayers B_{hf} is shifted from the bulk value by an amount ΔB_{hf} depending on t_{Co}. Using that $\Delta B_{hf}/B_{hf}$ is proportional to the change in atomic volume $\Delta V/V$ (or, alternatively, strain) the data suggest that the strain in the Co layers as probed by the bulk atoms is proportional to $1/t_{Co}$ from 6 Å up to bulk. The sign and magnitude of the strain, however, clearly depend on the chemical composition of the multilayer (Fig. 7). For Co/Ni the strain in the Co layers is compressive while for Co/Cu the strain is tensile (and depends on t_{Cu}). Since the lattice mismatch between Co and Ni is about +0.6% and between Co and Cu about -2.0% this strongly suggests that the strain in the Co layers is related to the lattice mismatch. If we compare the magnitude of the strain in Co/Cu for t_{Cu} = 42 Å with the magnitude of the strain in Co/Ni (t_{Ni} = 42 Å) we even find a perfect agreement between the ratio of the strains and the ratio of the mismatches.

For the interface atoms the same analysis can be performed to probe the strain at the interface. This provides us with the unique opportunity to probe the *variation of strain in the whole Co layer*.

CONCLUDING REMARKS

In multilayers and thin films the structure of the layers is extremely important for the observed physical properties. In this article we demonstrated that NMR is a promising technique which is able to probe locally the structure of the layers *within* a multilayer. The principles governing the application of NMR to magnetic multilayers were discussed and it was shown that in a relatively simple and straightforward way information can be obtained concerning the local structure and symmetry (fcc, bcc, hcp, stacking faults), the strain within the layers and the interface roughness. In favourable cases even the topology of the interfaces can be determined. The major drawback of NMR is its sensitivity. For most elements more than 1000 ML on 1 cm^2 are needed to get a reasonable signal to noise ratio.

In view of the current interest in the exchange coupling between the magnetic layers in a multilayer, it is interesting to note that the hyperfine field of the atoms in the non-magnetic spacer layers also reflects the details of this coupling. For these atoms the hyperfine field is namely mainly determined by the (induced) valence electron polarization (the "transferred" hyperfine field). Determination of the hyperfine field spectrum therefore should allow a quantitative as well as a qualitative test of the existing theories on a fundamental level.

Finally, we note without further discussion that NMR is not only sensitive to static properties through the positions, splitting, widths of the NMR spectra, but also to the dynamic properties of the electronic spin system through the nuclear spin-lattice relaxation time. In principle, by combination of a static and dynamic NMR study, this offers the opportunity to study locally the electronic structure within the multilayer.

ACKNOWLEDGEMENTS

We acknowledge the cooperation with Philips Research in Eindhoven on the research of magnetic multilayers. We would like to thank P. Panissod, F. J. A. den Broeder, P. J. H. Bloemen, K. Kopinga and E. A. M. van Alphen for their cooperation and valuable discussions. Part of this research was sponsored by the EEC in project GP^2M^3.

REFERENCES

1. F. J. A. den Broeder, D. Kuiper, A. P. van den Mosselaer and W. Hoving, Phys. Rev. Lett. **60**, 2769 (1988).
2. M. N. Baibich, J. M. Broto, A. Fert, F. Nguyen Van Dau, F. Petroff, P.Etienne, G. Creuzet, A. Friederich and J. Chazelas, Phys.Rev.Lett.**61**, 2472 (1988).
3. M. T. Johnson, R. Coehoorn, J. J. de Vries, N. W. E. McGee, J. aan de Stegge and P. J. H. Bloemen, accepted for publication in Phys. Rev. Lett.
4. H. A. M. de Gronckel and W. J. M. de Jonge, in Magnetic Multilayers, edited by L. H. Bennett and R. E. Watson (World Scientific, New Jersey 1992).
5. H. Yasuoka, in Metallic Superlattices, edited by T. Shinjo and T. Takada (Elsevier, 1987) p. 151 ff.
6. E. Fermi, Z. Phys. **60**, 320 (1930).
7. H. Akai, M. Akai, S. Blügel, B. Drittler, H. Ebert, K. Terakura, R. Zeller and P. H. Dederichs, Prog. Theor. Phys. Suppl. **101**, 11 (1990).
8. S. Blügel, H. Akai, R. Zeller and P. H. Dederichs, Phys.Rev.B **35**, 3271 (1987).
9. H. Brömer and H. L. Huber, J. Magn. Magn. Mater. **8**, 61 (1978).
10. A. J. Freeman, C. Li and R. Q. Wu, in Science and Technology of Nanostructured Magnetic Materials, edited by G. C. Hadjipanayis and G. A. Prinz (Plenum Press, New York 1991) p. 1 ff.
11. T. M. Shavishvili and I. G. Kiliptari, Phys. Stat. Sol. B **92**, 39 (1979).
12. P. C. Riedi and R. G. Scurlock, J. Appl. Phys. **39**, 1241 (1968).
13. S. Nasu, H. Yasuoka, Y. Nakamura and Y. Murakami, Acta Metall. **22**, 1057 (1974).
14. K. Le Dang, P. Veillet, Hui He, F. J. Lamelas, C. H. Lee and R. Clarke, Phys. Rev. B **41**, 12902 (1990).
15. J. F. Janak, Phys. Rev. B **20**, 2206 (1979).
16. P. C. Riedi, Hyp. Int. **49**, 335 (1989).
17. M. A. H. McCausland and I. S. Mackenzie, Adv. Phys. **28**, 305 (1979).
18. E. L. Hahn, Phys. Rev. **80**, 580 (1950).
19. A. M. Portis and A. C. Gossard, J. Appl. Phys. (Suppl.) **31**, 205 (1960).
20. E. A. Turov and M. P. Petrov, Nuclear Magnetic Resonance in Ferro- and Antiferromagnets (John Wiley & Sons, Inc., New York 1972).
21. M. B. Stearns, Phys. Rev. **162**, 496 (1967).
22. M. A. Butler, Int. J. Magn. **4**, 131 (1973).
23. A. Blacha and H. Brömer, Z. Physik B-Condensed Matter **41**, 9 (1981).
24. D. I. Hoult and R. E. Richards, J. Magn. Reson. **24**, 71 (1976).

25. K. Lee and W. Anderson, in CRC Handbook of Chemistry and Physics, edited by R. C. Weast, M. J. Astle and W. H. Beyer (CRC Press, Boca Raton 1988) p. E80-E85.
26. D. Hone, V. Jaccarino, T. Ngwe and D. Pincus, Phys. Rev. **186**, 291 (1969).
27. D. Fekete, H. Boasson, A. Grayevski, V. Zevin and N. Kaplan, Phys. Rev. B **17**, 347 (1978).
28. C. R. Houska, B. L. Averbach and M. Cohen, Acta Metall. **8**, 81 (1960).
29. H. A. M. de Gronckel, P. J. H. Bloemen, A. S. van Steenbergen, E. A. M. van Alphen and W. J. M. de Jonge, to be submitted to Phys. Rev. B.
30. H. A. M. de Gronckel, J. A. M. Bienert, F. J. A. den Broeder and W. J. M. de Jonge, in Magnetic Thin Films, Multilayers and Superlattices, Proc. of Symp. C of the 1990 E-MRS spring conference (Strasbourg, France, 29 may-1 june 1990), edited by A. Fert, G. Güntherodt, B. Heinrich, E. E. Marinero and M. Maurer (North-Holland, Amsterdam 1991) p. 457 ff.; also published in J. Magn. Magn. Mater. **93**, 457 (1991).
31. H. A. M. de Gronckel, K. Kopinga, W. J. M. de Jonge, P. Panissod, J. P. Schillé and F. J. A. den Broeder, Phys. Rev. B **44**, 9100 (1991).
32. H. A. M. de Gronckel, B. M. Mertens, P. J. H. Bloemen, K. Kopinga and W. J. M. de Jonge, J. Magn. Magn. Mater. **104-107**, 1809 (1992).
33. H. A. M. de Gronckel, C. H. W. Swüste, K. Kopinga and W. J. M. de Jonge, Appl Phys. A **49**, 467 (1989).
34. H. A. M. de Gronckel and J. A. M. Bienert, unpublished results.
35. C. Cesari, J. P. Faure, G. Nihoul, K. Le Dang, P. Veillet and D. Renard, J. Magn. Magn. Mater. **78**, 296 (1989).
36. Ph. Houdy, P. Boher, F. Giron, F. Pierre, C. Chappert, P. Beauvillain, K.Le Dang, P. Veillet and E. Velu, J. Appl. Phys. **69**, 5667 (1991).
37. K. Le Dang, P. Veillet, H. Sakakima and R. Krishnan, J. Phys. F: Met. Phys. **16**, 93 (1986).
38. E. A. M. van Alphen, H. A. M. de Gronckel and R. Coehoorn, unpublished results.
39. C. M. Schneider, J. J. de Miguel, P. Schuster, R. Miranda, B. Heinrich and J. Kirschner, in Science and Technology of Nanostructured Magnetic Materials, edited by G. C. Hadjipanayis and G. A. Prinz (Plenum Press, New York 1991) p. 37 ff.
40. H. A. M. de Gronckel and A. S. van Steenbergen, unpublished results.
41. K. Takanashi, H. Yasuoka, K. Kawaguchi, N. Hosoito and T. Shinjo, J. Phys. Soc. Japan **51**, 3743 (1982); ibid **53**, 4315 (1984).
42. N. Hamada, K. Terakura, K. Takanashi and H. Yasuoka, J. Phys. F: Met. Phys. **15**, 835 (1985).
43. K. Takanashi, H. Yasuoka, K. Takahashi, N. Hosoito, T. Shinjo and T. Takada, J. Phys. Soc. Japan **53**, 2445 (1984).
44. K. Takanashi, H. Yasuoka, N. Nakayama, T. Katamoto and T. Shinjo, J. Phys. Soc. Japan **55**, 2357 (1986).
45. Y. Suzuki, T. Katayama and H. Yasuoka, J. Magn. Magn. Mater. **104-107**, 1843 (1992).
46. M. Yudkowski, W. P. Halperin and I. K. Schuller, Phys. Rev. B **31**, 1637 (1985).

MAGNETO-TRANSPORT PHENOMENA OF MULTILAYERS WITH TWO MAGNETIC COMPONENTS

Teruya Shinjo

Institute for Chemical Research
Kyoto University
Uji, Kyoto-fu 611, Japan

INTRODUCTION

Interlayer coupling and magnetoresistance(MR) properties of magnetic multilayers are argued in many acticles of this volume. An antiferromagnetic interlayer coupling makes magnetizations of adjacent layers antiparallel and then the electric resistivity is greatly enhanced because of the spin–dependent scattering. Well–known examples of antiferromagnetically "coupled" type multilayer are Fe/Cr [1)] and Co/Cu.[2,3)] The mechanism of interlayer coupling is one of the most attractive subjects in recent years, since it has been found that the exchange interactions between magnetic layers generally depend oscillatorily on the thickness of non–magnetic intervening layer.[3,4)]

The subject of this article, on the other hand, is "non–coupled" type multilayers. If a multilayer includes two magnetic components with different coercive forces, magnetizations are aligned antiparallel in the course of field sweeping and then the resistivity is remarkably enhanced. Therefore, it is apparent that the existence of strong antiferromagnetic interlayer coupling is not a necessary condition for the giant MR effect of multilayers.

Multilayers exhibiting giant MR effect are classified into the following three groups.
(a) The interlayer coupling is strongly antiferromagnetic.
(b) The coupling is weakly antiferromagnetic.
(c) Two magnetic components are included in the multilayer and the spacing non–magnetic layer is so thick that the coupling between magnetic layers is negligibly weak. Magnetization and resistivity curves for each cases are schematically illustrated in Fig.1. In the case (a), magnetizations are aligned antiparallel as far as an external field is not strong, and the resistance vs. field curve shows the maximum in the vicinity of zero field. If the interlayer coupling is rather weak, an antiparallel alignment cannot be spontaneously established but occurs associated with domain wall formation. The maxima of resistivity appear at around the coercive fields, where adjacent layer magnetizations are the closest to antiparallel to each other. Such curves as the case (b) have been observed in many multilayer systems. Very recently, it has been reported that granular systems can also exhibit the giant MR effect [5,6)]. Their features are similar to the case (b). Therefore, it is suggested that the interaction

Magnetism and Structure in Systems of Reduced Dimension
Edited by R.F.C. Farrow *et al.*, Plenum Press, New York, 1993

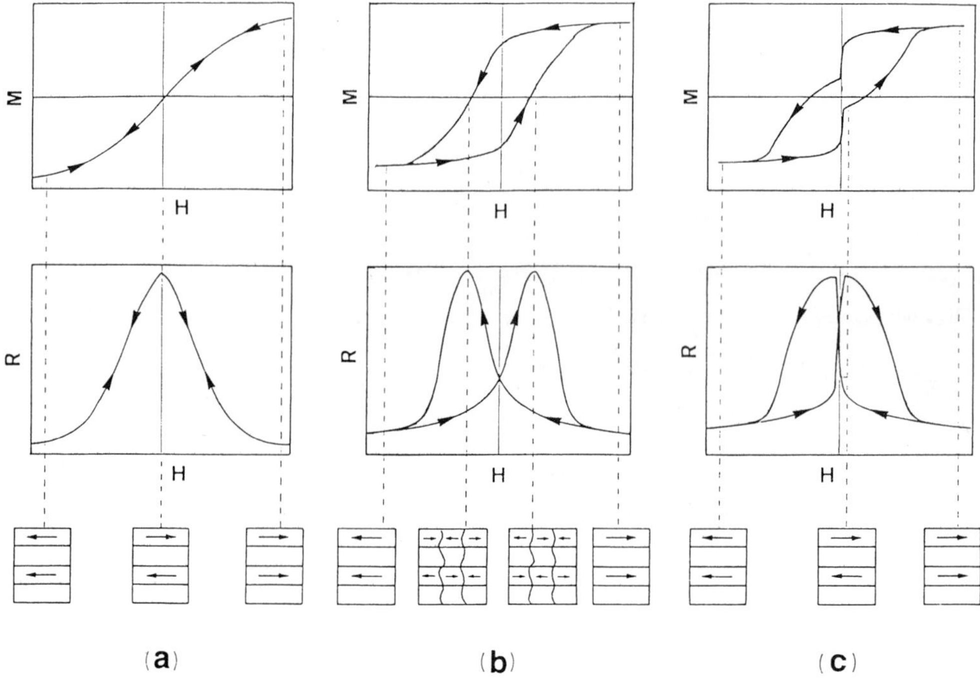

Fig.1. Schematic illustrations of magnetic hysteresis and magnetoresistance(MR) curves of multilayers,
(a) multilayers with strong antiferromagnetic coupling,
(b) the interlayer coupling is weakly antiferromagnetic,
(c) non–coupled type multilayers including two magnetic components with different coercive forces.

between ferromagnetic particles is weakly antiferromagnetic. In the case (c), the magnetization curve has two steps, corresponding to the independent flippings of soft and hard magnetic components. The enhancement of resistivity due to spin–dependent scattering occurs in the field region between the two coercive fields. This condition was experimentally realized in multilayers consisting of a non–magnetic component, (Cu, Au or Ag), and two magnetic ones, (Co and NiFe alloy), as will be described in the next section.[7–9)]

MR PROPERTIES OF NON–COUPLED TYPE Cu–BASED MULTILAYERS

Samples were prepared by successively depositing NiFe alloy, Cu, Co and Cu under ultrahigh vacuum (10^{-10} Torr during the deposition) on glass substrates at room temperature (RT). A nominal structure of multilayer sample for instance is;

Cr(50Å)/[Co(15Å)/Cu(35Å)/NiFe(15Å)/Cu(35Å)]×20.

For the deposition of so–called permalloy layer, Ni80%–Fe20% alloy was used as the evaporation source. Namely, Co and permalloy were adopted as hard and soft magnetic components, respectively, and the magnetic layers were separated by spacing Cu layers with an enough thickness to reduce the interlayer coupling to be negligible. The artificially layered structure with fcc(111) texture in the prepared samples was confirmed by using X–ray diffraction and electron microscope.[10)]

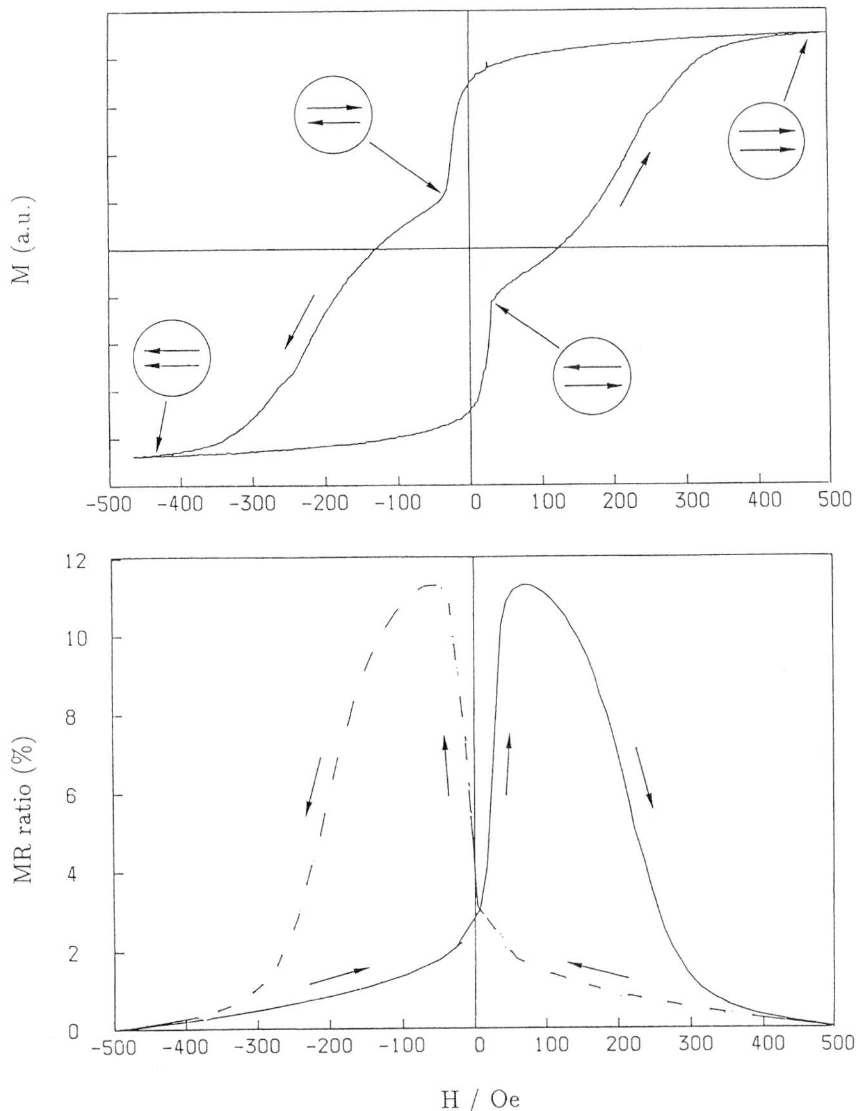

Fig.2. Magnetization and MR curves at RT of $Cr(50Å)/[Co(15Å)/Cu(35Å)/Ni_{80}Fe_{20}(15Å)/Cu(35Å)]\times 20$.

An example of MR curve at RT is shown in Fig.2, together with the hysteresis curve of the same sample. As clearly indicated by the two–step magnetization curve, two magnetic components change the directions independently. The permalloy layer changes the direction of magnetization easily, in a small field region, 0–30 Oe. On the other hand, much larger fields, 100–300 Oe, are required for the reorientation of Co layer magnetizations. Therefore, two magnetizations are oriented antiparallel in the field region between 0–300 Oe, and then the enhancement of resistivity is caused by the spin–dependent scattering. If the resistivity at H=500 Oe is taken as the standard, the maximum increase of resistivity is obtained as 11% at RT.

For the purpose to enhance the MR effect, in the case of non–coupled type multilayers, it is desirable to reduce the non–magnetic layer thickness. On the other hand, since the interlayer coupling has to be kept negligibly small, the spacing non–magnetic layer thickness is required to be larger than a certain value. Therefore to be looked for is the minimum thickness to keep magnetic layers uncoupled. which should be the optimum thickness to obtain the largest MR effect. In a previous work, the dependence of MR effect on Cu layer thickness was investiagted and 55Å was reported to be the minimum (optimum) thickness.[8] Afterwards, it was found that larger MR values are obtained from samples prepared on Cr buffer layers. It has been generally known that a buffer layer deposition is essential to obtain a flat surface. To avoid magnetic pin holes, interfaces are desirable to be flat. However, a thick buffer layer with a high conductivity (e.g.Au) is not benefitial to enhance the MR effect. A Cr buffer layer with 50Å thickness was found to be an appropriate choice. The conductivity of 50Å Cr layer is not high but a fairly smooth surface is formed. Then, the thickness of spacing Cu layer was able to be reduced down to 35Å, and the MR effect at RT was obtained to be 14%, which is the largest value at the present stage, among the MR values of non–coupled type multilayers.[11] If the Cu layer thickness is less than 35Å, magnetic layers are ferromagnetically coupled and behave as one magnetic system. Antiparallel alignment of magnetizations is no more realized in the course of field sweeping. Probably, the origin of this ferromagnetic coupling is magnetic pin holes. The intrinsic interlayer couling as a function of the spacing would be oscillatory but no systematic investigation has been made yet. Interlayer couplings between different magnetic materials also are interesting subjects.

TEMPERATURE DEPENDENCE

With a decrease of temperature, the MR effect, $\Delta\rho/\rho_o$ increases. The MR value for the sample in Fig.2 becomes more than 40% at 4.2K. Since the back–ground resistivity, ρ_o, of metals and alloys decreases with a decrease of temperature, the MR effect appears to increase even if the magnetic contribution, $\Delta\rho$, is constant. In order to study the spin–dependent scattering, the specific resistivity of $\Delta\rho$ of a Cu–based sample is shown in Fig.3, as a function of temperature. It is apparent that $\Delta\rho$ increases remarkably with a decrease of temperature and the relation is roughly expressed by a linear relation. Such feature is common for Au– or Ag–based samples. As shown in the figure, for all these cases, by extrapolating the $\Delta\rho$ vs T curves, the temperatures where the MR vanishes are estimated to be 500 ~ 600K, which is much lower than the Curie temperatures of Co or permalloy.

It is plausible that the spin–dependent scattering mainly occurs at interface sites and therefore the probability depends on the local magnetization at interface, whose temperature dependence may be much steeper than the bulk behavior. However, to our knowledge, an interface effect on local magnetization is only significant in a high temperature region near the Curie temperature.[12] Local magnetizations can be studied by using microscopic methods, like Mössbauer and NMR spectroscopies. A preliminary Mössbauer study on a Cu– based multilayer showed that the temperature dependence of interface hyperfine field, below RT, does not greatly differ from that of inside.[13] The temperature dependence of MR effect therefore does not correspond to the static magnetic behaviors. To account for the observed increase of MR effect even at very low temperatures, it seems necessary to take dynamical behaviors into account, such as magnon excitations. For a quantitative argument of MR effect, it is necessary to know the order degree of antiparallel alignment of magnetizations, at each temperature. Neutron diffraction is useful for this purpose and temperature dependent measurements on multilayers with two magnetic components are in progress.[14] Without the characterization of magnetic structure, experimentally obtained values of the MR effect cannot be analyzed by comparing with theories.

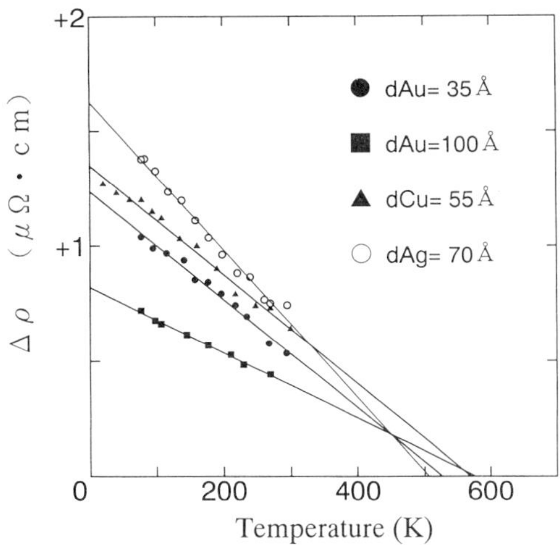

Fig.3. Temperature dependence of specific resistivity increase due to magnetic structure, in Cu, Au and Ag based multilayers. The thicknesses of Co and NiFe layers are always 25Å and the number of repetition of the four-layer unit deposition is 15, for each sample.

COMPARISON WITH Au- AND Ag-BASED MULTILAYERS

Instead of Cu, Au and Ag were adopted as the material for the non-magnetic spacing layer and multilayers including two magnetic components were prepared. The comparison between their MR properties is important for the understanding of the spin dependent scattering mechanism and it is of particular interest to look for the condition to enhance the MR effect.

As well as Cu-based films, non-coupled type multilayers exhibiting large MR effect can be prepared by using Au or Ag for the intervening layer. The MR results at RT for Cu-, Au- and Ag-films with similar structures are shown in Fig.4.[15] The number of repetition of four-layer unit deposition was 5 for each sample. The observed MR ratio for each system is plotted as a function of the spacing layer thickness, *d*. The multilayered structures in these samples were confirmed by using X-ray diffraction. If *d* is larger than 50Å, the crystallographic quality is excellent. However, for the samples of d<50Å, the quality is not always good and depends on the non-magnetic materials.

The result on the Cu-based films is a typical example of the dependence of MR effect on the spacing layer thickness. Because the spacing layers are highly conductive, the effect of spin dependent scattering becomes less significant if the thickness. *d*, is too large. Therefore, the MR effect increases with the decrease of *d*. If *d* becomes less than a critical value, magnetic layers are coupled by an interlayer exchange interaction and each layer magnetization cannot be oriented antiparallel and eventually the MR effect decreases. As shown in the figure, the critical value of *d*, that is the value for the largest MR ratio, for the case of Cu is about 35Å. If *d* is very small, the crystallographic quality becomes worse and then ρ_0 increases because of lattice imperfections, which is a factor to reduce the MR effect with decreasing of *d*.

Au-based films exhibit similar MR properties with those of Cu-based ones. The observed MR ratios show the maximum when *d* is about 50Å. if *d* is less than 50Å, an interlayer coupling begins to work. However, the coupling is antiferromagnetic, in contrast to the case of Cu. With the decrease of *d*, the MR behaviors of the Au-based films gradually change from non-coupled type to coupled one. Namely, the MR and magnetization curves for d > 50Å are like Fig.1(c), while those for d < 50Å are like Fig.1(a) or (b). For the

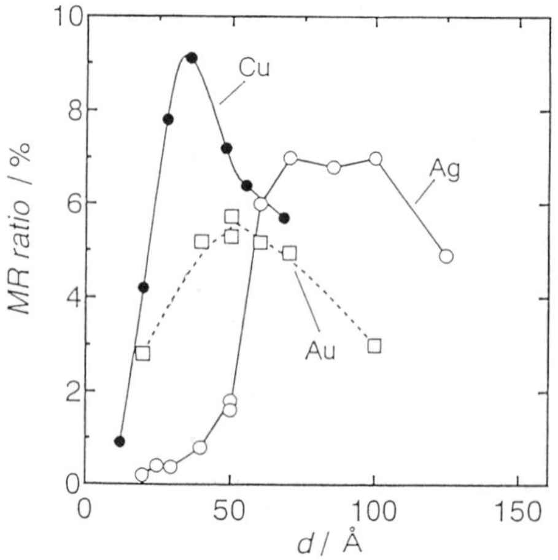

Fig.4. Comparison of MR effect, in Cu–, Au– and Ag–Based films.
d expresses the thickness of the non–magnetic layer (Cu,Au or A).
The thicknesses of magnetic layers are 25Å, for Au and Ag–based multilayers but are 10Å, for Cu–based samples. The number of repetition was 5, for all the cases.

samples with $d < 50$Å, the specific resistivity of $\Delta\rho$ does not change greatly. The decrease of MR ratio, with decrease of *d*, is mainly due to the increase of ρ_o.

In the case of Ag–based films, *d* is required to be larger than 60Å to exhibit non–coupled type MR effect. An interlayer coupling already exists at $d < 50$Å, whose sign is ferromagnetic in this case. In the region for $d > 60$Å, the Ag–based samples show larger MR ratios than those of the Cu–based samples, mainly because Ag–based samples have smaller ρ_o values.

The above results suggest that the interlayer couplings between Co and NiFe layers, through Cu or Ag intervening layer are ferromagnetic, while those through Au are antiferromagnetic. The length of coupling extends to the longest in the case of Ag and the shortest in Cu, among the three metals. However, these results strongly depend on the crystallographic qualities of samples and therefore the results for the samples with $d < 50$Å cannot be regarded as the quantitatively conclusive ones. On the other hand, if *d* is larger than 50Å, the crystallographic quality of the samples is fairly good in all these cases and the results can be compared quantitatively. It is to be noticed that the observed MR ratios are very similar for the samples with d=70Å, in the three systems. The values of $\Delta\rho$ are roughly the same, independently of the non–magnetic metal species. This result suggests that the probability of spin–dependent scattering at all interfaces, between magnetic layers and Cu, Au or Ag layers, is very similar.

In order to enhance the MR effect, it is favorable that the probability of spin–dependent scattering at the interface between magnetic and non–magnetic layers is high. Concerning the probability, however, the difference among Cu, Au and Ag is found to be small. From the selection of the non–magnetic metal, a drastic enhancement of the MR effect is not achieved. The reason why the largest MR effect was obtained in the Cu–based series is that the critical thickness of Cu for non–coupled type multilayer is smaller than that of Au or Ag.

MR OF NON-COUPLED TYPE MULTILAYERS IN ROTATING FIELDS

In the case of antiferromagnetically coupled multilayers, the magnetic structure being antiparallel in the absence of external field, can be transformed into parallel by applying an enough strong field. However, it is almost impossible to cotrol the spin directions in the intermediate stages. In contrast, in the non-coupled type multilayers, the relative angle between two magnetic components is able to be controlled.

Samples used for the rotating field measurements are the followings: [Cu(*d*)/Co(25Å)/Cu(*d*)/NiFe(25Å)]×15, where the Cu layer thickness, *d*, was chozen as 100 and 55Å.[16] After applying a field in the negative direction high enough to saturate both magnetizations, the field was swept back to a certain field and then rotated. The direction of magnetization in the hard component, Co, has no change, while that of the soft component, NiFe, follows the direction of external field. Thus, the relative angle of the two magnetizations can be varied as desired. Figure 5 shows the results at RT for the two samples. Initially applied field was 3 kOe. It was reduced to 20 Oe and then, rotated. The angle, θ, means the direction of the external field. If the NiFe layer magnetization is oriented always exactly to the direction of external field while the Co layer magnetization is invariable, θ is equal to the relative angle between two magnetizations.

The resistivity of the sample with *d*=100Å as a function of the angle, θ, of the external field direction is shown in Fig.5(a). The obtained curve is very close to an ideal cosine function. The curve was analyzed by assuming the following formula.

$$R=A_1\cos\theta + A_2\cos 2\theta + R_o \qquad (1)$$

The first term corresponds to the resistance due to the spin-dependent scattering and the second to the anisotropic MR effect. As the result of fitting, A_1 and A_2 are obtained as -6.1×10^{-2} and 1.1×10^{-3}, respectively. The value of A_2 is only 1/50 of A_1. Thus it is apparent that the contribution from anisotropic (classic) MR effect is negligible compared to the new MR effect due to the spin-dependent scattering. This result means that the NiFe layer magnetization is always oriented to the external field direction while the Co layer magnetization is not changed, and the resistivity caused by the spin dependent scattering is expressed by a cosine function. This functional form was originally proposed by Slonczewski for the spin valve effect in a tunneling junction between ferromagnetic layers.[17] Although the current direction in the present case is in the plane, the behaviors are understandable analogously with the spin valve effect.

The result for the case of *d*=55Å is shown in Fig.5(b). The profile is remarkably distorted from the cosine formula. This deviation is explained by assuming an interlayer coupling. When *d*=55Å, the interlayer coupling cannot be regarded as zero. If the direction of Co layer magnetization is not varied, a magnetic field caused by the exchange interaction is supposed to exist at the NiFe layer, which is always parallel (or antiparallel) to the direction of Co layer magnetization. The resistivity is again assumed to be expressed by a cosine function of the relative angle between the two magnetizations. However, in this case, the direction of the NiFe layer magnetization is not the same as that of external field but is determined as the vector sum of the external field and the exchange field. Theoretical curves for various magnitudes of external field, between 0 and 10 Oe, are shown in the same figure. A satisfactorily good fit is obtained when the exchange field is 7 Oe. As an exchange field over a Cu layer of 55Å, 7 Oe seems to be a reasonable magnitude.

An MR measurent under rotating fields was also reported by Dieny et al.[18] Their samples have a four-layer structure as NiFe/Cu/NiFe/FeMn. Two magnetic layers are

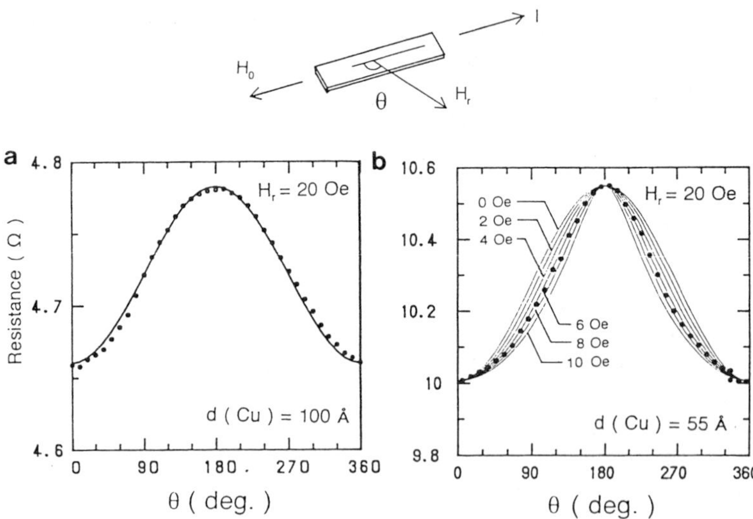

Fig. 5. Resistance at RT of [Cu(d)/Co(25Å)/Cu(d)/NiFe(25Å)]x15 under a rotating field, (20 Oe). The angle, θ, means the direction of the external field. (a) d = 100Å, (b) d = 55Å.

permalloy, but by depositing an antiferromagnetic FeMn layer, coercive forces are defferentiated. The MR properties are very similar to those of non–coupled type multilayers described here.

MERITS OF NON–COUPLED TYPE MULTILAYERS

For the study of interlayer coupling, a three–layer (sandwich) sample is much more convenient than a multilayer sample. Especially a sandwich sample where magnetic layers are separated by a wedge–shape non–magnetic spacing layer can furnish a systematic result at once.[19-22] On the other hand, for the study of transport properties, such a wedge–type sandwich is not useful. Concerning physical meanings, an information from sandwich sample is the same as that from multilayer sample. Actually many reports have been already published on the MR properties of sandwich samples. Sandwich samples of non–coupled type also have been investigated.[18,23,24] However, the very large field dependence of transport properties, so–called giant MR effect, can be observed only in multilayer samples. Therefore, to study magneto–transport properties of multilayers is of great importance both from fundamental and technological viewpoints.

It is certainly a merit of non–coupled type multilayers that the magnetic field to induce the MR effect is small. For technological applications, a high sensitivity for very weak fields is required. In the case of non–coupled type multilayers, the field dependence can be made very sharp as far as the soft magnetic component behaves indeed soft. For this purpose, the non–magnetic layer should be sufficiently thick, Then, the asbolute MR value has to be limited. However, even if the non–magnetic layer is enough thick, an MR effect larger than the anisotropic MR effect can be obtained. Therefore, to apply a non–coupled type multilayer for MR sensor or head seems to be promising. A goal of the MR studies is to find a condition for larger MR effect in smaller magnetic field.

Not only for applications, but also for fundamental researches, multilayers with two components are interesting subjects. The rotating field measurements mentioned in the preceding section is an example of experiment applicable for non–coupled type multilayers but impossible for coupled–type. In the case of non–coupled type, the magnetic structure can be modified by a weak field, which is sometimes very convenient for a practical measurement. Recently we have succeeded in observing the spin–dependent scattering effect in thermopower and also thermal conductivity.

The electric resistivity in a state with antiparallel magnetizations is much bigger than that in a parallel (ferromagnetically saturated) one. This difference, which was found to be much larger than expected, is the meaning of the giant MR effect. Such a difference with the same meaning is of course expected to appear not only in electric resistivity but also in other transport properties. Thermopower of non–coupled type multilayer, Cu/Co/Cu/NiFe, was measured by Sakurai et al.[25] As shown in Fig.6, the field dependence clearly shows the difference of the thermopower between spin parallel and antiparallel states. The profile of thermopower vs. magnetic field curve is consistent with that of electric resistivity and therefore it is no doubt that the observed peaks are caused by the magnetic structure of antiparallel alignment The sign of the thermopower is negative, and the absolute value of antiparallel state is smaller than that of parallel.

The field dependence of thermal resistivity also shows a very similar profile [26]. The spin dependent scattering affects only the conductivity due to electrons. Therefore, the magnetic field effect should be large at low temperatures, where the conductivity due to phonons is not dominant. Actually a clear differnce of thermal conductivity between spin parallel and antiparallel states was observed in a measurement at 10K. The result will be published elsewhere soon.

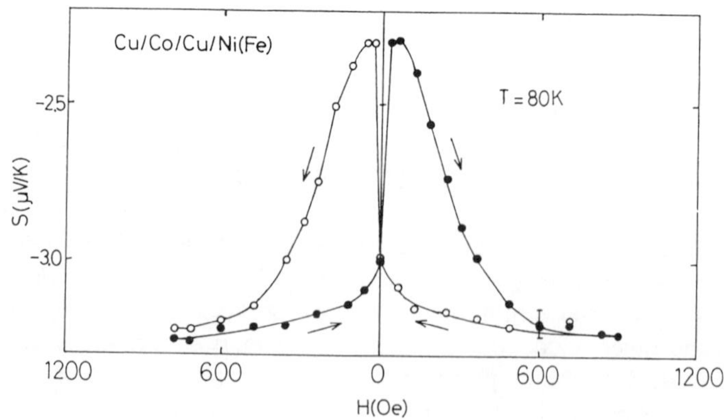

Fig.6. Thermopower at 80K of [Cu(50Å)/Co(25Å)/Cu(50Å)/NiFe(25Å)] x 15 as a function of magnetic field.

In this article, magneto–transport phenomena of non–coupled type multilayers were described. Interlayer couplings were not the subject here but the length and sign of interlayer exchange interaction is an indispensable knowledge to improve the non–coupled type MR effect. If the interlayer interaction between two homogeneous magnetic layers has been understood, attention will be paid to the interlayer interactions between heterogeneous combinations. Coupled type multilayers with two magnetic components will also be very interesting subjects.

ACKNOWLEDGEMENTS

The author would like to express sincere thanks to the collaborators; H. Yamamoto, T. Okuyama, T. Anno and J. Sakurai. The main part of the works introduced here has been supported by a Grant-in-Aid for Scientific Research on Priority Areas from the Ministry of Education, Science and Culture of Japan.

REFERENCES

1) M.N.Baibich, J.M.Broto, A.Fert, F.Nguyen Van Dau, F.Petroff, P.Etienne, G.Creuzet, A.Friederich and J.Chazelas, Giant magnetoresistance of (001)Fe/(001)Cr magnetic superlattices, Phys.Rev.Lett.61,2472(1988).
2) S.S.P.Parkin, Z.G.Li and D.J.Smith, Giant magnetoresistance in antiferromagnetic Co/Cu multilayers, Appl.Phys.Lett.58,2710(1991).
3) D.H.Mosca, F.Petroff, A.Fert, P.A.Schroeder, W.P.Pratt Jr. and R.Laloee, Oscillatory interlayer coupling and giant magnetoresistance in Co/Cu multilayers, J.Magn.& Magn.Mater.94,L1(1991).
4) S.S.P.Parkin, R.Bhadra and K.P.Roche, Oscillatory magnetic exchange coupling through thin Cu layers, Phys.Rev.Lett.66,2152(1991).
5) A.E.Berkowits, J.R.Mitchell, M.J.Carey, A.P.Young, S.Zhang, F.E.Spada, F.T.Parker, A.Hutten and G.Thomas, Giant magnetoresistance in heterogeneous Cu–Co alloys, Phys.Rev.lett.68,3745(1992).
6) J.Q.Xiao, J.S.Jiang and C.L.Chien, Giant magnetoresistance in nonmultilayer magnetic systems, Phys.Rev.Lett.68,3749(1992).

7) T.Shinjo and H.Yamamoto, Large magnetoresistance of field–induced giant ferrimagnetic multilayers, J.Phys.Soc.Jpn.59,3061(1990).
8) H.Yamamoto, T.Okuyama, H.Dohnomae and T.Shinjo, Magnetoresistance of multilayers with two magnetic components, J.Magn.& Magn.Mater.99,243(1991).
9) T.Shinjo, H.Yamamoto, T.Anno and T.Okuyama, Magnetism and magnetoresistance of metallic multilayers, Appl.Surf.Sci., in press.(1992).
10) T.Okuyama, H.Dohnomae, H.Yamamoto and T.Shinjo, Magnetism and magneto–resistance of multilayers with two magnetic components II. Au–based films, Mat.Res.Soc.Symp.Proc.231,223(1992).
11) H.Yamamoto, T. Anno and T.Shinjo: to be presented at the 1st International Symposium on Metallic Multilayers, to be held in Kyoto, 1993.
12) See a review for instance, T.Shinjo, Interface Magnetism, Surf.Sci.Repts.12.49(1991). References are cited therein.
13) T.Shinjo, H.Yamamoto, T.Okuyama, S.Araki, K.Mibu and N.Hosoito, Resistivity and magnetic structure in multilayers, Hyper.Int.68,333(1991).
14) N.Hosoito, in preparation.
15) T.Anno, H.Yamamoto and T.Shinjo, to be submitted to J.Magn.& Magn.Mater.
16) T.Okuyama, H.Yamamoto and T.Shinjo, Magneto–transport phenomena of multilayers with two magnetic components, J.Magn.& Magn.Mater. 11, in press,(1992).
17) J.C.Slonczewski, Conductance and exchange coupling of two ferromagnets separated by a tunneling barrier, Phys.Rev.B 39,6995(1989).
18) B.Dieny, V.S.Speriosu, S.S.P.Parkin, B.A.Gurney, D.R.Wilhoit and D.Mauri, Giant magnetoresistance in soft ferromagnetic multilayers, Phys.Rev.B 43 1297(1991).
19) J.Ungaris, R.J.Celotta and D.T.Pierce, Observation of two different oscillation periods in the exchange coupling of Fe/Cr/Fe(100), Phys.Rev.Lett.67,140(1991).
20) Z.Q.Qui, J.Pearson, A.Berger and S.D.Bader, Short–period oscillations in the interlayer magnetic coupling of wedged Fe(100)/Mo(100)/Fe(100) grown on Mo(100) by molecular beam epitaxy, Phys.Rev.Lett.68,1398(1992).
21) A.Fuss, S.Demokritov, P.Grünberg and W.Zinn, Short– and long period oscillations in the exchange coupling of Fe across epitaxially grown Al– and Au–interlayers, J.Magn.& Magn.Mater.103,L221(1992).
22) S.T.Purcell, M.T.Johnson, N.W.E.McGee, R.Coehoorn and W.Hoving, Two–monolayer oscillations in the antiferromagnetic exchange coupling through Mn in Fe/Mn/Fe sandwich structures, Phys.Rev.B 45,13064(1992).
23) C.Dupas, P.Beauvillain, C.Chappert, J.P.Renard, F.Trigui, P.Veillet, E.Velu and D.Renard, Very large magnetoresistance effects induced by antiparallel magnetization in two ultrathin cobalt films, J.Appl.Phys.67,5680(1990).
24) P.Grünberg, J.Barnas, F.Saurenbach, J.A.Fuss, A.Wolf and M.Vohl, Layered magnetic structures: antiferromagnetic type interlayer coupling and magnetoresistance due to antiparallel alignment, J.Magn.& Magn.Mater. 93,58(1991).
25) J.Sakurai, M.Horie, S.Arai, H.Yamamoto and T.Shinjo, Magnetic field effects on thermopower of Fe/Cr and Cu/Co/Cu/NiFe multilayers, J.Phys. Soc.Jpn.60,2522(1991).
26) H.Sato, Y.Aoki, Y.Kobayashi, H.Yamamoto and T.Shinjo, Thermal conductivity and Hall effect in Cu/Co/Cu/NiFe superlattce, in preparation.

THERMOELECTRIC POWER OF MAGNETIC MULTILAYERS

L. Piraux,[1] A. Fert,[2] P.A. Schroeder,[3] R. Loloee,[3] P. Etienne,[4] and T. Valet [4]

1 Unité de Physico-Chimie et de Physique des Matériaux, place Croix du Sud 1, B-1348 Louvain-la-Neuve, Belgium
2 Laboratoire de Physique des Solides, Université Paris-Sud, 91405 Orsay, France
3 Physics Department and Center for Fundamental Materials Research, Michigan State University, East Lansing, MI 48824, USA
4 Laboratoire Central de Recherche Thomson-CSF, 91404 Orsay, France

INTRODUCTION

Since the discovery of "giant magnetoresistance (MR) effects" [1,2], a large amount of work on the electrical resistivity of magnetic multilayers in applied fields has been done over the past few years. However, it is only recently [3-6] that the thermoelectric power of these systems has been investigated.

Sakurai et al [3] have measured the thermoelectric power of Fe27Å/Cr8Å and Cu50Å/Co24Å/Cu50Å/Ni(Fe)25Å multilayers in a magnetic field. They found very large magnetothermoelectric power (MTEP) effects correlated with the magnetoresistance. Conover et al [4] have carried out MTEP measurements at room temperature on Fe(32Å)/Cr(x) superlattices with x ≈ 5-50Å. They have found that the MTEP presents the oscillatory behavior as a function of the Cr layer thickness previously observed for the MR [7]. Also, the temperature dependence of the thermopower of the Al/Ni multilayer system in zero magnetic filed has been reported [5].

In this paper, we first present the results of experimental studies of the thermoelectric power in Co/Cu, Fe/Cu, Fe/Cr and Cu/Co/Cu/NiFe multilayers, both in the presence and absence of a magnetic field. Next, we propose a possible explanation based on a major contribution from spin-dependent electron-magnon scattering to the diffusion thermopower.

EXPERIMENT

Co/Cu and Fe/Cu multilayers were prepared by sputtering at Michigan State University while Fe/Cr and Cu/Co/Cu/NiFe multilayers were grown by MBE and sputtering respectively at the LCR Thomson CSF (NiFe=permalloy). Both (Co11Å/Cu9Å) x 30 and (Fe15Å/Cu15Å) x 60 multilayers are deposited on Si(001) with a buffer layer of iron. The preparation and characterisation are described elsewhere [8,9]. In Co/Cu, the Co and Cu layers are fcc with a pronounced (111) texture [8]. For Fe15Å/Cu15Å, the Fe and Cu layers are bcc with (110) texture [9,10]. For the (001)bcc Fe26Å/Cr9Å multilayer grown on a MgO(001) substrate, the thickness of the non-magnetic layers corresponds with maximum antiferromagnetic coupling between the magnetic layers at H=0. We have also studied a non-coupled type multilayer prepared by successively depositing Co(10Å), Cu(50Å) and permalloy (50Å) with a total tickness of 2000Å. In this multilayer, the coercive field is much smaller in permalloy than in Co and a giant MR (~20% at 4.2K) is induced in a small field (~80 Oe).

Thermoelectric power measurements were performed using a static heat and sink four-probe method. The multilayers were glued on a copper block whose temperature could be precisely controlled over the range 1.3-200K. Heating currents of 10-200 mW through a 10 kΩ metal-film resistance were applied to the sample producing temperature gradients of up to ~ $2Kcm^{-1}$. Two carbon glass resistors carefully attached to the sample with GE varnish were used as thermometers to measure both the absolute temperatures and the temperature differences. The power dissipated in each carbon glass sensor was limited to 10^{-9}W in order to avoid self-heating. At T~200K, the thermometers are becoming too insensitive to be useful while radiation problems become increasingly important, so these determine our upper limit. Electrical contacts to the sample for electrical conductivity and thermoelectric power measurements were made with platinum paste. Chromel P, copper and superconducting NbTi wires were used independently for voltage measurements. Voltage differences were measured using a Keithley K181 nanovoltmeter with a resolution of about 10^{-8} V and corrected for any offsets present when the sample had no temperature gradient. Most of the random error is due to the measurement of the temperature gradient. The thermopower of the chromel-P exhibits very weak magnetic field dependence while the field has relatively little influence on the resistance of the carbon glass thermometers. However, this last point was not really useful in this experiment since the temperature gradient is entirely due to the substrate so that it is magnetic field independent. Magnetic fields up to 5 tesla were provided by a superconducting magnet. Co/Cu, Fe/Cu and Fe/Cr multilayers were mounted with the layers perpendicular to the applied magnetic field while for Cu/Co/Cu/NiFe the field was parallel to the layers.

We define the MTEP as $\partial S(H) = S(H) - S(H=0)$. We call ΔS the total change in thermoelectric power between the F and the AF states, i.e. $\Delta S = S_F - S_{AF}$, where S_{AF} is the thermoelectric power for an antiferromagnetic arrangement and S_F is the thermoelectric power when the magnetizations of adjacent layers are parallel.

EXPERIMENTAL RESULTS

For Co/Cu, Fe/Cu and Cu/Co/Cu/NiFe, the field dependence of the MTEP is similar to that of the MR with saturation occuring at the same magnetic field. This behavior is illustrated in Fig.1a-b for a Co11Å/Cu9Å sample. The main difference between the MTEP and the MR is in their temperature dependence. With increasing temperature, the MR decreases, rapidly for Fe/Cu and Cu/Co/Cu/NiFe multilayers, less rapidly for the Co/Cu system. In contrast, the MTEP is enhanced with increasing temperature. This appears clearly in Fig.2a-c where the total change in thermoelectric power, $\Delta S=S_F-S_{AF}$, is plotted versus temperature. For Co/Cu, ΔS is positive and very small at low temperature, then becomes negative and increases steeply above 30-40K. For Fe/Cu, the MTEP is very small at low temperatures and also increases steeply above 30-40K. In Cu/Co/Cu/NiFe, the increase of the MTEP is more progressive.

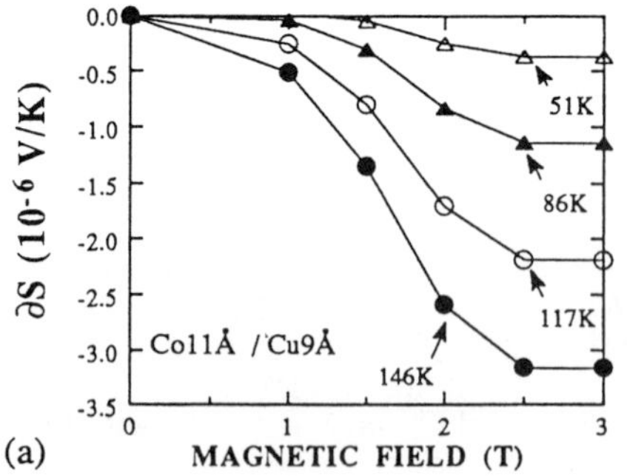

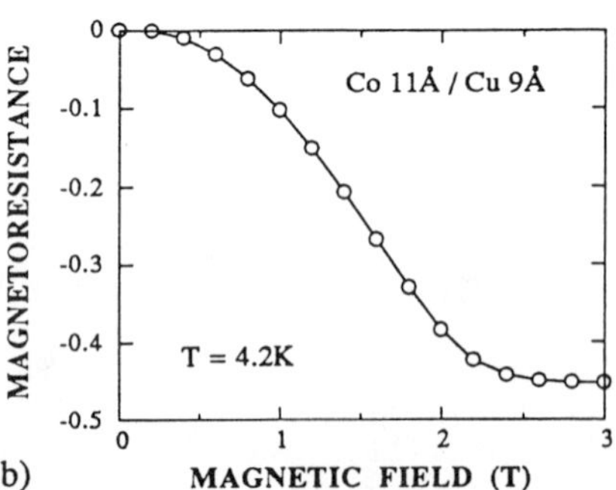

Figure 1. (a) MTEP as a function of magnetic field for a Co11Å/Cu9Å multilayer at selected temperatures, as indicated. In (b) the magnetoresistance behavior at 4.2K is shown for comparison. In both cases, the magnetic field is perpendicular to the layers; the solid lines are guides for the eyes.

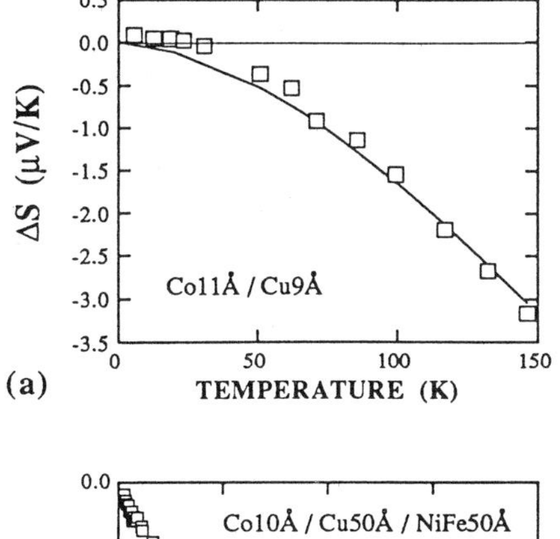

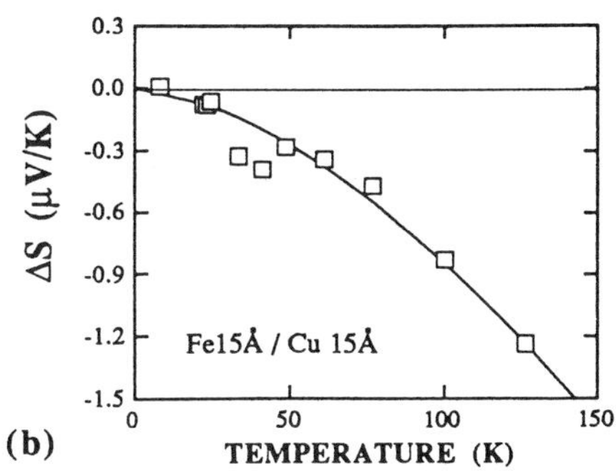

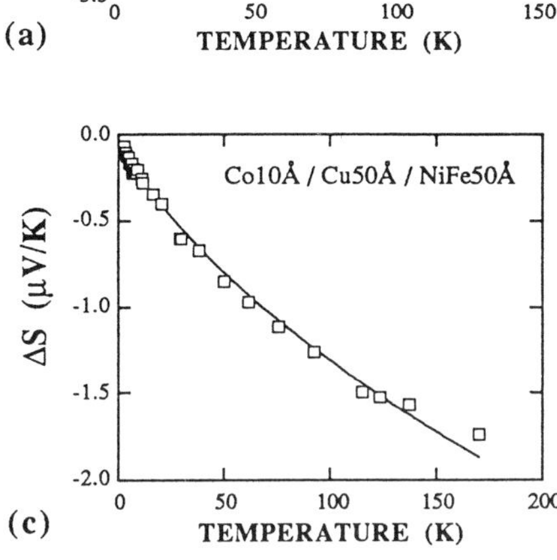

Figure 2 . Temperature dependence of the MTEP at the saturation field, ΔS, for Co11Å/Cu9Å (a), Fe15Å/Cu15Å (b) and Cu50Å/Co10Å/Cu50Å/NiFe50Å (c) multilayers. The solid curves correspond to the fit using Eq.(4), as described in the text.

The experimentally observed behavior for the MTEP is less simple in Fe/Cr than in the other studied multilayers. The MTEP ∂S of a Fe26Å/Cr9Å multilayer is plotted vs H at several temperatures in Fig.3a. The saturation field agrees with that of the MR, but the field dependence seems to be complicated by some competition between negative and positive contributions. The total change ΔS, shown in Fig.3b, is positive below 50K then becomes negative between 50K and 120K and changes again its sign at higher temperature. Experimental MTEP data from Sakurai et al [3] on a Fe27Å/Cr8Å multilayer are also reported in Fig.3b. Though in their experiment the magnetic field was parallel to the layers, we see that the results are very similar.

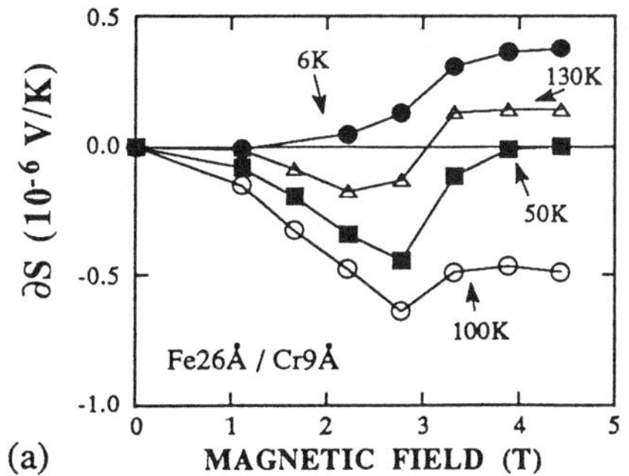

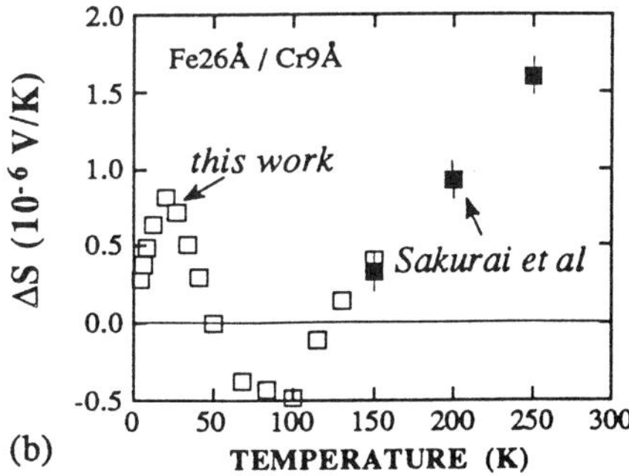

Figure 3 . Magnetic field dependence of the MTEP for a Fe26Å/Cr9Å multilayer at various temperatures (a); the inset shows the MR behavior at 4.2K. (b) MTEP at the saturation field, ΔS, plotted against temperature; the data reported in Ref.3 for a Fe27Å/Cr8Å multilayer at higher temperature are also indicated in Fig.3b.

In Fig.4a-c, we show separately the temperature dependences of S_{AF} and S_F for Fe/Cr, Cu/Co/Cu/NiFe and Co/Cu multilayers. We see that both S_{AF} and S_F are positive in Fe/Cr and negative in Cu/Co/Cu/NiFe with an almost linearly temperature variation for the latter specimen. These results are in agreement with the data of Sakurai et al [3] on similar multilayers. For Co/Cu, the thermopower is positive at low temperature then becomes negative at higher temperature. Consequently, for such multilayers there is a small temperature region where S_{AF} is positive while S_F is negative. Fig4c illustrates this feature for a Co10Å/Cu9Å multilayer. For most multilayers, we found that the sign of the thermoelectric power can be qualitatively understood combining the associated thermopower for pure metals (S is positive in pure Cu, Fe and Cr and negative in Co).

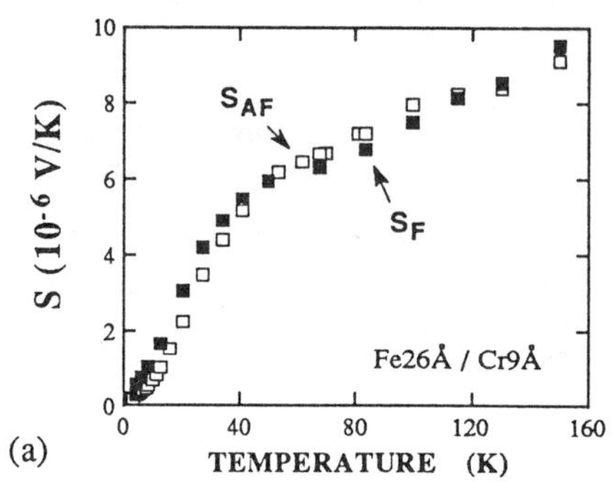

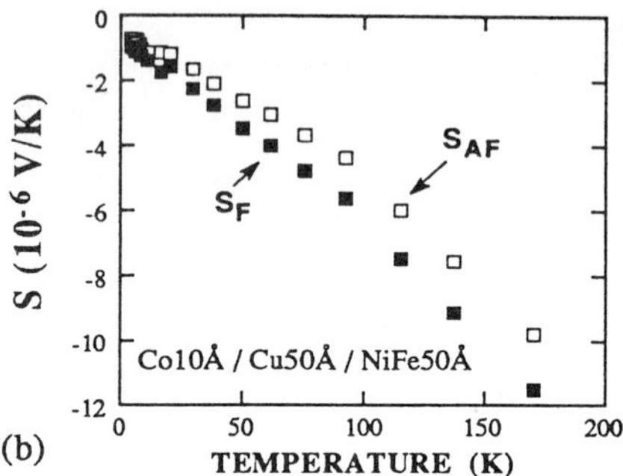

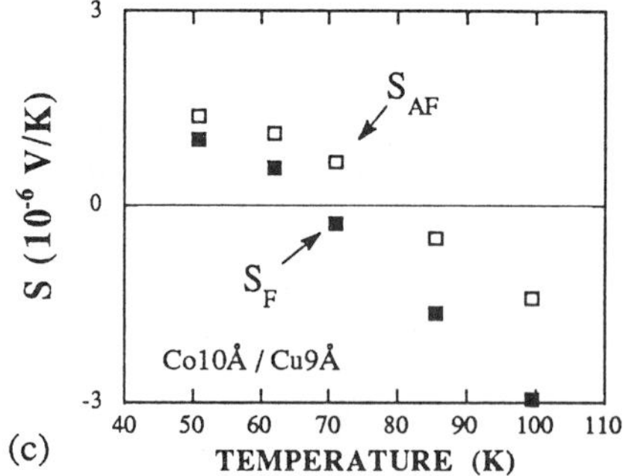

Figure 4 . Temperature dependence of the thermoelectric power coefficient in the AF (S_{AF}) and F (S_F) states for Fe26Å/Cr9Å (a), Cu50Å/Co10Å/Cu50Å/NiFe50Å (b) and Co10Å/Cu9Å (c) multilayers.

THEORETICAL MODEL

In general, the thermopower of pure metals consists of two parts : a diffusion part and a phonon-drag part. In magnetic multilayers, the phonon-drag contribution is rendered negligible due to the short electron mean free path so that the thermopower is dominated by the diffusion term. For a metal, the usual expression for the diffusion thermoelectric power is given by the Mott relation :

$$S = \frac{\pi^2 k_B^2 T}{3 e} \left(\frac{\partial \ln\sigma}{\partial \varepsilon}\right)_{\varepsilon_F} \qquad (1)$$

where the derivative term expresses the energy variation of the conductivity σ at the Fermi energy. The model we propose to account for the large enhancement of the MTEP with temperature is based on a significant contribution from electron-magnon scattering to the diffusion thermoelectric power. Previous theoretical calculations [11] have demonstrated that a diffusion thermoelectric power approaching the classical limit of k_B/e (~ 86μV/K) can be obtained in ferromagnetic systems as the two following conditions are simultaneously satisfied:

- There are two electron channels (spin ↑ and spin ↓) with different conductivities σ_1 and σ_2 (the difference is mainly due to spin dependent elastic scattering).

- In addition, the electrons are submitted to inelastic scattering by magnons with gain and loss of energy for the spin ↑ and spin ↓ electrons respectively (or annihilation and creation of magnons).

With magnon energies of the order of k_BT, this leads to $\frac{1}{\sigma}\left(\frac{d\sigma}{d\varepsilon}\right)_{\varepsilon_F} \sim \frac{1}{k_BT}$ and $S \sim \frac{\pi^2}{3}\frac{k_B}{e}$. This is much larger than the usual thermopower associated to scattering by defects, impurities or phonons with $\frac{1}{\sigma}\left(\frac{d\sigma}{d\varepsilon}\right)_{\varepsilon_F} \sim \frac{1}{\varepsilon_F}$ and $S \sim \frac{\pi^2}{3}\frac{k_B}{e}\left(\frac{k_BT}{\varepsilon_F}\right)$.

A straighforward calculation based on the Mott formula of the TEP and assuming free electrons (with an exchange gap in k-space between the Fermi surfaces) leads to [6,12] :

$$S_{\uparrow(\downarrow)} = -(+)\frac{\pi^2 k_B}{3|e|} R(T) \qquad (2)$$

R(T) is a coefficient of the order of unity varying slowly with temperature (R(T) ~ 1/lnT except at very low temperature where R vanishes exponentially). Since, in the scattering by magnons, there is gain and loss of energy for spin ↑ and spin ↓ electrons respectively, the derivative of the relaxation rate at the Fermi energy has opposite signs for the two spin directions, and consequently $S_\uparrow$ and $S_\downarrow$ have opposite signs.

Proceeding now to the case of multilayers, we call i_+ (i_-) and S_+ (S_-) the current density and TEP coefficient for the electrons with $s_z = +1/2$ (-1/2). The resulting TEP is given by :

$$S = \frac{i_+ S_+ + i_- S_-}{i_+ + i_-} \qquad (3)$$

We first suppose, for simplicity, that the only contribution to the TEP comes from electron-magnon scattering. In the ferromagnetic state, with, say, all the magnetic layers having $s_z = +1/2$ as majority spin direction, we have $i_+ \sim \rho_\uparrow^{-1}$, $i_- \sim \rho_\downarrow^{-1}$, $S_+ = S_\uparrow$ and $S_- = S_\downarrow$. If for example, $\rho_\uparrow >> \rho_\downarrow$, we expect $S_F \approx S_\uparrow$. In contrast, in the AF state and for thin enough layers, the spin ↑ and spin ↓ characters are symmetrically mixed in each channel. This leads to a balance between the positive and negative contributions to the TEP and to $S_{AF} \approx 0$. Consequently, we are expecting $\Delta S \approx -\pi^2 k_B R(T)/3|e|$ for $\rho_\uparrow << \rho_\downarrow$ (+ sign for the opposite case with $\rho_\uparrow >> \rho_\downarrow$).

We next consider a more realistic situation where both elastic scattering and inelastic electron-magnon scattering are taken into account so that the magnon contribution to the thermoelectric power must be reduced in proportion. By applying appropriate rules to calculate the resulting thermoelectric power when each set of carriers experiences distinct scattering mechanisms, one finds [12] for the total change in thermoelectric power ΔS

$$\Delta S = -\frac{\pi^2 k_B R(T)}{3|e|} \frac{(\rho_\downarrow - \rho_\uparrow)(\rho_\downarrow + \rho_\uparrow)}{\left[\rho_\uparrow \rho_\downarrow + \rho_{\uparrow\downarrow}(\rho_\uparrow + \rho_\downarrow)\right]\left[\rho_\uparrow + \rho_\downarrow + 4\rho_{\uparrow\downarrow}\right]} \rho_{\uparrow\downarrow} \qquad (4)$$

where $\rho_{\uparrow\downarrow}$ is the so-called spin-mixing resistivity, proportional to the electron-magnon scattering rate [13]. From relation (4) we see that the sign of the MTEP depends on whether the ratio between the resistivities for the two spin directions ($\alpha = \rho_\downarrow / \rho_\uparrow$) is greater or less than unity, i.e., ΔS is negative for $\rho_\downarrow > \rho_\uparrow$ and positive for $\rho_\downarrow < \rho_\uparrow$. As a function of temperature, $|\Delta S^m|$ is zero at T=0K ($\rho_{\uparrow\downarrow}=0$), increases with $\rho_{\uparrow\downarrow}$, goes through a maximum for $\rho_{\uparrow\downarrow} = 0.5\,(\rho_\uparrow\rho_\downarrow)^{1/2}$ and decreases to zero for $\rho_{\uparrow\downarrow} >> (\rho_\uparrow\rho_\downarrow)^{1/2}$. Fig.5 illustrates the theoretically predicted behavior of ΔS(T) over a wide temperature range for $\alpha << 1$ and $\alpha >> 1$ (a quadratic temperature dependence has been assumed for $\rho_{\uparrow\downarrow}$).

We note that at low temperature, where $\rho_{\uparrow\downarrow}$ is much smaller than $\rho_\uparrow$ and $\rho_\downarrow$, Eq.(4) can be reduced to the following approximate form

$$\Delta S \approx - \; 285 \; R(T) \; (\alpha - 1) \; \frac{\rho_{\uparrow\downarrow}}{\rho_\downarrow} \qquad (5)$$

where ΔS is expressed in $\mu V/K$.

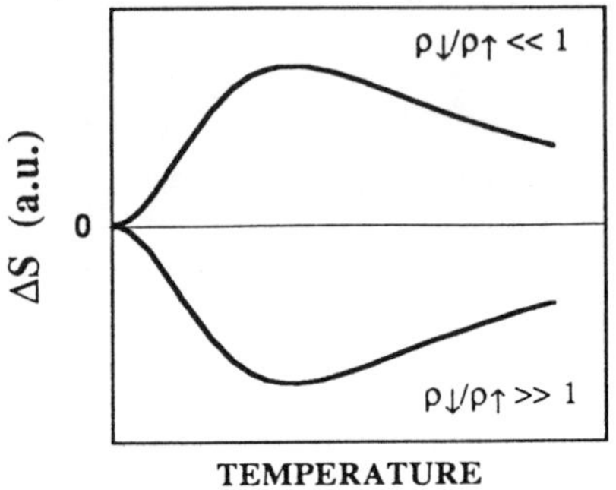

Figure 5 . Variation of ΔS with temperature as calculated from relation (4) for different signs of $(\alpha$-1).

DISCUSSION

We turn now to an examination of experimental MTEP data in the light of the model developped in the previous section. We suggest that MTEP at not too low temperature is dominated by a major contribution from electron-magnon term. At very low temperature when the magnon contribution vanishes there should be additional contributions arising from spin dependent elastic scattering. Our discussion will mainly focuse on the temperature dependence of the total MTEP, ΔS, and on its sign.

a) MTEP due to spin dependent electron-magnon scattering

We have fitted the MTEP data of Figs.2a-c to the above mentioned expression (4) for the contribution from electron-magnon scattering to the diffusion thermoelectric power. For Fe/Cu and Co/Cu, the experimental ΔS vs T variation can be reproduced assuming a T^2 variation for $\rho_{\uparrow\downarrow}$ (see Figs 2a and b) in agreement with the existing data on $\rho_{\uparrow\downarrow}$ in bulk ferromagnets [13]. The fit gives $\rho_{\uparrow\downarrow}/T^2 \sim 4.9 \; 10^{-5} \; \mu\Omega cm/K^2$ for Co/Cu and $\rho_{\uparrow\downarrow}/T^2 \sim 8.6 \; 10^{-5} \; \mu\Omega cm/K^2$ for Fe/Cu which is also of the same order of magnitude with the values previously reported on bulk ferromagnets [13]. The larger $\rho_{\uparrow\downarrow}$ in Fe/Cu is in agreement with a smaller value of the D constant of the magnon dispersion curve in Fe (in agreement also with the more pronounced temperature dependence of the MR in Fe/Cu). In contrast, for the Cu/Co/Cu/NiFe multilayer a good fit was obtained assuming a linear T dependence for $\rho_{\uparrow\downarrow}$ (see Fig.2c) . This different temperature dependence of ΔS in Cu/Co/Cu/NiFe is an interesting result. In bulk materials, the resistivity $\rho_{\uparrow\downarrow}$ of disordered ferromagnetic systems includes an additional contribution from "incoherent" magnon scattering proportional to $T^{3/2}$ [14]. In permalloy layers, as the main exchange scattering should arise from the disordered system of Fe moments, the onset of incoherent terms in $\rho_{\uparrow\downarrow}$ might account for the different temperature dependence of ΔS.

We point out that in the temperature range investigated in this work (with $\rho_{\uparrow\downarrow}$ relatively small compared to $\rho_\downarrow$, $\rho_\uparrow$), the initial increase of $\Delta S(T)$ is almost proportional to the spin-mixing term $\rho_{\uparrow\downarrow}$. Using our results for $\rho_{\uparrow\downarrow}$ and for the resistivity of our samples, a maximum well above room temperature is expected.

In Fe/Cr, ΔS becomes positive and starts increasing at about 120K. In the data of Sakarai et al [3] on Fe27Å/Cr8Å, this increase continues up to room temperature. The positive sign is predicted by Eq.4 for $\rho_\uparrow > \rho_\downarrow$ which is actually in agreement with the strong resonant scattering of the spin ↑ conduction electrons by Cr atoms in Fe ($\rho_\downarrow / \rho_\uparrow \approx 0.17$ in bulk alloys). This is also in agreement with the theoretical arguments developed for multilayers by Inoue et al [15]. In contrast, the negative sign of the MTEP for Co/Cu, Fe/Cu and in Cu/Co/Cu/NiFe corresponds to $\rho_\uparrow < \rho_\downarrow$, as expected from electronic structure arguments [9,16].

b) MTEP due to spin dependent elastic scattering

At low temperature, the MTEP in Co/Cu becomes positive and very small (see Fig.2a). As the contribution from electron-magnon scattering vanishes at low temperatures, the positive residual MTEP is probably due to the contribution from elastic scattering. From Eq.3, and as for the magnon term, the crossover from spin-mixed channels at low field to separated spin ↑ and spin ↓ channels at high field should also change the contribution from elastic scattering to the thermoelectric power. This MTEP term has been calculated by Inoue et al [16] for TM/Cu multilayers and is expected to be proportional to T while its sign depends on the nature of the magnetic layers. The predicted sign is positive for Co/Cu which is in agreement with our sign for ΔS below 20K. A negative sign is predicted for Fe/Cu [16]. In our measurements on Fe/Cu, the magnetic field-induced emf signals below 20K fall almost within the experimental uncertainties (see results from Fig. 2b) so that it was not possible to determine accurately the MTEP values. So, at least for Co/Cu, it is possible to understand the small positive MTEP observed at low temperature in terms of a positive "elastic term", the much larger negative "magnon term" appearing above 40K approximately.

The behavior of the MTEP for Fe/Cr is more complex than for Co/Cu and Fe/Cu. Fig.3a suggest some competition between a negative contribution (generally predominant at low field) and a positive contribution saturating at the saturation field of the magnetoresistance. This complexity is also reflected by the two changes of sign observed in the low temperature range, as shown in Fig. 3b. We believe that this behavior is due to additional spin dependent scatterings by the magnetic moments induced on the Cr atoms near the Fe/Cr interfaces.

CONCLUSIONS

The thermoelectric power of magnetic multilayers (Co/Cu, Fe/Cu, Fe/Cr and Cu/Co/Cu/NiFe) has been investigated as a function of temperature and magnetic field. Most specimens were found to exhibit pronounced magnetothermoelectric power (MTEP) effects correlated with their giant negative magnetoresistance. However, whereas the magnetoresistance is a decreasing function of temperature, the MTEP displays the opposite behavior, i.e. the MTEP increases rapidly with temperature. In the high temperature range, the MTEP is negative for Co/Cu, Fe/Cu and Cu/Co/Cu/NiFe and positive for Fe/Cr.

Our theoretical model based on the contribution from electron-magnon scattering to the diffusion thermoelectric power explains the rapid increase of MTEP with temperature. The experimental results can be accounted for with reasonable assumptions for the electron-magnon resistivity (magnitude and temperature dependence of $\rho_{\uparrow\downarrow}$). The sign of the MTEP corresponds to $\rho_\uparrow$ smaller than $\rho_\downarrow$ for Fe/Cu, Co/Cu and Cu/Co/Cu/NiFe and $\rho_\uparrow$ larger than $\rho_\downarrow$ for Fe/Cr, as expected from electronic structure arguments.

ACKNOWLEDGEMENTS

This work was partly funded by the United States National Science Foundation under grant No. DMR-88-013287 and by the Michigan State University, Center for Fundamental Materials Research. The work performed in Louvain-la-Neuve was in the framework of the programme "Action de Recherche Concertée" sponsored by the Belgian State (Ministry of Scientific Policy). L.P. is Research Associate of the National Fund for Scientific Research (Belgium).

REFERENCES

[1] M. N. Baibich, J. M. Broto, A. Fert, F. Nguyen Van Dau, F. Petroff, P. Etienne, G. Creuzet, A. Friederich and J. Chazelas, Phys. Rev. Lett. 61 (1988) 2472
[2] G. Binasch, P. Grunberg, F. Saurenbach and W. Zinn, Phys. Rev. B39 (1989) 4828
[3] J. Sakurai, M. Horie, S. Araki, H. Yamamoto and T. Shinjo, J. Phys. Soc. Jpn 60 (1991) 2522
[4] M. J. Conover, M. B. Brodsky, J. E. Mattson, C. H. Sowers and S. D. Bader, J. Magn. Magn. Mat. 102 (1991) 15
[5] H. Sato, I. Sakamoto and C. Fierz, J. Phys. : Condens. Matter.3 (1991) 9067
[6] L. Piraux, A. Fert, P.A. Schroeder, R. Loloee and P. Etienne, J. Mag. Mag. Mat. 110 (1992) L247
[7] S. S. P. Parkin, N. More and K. P. Roche, Phys Rev. Letters 64 (1990) 2304
[8] D. H. Mosca, F. Petroff, A. Fert, P. A. Schroeder, W. P. Pratt and R. Loloee, J. Magn. Magn. Mat. 94 (1991) 1
[9] F. Petroff, A. Barthélémy, D. H. Mosca, D. K. Lottis, A. Fert, P. A. Schroeder, W.P. Pratt, R. Loloee and S. Lequien, Phys. Rev. B44 (1991) 5355; for the preparation and properties of Co/Cu see also S. S. P. Parkin, Z. G. Li and D. Smith, Appl. Phys. Lett. 58 (1991) 2710
[10] S. Pizzani, F. Baudelet, D. Chanderris, A. Fontaine, H. Magnan, J. M. Georges, F. Petroff, A. Fert, R. Loloee and P.A. Schroeder, Phys. Rev. B (in press)
[11] T. Kasuya, Prog. Theor. Phys. 22 (1959) 227; I. Y. Korenblit and Y. P. Lazarenko, Sov.Phys. JETP 33 (1971) 837
[12] A. Fert and L. Piraux, to be published
[13] I. A. Campbell and A. Fert, Ferromagnetic Materials, Vol.3 (Edited by E. P. Wohlfarth, North Holland, 1982), p.747; A. Fert and I. A. Campbell, J. Phys.F 6 (1976) 849
[14] D.L. Mills, A. Fert, I.A. Campbell, Phys. Rev. B4 (1971) 196
[15] J. Inoue, A. Oguri and S. Maekawa, J. Phys. Soc. Jpn 60 (1991) 376
[16] J. Inoue, H. Itoth and S. Maekawa, preprint

MAGNETORESISTANCE INTERPRETATION AND MAGNETIZATION PROCESSES IN Co-Re AND Co-Cu MULTILAYERS

P.P.Freitas[1,2], I.G.Trindade[1], L.V.Melo[1,2]
N.Barradas[3], and J.C.Soares[3]

1-INESC,Rua Alves Redol,9,1000 Lisboa-Portugal
2-IST,Av. Rovisco Pais,1000 Lisboa-Portugal
3-CFNUL,Av.Prof.Gama Pinto 2,1699 Lisboa-Portugal

1-INTRODUCTION

Co-Re and Co-Cu multilayers show oscillatory exchange coupling together with magnetoresistance enhancement when the magnetic layers are antiparallel to one another[1,2]. The main mechanism proposed to account for the observed magnetoresistance is spin-dependent scattering either at the interfaces between the magnetic and the non-magnetic layers or within the ferromagnetic layers[3]. The magnetoresistance arises from the assymetry of the scattering rates for the spin-down and spin-up channels. If only interface scattering is considered this assymetry can be linked with the known different values of residual resistivity for non-magnetic impurities in a ferromagnetic matrix. For example, if we consider Co-Re multilayers, then interface spin-dependent scattering would arise from the fact that $\gamma = \rho_{\downarrow} / \rho_{\uparrow} << 1$ for Re impurities in a Co matrix. Since this is an interface effect, we expect the effect to get smaller as the thickness of the ferromagnetic layer increases and the relative weight of the interface gets reduced. If we take bulk scattering only, then the assymetry in scattering rates means different conductivities and mean-free paths for the spin-up and spin-down channels within the ferromagnetic layer. From the nature of band structure in transition metals and their alloys, this assymetry is strong in hcp or fcc Co, or in Permalloy. It is very weak in bcc Fe. As the thickness of the ferromagnetic layer increases relative to the total multilayer thickness, we expect the MR effect first to increase as the number of scattering centers increases but then to decrease as the thickness of the magnetic layer gets larger than the mean-free path in the layer. The regions of the ferromagnetic layer which are at a distance from the interface greater than this mean-free path will not contribute to the magnetoresistance. Checks on the effect of interface scattering can be done by adding very thin third element layers at the interfaces without changing the structure of the multilayer. Checks for spin-dependent bulk scattering can be done by adding small amounts of Fe to Co-based multilayers, or by looking for a maximum in the MR dependence versus thickness of the ferromagnetic layer.

We use a semi-classical model to solve for the electrical resistivity of the multilayer structure, extending the Fuchs-Sondheimer formalism based on Boltzmann equation, to a multilayer structure with spin-dependent scattering either at the interfaces or within the magnetic layers[4].

This model was already applied to Fe-Cr[5] and Co-Cu [6], and spin-valve structures[7]. In this paper we report a study of the MR of two different systems having the same ferromagnetic layer(Co) but different non-magnetic spacers(Re or Cu). We used the semi-classical model extended to incorporate both interface and bulk spin-dependent scattering. Both these systems should present similar contributions from bulk scattering from the Co layers, but quite different interface scattering since $\gamma <<1$ for Re, while $\gamma >>1$ is expected for Cu. In the case of Co-Re, interface scattering was further studied by adding a third element (Fe,Ni,Mo,Cr) at the interface. Both multilayer systems were prepared under the same experimental conditions and using the same deposition system.

The second part of the paper deals with the study magnetization processes in antiferromagnetically coupled magnetic multilayers . The competitive role of the anisotropy, antiferromagnetic exchange, and Zeeman terms have been analised by comparing the experimental magnetization curves with predicted phase diagrams obtained from total energy minimization. In Fe/Al/Fe, Fe/Au/Fe, Co/Cu/Co sandwiches[8,9] a biquadratic exchange energy term was introduced to account for the magnetization behavior. In Fe/Cr bilayers deposited on Fe wiskers both components of the magnetization, parallel and perpendicular to the applied field were monitored using suitable magneto-optic techniques, and the results compared with the phase diagram obtained again by total energy minimization[10,11]. In this paper we report a study of magnetization reversal processes in $(Co\text{-}Re)_{xn}$ multilayers. We study in particular structures with two, three and five antiferromagnetically coupled Co layers. Both components of the magnetization parallel and perpendicular to the applied field are measured by magneto-optic techniques. We find two types of hysteresis cycles respectively stair-case cycles where magnetization reversal occurs by spin-flop or domain-wall propagation processes, or s-like cycles where the magnetizations of all three individual layers rotate simultaneously but gradually into the applied field direction.

The experimental results are compared with a Monte Carlo (MC) simulation of the magnetization cycle using the XY model. The behavior of each individual layer as well as the spin-configurations in the remanent state and other applied fields are discussed. Previous MC studies of magnetic multilayers were performed using the Ising model and addressed two different problems. The first study dealt[12] with the analysis of the phase diagram and critical behavior of a magnetic bilayer with competing coupling constants J1 and J2 respectively the coupling constants within the ferromagnetic layers and between the ferromagnetic layers. In the second and more recent study[13] a $(Ni\text{-}Fe)_{xn}$ multilayer was discussed starting from the atomic exchange integrals between the constituent atoms. The nature of the magnetic coupling between the Fe layers was found to depend on the number of atomic Ni planes in the spacer layer.

2-EXPERIMENTAL METHOD

The Co-Re and Co-Cu multilayers were prepared in a load-locked high-vacuum sputtering system under conditions previously described[1,14]. Base pressure is $5\text{x}10^{-8}$ Torr and deposition rates were monitored by in-situ quartz-crystals calibrated by Rutherford Backscatering(RBS). Co was deposited at 2 to $2.5\AA/s$ while Re and Cu were deposited at a rate of 0.4 to $0.5\AA/s$. Full structural information on the Co-Re multilayers was published elsewhere. Fig.1a) shows RBS results for two multilayer structures consisting of three Co layers 30Å thick separated by Re spacers 21Å and 7Å thick respectively. A glazing angle geometry was used that allowed the resolution of the individual Re spacers. The solid lines through the data are theoretical fits assuming bulk densities for Co and Re and no interface mixing. The Co-Re superlattices grown on the Re buffer are hcp and highly textured with the [001] axis perpendicular to the plane of the sample. We used either Corning glass or Si[111] or Si[100] substrates. Perturbed Angular Correlation measurements (PAC)indicate a higher content of stacking faults in films deposited on glass substrates. This technique was also used to determine the structure of Co-Cu multilayers. We found that a for Co thicknesses larger than 30Å both fcc and hcp phases coexist, apart from a large number of stacking faults(Fig.1b). All samples were deposited at room temperature.

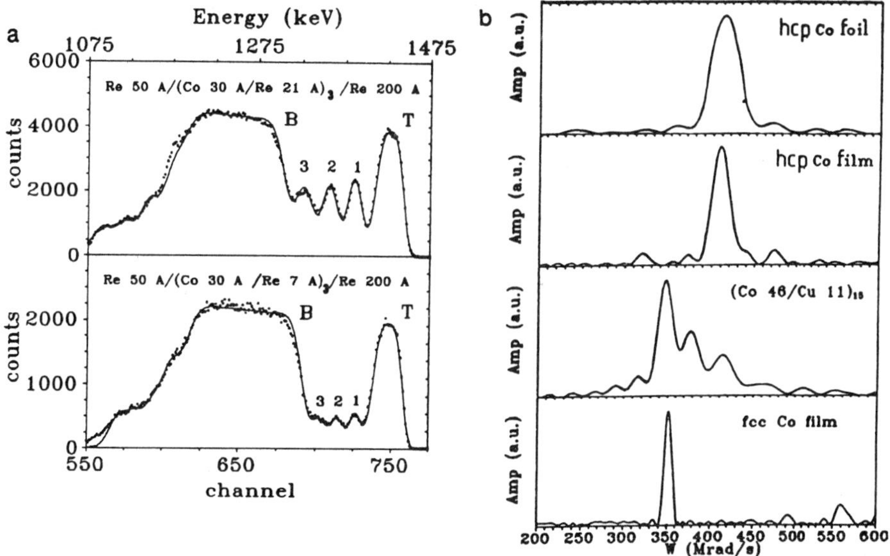

Fig.1-a) RBS results and theoretical fit(solid line) for two $(Re_{t_{Re}}Co_{38\AA})\times 3$ structures with t_{Re}=21Å and 5Å respectively. The incident beam was at an 80 degree angle to the normal to the plane of the sample. The good agreement between experimental and calculated spectra means no interface mixing is occurring. b) PACS results showing hcp structure for a Co film on top of a Re buffer, while fcc is predominant for the Co in a Co-Cu superlattice grown on a Fe buffer.

The magnetization behavior was analised using a differential magneto-optics set-up in the transverse Kerr effect geometry where the applied magnetic field is perpendicular to the plane of incidence[15,16] . To measure the magnetization component parallel to the applied field we use p polarized light(θ_p=0, measured with respect to the plane of incidence) and the analiser aligned with the polarizer(θ_a=0). We are indeed just making a reflectivity measurement which is not sensitive to the analiser angle. This transverse Kerr effect measurement then gives the expected hysteresis cycle for a ferromagnetic film with in-plane magnetization. The magnetization component perpendicular to the applied field direction is obtained simply by rotating the analiser to a position close to extinction ($\theta_a=\pi/2$).

This perpendicular component can also be observed with s-polarized light and with the analiser close to extinction ($\theta_p=\pi/2, \theta_a=0$).

3-EXTENSION OF THE CAMLEY AND BARNAS MODEL

Our approach was that of Camley and Barnas[4] generalizing their initial model to include spin-dependent scattering in the ferromagnetic layers. In the scope of the model this means having different mean-free paths for the spin-up and spin-down electrons in the magnetic layer. The assymetry in bulk spin-dependent scattering is given by the parameter $\beta=\lambda_\uparrow/\lambda_\downarrow$. The assymetry in the interface transmission coefficients $T_{\uparrow,\downarrow}$ is measured by the parameter $\alpha= D_\downarrow/D_\uparrow$ where $D_{\uparrow,\downarrow}=1-T_{\uparrow,\downarrow}$. Here we assumed perfect transmission at the internal interfaces.

We first solved the Boltzmann equation for a five layer structure to simulate the behavior of real structures with two magnetic layers, a spacer, a buffer and a top protective layer. In this case, diffuse scattering is introduced in the buffer-substrate and top-air interfaces, and perfect transmission is assumed for all internal interfaces.

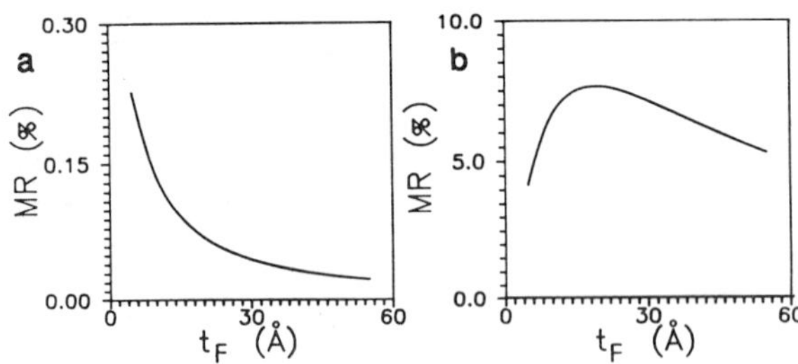

Fig.2-Calculated MR dependence on the thickness of the ferromagnetic layer t_F for an infinite antiferromagnetically coupled superlattice using interface scattering only (a) and assuming bulk spin dependent scattering only (b).

The infinite superlattice is simulated just by taking the three internal layers(F/NM/F) and assuming specular reflection in the middle plane of the top and bottom magnetic layers. In this case the external layers play no role.

Finite structures with n bilayers of Co-Re or Co-Cu were also studied by iteratively solving the boundary conditions in all the interfaces.

Fig.2a) shows typical results for the MR dependence of an infinite superlattice on the thickness of the magnetic layer, for interface scattering only. We used $\lambda_F=\lambda_\uparrow+\lambda_\downarrow=49$Å Å, $\lambda_{NM}=24$Å,$\beta=1$, $\alpha=0.4$, and $D_\uparrow=0.13$. As we are considering interface spin-dependent scattering only, when the thickness of the ferromagnetic or non-magnetic layers increase, the MR decreases due to a simple dilution effect.

Fig.1b) shows equivalent results but for bulk scattering only. We used $\beta = 6$ with $\lambda_\uparrow$ =42Å, $\lambda_\downarrow$=7Å, α=1, $D_\uparrow$=0.13, λ_{NM}=24Å. The MR dependence on the thickness of the non-magnetic layer is similar, but a quite different behavior occurs as a function of the thickness of the magnetic layer. As the thickness of the ferromagnetic layer increases, the bulk spin-dependent scattering contribution to the MR first increases since the number of spin-dependent scattering centers in the total sample is proportional to the relative volume of magnetic material. A maximum MR occurs close to 30Å and the MR starts to decrease due to a current shunting effect through regions of the ferromagnetic layer that are away from the interface by a distance greater than the electron mean-free path in the ferromagnetic layer, and that therefore do not contribute to the magnetoresistance[7].

4-MONTE CARLO SIMULATION

A Monte Carlo simulation was performed using the XY model and the Metropolis algorithm[17]. For the results shown in this article we used a hcp lattice with 48 spins in each hcp layer. Fig.3a) shows the cell structure used to simulate the behavior of antiferromagnetically coupled hcp bilayers. In this very simple model, each experimental ferromagnetic layer of thickness d is simulated by two basal planes (A_1/B_1,A_2/B_2) that interact ferromagnetically with each other. The antiferromagnetic interaction between two experimental magnetic layers is represented by the interaction between the interfacial basal planes B_1 and A_2.

The Hamiltonian reflects the ferromagnetic exchange interaction J between spins in each ferromagnetic layer (this involves the two basal planes for the hcp lattice), the antiferromagnetic exchange interaction between spins belonging to different ferromagnetic layers, J_{af} and the Zeeman and crystal field interaction terms . No magnetostatic term was used. We consider only nearest-neighbors for the exchange interactions. The energy per spin at each site of the lattice takes the form,

$$E_i = -J\vec{S_i}.\sum \vec{S_{nn}} - J_{af}\vec{S_i}.\sum' \vec{S_{nn}} - m_s\vec{S_i}.\vec{H} - D(\vec{S_i}.\hat{n})^2 \qquad (1)$$

where $\vec{S_i} = \cos\theta_i \hat{u_x} + \sin\theta_i \hat{u_y}$ and θ_i is the angle between the moment at lattice site i and the magnetic field. D is the local anisotropy energy and $\hat{n}$ the unit vector representing the easy axis direction. $\vec{S_{nn}}$ corresponds to the nearest neighbor spins, and ms is the atomic moment. $\sum$ runs over nearest neighbors on the ferromagnetic layer and $\sum'$ runs over nearest neighbors in different ferromagnetic layers.

Consider for the moment J_{af}=D=0. For Co atoms on a hcp structure $J_{Co-Co} \approx$ 1.4e^{-14}erg[18]. Comparing the Zeeman and the exchange terms in the Hamiltonian this would lead to a saturation field in excess of 10T which is obviously quite different from the observed experimental behavior (our thin Co films saturate at a couple of hundred Oe). This occurs since in actual ferromagnetic films, magnetization processes involve the coherent rotation of large groups of spins rather than incoherent and individual spin rotations.

We then introduce a cell consisting of 10^4 spins in such a way that the Zeeman term leads to saturation fields of hundred of Oe. The volume of such a cell is about $1x10^{-19}cm^3$ and for a $10\mathring{A}$ thick film this corresponds to a domain $110\mathring{A}$ wide. This unit volume can be associated with the characteristic magnetic volume being flipped at each Monte Carlo step. Equation 1 then becomes

$$E_{cell} = -J'\vec{S}_i.\sum\vec{S}_{nn} - (J'_{af}/d)v\vec{S}_i.\sum\vec{S}_{nn} - M_s v\vec{S}_i.\vec{H} - Kv(\vec{S}_i.\hat{n})^2 \tag{2}$$

where we have introduced the actual measured antiferromagnetic exchange interaction per unit area betweeen two ferromagnetic layers J'_{af}, the saturation magnetization of the ferromagnetic layer Ms, the anisotropy energy per unit volume,K, and the ferromagnetic exchange between two cells J', and the film thickness d. The effective antiferromagnetic interaction between two cells is $J''_{af}=J'_{af}\times v/d$. For a fixed cell volume v, J' is chosen in order to produce square hysteresis cycle with a certain coercive field. The antiferromagnetic interaction is now turned on and the Monte Carlo simulation compared with the experimental results. The remanent state spin configurations depend on the interplay between J',J'_{af} and K. The saturation field depends essentially in the interplay between the Zeeman term, J'_{af} and K.

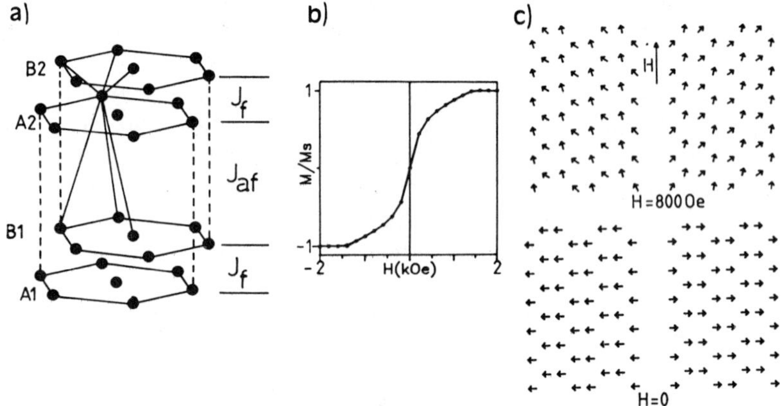

Fig.3-a) Cell structure for the hcp lattice used in the MC simulations. Notice the ferromagnetic exchange interactions inside the ferromagnetic layers, but the antiferromagnetic interaction between the two interface layers. b) Calculated hysteresis cycle obtained by the MC simulation for an antiferromagnetically coupled bilayer with J'_{af}=-0.163 erg/cm^2, K=0 and d=$38\mathring{A}$ and J'/J''_{af}=0.4. c) Spin configurations for the two layers in the remanent state and for H=800 Oe.

Fig.3b) shows the simulated hysteresis cycle obtained for an antiferromagnetically coupled hcp bilayer using the following parameters, M_s=1432emu/cm^3, J'_{af}=-0.163erg/cm^2, and $J'/\ J''_{af}$=0.4, d=$38\mathring{A}$. We also show the spin configurations for the remanent state, and for an applied field of 800 Oe . Notice that at the remanent state the magnetization is aligned at 90 degrees to the field direction(Fig.3c).

5-RESULTS

Fig.4a) shows experimental MR and resistivity dependences on the Re thickness for Co-Re superlattices with the following structure, glass/$Re_{50Å}$/$Co_{20Å}$ /($Re_{t_{Re}}$ $Co_{20Å}$)x15/$Re_{50Å}$. The MR was measured with the applied magnetic field in-plane and perpendicular to the current . The solid line through the MR data points corresponds to the calculated MR for antiparallel alignement of the Co layers using interface scattering only. The parameters used for simulation were: λ_{Co}=76Å, λ_{Re}=24Å, $\beta=\lambda^{\uparrow}_{Co}/\lambda^{\downarrow}_{Co}=1$, $D_{\uparrow}=0.3$, $\alpha=D_{\downarrow}/D_{\uparrow}=0.3$. The Co mean-free path corresponds to a room-temperature resistivity of 16±1 $\mu\Omega$.cm for 500Å thick films. The Re mean-free path is obtained from the ratio of the measured room-temperature resistivities of Co and Re with ρ_{Re}=45± 5 $\mu\Omega$.cm (500Å thick film).

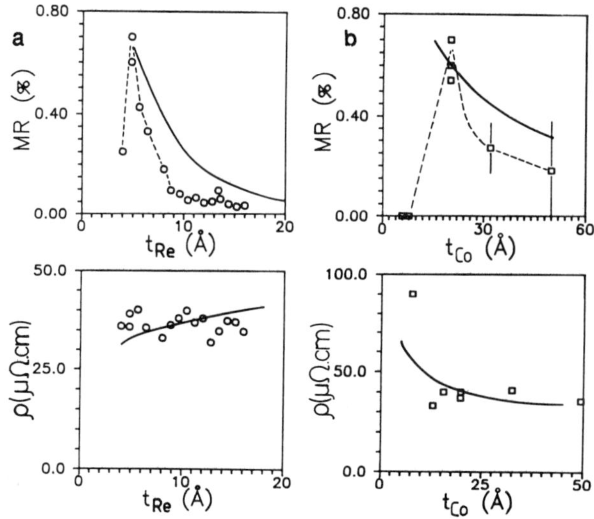

Fig.4-Magnetoresistance and resistivity dependences on Re (a) and Co (b) thicknesses for Co-Re superlattices with the following structure, glass/50ÅRe/$Co_{t_{Co}}$/($Re_{t_{Re}}Co_{t_{Co}}$)x15/50ÅRe. All data is taken at room temperature. The solid lines through the data are theoretical fits with the parameters indicated in the text.

The solid line through the resistivity data corresponds to the calculated resistivity per trilayer(Co1/Re/Co2) normalized to the experimental resistivity for the superlattice with t_{Re}=17Å. The essentially constant value observed for the resistivity of the Co-Re superlattices as t_{Re} varies from 5Å to 18Å is due to the very high resistivity of the Re layers when compared with the Co layers.

Fig.4b shows the MR and resistivity dependences on Co thickness for glass/$Re_{50Å}$ /$Co_{t_{Co}}$/($Re_{5Å}$ $Co_{t_{Co}}$)x15/$Re_{50Å}$ superlattices. The MR decrease with increasing Co thickness is due to current shunting through the thicker Co layers. For Co thicknesses below 8Å, the measured MR value is below $1x10^{-4}$.

We think this is evidence for pinholes or loss of continuity in the Co layer. The error bars for t_{Co}=30 and 50Å arise from the uncertainty in substracting the large anisotropic magnetoresistance contribution since the remanent state domain configuration is not known. For Co thicknesses below 20Å the spin-valve effect dominates for antiferromagnetically coupled films. The strong decrease of the MR vs t_{Co} cannot be accounted for if we include bulk scattering with $\beta \approx 5$ or larger as a typical value for Co. We therefore conclude that in the Co-Re multilayer system interface scattering has a dominant role as the spin-dependent scattering mechanism giving rise to the enhanced MR. The set of fitting parameters found for the Co-Re multilayers was then used to calculate the MR for sandwiches and structures consisting of a finite number of bilayers. Fig.5 shows the calculated MR for a set of $Co_{20\AA}(Re_{5\AA}Co_{20\AA})_{\times n}$ multilayers with n varying from 1(trilayer without buffer and top layers) to 8.

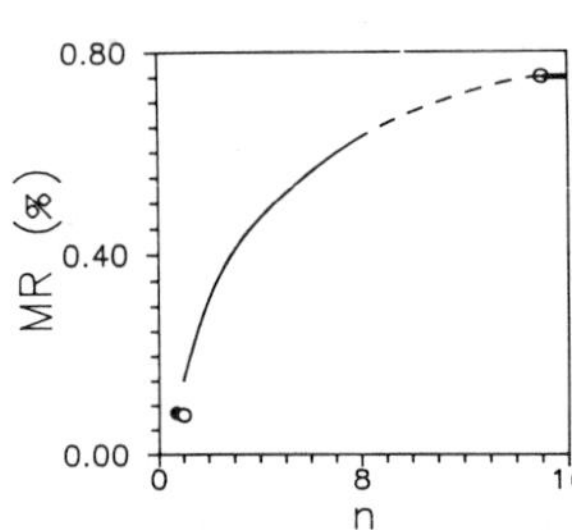

Fig.5-Calculated MR dependence on the number of (Co-Re) bilayers n for a multilayer with the structure $Co_{20\AA}(Re_{5\AA}Co_{20\AA})_{\times n}$(no top or buffer layers). Two experimental data points are shown one for a n=15 structure, and another for a n=1 structure, both with 50Å Re buffer and top layers. The solid dot corresponds to the calculated MR for n=1 but including buffer and top layers.

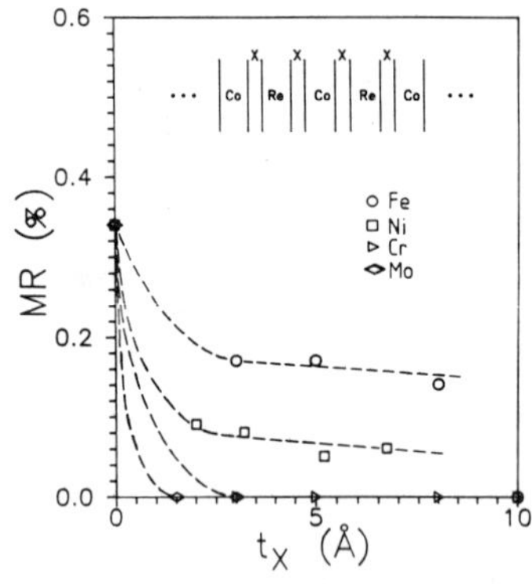

Fig.6-Magnetoresistance dependence on the thickness of a third element x=Fe,Ni,Cr or Mo added at each Co-Re interface. While Cr and Mo kill the antiferromagnetic coupling, the data with Fe or Ni can be explained just by changing the interface transmission coefficients.

The solid line through points of abcissa 15 and 16 correspond to the infinite superlattice. Two experimental data points are included for comparison(open circles). One for a multilayer with n=15 and another for n=1, but in both cases with Re top and buffer layers 50Å and 150Å thick respectively. This last sandwich structure has a MR =0.083% at room temperature. If we just take the calculated solid line(no Re buffer , no Re top) we overshoot and predict a MR value of 0.16%. If we solve the Boltzmann equation for this 5 layer structure using diffuse scattering at the outer interfaces and the same parameters used in the fitting of the Co-Re multilayers we obtain $(\rho_{\uparrow\uparrow}$-$\rho_{\uparrow\downarrow})/\ \rho_{\uparrow\uparrow}$=0.08% in striking agreement with the experimental result(solid circle).

The study of interfacial spin-dependent scattering in these samples was further extended by adding a third element X=Fe, Ni, Cr or Mo at the Co-Re interfaces as a very thin layer of thickness t_X. The observed MR dependence on t_X is shown in Fig.6. We found that adding 3Å of Cr (about 1ML) or 1.5Å of Mo (about 0.5 ML) at each interface is enough to completely destroy the antiferromagnetic coupling between the ferromagnetic layers, drastically reducing the MR value. However this coupling is still present when 3Å of Fe(1ML) or 2Å of Ni(0.6ML) are added at each interface. Furthermore the AF coupling persists even for Fe or Ni thicknesses up to 10Å. In this case , these layers are exchanged coupled to the Co layers and the antiparallel alignement persists throughout the magnetic layer.

The lower MR value can be attributed to a smaller assymetry in the spin-dependent transmission coefficients for Ni/Re and Fe/Re interfaces when compared with Co/Re interfaces. In all these cases, the ratios $\rho_\downarrow/\rho_\uparrow$ for the respective impurities in Ni, Fe, or Co matrices range between 0.3 and 0.4[19]. It is also known that $\beta \approx 1$ for Fe , and therefore if bulk spin-dependent scattering in the magnetic layers were important , adding 10Å of Fe should significantly decrease the MR. This is not observed. The weak t_X dependence of the MR for t_X thicknesses above 3Å can be fully explained as a dilution effect only. These results confirm our analysis of the Co-Re multilayer data reinforcing the argument in favour of interfacial spin-dependent scattering.

Figs.7a) and b) show the experimental MR and resistivity dependence on Cu thickness for Co-Cu multilayers with the following structure, glass/$Fe_{50Å}$ /($Co_{11Å}$ $Cu_{t_{Cu}}$)$_{\times 16}$. The Fe buffer imposes fcc texture and was found to improve the structural quality of the multilayer growth[2]. A MR effect of 45% was obtained at room temperature for a 30-bilayers sample deposited on a glass substrate with t_{Co}=11Å and t_{Cu}=10Å. We observed MR maximum for t_{Cu} near 10, 22 and 32Å corresponding to periodic antiferromagnetic coupling between the Co layers. Based on a measured Cu room temperature resistivity of 4.5±0.5 $\mu\Omega$.cm for a 500Å thick film we predict a mean free path for copper λ_{Cu}=320Å, and we kept the value of λ_{Co}=76Å as used for Co-Re samples. The very large MR value and the fact that Cu stands to the right of Co in the same period of the periodic table requires $\alpha \gg 1$. Using these parameters and interface scattering only we failed to get a good fit for the experimental MR data. Bulk scattering is then added to account for the observed Cu and Co thickness dependences. The parameters used for the best simulation were β_{Co}=6, $\lambda^{\uparrow}_{Co}$=60Å, λ_{Cu}=420Å, $D^{\uparrow}$=0.15 and α=5. The solid line through the MR data corresponds to a fit to antiferromagnetically coupled samples. The solid line through the resistivity data corresponds to the calculated resistivity per trilayer (Co1/Cu/Co2) normalized to the experimental resistivity for the superlattice with t_{Cu}=40Å.

In Fig.7c) we show the separate contributions of interface scattering and of bulk scattering to the total MR. From these results we conclude that both contributions are of the same order. One Co-Cu interface has a contribution to the MR equivalent to 8Å of bulk Co.

The MR and resistivity dependences on t_{Co} are shown in Figs.7d) and e) for a set of glass/$Fe_{50Å}$ /($Cu_{22Å}$$Co_{t_{Cu}}$)$_{\times 16}$ superlattices. Notice that we have moved towards the second antiferromagnetic peak in order to avoid working with a very sharp MR peak. There is qualitative agreement between the experimental and theoretical data although there is a constant mismatch between the MR amplitudes (11% to 16%) probably due to a small deviation in the Cu thickness.

The approximately constant value of the MR versus t_{Co} cannot be explained if we use interface scattering only.

The set of parameters found for the Co-Cu superlattices was then used to calculate the MR for selected sandwich structures. Fig.8a) shows the MR for a glass/$Fe_{50Å}$/$Co_{11Å}$/$Cu_{10Å}$/$Co_{11Å}$/$Cu_{10Å}$ structure reaching 8% at room temperature. The calculated MR for the five-layer structure is of 5%. From the experimental point of view, we should point out that it is not trivial to obtain giant magnetore-

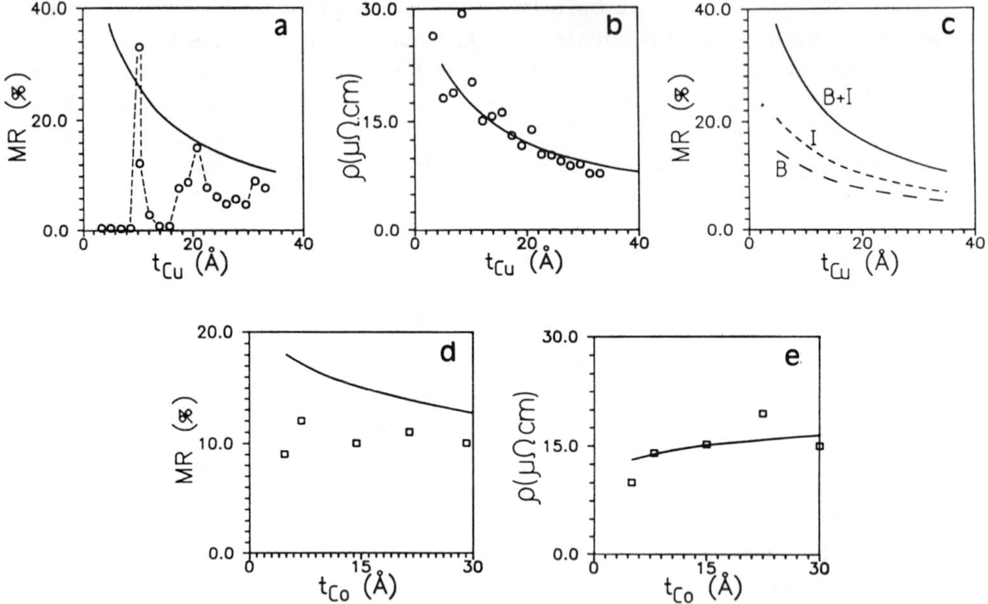

Fig.7-Magnetoresistance (a, c and d) and resistivity (b, e) dependences on Cu and Co thicknesses for Co-Cu superlattices with the following structure, glass/50ÅFe/($Cu_{t_{Cu}}Co_{t_{Co}}$)$_{\times16}$. All data is taken at room temperature. The solid lines through the data are theoretical fits with the parameters indicated in the text (interface plus bulk spin-dependent scattering). The separate contributions of bulk and interface scattering are shown in (c).

sistance (GMR) in sputtered Co-Cu-Co sandwiches for t_{Cu}=10Å. Several authors have reported the inability for producing this result, being only able to find GMR for t_{Cu}=22Å. This is clearly a structural problem arising from the initial growth conditions of the multilayers. The GMR at t_{Cu}=10Å is readily obtained when the number of bilayers starts to increase. A sandwich structure grown with t_{Cu}=22Å, t_{Co}=60Å shows a MR of 6.4% with a saturation field of 200 Oe.

Fig.8b) shows the hysteresis cycles for Co-Cu multilayers with t_{Co}=10Å and t_{Cu}=10 and 22Å respectively. Notice that a perfect antiferromagnetic signal with zero remanence is obtained for t_{Cu}=10Å but a square hysteresis cycle is obtained for t_{Cu}=22Å.

In the latter case, this means that the coupling is too weak giving rise to a saturation field which is smaller than the coercive field of the ferromagnetic layer. The peak in the magnetoresistance probably arises from scattering between antiparallel aligned domains.

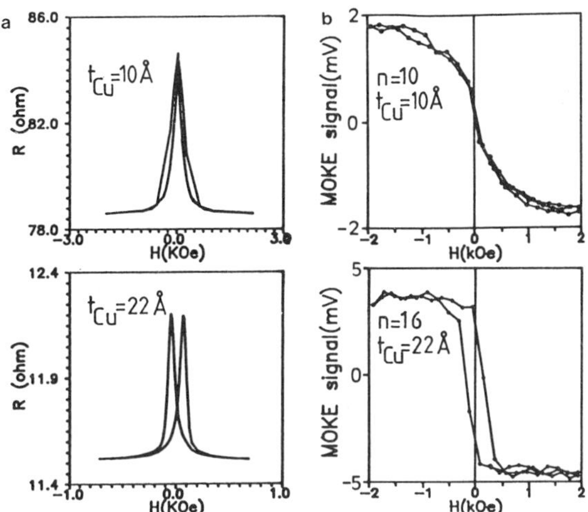

Fig.8-a) Magnetoresistance at room temperature, for two glass/$Fe_{50\mathring{A}}$/$Co_{11\mathring{A}}$/$Cu_{t_{Cu}}$/$Co_{11\mathring{A}}$/$Cu_{t_{Cu}}$ structures with $t_{Cu}=10\mathring{A}$ and $t_{Cu}=22\mathring{A}$. b) Magnetization hysteresis cycles obtained by longitudinal Kerr effect for two superlattices with the following structure, glass/$Fe_{50\mathring{A}}$/($Co_{11\mathring{A}}$/$Cu_{t_{Cu}}$)x16 with the same copper thicknesses.

We will now concentrate on the results for the magnetization reversal processes in $(Co_{t_{Co}}Re_{5\mathring{A}})_{xn}$ multilayer structures. Fig.9 shows the parallel a) and perpendicular b) components of the magnetization for a $Re_{50\mathring{A}}/(Re_{5\mathring{A}}Co_{38\mathring{A}})x3/Re_{150\mathring{A}}$/glass antiferromagnetically coupled structure. First we notice that the parallel component hysteresis cycle has three subloops and a remanence slightly below 1/3 of the magnetization saturation value. This indicates a remanent state with the magnetization of the top and bottom layers aligned in the field direction while the middle layer is antiparallel with the other two. The perpendicular component is fifty times smaller than the parallel one. This reflects the fact that the total perpendicular component is almost null, implying that the individual perpendicular components in each magnetic layer tend to cancel each other. For comparison, for a pure Co film 500Å thick on glass, we measure parallel and perpendicular component signals of the same order of magnitude. Fig.9b) shows that the perpendicular component reaches a maximum value just after reversal of the applied magnetic field. This indicates that magnetization reversal occurs either through rotation processes or 90 degree domain wall propagation processes affecting all three magnetic layers.

This magnetization-reversal behavior can be further analised by performing a Monte Carlo simulation of three antiferromagnetically coupled layers to check the behavior of the magnetization in each layer during the hysteresis cycle. We assume at first no anisotropy, J'_{af}=-0.163erg/cm^2, J'/J''_{af}=0.1 and d=38Å.

Fig.10 shows the calculated parallel magnetization components for the trilayer structure. Fig.10-I) presents the parallel magnetization component for three layers a), for the top or bottom layers b) and for the middle layer c). We see immediately that the three-loop structure observed experimentally was not reproduced although the total parallel component shows the expected 1/3 remanence while the perpendicular component(not shown) is found to be about 100 times smaller than the parallel one . This agrees with our experimental observation that the perpendicular component of the three layers just cancels out.

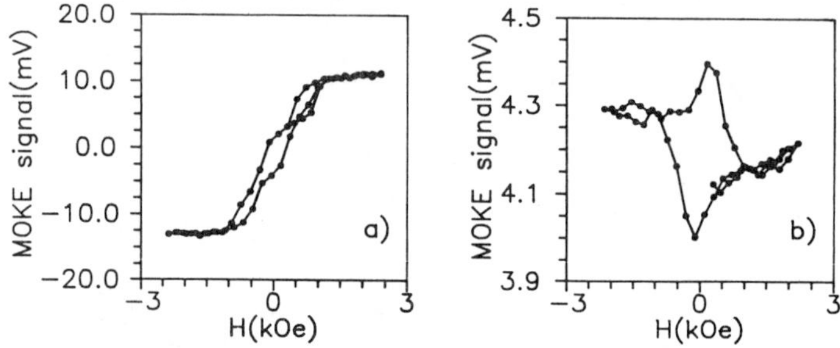

Fig.9-Measured parallel (a) and perpendicular (b) magnetization components for a ($Co_{38Å}Re_{5Å}$)x3 multilayer. Notice that the perpendicular component signal is 50 times smaller than the parallel component, which has a remanence of 1/3 of the saturation magnetization. The maximum in the perpendicular component occurs just after field reversal where 180 degrees magnetization reversal processes are ocurring in all the layers.

Starting from the saturated state and down to 200 Oe, and while the parallel component of the top and bottom layers keeps essentially aligned with the magnetic field,the middle layer gradually rotates by 180 degrees due to the antiferromagnetic coupling interaction that is dominant at small applied fields. In reality, the top and bottom layers deviate about 30 degrees from the applied field direction in order to compensate the perpendicular component coming from the rotation of the middle layer.

Between 200 and -200 Oe the magnetization of the top and bottom layers reverses following the applied field while the middle layer sweeps back into the original orientation to maintain the antiparallel arrangement. These three 180 degrees magnetization reversal processes give rise to the observed maximum perpendicular component just after the field has been reversed. From -200Oe to -1200Oe, the top and bottom layers remain pratically aligned with the field and the middle layer gradually rotates into the field direction. The small decrease in the magnetization of the top and bottom layers observed between 1200 and 200 Oe is due to a canting of the magnetization within the ferromagnetic layers. This can be seen in Fig.10-II where the spin configurations for the remanent state and H=600 Oe are presented.

The observed canting of the magnetization in the ferromagnetic layer can be explained just in terms of energy minimization. The external basal planes(A_1,B_2) only feel the Zeeman term and align with the field. The interface basal planes (B_1,A_2) feel both the Zeeman term and the antiferromagnetic interaction with the middle layer that wants to sit at 90 degrees towards the applied field direction. With respect to the remanent state, notice that contrary to what happens in the antiferromagnetically coupled bilayer(Fig.3) the spins are all aligned either parallel or antiparallel to the field direction.

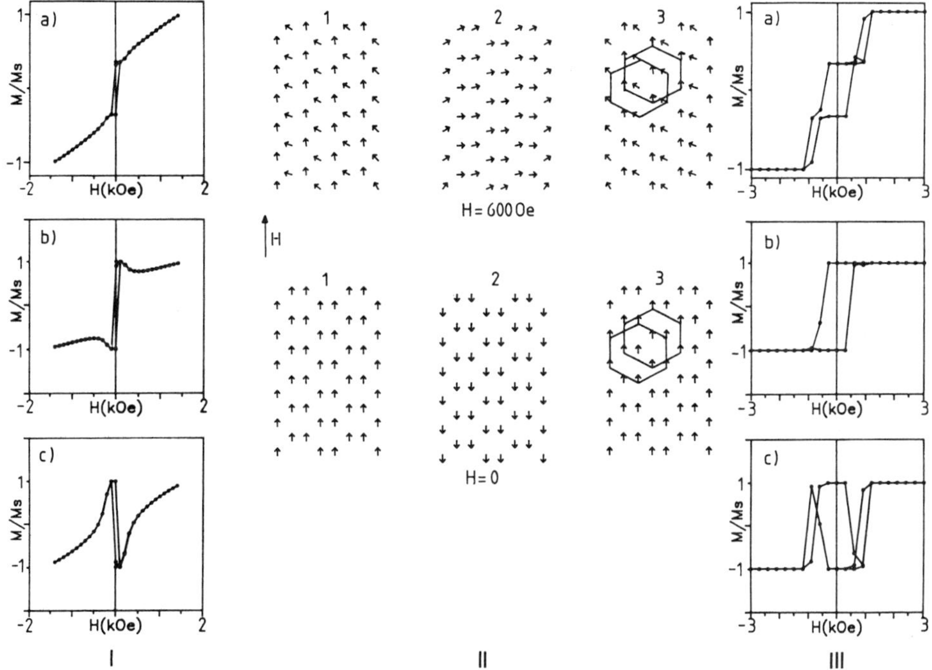

Fig.10-I) Simulated behavior for an antiferromagnetically coupled trilayer. a) total magnetization, b) bottom or top layers, and c) middle layer. II) Spin configurations for the antiferromagnetically coupled trilayer at the remanent state and at an applied field of 600 Oe. III) Simulated behavior for an antiferromagnetically coupled trilayer after introducing uniaxial anisotropy. a) total magnetization, b) bottom or top layers, and c) middle layer.

The experimentaly observed three-loop structure can be simulated introducing uniaxial anisotropy in the Hamiltonian, with the magnetic field applied along the easy axis. This probably means that stress due to mismatch and growth mechanisms induces a preferential magnetization direction through magnetostriction. Fig.10-III shows the simulated hysteresis cycle after introducing an uniaxial anisotropy term ($K=3x10^5$ erg/cm^3). The three loop structure is obtained and the cycle closely resembles the obtained experimental result. In a) we show the parallel component of the magnetization for the three layers, in b) the parallel component of the magnetization for the top or bottom layers , and in c) the parallel component for middle layer . Comparing with Fig.10-I, we see that the main difference is that we observe no gradual rotation processes .

Fig.10-III shows that the bottom and top layers reverse its magnetization coherently at the coercive field H_c. On the other hand, the middle layer performs three 180 degrees sudden magnetization reversal processes during half of the hysteresis cycle.

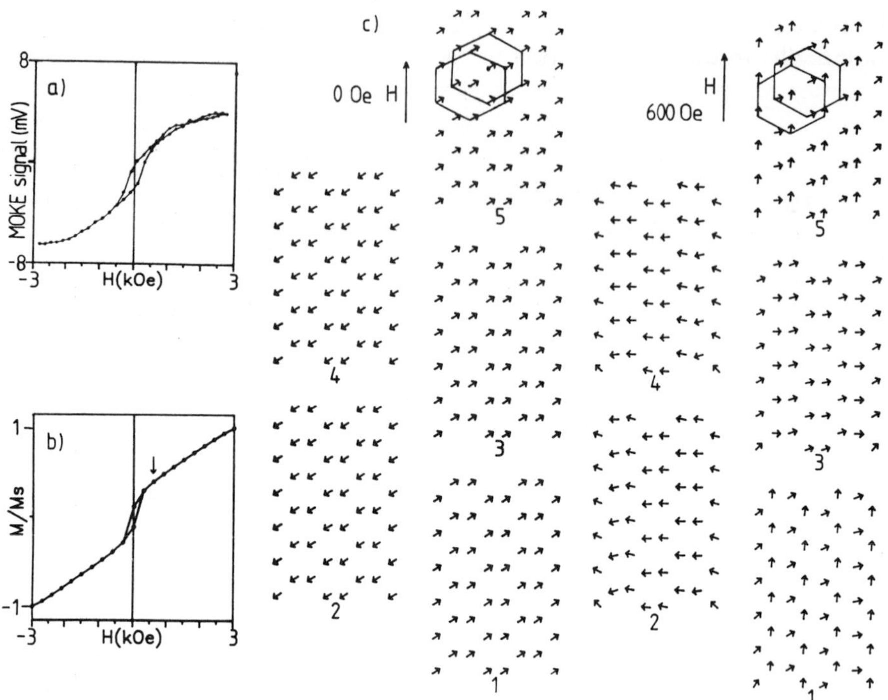

Fig.11-a)Parallel component of the magnetization for the $(Co_{20\AA}Re_{5\AA})\times 5$ structure. Notice that at the remanent state the magnetization is 1/5 of the saturation magnetization. The multiloop structure found for the $(Co_{38\AA}Re_{5\AA})\times 3$ structure no longer occurs. b) Calculated hysteresis cycle for a system of five antiferromagnetically coupled layers. c) Spin-configuration for the five layers at the remanent state and for a 600 Oe applied field.

We did a similar analysis for a $Re_{20\AA}/(Co_{20\AA}Re_{5\AA})\times 5$ $Re_{150\AA}$/glass multilayer . Fig.11 show the magnetization hysteresis cycle for this structure a) together with the Monte Carlo simulation for the parallel component b) and spin configurations for the five layers c). In this case the experimental remanence is at 1/5 of the magnetization saturation value and no hysteresis occurs apart from the small loop near zero applied field. Notice the absence of a multi-loop structure. This seems to indicate that existing stresses have relaxed and that the anisotropy effect is no longer observed.

The Monte Carlo simulation qualitatively reproduces our result. We used M_s= 1432emu/cm^3, J'_{af}=-0.163erg/cm^2, and J'/J''_{af}=0.05, d=20Å. A careful analysis of the remanent state data shows a remanence lower than 1/5. By analising the remanent spin configuration one understands that the system has chosen a remanent state with the magnetization at a 60 degree angle to the applied field direction. The origin for this lies in the precedent spin configuration taken at a field of 600 Oe (Fig.11c). While the internal three layers sit at 90 degrees to the applied field direction, the top and bottom layers show again canted ferromagnetism with the very top and bottom basal planes aligned with the field but with the interface planes at 60 degrees to the field direction.

It is therefore easier for the internal three layers to rotate 30 degrees for full antiparallel alignment at zero field, than obliging the 5 layers to a 90 degrees rotation. Introduction of a small uniaxial anisotropy is enough to destroy the canting effect and to reinstaure the 1/5 remanence.

6-CONCLUSIONS

Spin-valve magnetoresistance was studied in two superlattice systems $(Co\text{-}Re)_{xn}$ and $(Co\text{-}Cu)_{xn}$. The MR and resistivity data versus t_{Co} and t_{NM} were successfully analised using an extension of the Camley and Barnas model including interface and bulk spin-dependent scattering. We find that while interface spin-dependent scattering is the dominant scattering mechanism for the MR in Co-Re superlattices, bulk spin- dependent scattering is needed to account for the MR dependence on Co thickness in the Co-Cu superlattices. The effect of interface scattering in Co-Re multilayers was further studied by adding a third element x=Fe,Ni,Cr or Mo at the Co-Re interfaces. 1ML of Cr or Mo were found to be sufficient to kill the antiferromagnetic coupling while Ni or Fe only alter the transmission coefficients spin-assymetry at the Ni/Re or Fe/Re interfaces when compared with the Co/Re interfaces.

Magnetization reversal processes were analised in $(Co_{t_{Co}}Re_{5\mathring{A}})_{xn}$ antiferromagnetically coupled films(n=2,3 and 5). Multiloop hysteresis cycles where the magnetization reversal occurs through spin-flop or domain wall propagation processes appear for structures with n=3, while single loop hysteresis cycles where gradual rotation processes occur simultaneously in all layers are found for n=5. Monte Carlo simulations suggest that the multiloop structure reflects uniaxial anisotropy effects probably related to magnetostricion and stresses. As the number of layers increase, this effect becomes less important and simultaneous rotation processes in all layers start to occur. The two magnetization components were observed using a suitable magneto-optical set-up. Maximum in the perpendicular component occurs just after field reversal where all layers are reversing their magnetizations by 180 degrees. The very small value of the total perpendicular component means that the three magnetic layers orientate themselves in such a way that their perpendicular components cancel each other.

REFERENCES

1-P.P.Freitas, L.V.Melo, I.G.Trindade, M.From, J.Ferreira and P.Monteiro, Phys.Rev.B45,2495(1992)
2-S.S.Parkin, R.Bhadra, and K.P.Roche, Phys.Rev.Lett.66,2152(1991)
3-M.N.Baibich, J.M.Broto, A.Fert, F.Nguyen Van Dau, F.Petroff, P.Etienne, G.Creuzet, A.Friederich, and J.Chazelas, Phys.Rev.Lett.61, 7472(1988)
4-R.E.Camley and J.Barnas, Phys.Rev.Lett.63,664(1989)
5-J.Barnas, A.Fuss, R.E.Camley, P.Grunberg, and W.Zinn, Phys.Rev.B42,8110(1990)
6-D.H.Mosca, D.Lottis, P.A.Schroeder, and A.Fert, to be published
7-B.Dieny, Europhys.Lett.17,261(1992)
8-P.Grunberg, J.Barnas, F.Saurenbach, J.A.Fuβ, A.Wolf, and M.Vohl, J.Magn.Magn.Mater.93,58(1991)

9-A.Fuβ, S.Demokritov, P.Grunberg, and W.Zinn, J.Magn.Magn.Mater.(1992)
10-W.Folkerts, J.Magn.Magn.Mater.94,302(1991)
11-W.Folkerts, and S.T.Purcell, unpublished=
12-A.M.Ferrenberg, and D.P.Landau, J.Appl.Phys.70,6215(1991)
13-M.B.Taylor, and B.L.Gyorffy, in Digests of the International Colloquium on Magnetic Films and Surfaces, Glasgow(1991)
14-L.V.Melo, I.G.Trindade, M.From, P.P.Freitas, N.Teixeira, M.F.da Silva, and J.C.Soares, J.Appl.Phys.70,7370(1991)
15-J.M.Florczak, and E.Dan Dahlberg, J.Appl.Phys.67,7520(1990)
16-J.M.Florczak, E.Dan Dahlberg, J.N.Kuznia, A.M.Wowchak, and P.I.Cohen, J.Appl.Phys.69,4997(1991)
17-D.P.Landau, Phys.Rev.B13,2997(1976)
18-C.Kittel, in Introduction yo Solid State Physics,ed.John Wiley& Sons,Inc.,New York,p.426,(1986)
19-I.A.Campbell, and A.Fert, in Ferromagnetic Materials, edited by E.P.Wolfarth (North-Holland, Amsterdam,1982) Vol.3,p.747

MAGNETIC NANOPARTICLES: DIFFERENT SYSTEMS FOR A PATCHWORK OF NEW EFFECTS IN MAGNETISM

B. Barbara, A. Marchand, and L.C. Sampaio

Laboratoire de Magnétisme Louis Néel, CNRS, BP. 166
38042-Grenoble, France

INTRODUCTION

Many of the important problems of layers and multilayers, such as the nature of the interfaces (chemical gradient...), their topology (roughtness lenghscale...),... and their effects on physical properties (modulus and direction of the interfacial magnetization, interfacial anisotropies, coupling through non magnetic layers, polarized conduction...) are also of prime importance in small magnetic particles. Several types of small particles can be produced using different techniques e.g. size-distributed and randomly dispersed particles (metallurgy...) or artificial 2-D lattices (lithography...). The main difference with layers and multilayer obviously comes from the fact that, in these lattices, the "nano-periodicity" is not limited to one dimension. This pushes the layers and multilayers systems towards a new field of "well controlled complex systems".

In this paper we describe two of the most recent effects involving magnetic nanoparticles : (i) the quantum tunneling of the magnetization, coherent at lengthscales much larger than atomic ones, and (ii) the "giant" magnetocaloric effects of ferromagnetic nanoparticles. The first, may be the most fascinating, is not yet really proved. However very carefull magnetic after-effect experiments performed on an assembly of particles of $TbCeFe_2$, give strong arguments in favour of this effect below 1 Kelvin[1]. Our main result consists in the observation of a change, in decreasing temperature, of the magnetic relaxation from a Thermally Activated (TA) regime to a regime in which the relaxation is nearly independent of temperature (first section after the introduction). Our contribution to the "giant" magnetocaloric effect of small particles, recently discovered at NIST[2] is to show, by

Magnetism and Structure in Systems of Reduced Dimension
Edited by R.F.C. Farrow *et al.*, Plenum Press, New York, 1993

numerical calculations, the role of particle size distributions (second section). the main effect of a particle size distribution is to broaden the field(H) and temperature(T) variations of the magnetic entropy and therefore to diminish the magneto-caloric effect. However this conclusion, which could easily be expected, appears not to be true in low fields (few to several tens of Oe) and large (nitrogen to room) temperatures where interesting features can be observed. A realisation of some of these results is given with the study of the magnetic entropy of different sets of particles of Fe embedded in an alumina matrix.

QUANTUM TUNNELING OF THE MAGNETIZATION AT THE NANOSCOPIC SCALE IN PARTICLES OF $CeTbFe_2$

Several years ago, A.J. Leggett et al[3] predicted that one of the most intriguing effect of quantum mechanics, namely Quantum Tunneling, could take place on the Macroscopic scale, provided that dissipative interaction with the rest of the World (thermal bath, vibrations, etc...) were small enough. This fascinating prediction was șuccessfully tested for the first time on Josephson junctions at IBM Yorktown Heights[4]. It is extremely important to develop other systems in which Macroscopic Quantum Tunneling (MQT) could easily be observed and analyzed. Concerning magnetic systems, early experiments by Uehara and Barbara[5,6] presented indirect evidence for quantum tunneling of very small sections of domain wall in a large single crystal of $SmCo_{3.5}Cu_{1.5}$. However the interpretation of such a result was rather difficult due to the very large number of domain walls that certainly do not move independently. In order to reduce domain wall interactions we have repeated the same magnetic relaxation experiments, but on small particles (of $Tb_{0.5}Ce_{0.5}Fe_2$) of mean diameter 150 Å, obtained by reaction with H_2 gas under pressure. The average grain diameter was verified by X-ray measurements. The particles were randomly dispersed in a polymer matrix such that interactions between different particles are negligible. The large cubic anisotropy of the Tb^{+3} ions increases the low temperature range in which MQT can be observed[7] and minimizes the effects of long range dipolar interactions[1*]. In average, each particle contains, near the coercive field, a single domain wall (fig.1a). In the TA regime, the temperature dependence of the relaxation rate τ for a small volume of spins V is given by :

$$1/\tau = 1/\tau_0 \exp(-E(H)/k_B T) \qquad (1)$$

[1*] The maximum residual field near the state M = 0 is roughly given by $H'_d \approx M_s / \sqrt{Z}$ where Z is the number of neighbouring particles and M_s the magnetization per unit volume. In the less favorable hypothesis where the distance between particles were equal to their radius, $M_s \approx 200$ emu / cm^3. Assuming that the dipolar interactions should be summed up to 4th neighbouring particles we get $H'_d \approx 200 / \sqrt{100} =$ 20 Oe. This field being more than two orders of magnitude smaller than our measuring field $H \approx H_c \approx 5$ kOe, any eventual dipolar-glass transition will be easily distroyed by our magnetic field (H_c is due to the large anisotropy of $Tb_{0.5}Ce_{0.5}Fe_2$).

where $1/\tau_o$ is the characteristic attempt frequency and $E(H) = kVf(H)$ is a field dependent energy barrier. The constant k takes into account the total anisotropy of the sample. In the TA regime the relaxation rate diverges as T tends to zero. However, if MQT[3,6,7,8-11] is present, τ is given by

$$1/\tau = 1/\tau_o \exp(-E(H) / k_B T_c(H)) \quad (2)$$

where $T_c(H)$ is temperature independent (defined as the crossover temperature into the MQT regime) and thus the relaxation rate remains finite.

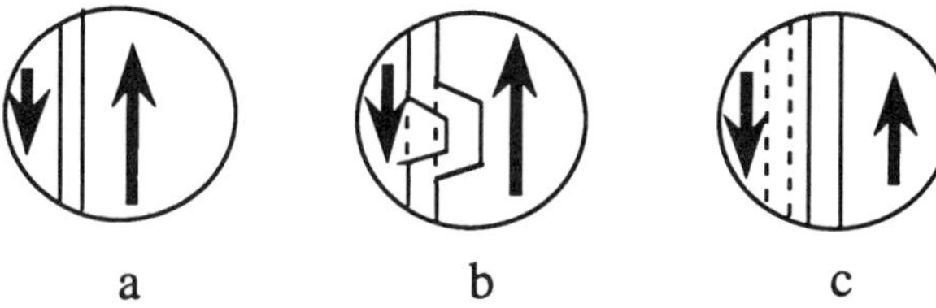

Figure 1. Shematic representation of the supposed motion of a domain wall in our particle. In (a) the domain wall divides the particle in two Weiss domain. In (b) a 2-Dimensional nucleation takes place on the domain wall. In (c) thie nucleus (local domain wall deformation) moves from the center of the particle to its boundaries (soliton) In (b) and (c) the dashed lines represent the initial position of the domain wall in (a).

The low temperature measurements were made using a high-field low-temperature SQUID magnetometer. For each relaxation curve, the magnetization of the sample was first saturated in a field of 8 Tesla. The field was then decreased to zero, reversed and stabilized at a given value of H close to the sample coercive field H_c (when T increases from 50 mK to 10 K, H_c decrease from 6.3 to 1.7 kOe). Measurements of the absolute value of M were made by displacing the sample (which is attached to a miniature dilution refrigerator) within the pickup coils of the SQUID. The recording of the time decay of the magnetization near H_c as a function of time was made over a period of 1 to 2 hours (fig.2).

Interpretation of the results of our powder sample is made in terms of a distribution of field dependent energy barriers n(E(H)) that takes into account the different sizes of the particles and the random orientation of the particles in the matrix. The decay of the magnetization for a distribution of energy barriers is given[12,13] by

$$(1/2M_s)(dM/dt) = k_B T \int n(E(H))\ e^{-\lambda t}\ d\lambda \quad (3)$$

where M_s is the saturation magnetization and $\lambda = 1/\tau = 1/\tau_o \exp(-E(H)/k_B T)$. By measuring M(t) close to H_c, this problem can be further simplified. This is because near H_c, n(E(H)) is at its maximum. Indeed the maximum slope of dM/dH at H_c is a direct consequence of being at the distribution maximum $n_{max}(E)$. Therefore, n(E(H)) varies slowly in comparison to $e^{-\lambda t}$ and eq.3 may be approximated[5] as

$$1/\tau = 1/\tau'_o \exp(-E(H)/k_BT) \qquad (4)$$

where E(H) and τ are respectively a most probable activation energy and the corresponding most probable relaxation time. Expression (4) has been derived from expression (3) in ref. 5, where it was also shown that $\tau'_o = \tau_o / k_BT\, n(E)$. If the experiments are performed near the coercive field, τ is related to the measured magnetization through the relation :

$$1/\tau = (1/2M_S)(dM/dt)_{M=0} \qquad (5)$$

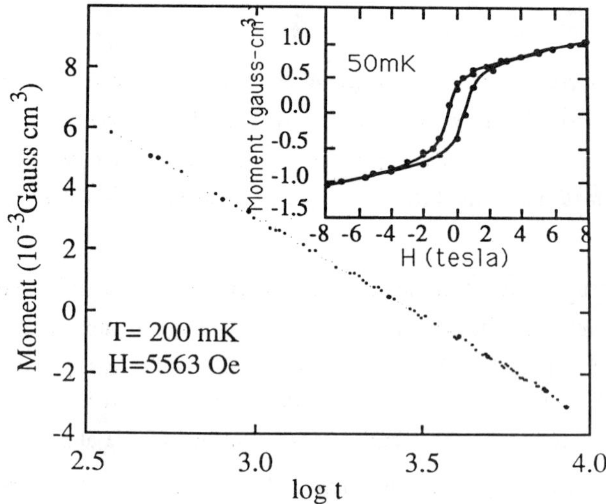

Figure 2. Decay of the sample magnetic moment at T = 200mK and H = 5563 Oe. Inset : Hysteresis loop measured at 50 mK.

Physically this relaxation time corresponds to the time needed to reverse one half of the magnetic moments of the energy barrier distribution (the magnetization, equal to the saturated one M_S at the time $t = \tau o$, is measured till it decays down to M = 0). Experimentally the relaxation time τ is directly related to the slope $(dM/dt)^{-1}$ of the M(t) curve at its intercept at M = 0. Furthermore, near H_c, $n_{max}(E)$ is only slightly dependent on the temperature and magnetic field. The measurements therefore are not particularly sensitive to the n(E) distribution and always take into account roughly the same number of energy barrier sites. This procedure is also interesting for other reasons : (i) The relaxation is faster. (ii) The sample demagnetizing field is negligible. (iii) Eventual demagnetizing fields of dipolar origin, as well as dipolar glass effects between particles, are negligible compared to the applied field

(H = 1.7 to 6 kOe). More than 70 different values of $\tau(H,T)$ have been measured which allows us to determine the field and temperature dependence of the corresponding energy barriers. It is interesting that MQT could in principle be observable in zero field. However this would then require extremely weak zero field energy barriers E(0). In such cases other energy terms such as dipolar fields or spin waves would overcome the effect of MQT, even if they are weak. For that purpose an assembly of very well separated particles is required. Regarding the field dependence of the energy barriers, Néel[13] found for applied fields H close to the maximum coercive field H_0 (defined as the coercive field in the limit T = 0 for TA and in absence of MQT), $E \propto (H_0-H)^{\alpha}$ with $\alpha > 0$. On the other hand, in several highly anisotropic systems the form $E \propto (1/H - 1/H_0)^{\beta}$ with $\beta > 0$ has been proposed and observed[5,6,14,15]. To chose between these forms in the present case, we have plotted $(\log\tau_0/\tau)^{1/\alpha}$ vs. H and $(\log\tau_0/\tau)^{1/\beta}$ vs. H^{-1} for several values of α and β. Only for $\beta = 1$ did we get a coherent set of aligned data points extrapolating to nearly the same point in the plane $(\ln(1/\tau), 1/H)$. A plot of the data points is given fig.3. The straight lines in the figure are the values of $1/\tau$ calculated from the sum of eq.1 and 2 (using the same temperatures as the data points for comparison) with the respective modifications of eq.4. The best fitting parameters $1/H_0 = 0.15 \pm 0.02$ kOe^{-1} and $\log(1/\tau_0') = 8.5 \pm 0.5$ (coordinates of the focal point) were used for the calculation lines.

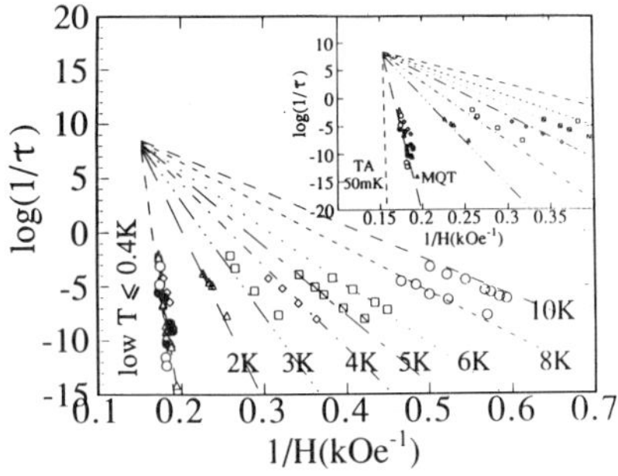

Figure 3 Log($1/\tau$) vs. 1/H in the range $0.05 \leq T(K) \leq 10$. The dashed lines are calculated from eq.1 and 2. Low temperature data: T = 400, 200, 100, 66 and 50 mK.

These parameters were obtained by a rather sensitive scaling plot of the data as $\log((1/2M_s)$ dM/dt) vs. E(H)/T.). This plot is shown in fig.4. The fit to the high temperature data is quite good considering the approximations in the model. The data points obtained near 1.5 K to 3 K deviate slightly from the calculated lines which may be caused by the proximity to the crossover into the MQT regime (see below). Nevertheless, we conclude that the energy barrier in our system E(H) is proportional to $1/H - 1/H_0$, where $H_0 = 6.5 \pm 0.5$ kOe is the maximum coercive field. It is noteworthy that these plots and their extrapolations are very similar to those which establish our present belief of MQT in Josephson junctions, see eg ref.4,16 and fig. 5. This is also similar to the energy barrier, proportional to 1/E, observed in charge density waves, where E is the electric field (see eg ref. 17) and for 3D-ferroelectrics domain walls or dislocations[15]. The 1/H law has been observed in bulk magnetic systems only if the anisotropy energy is large enough and can be interpreted in terms of nucleation 2-D "kinks" on narrow domain walls[5,14].

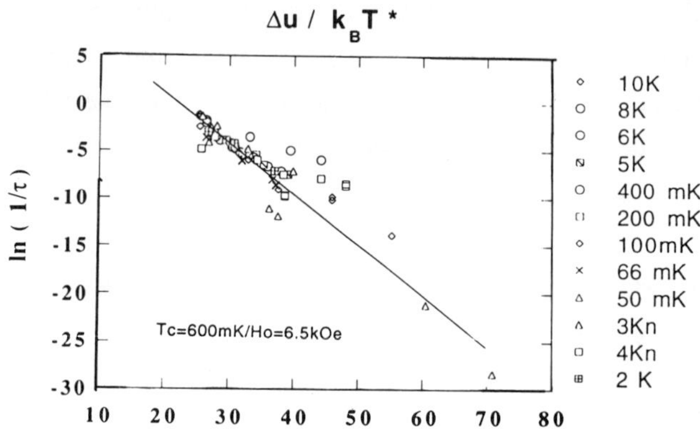

Figure 4. Scaling plot of the most probable relaxation time τ, according to expression (5) with $E(H) = 1/H - 1/H_0$

An important aspect of the calculated behavior of $1/\tau$ is that below the crossover temperature into the MQT regime the relaxation becomes independent of temperature. This is shown in the inset of the fig.3 by the line marked MQT (any calculated line below T_c is essentially on top of this one). The field dependence $T_c(H)$ of the crossover temperature for domain wall tunnling is expected to be weak[11]. It has been used in the calculation with the value $T_c(0) = 600$ mK which was obtained on the same scaling plot mentioned, with the modification for the low temperature data of $\log(1/2M_s.dM/dt)$ vs. $E(H)/T_c(H)$. The measured data points for $T \leq 400$ mK are very close to the calculated limit and show the tendency toward temperature independence. This result can be compared to what one would expect if only TA were present, as shown in the insert for T = 50 mK (line labeled TA 50 mK).

The slopes of the lines in fig.3 are related to the temperature dependence of eq. 1 and 2. If we define $1/T^* = d\log(1/\tau)/d(1/H)$, then $T^* \alpha T$ in the high temperature TA regime. If we neglect the field dependence of the cross over temperature, then $T^* \alpha T_c$ in the MQT regime. This expression applied to fig.3 leads to the variations of $1/T^*$ with the measured temperature $1/T$ given in fig.6. (The solid line follows the points obtained from the least squares fit to the data at each temperature. The dashed line has been calculated from eq.1 and 2, by taking the sum of the TA and QT probabilities, as in fig. 3). A least squares fit to the high temperature data is shown as the straight line in fig.6 where thermal activation is clearly obeyed (see inset). The extrapolation of the TA behavior to low temperature, and the strong deviation of the measured results from this line which, tends toward a constant value, is our main conclusion. We believe it is important because it corresponds well with the general predictions by Leggett[3] for Macroscopic Quantum Tunneling, and more particularly to the recent predictions for MQT in magnetic systems[6,7,8-11].

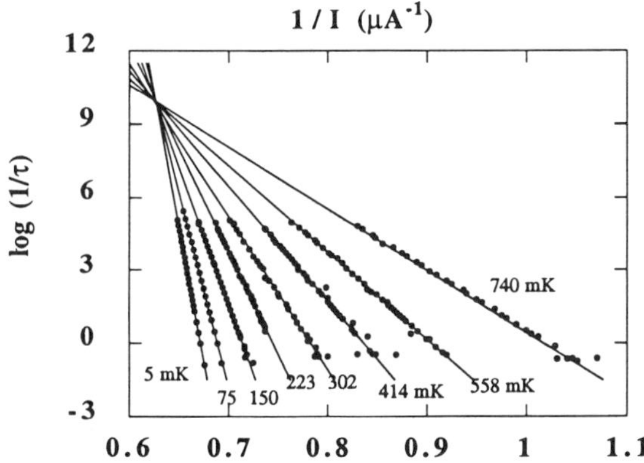

Figure 5. Plot similar to the one of figure 2, but for a Josephson junction (see ref.4 and 16).

A rough estimate for the volume of the tunneling entities can be obtained by relating the expression for the energy barriers $E(H)=k_BT^* \ln(\tau/\tau'_0)$ to the Zeeman energy $E=M_sVH$ (H is the measuring field and M_S = 200 emu/cm^3 is the saturation magnetization). The slope of $\log\tau$ vs. H at a given temperature gives a value of V that decrease from 1.2 .10^5 Å^3 to 2.10^4 Å^3 when T decrease from 10 K to 50 mK. Another way to evaluate V is to consider that in the TA regime T^* should be equal to T. If one asumes that (ie if the $1/T^*$ and $1/T$ scales of the inset of Fig.6 are the same) we get V = 3.10^4 Å^3 and V=1.2.10^3 Å^3 at 10 K and 50 mK respectively. These values are comparable to the precedent one, especially if one considers the rough character of this approach. These values of the TA or QT volumes are much smaller than the particle volume (10^6 Å^3). We conclude that in this system the QT and the TA reversals of the magnetization are not uniform. In order to explain these results, we propose the following : the mean particle size (150 Å) is smaller than the critical single domain particle size (10^3 - 10^4 Å) but larger than the thickness of one domain wall (≈ 30 Å), therefore it is likely that near the coercive field, many of the particles contain one domain wall in a metastable equilibrium (two or more domain walls would cost too much energy because at the thermodynamical equilibrium and in zero field a particle of $TbCeFe_2$ should be single domain). The domain walls should be pinned essentially at the particle boundaries due to surface roughness or disorder, etc. Each elementary event can be described by an irreversible jump of the domain wall due to the passage of its energy barrier by TA at high temperature or by QT at low temperature. Concerning the crossover temperature, the value $T_c = 0.6$ K compares very well with $T_c = 0.3$ K calculated for a single domain wall of area $R = 4\pi.75^2$ Å^2 using the theory of Stamp[11]. Interestingly the expression of Chudnovsky and Gunther[7] for homogeneous reversal of the magnetization (at the unison), gives the same value $T_c = 0.3$ K (here we have to take into account the value of the magnetocrystalline anisotropy $K = 10^7$ erg/cm^3).

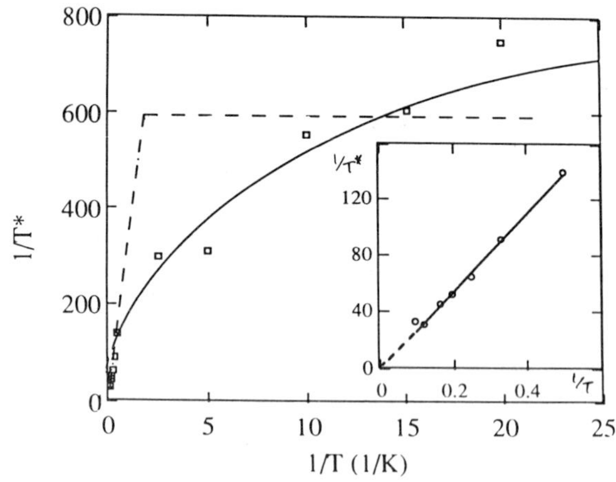

Figure 6. Effective reciprocal temperature 1/T* vs. measured temperature 1/T : Dashed lines are calculated for $T_c = 0.6$ K, the solid line (guide to the eye) follows the experimental values. Inset: High temperature TA regime fit to data.

Is it possible to find other explanations? One other possibility could be related to the existence of some dissipation leading to sample heating, and therefore to an increase of T*. Dissipation can result from inelastic scattering of the "tunneling particle" within its own medium leading to the creation of quasi-particles. Phonons and or spin-waves can be generated by each quantum event through the highly non-symmetrical energy wells (effect of the applied field). Eddy current also can produce dissipation. Sample heating has already been observed in highly dissipative bulk ferromagnets[1, 18,19] (see also fig. 7). This phenomenon leads to non-reproducible dynamics, and at lower temperature, to a staircase behavior of the magnetization. In the present study our results were, on the contrary, extremely well-reproducible and did not show any indication for a staircase behavior at the lowest temperatures. This difference between bulk and small particle samples can easily be understood if one considers that quasi-particles can much more easily be absorbed by the thermal bath, through sample boundaries, in small (nano-) particles than in bulk samples.

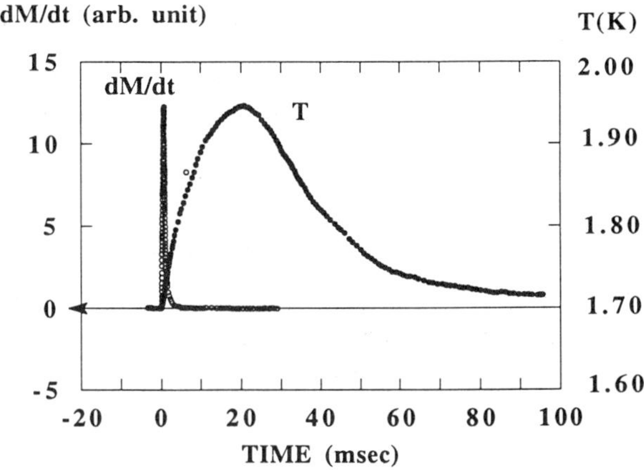

Figure 7. Time dependance of the measured magnetization change dM/dt and sample temperature T in the single crystal of $SmCo_{3.5}Cu_{1.5}$ after a magnetization jump at 1.5 K. As mentionned in the text the real increase of temperature is probably larger than the measured one (0.2 K). However the magnetization jump vanishes while the increase of temperature is still very small. This suggests that sample heating does not affect in a crucial way the quantum event which initiated the jump, but only further quantum events.

For a given amount of dissipation, self-heating should roughly depend on the ratio between the number of MQT events per unit time and the sample surface. It will therefore increase with the volume V of the tunneling entity. Due to the relatively modest size of the tunneling volume in our case, we conclude the probable weakness of self-heating dissipation in this study. Futheremore, according to the theory, phonons dissipation is always negligible[3,6,7,8-11] whereas spin-waves dissipation can only be neglected if the temperature is smaller than the spin-waves gap Δ[6,11]. This last is of the order of magnitude of the anisotropy energy per atom. For $Tb_{0.5}Ce_{0.5}Fe_2$, we find $\Delta = 7$ K, a value larger than the cross-over temperature by one order of magnitude. This is also true in $SmCo_{3.5}Cu_{1.5}$[5] or Dy_3Al_2[20]. One can then expect spin-waves dissipation to be negligible in the low temperature regime of these three systems.. It is however possible that domain wall excitations with free soliton motions could deeply modify our present understanding of these phenomena [5,6,14] (see also fig. 1 and next paragraph).

1n summary, we have shown that our experimental results on the time decay of the magnetization of small ferromagnetic particles of $Tb_{0.5}Ce_{0.5}Fe_2$ can be explained on a very consistent way in terms of Quantum Tunneling of the Magnetization. Each QTM event seems to result from the Coherent Quantum Jump of a ferromagnetic domain wall (Bloch wall) across a particle (planar at the scale of the particle). The sum of all CQJ leads to quantum irreversible noise, analogous to the Barkhausen noise in thermally activated processes. We have indicated in the fig 1 how such a CQJ takes place within a given particle. The observation of an energy barrier proportionnal to 1/H suggests that the domain motion process through 2D-nucleation processes. Such processes are possible within a particle of diameter $D \cong 150$ Å because the critical radius for 2D-nucleation is the domain wall surface[5,21] $R_C \cong \alpha\gamma / M_S H_C$ is smaller than D. In general $\alpha \cong 0.1$[15,21] and for $TbCeFe_2$, $\gamma \cong 20$ ergs/cm^3, $M_S \cong 200$ emu/cm^3 and $H_C \cong 6$ kOe, hat gives $R_C \cong 160$ Å which is precisally the order of the mean particle diameter. Statistical fluctuation of α should lead to distribution R_C values but in any case $R_C \geq \delta$ the Bloch wall thickness and in $TbCeFe_2$ $\delta \cong 50$ Å see the fig1(a, b, c). Quantum resonance effects of antiferromagnetic particles which constitute a different (more delicate but nicer) way to look at quantum effects have been predicted[22] and recently observed in particles of ferritine by Awschalom et al[22].

GIANT MAGNETOCALORIC EFFECT IN FERROMAGNETIC NANOPARTICLES. ROLE OF SIZE DISTRIBUTIONS

Upon a change of the applied magnetic field, the magnetic entropy of a set of magnetic moments is modified. If this field change is performed adiabatically, it induces a change of the sample temperature. This well known phenomenon called "magnetocaloric effect", has been a long ago (and still is) applied to refrigeration by adiabatic demagnetization of paramagnetic samples. Its physical origin is simple : the entropy S_H of a set of paramagnetic moments magnetically saturated by application of a strong magnetic field, is small. It must increase if the magnetic moments disorder upon some field reduction (to say zero, $S_0 > S_H$). In order to do so, the system of magnetic moments has to take the disordering energy in the phonons bath and this leads to a decreasing of the sample temperature. The strengh of the entropy change, $\Delta S = S_H - S_0$, depends upon magnetic parameters and in particular upon the temperature dependence of the magnetization M(T). As shown in elementary textbooks (see eg ref. 23) the Maxwell relation :

$$(\delta S / \delta H)_T = (\delta M / \delta T)_H \tag{6}$$

allows, after integration, to get the entropy change from the magnetization curves M(H) :

$$\Delta S = \int (\delta M / \delta T)_H \, dH. \tag{7}$$

For a classical paramagnet, the magnetization curve is described by the Langevin function, $M = NM_SL(X)$ of argument $X = M_SH/k_BT$, where N represents the number of magnetic moments of amplitude M_S. In low fields, $M = NM_S^2H / 3k_BT$, and therefore $\Delta S_V = -NM_S^2H^2 / 6k_BT$. These expressions are also valid for a set of n identical superparamagnetic particles of volume V and magnetic moment $m_S = M_SV$. Recently R. Shull pointed out that the entropy change should be larger for such a set of superparamagnetic particles than for an assembly of paramagnetic atoms having the same number of magnetic atoms, ie such as $nV = Nv$, where v represents the atomic volume[2,24]. As a matter of fact it is easy to show that in low fields the entropy change should be V/v times larger for the set of small particles. Although the particles volume V should be small enough so that the blocking temperature remains smaller than the temperature at which the experiment is performed, this number can be fairly large (typically of the order of 10^3). In a previous work we have shown that the magnetization curves of different sets of superparamagnetic particles of Fe (of the order of few nanometers) epitaxed on alumina grains of about 10 μm, can fairly well be interpreted in terms of scaling, by assuming powerlaw particles size distributions with a lower size cut-off[25]. Such distributions were in fact similar to those (log-normal) directly derived from TEM observations. In order to investigate the magnetocaloric effect in our Fe particles, we have first considered the effects of a size distribution in the calculation of the entropy change.

The entropy change per volume unit can be written :

$$\Delta S = \int dV\, n(V)\, \Delta S_V(H/T) \,/ \int dV\, V\, n(V) \qquad (8)$$

where n(V) represents the size distribution and ΔS_V the entropy change for a particle of volume V. If M(H,T) is given by a Langevin function,

$$\Delta S_V = k_B(1 - x \coth x + \ln((\sinh x)/x)) \qquad (9)$$

with $x = m_SH / k_BT$.

In general the particles size distribution is a log-normal one. We have used this distribution in our numerical calculations. Before to consider this case let us consider a very simple analytical case in which the increasing and decreasing parts of n(V) are represented by two powerlaws, one with a positive exponent (smallest volumes) and the other one with a negative exponent (largest volumes) [25] :

$$n(V) = C_1V^{\alpha} \quad \text{and} \quad n(V) = C_2V^{-\beta} \qquad (10)$$

α and $\beta > 0$.

Inserting (10) into (8) and taking the limits $V \cong 0$ and $V \cong \infty$ for the integrals give for ΔS the following forms, dependent on α or β :

$$\Delta S \propto (M_S H/kT)^{-(\alpha+1)} \qquad \text{for large } H/T,$$

and (11)

$$\Delta S \propto (M_S H/kT)^{\beta-1} \quad \text{for small } H/T$$

Using the same limits ($V \cong 0$ and $V \cong \infty$) for the log-normal distribution,

$$n(\ln V) = (1/\sigma\sqrt{2\pi})\, \exp(-\ln^2(V/V_0)/2\sigma^2) \qquad (12)$$

(where V_0 is such as $\langle \ln V \rangle = \ln V_0$) expression (8) leads to scaling form

$$\Delta S = (\sigma\sqrt{2\pi}\, \ln V_0)^{-1}\, F(\sigma, M_S H V_0/kT) \qquad (14)$$

where F is an unknown function which should be S-shaped, as shown qualitatively by expressions (11). The departure of the curve ΔS vs H/T, should have a positive or negative curvature according to the value of the curvature of the distribution of those particles which participate to the magnetocaloric effect. In particular, for the powerlaw distribution, the curvature of ΔS vs H/T should be negative for $\beta > 1$ and positive for $\beta < 1$ (note that $\Delta S = S_H - S_0$ is always negative). A very rough but enlightening comparizon could be made between the powerlaw and log-normal distribution, if this last is written :

$$n(V) \propto (V/V_0)^{-(\ln(V/V_0))/2\sigma^2} \qquad (15)$$

Assuming the exponent not to be too much dependent on V, this expression will be similar to (10) where $\beta \propto 1/\sigma^2$. We therefore expect the initial curvature of ΔS vs H/T to change from positive to negative when the distribution broadens.

In order to check more quantitatively these ideas, we have calculated numerically the entropy change using expressions (8) (9) and (12).For large values of H/T, ΔS is strongly reduced by the broadening of the distribution (fig.8). However this tendency to broadening and reduction is reversed when $M_S H V_{moy} / kT$ is smaller than 1 or 2 (fig.9) (ΔS is larger for $\sigma = 1$ than $\sigma = 0.5$ or 0.1) (V_{moy} is identical to $V_0 = 4\pi R^3_{moy} / 3$). For Rmoy = 25 Å and $M_S = 1800$ emu/cm^3, we get $H/T < 5$ (ie H = 1500 Oe at room temperature). It is interesting to see that the expected change of curvature for broad enough distributions is observed when σ is larger than 1. In order to show more clearly this effect we have plotted ΔS vs σ for different H/T (fig.10.11).

The entropy change is maximum for 1< σ <2. Its value with a distribution σ = 1 is 1.5 larger the maximum value obtained in the absence of distribution (σ → 0, fig. 7). This ratio comes out to be larger for smaller H/T ratio. Fig. 10 and 11 show that ΔS(σ = 1 or 2) is larger than ΔS (σ → 0) by order of magnitude. The effective increasing of ΔS by distribution broadening seems to be due to the existence of small particles when the distribution broadens.

Let us now give some experimental results on our particles of Fe. These particles have been prepared at Toulouse (Laboratoire de Chimie des Matériaux Inorganiques), in narrow connection with the Louis Néel Laboratory concerning magnetic characterizations[26]. These

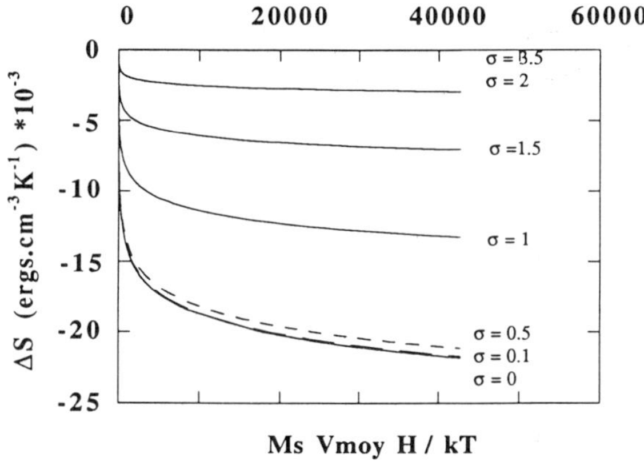

Figure 8. Entropy change ΔS vs H/T calculated for Fe particles distributed according to a log-normal distribution with mean particles volume Vmoy and different standard deviations σ.

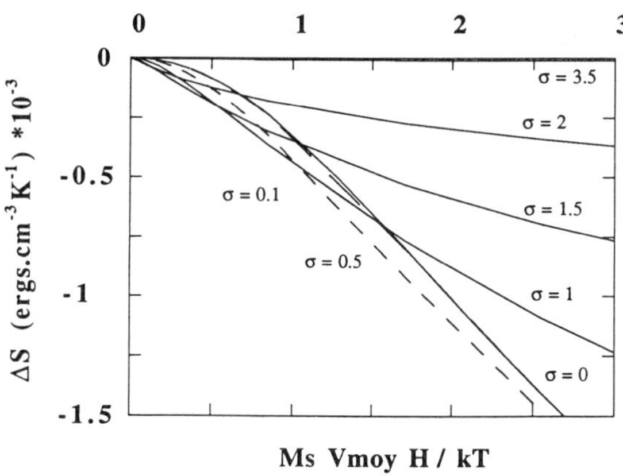

Figure 9. Same plot as fig 8 for smaller scales.

Fe particles of mean sizes of several nanometers are epitaxied on much larger alumina particles (several tens of μm). Mean particles sizes and standard deviations are larger the larger the Fe concentration. Particles sizes distributions shift with Fe concentrations. The influence of these shifts on magnetization curves have been studied via the scaling behavior of the magnetization near the T = 0 Kelvin transition phase for non interacting particles[22]. Fig.12 shows an example of magnetization curves obtained in these systems (10 at.% of Fe).

The variations of entropy changes when the field varies from 0 to H, have been obtained from these curves using the expression (16) deduced from (7) for very small temperature differences ΔT :

$$\Delta S = (1/\Delta T) \int \Delta M \, dH \tag{16}$$

The scaling predicted by expression (14) is roughly verified (fig.13). The same type of curves have been obtained for three different specimens of 10at.% of Fe. The deviation

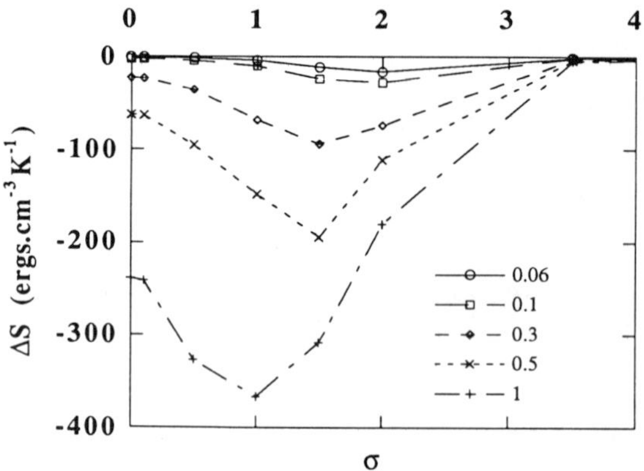

Figure 10. Entropy change ΔS vs σ for different values of H/T.

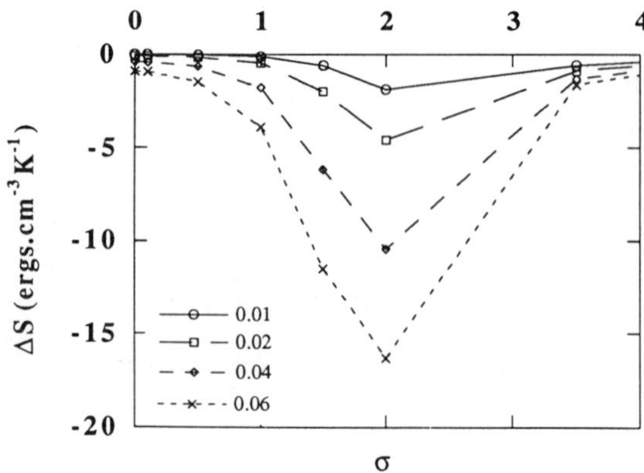

Figure 11. Same plot as fig. 12 for smaller scales.

from scaling can be due to several reasons : bimodal distribution, temperature and field dependent effective distribution intervening in the entropy change in particular due to the existence of blocking temperatures. The sensitivity of the entropy change to the distribution could be used to determine the distribution.

In any case the discussion given above shows that the standard size deviation of these samples is rather small, say $\sigma < 1$, in agreement with size distributions directly obtained by TEM. Let us now evaluate quantitatively the entropy changes and how "giant" it could be. For comparison let us take paramagnetic garnet GGG which is commonly used as refrigerant especially due to the large Gd moment. If used at the nitrogen temperature T = 77 K, it would roughtly give ΔS = 50 erg/gK for a field variation of 10 kOe. In the same conditions our particles give 20 times more. However if the weight of alumina is taken into account, this factor is reduced by 10 (we have 10% of Fe), and we come to the same order of magnitude. Now if instead of this particles we used Gd particles the effect would be multiplied by ratio $((7/2)^2 \cong 10$. Nevertheless it is clear from fig.14, that the same system with particles of mean volume 10 times smaller should give for the same H/T (about 100 Oe/K), an entropy change 50 times larger. Note that the mean radius of such particles would be only 2 times smaller ie about 12 Å.The effect of enhancement calculated numerically for broad distributions, and associated to a curvature change, should not play a significant role for such relatively large H/T values. As calculated above an enhancement is predicted for H/T < 1. As an example for H/T = 0.06, the entropy change of our system of particles of Fe should be multiplied by 20 (see fig.10,11), ie should reache values of the order of $2\ 10^4$ erg/gK, if the distribution were two or three times broader. Physically the effect of distribution broadening probably results from a correlative increase of smaller particles on the left side of the distribution.

As a summary, particles size distributions, always present in real systems, modify importantly the curves of variations of the entropy change vs H/T. Scaling behavior,

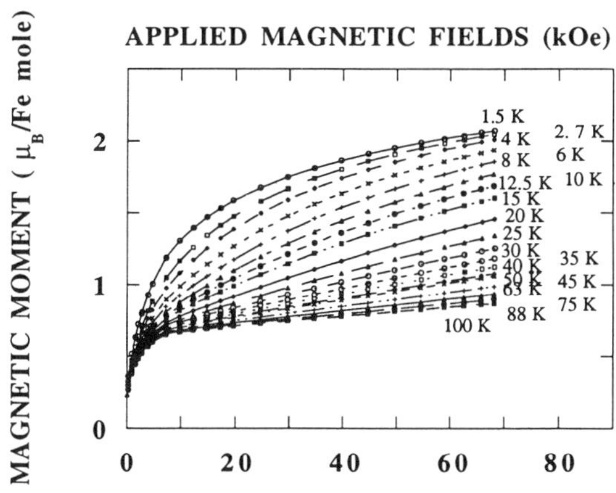

Figure 12. Magnetization of a composite Fe-Al_2O_3 with 10 at% of Fe.

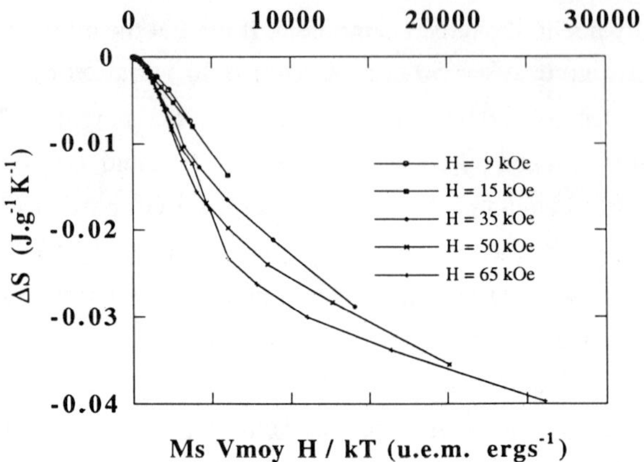

Figure 13. Entropy change measured in Fe particles in an alumina matrix.

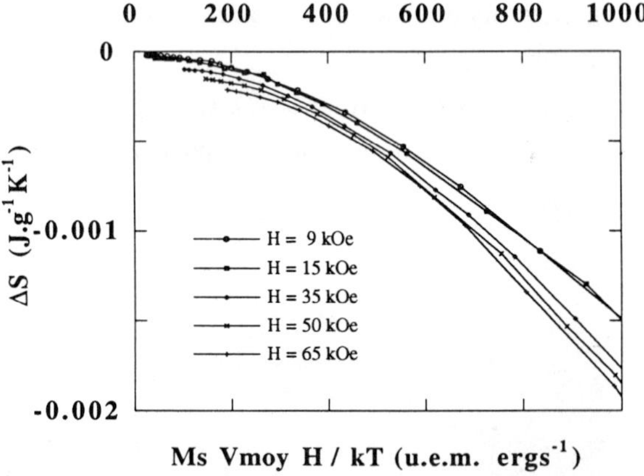

Figure 14. Same plot at fig. 13 for smaller scales.

although not very good, is observed and should allow to analyse in more details what are the particles which switch for given H and T and therefore participate to the magneto-caloric effect. Very large magneto-caloric effect is obtained with Fe particles in alumina and effects larger by one or two orders of magnitude are expected on the same system with broader distributions and smaller particles. For other systems it is important to get good dispersions of particles in matrixes as light as possible.

CONCLUSION

In this paper we have given two examples of our recent works on small particles. Measurements of the time dependence of the magnetization of nanoparticles of $TbCeFe_2$ of 150 Å of mean diameter give strong support for the existence of quantum tunneling of domain walls in these particles. Dissipation effects which generally distroy quantum coherence even at the short timescale of a single tunneling event, are extremelly small in such particles. If one assumes the well admitted theoretical result[3] according which the tunneling probability is reduced by dissipation by the factor $p = p_0 \exp(-V/V_0)$, then the effect of dissipation of a given material of macroscopic size will be reduced if this material is reduced to say 100 Å, to p' such as $p'/p_0 = \exp(-10^{-6} \ln(p/p_0))$, which is enormous. Futhermore in the particles of 150 Å the effect of spin waves which probably constitutes the most important source of dissipation[6,11] should not be important here : due to the large anisotropy of this system the cross-over temperature (0.6 K) between TA and QT is much smaller than the spin waves gap.

The second part of this paper was devoted to the effect of sizes distributions on the giant magnetocaloric effect of nanoparticles, recently discovered by R. Shull[2,24]. Our numerical calculations and magnetization experiments performed on Fe nanoparticles epitaxied in an alumina matrix, show that particles sizes distributions can modify on important ways the variations of the entropy changes with the ratio H/T. Broad distributions give much more important magnetocaloric effect. The magnetocaloric effect of smaller particles beeing larger, this observation can be attributed to the existence of much smaller particles on the lower particle side of broad distributions. The particles of Fe embedded in an alumina matrix give important entropy change which should provide interesting applications for refrigerator.

REFERENCES

1. C. Paulsen, L.C. Sampaio, B. Barbara, D. Fruchart, A. Marchand, J.L. Tholence and M. Uehara, Mesoscopic quantum tunneling in small ferromagnetic particles, Phys. Lett A, 161:319 (1991) and Macroscopic quantum tunneling effects of Bloch walls in small ferromagnetic particles. Europhys. Lett. 19:643 (1992).
   B. Barbara, C. Paulsen, L.C. Sampaio, M. Uehara, D. Fruchart, J.L. Tholence and A. Marchand, Observation of quantum tunneling of the magnetization vector in small particles with or without domain walls, Proc. Int. Workshop"Studies of magnetic properties of fine particles and their relevance to materials science", Rome, Italy (nov. 1991), Elsevier Publ., Amsterdam, The Netherlands.

B. Barbara, L.C. Sampaio, A. Marchand, M. Uehara, C. Paulsen, J.L. Tholence and D. Fruchart, Macroscopic quantum tunneling in ferromagnetic particles, Symposium on the foundation of quantum mecanics, Tokyo (August 1992).
L.L. Balcells, J.L. Tholence, S. Linderoth, B. Barbara and J. Tejada, Quantum tunneling of magnetization in metallic FeC ferrofluids, Z. Phys. B, to appear.

2. R.D. Shull, R.D. MCMichæl, L.D. Schartzendruber and L.H. Bennett, Magnetocaloric effect in fine magnetic particle systems, Proc. Int. Workshop"Studies of magnetic properties of fine particles and their relevance to materials science", Rome, Italy (nov. 1991), Elsevier Publ., Amsterdam, The Netherlands.
3. A.J. Leggett, S. Chakravarty, A.T. Dorsey, M.A. Fisher, A. Garg and W. Zwerger, Dynamics of the dissipative two-state system, Rev. Mod. Phys. 59:1 (1987).
4. R.F. Voss and R.A. Webb, Macroscopic quantum tunneling in 1 μm Nb Josephson junctions, Phys. Rev. Lett. 47:265 (1988).
5. M. Uehara and B. Barbara, Non coherent quantum effects in the magnetization reversal of a chemically disordered magnet : $SmCo_{3.5}Cu_{1.5}$, J. Phys. (France) 47:325 (1986) and Proceeding of the Rare Earth Conf. Durham, Inst. Phys. Conf. series 37:203 (1978) and therein references.
6. P.C.E. Stamp, E.M. Chudnovsky and B. Barbara, Quantum tunneling of magnetization in solids, Int. J. Mod. Phys. B6:1355 (1992).
7. E.M.Chudnovsky and L. Gunther, Quantum tunneling of magnetization in small ferromagnetics particles, Phys.Rev. Lett. 60:661 (1981).
8. L. Gunther, Quantum tunneling of magnetization, Phys World 3:28 (1990).
9. M. Enz and R. Shilling, Spin tunneling in the semiclassical limit, J. Phys. C, 19: 1765 (1986).
10. A. Garg and G.H. Kim, Dissipation in macroscopic magnetization tunneling, Phys. Rev. Lett. 63:2512 (1989).
11. P.C.E. Stamp, Quantum dynamics and tunneling of domain walls in ferromagnetic insulators, Phys. Rev. Lett. 66:2802 (1991).
12. R. Street and J.C. Wooley, A study of magnetic viscosity, Proc. Phys. Soc. A62:562 (1949).
13. L. Néel, Influence des fluctuations thermiques sur l'aimantation de grains magnétiques très fins, C. R. Acad. Sc. Paris, 228:664 (1949). L. Néel, Les œuvres scientifiques de Louis Néel, Editions du CNRS, 15 Quai Anatole France, 75700 Paris.
14. T. Egami, Theory of intrinsic magnetic after-effect . I. Thermally activated process, Phys. Status Solidi, A20:157 (1973), II. Tunneling process and comparison with experiments, Phys. Status Solidi, B57:211 (1973).
15. B. Barbara, G. Fillion, D.Gignoux and R. Lemaire, Magnetic after-effect associated with narrow domain wall in some rare earth based intermetallic compounds, Sol. State Comm. 10:1149 (1972) and references therein.
16. B. Barbara, P.C.E. Stamp and M. Uehara, Mesoscopic tunneling in disordered magnetics systems : a comparison with tunneling in squids, J Phys. 49:C8-529 (1988).
17. J. Bardeen, Superconductivity and other macroscopic quantum phenomena, Phys. Today 43(12):25 (1990).
18. M. Uehara, B. Barbara, B. Dieny and P.C.E. Stamp, Staircase behaviour in the magnetization reversal of a chemically disordered magnet at low temperature, Phys. Lett. 114A:23 (1986).
19. Y. Otani, J.M.D. Coey, B. Barbara, H. Myyajima, S. Chikazumi and M. Uehara, J. Appl. Phys. 67:4619 (1990).
20. B. Barbara, Homogenous and inhomogenous pinning in hightly anisotropic materials, Proc. 2nd Int. Symposium on Anisotropy and Coercivity, Univ. of Dayton 137 (1978).
21. B. Barbara, and E.M. Chudnovsky, Macroscopic quantum tunneling in antiferromagnets, Phys. Lett. A145: 205 (1990).
22. D.D. Awschalom, M.A. Mc Cord and G. Grinstein, Observation of macroscopic spin phenomena in nanometer scale magnets, Phys. Rev. Lett. 65:773 (1990).
23. Maxwell relations, Y. Rocard. "Thermodynamique", Masson et Cie, Paris (1952).
24. R.D. McMichæl, R.D. Shull, L.D. Schartzendruber, L.H. Bennett, and R.E. Watson, Magnetocaloric effect in superparamagnets, J.M.M.M. 111:29 (1991).
25. A. Marchand, B. Barbara, P. Mollard, X. Devaux and A. Rousset, On a serie of nanoparticles of iron epitaxed on Al_2O_3 : a new field, temperature and concentration (of Fe) scaling plot of the magnetization curves, to be published in J. Magn. Magn. Mat. (Nederlands). B. Barbara J. Filippi, A. Marchand, P. Mollard V.S. Amaral, X. Devaux and A. Rousset, High field magnetization of two shortly correlated systems : random anisotropy systems and nanoparticles, J. Phys. I (France) 2, 101-9, (1992).
26. A. Marchand, X. Devaux, B. Barbara, P. Mollard, M. Brieu and A. Rousset, Microstructural and magnetic characterization of alumina-iron nanocomposite, To appear in J. Mater. Sci.

MAGNETIC PROPERTIES OF SUB-MICRON PARTICLES

S. Schultz and D. R. Smith

Center for Magnetic Recording Research, and
Department of Physics, 0319
University of California, San Diego
La Jolla, California 92093-0319

INTRODUCTION

In 1985 the Center for Magnetic Recording Research (CMRR) at the University of California in San Diego, initiated a fundamental study of magnetic particle and thin film hysteresis as part of a comprehensive program dedicated to developing very high density tape and disc magnetic storage systems. Since magnetic hysteresis is the essential physical phenomonon underlying the entire fifty-billion dollar per year magnetic storage industry, it had obviously been extensively studied, but despite these prior efforts, a detailed quantitative microscopic description of even one real experimental system had never been achieved. Similarly, while there were a few theoretical model systems, such as a uniformly magnetized ellipsoid, which admit of a complete magnetostatic solution in a uniform applied field, there was no detailed quantitative description of even one model isolated sub-micron magnetic particle. Of course, what is ultimately desired, is not only a complete description of the static and dynamic magnetic properties of individual particles (or the grains of a thin film), but also a methodology for a quantitative description of a strongly interacting collection of such particles, i.e, a prediction of the macroscopic hysteretic properties of the media.

The CMRR effort has subsequently been supported via a Materials Research Group award from the National Science Foundation, and in this paper we provide references for much of the progress that has been made toward providing that quantitative body of experimental data and numerical simulation that is required in order to eventually describe the internal variables of a magnetic many-body hysteretic system. We will focus on the first phase of the program which is to quantify the behavior of single magnetic sub-micron particles.

What are the properties of the individual magnetic entities that one would like to be able to describe, in order to set the basis for the eventual complete description of the hysteresis of a classical many-body strongly interacting magnetic system? These are the internal time dependent magnetization distributions (over as fine a length scale as needed) everywhere throughout the object for any arbitrary applied field distribution in space and time. We can further subdivide this requirement (which is daunting even for a single isolated realistic sub-micron particle, let alone an interacting group of particles) into the following subset, which constitutes the basis for our theoretical and experimental program:

i. Determination of the particle's total (vector) magnetic moment as a function of a time independent uniform field applied at any angle.

Magnetism and Structure in Systems of Reduced Dimension
Edited by R.F.C. Farrow *et al.*, Plenum Press, New York, 1993

ii. For particles which have a large remanence, we may modify (i) to refer to only a determination of the switching field as a function of a uniform field applied at any angle.

iii. Recognizing that our eventual goal is to cope with interacting particles, for which the internal field gradients are very large, we desire to know the magnetostatic response to applied non-uniform magnetic fields.

iv Recognizing that real particles will have significant variations in geometry, morphology, and internal variations of material properties (both structural and magnetic) we desire to know the corresponding effects on the magnetostatic response of (i-iii) above.

v. Recognizing the very large surface to volume ratio of the sub-micron particles of interest, and that the change of symetry and boudary conditions at the particle surface can have profound local effects, which in turn can strongly change the magnetostatic response of (i-iv) we desire to know the corresponding consequences of changing the surface magnetic anisotropy and magnetization.

vi. Finally, while recognizing that a complete real time description and experimental verification at the atomic spatial and temporal scales may not be readily forthcoming, we nonetheless wish to be able to include the interaction of the particles with a thermal bath. Thus, we wish to be able to fully characterize and model magnetic viscosity, (i. e., thermally activated switching), and even reversible thermal fluctuations.

The experimental systems that have been investigated include:

i. Discrete column reversal in perpendicular Co-Cr. We measure the switching statistics as a function of applied field (i.e., the columnar "Barkhausen noise" during a hysteresis cycle),[1] as well as during thermal time decay.[2]

ii. Magnetization and hysteretic properties of individual sub-micron Permalloy particles produced by nanolithography.[3]

iii. The switching fields of isolated γ-Fe_2O_3[4] and Permalloy[5] particles, and the angular dependence of the switching field.[6,7]

iv. Determination of the thermal switching of both γ-Fe_2O_3,[6] and Permalloy particles.[7]

In order to make measurements of the magnetic properties of single sub-micron particles, several new techniques,[5,8] and extension of older ones had to be developed.[4] In parallel with the experimental programs, there is also an intensive theoretical and numerical simulation program at CMRR. Several appropriate references are provided in Ref.9.

ACKNOWLEDGEMENTS

The author greatly appreciates the large number of co-workers who have made substantive contributions to the body of work referenced. Their individual names are not listed here, but may be readily ascertained from the referenced citations. We gratefully acknowledge support from both the Center for Magnetic Recording Research and the National Science Foundation, grant number DMR-90-10908

REFERENCES

1. Webb, B. C. and Schultz, S., "Observation of discrete jumps allowing determination of the statistics for magnetization reversal of interacting magnetic particles in Co-Cr thin-films",Sol.St. Commun. **68**, #5, 437 (1988); and Webb, Bucknell C., and Schultz, S., "Detection of the Magnetization Reversal of Individual Interacting Single-Domain Particles within Co-Cr Columnar Thin-Films", (Vancouver) IEEE Trans. on Magns. **24**, #6, 3006 (Nov 1988).

2. Webb, Bucknell C., and Schultz, S., "Detection of individual particle switching during hysteresis and time decay in Co-Cr thin-films", published in J.de Physique, Colloque C8, Supp. **49**, #12, 1975 (1988) Supp. Series. (ICM'88 Paris).

3. Smyth, J. F., Schultz, S., Fredkin, D. R., Koehler, T. R., McFadyen, I. R., Kern, D. P., and Rishton, S. A., "Hysteresis in lithographic arrays of permalloy particles: theory and experiment" Invited talk presented at the 35th Ann. Conf. MMM'90, San Diego Oct. 29 - Nov. 1, 1990, J. Appl. Phys. **69**, #8, 5262 (1991).

4. Salling, C. T., Schultz, S., McFadyen, I. R., and Ozaki, M. "Measuring the coercivity of individual sub-micron ferromagnetic particles by Lorentz microscopy", 5th Joint MMM-Intermag Conf., Pittsburgh, PA, June 18-21,1991. IEEE Trans. Mag., **27**, #6, 5184 (1991).

5. Gibson, G. A., Schultz, S., Smyth, J. F., and Kern, D. P."Observation of the switching fields of individual Permalloy particles in nanolithographic arrays via magnetic force microscopy", 5th Joint MMM-Intermag Conf., Pittsburgh, PA, June 18-21, 1991. IEEE Trans. Mag., **27**, #6, 5187 (1991).

6. C. Salling, R. O'Barr, S. Schultz, M. Ozaki, and I. McFadyen, in preparation.

7. Lederman, M., and Schultz, S., "Observation of thermal switching of a single ferromagnetic particle", to be published in the Proc. of the 37th Ann. Conf. on Magnetism and Magnetic Materials, in J. Appl. Phys. (1992).

8. Tobin, V. M., Schultz, S., and Oseroff, S. B., "Pico-emu Magnetic Moment Measurements with a Millikan Oil Drop Magnetometer", presented at the APS March Meeting, 12-16 March 1990, Anaheim. Bull A.P.S., **35**, #3, 646 (1990); Gibson, G. A., and Schultz, S., "A high sensitivity alternating gradient magnetometer for use in quantifying magnetic force microscopy", presented at the 35th Ann. Conf. MMM'90, San Diego Oct. 29 - Nov. 1, 1990, J. Appl. Phys. **69**, #8, 5880 (1991).

9. Zhu, J-G., and Bertram, H. N., "Magnetization Reversal in Co-Cr Perpendicular Thin Films:, J. Appl. Phys., **66**, #3, 1291-1307 (1989); Schabes, M. E., and Bertram, H. N.,"Ferromagnetic Switching in Elongated γFe_2O_3 Particles", J. Appl. Phys. **64**, 5832-5834 (1988); Schabes, M. E., and Bertram, H. N., "Effects of Applied Field Inhomogeneities on the Magnetization Reversal of Elongated γFe_2O_3 Particles", IEEE Trans. Magn, **MAG-25**, #5, 3662-3664 (1989); Schabes, M. E., and Bertram, H. N., "Magnetization Reversal of Cobalt Modified γFe_2O_3 Particles", J. Appl. Phys., **67**, #9, 5149-5151 (1990); Yuan, S.-W., Bertram, H. N., Smyth, J. F., and S. Schultz, "Size effects of switching fields of thin Permalloy particles",to be published in the Proc. of INTERMAG'92, in the Sept. 1992 issue of IEEE Trans. of Magn.; Fredkin, D. R., and Koehler, T. R., "*Ab initio* Micromagnetic Calculations for Particles", J. of Appl. Phys., **67**, 5544 (1990); Invited paper at MMM'89; Fredkin, D. R., Koehler, T. R., Smyth, J. F., and Schultz, S., "Magnetization Reversal in Permalloy Particles - Micromagnetic Computations", J. Appl. Phys., **69**, 5276 (1991); and Chen, W-J., Fredkin, D. R., and Koehler, T. R., "Micromagnetic Studies of Interacting Permalloy Particles", to be published in the Proc. of INTERMAG'92, in the Sept. 1992 issue of Trans. of Magn..

GIANT MAGNETORESISTANCE IN GRANULAR MAGNETIC SYSTEMS

J. Samuel Jiang, John Q. Xiao, and C. L. Chien

Department of Physics and Astronomy
The Johns Hopkins University
Baltimore, MD 21218, USA

Since the discovery of giant magnetoresistance (GMR) in Fe/Cr multilayers [1], the magneto-transport properties of magnetic multilayers have received considerable attention over the past few years [2-12]. The GMR effect is of fundamental importance in that it reveals the interplay between electron transport and magnetism. It has been associated with the existence of antiferromagnetic exchange coupling between the ferromagnetic layers across the non-magnetic layers [2, 3]. Furthermore, the electrical resistivity of some of these multilayers can be reduced by as much as a factor of two upon the application of an external magnetic field [1]; the sheer size of this effect makes these materials potential candidates for magneto-resistive applications. A number of other magnetic multilayer systems, such as Co/Ag [4] and ($Ni_{81}Fe_{19}$)/Cu [5], have also been found to exhibit the GMR effect as the result of extensive research efforts.

Until very recently, the GMR effect has been realized only in layered structures. Planar structure is featured in all theoretical attempts aimed at explaining the mechanism of GMR[6-8]. Among the most discussed is the "two-current model", which describes the decrease in resistivity as the result of the "short-circuit effect" of one of the spin channels due to the alignment of the magnetic layers [9].

It is of fundamental interest to investigate how the electron transport is related to the magnetic structure in systems other than multilayers. Of practical concern is whether the layered structure, whose preparation requires MBE or sputtering with stringent deposition control, is essential to the existence of giant magnetoresistance.

We have chosen metallic granular magnetic media to study a number of crucial and hitherto unexplored aspects pertinent to GMR. Specifically, we wish to explore the extent to which GMR depends upon the layered structure, the consequence when the magnetic entities are far apart so that they are effectively decoupled, and the relationship between GMR and the magnetic alignment. We have observed GMR in granular Co-Cu, Co-Ag, Fe-Cu and Fe-Ag systems with magnitudes comparable to those in multilayers [14, 15]. Also

Magnetism and Structure in Systems of Reduced Dimension
Edited by R.F.C. Farrow *et al.*, Plenum Press, New York, 1993

noteworthy is that Berkowitz *et. al.* have simultaneously reported the observation of GMR in granular Co-Cu systems [13].

The metallic granular materials consist of ultra-fine magnetic particles embedded in conducting non-magnetic matrices. While Fe and Co form alloys with most of the metals, they are virtually insoluble in Cu and Ag under equilibrium condition. The mutual immiscibility between metal and insulator has long been utilized to promote phase separation in fabricating metallic granular materials [16]. If the magnetic granules are smaller than the critical size for single domain particles, all the moments within a particle are aligned ferromagnetically. Materials that contain single-domain magnetic particles in a *metallic* medium, such as granular Fe-Cu and Co-Cu, have recently been studied by Childress and Chien [17, 18]. As will be discussed later, the fact that the particle is single domain has important ramifications in the existence of GMR.

We have used a DC magnetron sputtering system with composite targets of (Fe or Co)/(Cu or Ag) to deposit films onto substrates held at constant temperatures (T_S) ranging from 77 K to 900 K. Deposition at low substrate temperatures (e.g. T_S=77 K) yields metastable alloys due to the high quenching rate [19]. Upon heat treatment, the magnetic component precipitates out as the result of phase separation. By varying the composition and the annealing temperature (T_A), the size and the density of the magnetic granules can be controlled. For depositions carried out at high substrate temperatures, the *in-situ* annealing also leads to the formation of granular structures.

The morphology of the samples has been studied using several methods. Fig. 1 shows the X-ray diffraction pattern of $Co_{20}Ag_{80}$ (20 vol. % Co) after undergoing different heat treatments. The as-prepared sample (T_S=77 K) is a metastable fcc alloy. After annealing, a second set of fcc diffraction peaks emerges, indicating the formation of small fcc Co particles in fcc Ag. At successively higher T_A, the Co precipitates grow lager, resulting in narrower diffraction peaks. These microscopic features are also verified by transmission electron microscopy. Convergence beam electron diffraction and energy dispersive analysis both confirm that the Co particles are surrounded by the Ag matrix.

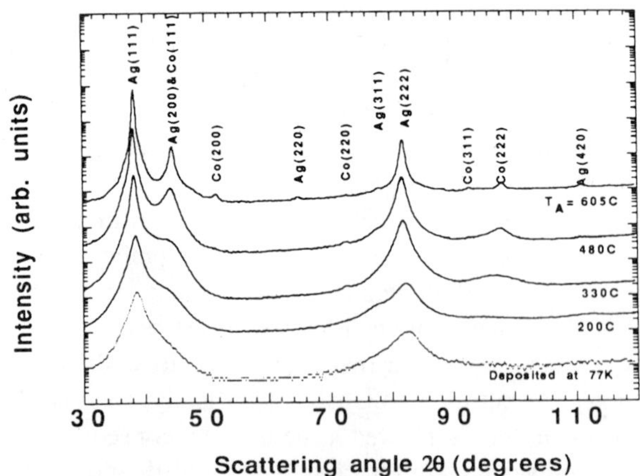

Figure 1. X-ray diffraction patterns of the $Co_{20}Ag_{80}$ (20 vol. % Co) sample deposited at 77 K (fcc alloy) and granular fcc Co in fcc Ag after annealing at 200 °C, 330 °C, 480 °C and 605 °C for 10 min. For clarity, the diffraction patterns have been shifted vertically.

The electrical resistivities have been measured using a standard four-terminal configuration with external magnetic fields up to 50 kOe. Longitudinal resistivity $\rho_{||}$ (with the magnetic field parallel to the current), perpendicular resistivity $\rho_{\perp}$ (with the field perpendicular to the film plane), and transverse resistivity ρ_T (with the field in the sample plane and perpendicular to the current) have been measured. A SQUID magnetometer has been employed in measuring the magnetic properties.

For each sample, we measured the resistivity starting from the initial unmagnetized state and then increasing the field to 50 kOe. Afterwards the field was cycled from +50 kOe to -50 kOe and then back to +50 kOe again. Fig. 2(a) depicts the magnetoresistance MR ($\Delta\rho/\rho$) at 5 K for the phase-separated $Co_{16}Cu_{84}$ (T_s=350 °C). All three MR (longitudinal, perpendicular and transverse) give very similar results. The initial curve is history-dependent, we will focus on the reproducible part of the hysteresis loop. Also shown in Fig. 2(b) is the magnetization hysteresis loop from SQUID measurements. It is seen that MR closely correlates with the magnetization. The resistivity attains a maximum at the coercive field (H_C) where the magnetization is zero, and asymptotically approaches the lowest value as the magnetization reaches saturation M_S. In fact, as shown in Fig. 3, the MR is found to be approximately proportional to the square of the magnetization, i. e.:

$$\frac{\Delta\rho}{\rho} \cong -A\left(\frac{M}{M_s}\right)^2, \tag{1}$$

where $A = 0.065$ for $Co_{16}Cu_{84}$ (T_s=350 °C).

It should be noted that within each of the single-domain particles, the moments are ferromagnetically aligned, although the magnetization axes of the particles are randomly oriented in the unmagnetized state. The external field coerces the magnetization axes of the particles to rotate toward alignment. The magnetization reaches saturation when a total alignment is achieved.

At low temperatures, scattering processes due to phonons and magnons are negligible. Therefore the variation in the resistivity is solely due to the elastic scattering of the conduction electrons by the magnetic particles. Eq. (1) suggests that the resistivity at low temperatures can be expressed as:

$$\rho = \rho_0 + \rho_m\left[1-\left(\frac{M}{M_s}\right)^2\right], \tag{2}$$

where ρ_0 is the residual resistivity due to the disorder scattering. The M^2 dependence in Eq. (2) can be understood as follows: The contribution from the elastic magnetic scattering can be represented by the average of $\mathbf{M_i}\cdot\mathbf{M_j}$ over the mean free path, where $\mathbf{M_i}$ and $\mathbf{M_j}$ the magnetization of the particles. Assuming the particles are identical and non-interacting, it is straightforward to show that $\langle\mathbf{M_i}\cdot\mathbf{M_j}\rangle=\langle\mathbf{M_i}\rangle\cdot\langle\mathbf{M_j}\rangle=\langle\mathbf{M}\rangle^2$. Therefore the M^2 dependence of GMR is the manifestation of the extra resistivity from the elastic magnetic scattering. Since each particle contains a large number of moments that scatter the electron coherently, the MR effect will be much larger than that due to small clusters in diluted alloys and spin glasses. The MR is thus a gauge of the alignment of the magnetization axis, it is not unexpected that different geometries (perpendicular, longitudinal and transverse)give similar MR results in an isotropic system such as granular materials.

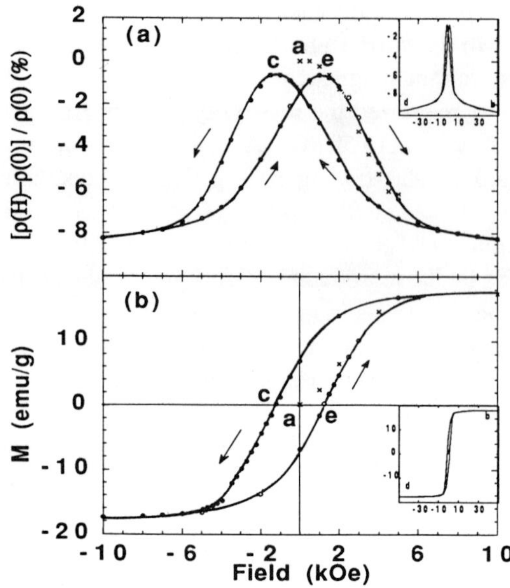

Figure 2. Transverse magnetoresistance $[\rho_\perp(H) - \rho_\perp(0)]/\rho_\perp(0)$ as a function of external field (H) and hysteresis loop at 5 K for phase-separated $Co_{16}Cu_{84}$ sample deposited at $T_s = 350$ °C. The crosses, open and closed circles denote the initial curve (a → b), and branches of decreasing (b → c → d) and increasing fields (d → e → b).

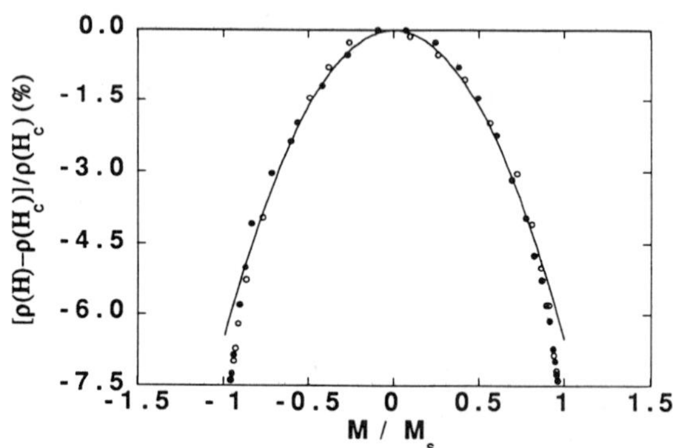

Figure 3 . Giant magnetoresistance $\Delta\rho/\rho = [\rho_\perp(H) - \rho_\perp(H_c)]/\rho_\perp(H_c)$ vs. global magnetization at 5 K for phase-separated $Co_{16}Cu_{84}$ sample deposited at $T_s = 350$ °C taken from Fig. 2. Open and closed circles are those with increasing and decreasing fields respectively. The solid curve is $\Delta\rho/\rho = -0.065\ [M/M_s]^2$.

The effect of particle size on GMR in $Co_{20}Ag_{80}$ (20 vol. % Co) is shown in Fig. 4. As mentioned above, a higher annealing temperature results in larger particle sizes. The values of $\Delta\rho/\rho$=35%, 43%, 38% and 22% have been observed at 5 K in samples with T_A=200 °C, 330 °C, 480 °C and 605 °C, respectively. The magnitudes of GMR in multilayers are more commonly reported as $(\Delta\rho/\rho) = [\rho_{AF}-\rho_F]/\rho_F$ in the literature, where ρ_{AF} and ρ_F refer to the resistivities in the antiferromagnetically and ferromagnetically aligned states in multilayers. In the case of granular systems, they represent the M=0 and $M=M_S$ states. This definition gives GMR's of 54%, 75%, 61% and 28% for the respective samples. These are among the largest GMR effects reported. The magnitude of GMR first increases with the particle size, and then decreases as the particles get even larger, demonstrating that the GMR effect can be maximized by optimizing the process condition [14]. As expected, the saturation field Hs also decreases progressively with the increasing T_A.

In general, the resistivity of the granular magnetic materials can be expressed as:

$$\rho(T,H) = \rho_0 + \rho_{ph}(T) + \rho_m(T)\left[1 - F\left(\frac{M(H,T)}{M_s}\right)\right], \tag{3}$$

where $\rho_{ph}(T)$ is the phonon contribution. The last term represents the contribution from the magnetic scattering. The function $F(x)$ has the limiting values of $F(1)=1$ and $F(0)=0$. The detailed form of $F(x)$ depends on the size and density of the particles, and whether the particles are interacting. In the case of $Co_{16}Cu_{84}$, $F(M/M_S) \cong (M/M_S)^2$. In order to achieve large GMR, the last term in Eq. (3) must be a significant fraction of the total resistivity. It is important to examine $\rho_m(T)$ itself, which can be deduced form the resistivity difference $\Delta\rho(T)=\rho(T,0)-\rho(T, 50\text{ kOe})$ for samples whose saturation field is less than 50 kOe. Fig. 5 shows the resistivity of the Co-Ag samples (T_A=330 °C, 480 °C and 605 °C) measured as a function of the temperature at zero field and with an external field H=50 kOe. It is seen that ρ_m is only weakly dependent on T, which might be due to the thermally activated magnetization fluctuations and inelastic scattering of the conduction electrons. However, the GMR effect dwindles as the temperature increases in general, since the resistivity due to phonon scattering increases with temperature.

Finally we compare the GMR effect with the anisotropic magnetoresistance (AMR) effect exhibited by ferromagnetic materials. As the name suggests, AMR depends upon the relative direction of **H** and **j**; the longitudinal resistivity increases with the external field while the transverse resistivity decrease with the field. It is the resistivity difference between the two geometries that has been employed in many magnetoresistive devices. In granular materials, GMR is nearly isotropic since it only depends upon the degree of alignment of the magnetization axis of the particles. The AMR effect, however, still exists in these materials. A close-up view reveals that $\rho_{||} > \rho_T$ in granular Co-Ag samples. This is the remanence of AMR in a system that contains ferromagnetic entities.

It is worth mentioning that in conventional magnetoresistive materials such as permalloy the anisotropic magnetoresistance at 5 K is about 0.7μΩ-cm [20], while for granular Co-Ag the values of ρ_m are 5.5, 3.5 and 1.6 μΩ-cm for samples with T_A=330 °C, 480 °C and 605 °C. The fractional change $\Delta\rho/\rho$ at room temperature, is 3% for permalloy and 12%, 17% 19% and 5% for Co-Ag annealed at 200 °C, 330 °C, 480 °C and 605 °C respectively.

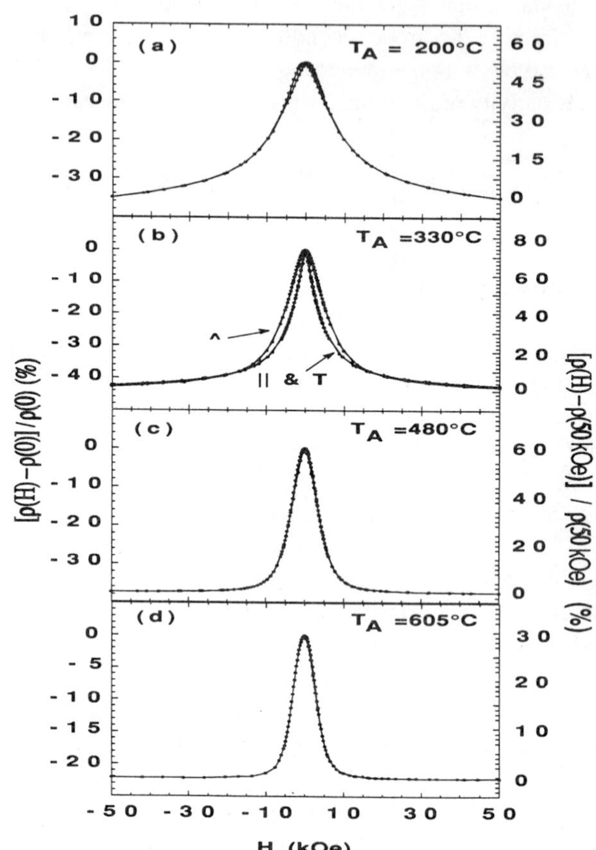

Figure 4. Perpendicular magnetoresistance of the $Co_{20}Ag_{80}$ (20 vol. % Co) samples annealed at (a) 200 °C, (b) 330 °C, (c) 480 °C, and (d) 605 °C. For the results in (b), the longitudinal ($\rho_{||}$) and transverse (ρ_T) are the inner curves which are indistinguishable, whereas the perpendicular ($\rho_{\perp}$) is the outer curve.

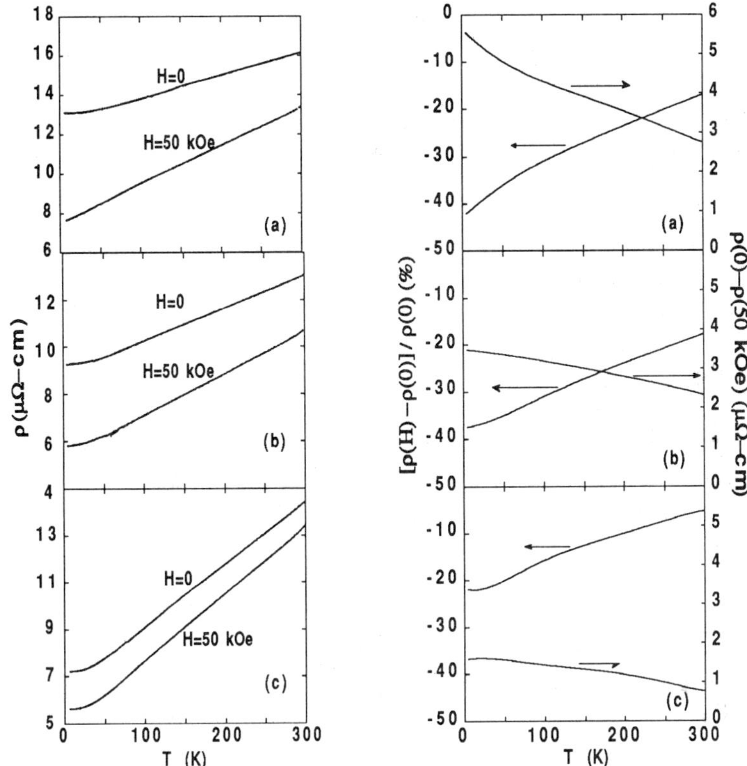

Figure 5. Temperature dependencies of the resistivity at $H = 0$ and 50 kOe (left panel), giant magnetoresistance $[\rho(H)-\rho(0)]/\rho(0)$ and $\rho(0,T) - \rho(50\ \text{kOe},T) \approx \rho_m(T)$ (right panel) of $Co_{20}Ag_{80}$ annealed at (a) 330 °C, (b) 480 °C, and (c) 605 °C.

In conclusion, we have observed giant magnetoresistance in granular magnetic materials. Layered structure is not a prerequisite for GMR. The magnetic scattering of the conduction electrons by the single-domain particles contribute to the extra resistance. GMR is determined by the particle size and density. Since the resistivity depends only upon the degree of alignment of the magnetization axis, GMR in granular systems is isotropic. Granular materials offer significantly larger magnetoresistance effect, yet the simplicity in their fabrication facilitates an easy alternative for exploiting this new phenomena.

ACKNOWLEDGEMENT

This work has been supported by Office of Naval Research under grant No. N00014-91-J-1633.

References

1. M. N. Baibich, J. M. Broto, A. Fert, F. Nguyen van Dau, F. Petroff, P. Etienne, G. Creuzet, A. Friederich, and J. Chazeles, Phys. Rev. Lett. **61**, 2472 (1988).
2. S. S. P. Parkin, R. Bhadra, and K. P. Roche, Phys. Rev. Lett. **66**, 2152 (1991).
3. F. Petroff, A. Barthelemy, D. H. Mosca, D. K. Lottis, A. Fert, P. A. Schroeder, W. P. Pratt, Jr., R. Loloee, and S. Lequien, Phys. Rev. **B44**, 5355 (1991).

4. W. P. Pratt, Jr., S. F. Lee, J. M. Slaughter, R. Loloee, P. A. Schroeder, and J. Bass, Phys. Rev. Lett. **66**, 3060 (1991).
5. S. S. P. Parkin, Appl. Phys. Lett. **60,** 512 (1992).
6. R. E. Camley and J. Barnes, Phys. Rev. Lett. **63**, 664 (1989).
7. P. M. Levy, S. Zhang, and A. Fert, Phys. Rev. Lett. **65**, 1643 (1990).
8. S. Zhang and P. M. Levy, J. Appl. Phys. **69**, 4786 (1991).
9. See e. g., A. Fert, *Science and Technology of Nanostructured Magnetic Materials*, edited by G. C. Hadjipanayis and G. A. Prinz, (Plenum Press, New York) 1991, p. 221.
10. B. Dieny, V. S. Speriosu, S. S. P. Parkin, B. A. Gurney, D. R. Wilhoit, and D. Mauri, Phys. Rev. **B43**, 1297 (1991).
11. A. Chaiken, G. A. Prinz, and J. J. Krebs, J. Appl. Phys. **87**, 4892 (1990).
12. P. Baumgart, B. A. Gurney, D. R. Wilhoit, T. Nguyen, B. Dieny and V. S. Speriosu, J. Appl. Phys. **69**, 4792 (1991).
13. A. Berkowitz, A. P.Young, J. R. Mitchell, S. Zhang, M. J. Carey, F. E. Spada, F. T. Parker, A. Hutten, and G. Thomas, Phys. Rev. Lett. **68**, 3745 (1992).
14. J. Q. Xiao, J. S. Jiang, and C. L. Chien, Phys. Rev. Lett. **68**, 3749 (1992).
15. J. Q. Xiao, J. S. Jiang, and C. L. Chien, Phys. Rev. **B** (Oct. 1, 1992).
16. See e.g., C. L. Chien, J. Appl. Phys. **69**, 5267, (1990) and references therein.
17. J. R. Childress, C. L. Chien, and M. Nathan, Appl. Phys. Lett. **56**, 95 (1990).
18. J. R. Childress and C. L. Chien, J. Appl. Phys. **70**, 5885 (1991).
19. J. R. Childress and C. L. Chien, Phys. Rev. **B43**, 8089 (1991).
20. See e.g., T. R. McGuire and R. I. Potter, IEEE Trans. Mag. **MAG-11**, 1018 (1975).

INTERLAYER EXCHANGE COUPLING: RKKY THEORY AND BEYOND

P. Bruno and C. Chappert

Institut d'Electronique Fondamentale, CNRS UA 022
Bât. 220, Université Paris-Sud, F-91405 Orsay CEDEX, France

1 INTRODUCTION

The problem of exchange coupling between ferromagnetic transition-metal layers separated by a non-magnetic metal spacer is currently attracting considerable attention. The impetus has been given by the discovery, for Fe/Cr/Fe and Co/Ru/Co systems, that the coupling oscillates periodically in sign and magnitude as a function of the non-magnetic spacer thickness [1]. It has been subsequently found that oscillatory coupling occurs indeed for almost all transition and noble metals (as spacer material) [2]. The explaination of this remarkable phenomenon is a challenge to the theory and has motivated a number of theoretical investigations.

The oscillatory behavior bears much resemblance with the one observed for Ruderman-Kittel-Kasuya-Yosida (RKKY) interactions between magnetic impurities. Thus the RKKY interaction appears a good candidate for the mechanism of oscillatory interlayer coupling. However, when applied in its simplest version (i.e. making a free-electron approximation and assuming a uniform continuous spin distribution within the ferromagnetic layers) [3], the RKKY theory predicts a period $\Lambda = \lambda_F/2 \approx 1$ monolayer (ML), which is much shorter than the experimental ones. It has been shown then that long periods can indeed be obtained within a simple RKKY theory, provided that the *discreteness* of the spacer thickness is taken into account [4]. Recently, we have presented a general theory of RKKY interlayer exchange coupling [5]; in this approach, the coupling is related in a physically transparent manner to the topological properties of the Fermi surface of the spacer material. In particular, we showed that *multiple periods* may occur, depending on the in-plane atomic structure. For the first time, quantitative predictions were obtained for the oscillation periods in the case of noble-metal spacers; these predictions have been well verified by the experiment. The RKKY theory of interlayer coupling will be reviewed in Sec. 2.

In spite of its success for predicting oscillation periods, the RKKY theory is not capable of describing correctly the *magnitude* and the *phase* of the coupling oscillations, essentially because it relies on a poor description of the interaction between the

Magnetism and Structure in Systems of Reduced Dimension
Edited by R.F.C. Farrow *et al.*, Plenum Press, New York, 1993

ferromagnetic layers and the conduction electrons. In order to improve over the RKKY approach and be able to describe consistently the coupling *amplitude* and *phase* of the oscillations, one has to treat explicitely the hybridization between the ferromagnetic layers and the spacer. This can be done in terms of *virtual bound states*, by analogy with the Friedel-Anderson theory of dilute magnetic alloys [6]. This model yields *periods* consistent with the RKKY theory, which is satisfactory. In addition, the *amplitude* and *phase* of the oscillatory coupling are found to be determined, respectively, by the *spin polarization* and the *occupancy* of the *virtual bound states* at some specific points of the two-dimensional Brillouin zone. This model will be presented in Sec. 3.

2 RKKY THEORY

2.1 Model

We consider two ferromagnetic monolayers F1 and F2 embedded in a non-magnetic metal. The distance between F1 and F2 is $z = (N+1)d$, where d is the spacing between atomic planes and N the number of atomic planes of the spacer. The magnetic layers considered here consist of a single atomic layer, whereas those used for experimental studies are usually thicker; nevertheless, it has been found experimentally that the interlayer exchange coupling is roughly independent of the thickness of the magnetic layers, so that most of the coupling can be ascribed to the outmost magnetic planes. Thus, as a first approximation, a model with monoatomic magnetic layers is expected to describe correctly the coupling phenomenon. The magnetic layers are assumed to consist of spins $\mathbf{S}_i$ located on the atomic positions $\mathbf{R}_i$ of the host metal. The magnetic layers are thus *coherent* with the spacer; this assumption plays an important rôle in the problem. It is expected to be relevant for epitaxially grown systems.

A magnetic layer (say F1) interacts with the conduction electrons of the host material, and induces a spin-polarization around it. This polarization is propagated accross the spacer and eventually interacts with F2, giving rise to an effective exchange interaction between F1 and F2. The problem of the exchange coupling between F1 anf F2 can thus be splitted into two aspects: (i) the interaction between a ferromagnetic layer and the host conduction electrons, and (ii) the way the spin-polarization is propagated across the host material. For the case of transition-metal magnetic impurities, aspect (i) is usually ascribed to the so-called s-d mixing interaction [7–9]. Investigations of the interlayer exchange coupling on the basis of s-d mixing have been done by various authors [6, 23]; this approach is more sophisticated than the RKKY one. It will be discussed in Sec. 3.

In the present Section, we aim to focus on aspect (ii) of the problem, in particular on the selection of the oscillation periods. For this purpose, it is sufficient to approximate the coupling between the spins $\mathbf{S}_i$ and the conduction electrons (spin $\mathbf{s}$, position $\mathbf{r}$) by a contact potential

$$\mathcal{V}_i(\mathbf{r}, \mathbf{s}) = A\,\delta(\mathbf{r} - \mathbf{R}_i)\;\mathbf{s} \cdot \mathbf{S}_i\;; \tag{1}$$

this is the form originally used by Ruderman and Kittel for investigating the indirect exchange coupling between nuclear spins, and extended later by Kasuya and Yosida [11]. The contact interaction (1), applied to transition-metal spins, is a rather crude approximation; it usually predicts incorrect phases for the oscillatory coupling, and the coupling strength is described by an adjustable parameter, A. These limitations of the RKKY model should be kept in mind when comparing its predictions with experimental results. On the other hand, it yields correct results for the oscillation periods, which are the quantities of interest here.

The RKKY interaction between two spins $\mathbf{S}_i$ and $\mathbf{S}_j$ is [11]

$$\mathcal{H}_{ij} = J(\mathbf{R}_{ij})\ \mathbf{S}_i \cdot \mathbf{S}_j\ , \tag{2}$$

where the exchange integral $J(\mathbf{R}_{ij})$ is essentially the Fourier transform of the non-uniform susceptibility $\chi(\mathbf{q})$. The interlayer coupling is obtained by summing $\mathcal{H}_{ij}$ over all the pairs ij, i and j running respectively on F1 and F2. The coupling energy per unit area can be written

$$E_{1,2} = I_{1,2} \cos\theta_{1,2}\ , \tag{3}$$

where $\theta_{1,2}$ is the angle between the magnetizations of F1 and F2. The interlayer coupling constant $I_{1,2}$ is given by

$$I_{1,2} \sim \sum_{j \in \mathrm{F2}} J(\mathbf{R}_{Oj})\ , \tag{4}$$

where O labels one site of F1 taken as the origin. Note that within the present sign convention, positive (respectively negative) values of $I_{1,2}$ correspond to antiferromagnetic (resp. ferromagnetic) coupling.

2.2 Free-Electron Approximation

Before attempting to calculate the coupling in a general case, it is instructive to first examine it within the free electron approximation. The calculations can then be performed in an almost completely analytical manner, so that the results are physically transparent; moreover, many features of the coupling thus obtained remain qualitatively valid in more realistic situations. In all this Section, the host material will be approximated by a free-electron gas with the same density; since the model will be applied to noble metals in this paper, we consider fcc host materials with one conduction electron per atomic cell, so that the Fermi vector is $k_F = (12\pi^2)^{1/3}/a$, where a is the lattice parameter. The spins $\mathbf{S}_i$ are still supposed to be located on the atomic sites of the fcc lattice. The exchange integral within the free-electron approximation is given by the well known expression [11]

$$J(\mathbf{R}) \sim\ F(2k_F R)\ , \tag{5}$$

with

$$\begin{aligned} F(x) &= \frac{x \cos x\ -\ \sin x}{x^4} \\ &\approx \frac{\cos x}{x^3} \qquad \text{for } x \to +\infty\ ; \end{aligned} \tag{6}$$

it oscillates with a period $\Lambda = \lambda_F/2$ and decays as R^{-3}. The oscillatory behavior arises from the logarithmic singularity at $q = 2k_F$ of the susceptibility $\chi(\mathbf{q})$ for a free-electron gas.

As a first approximation, one may try to replace the actual ferromagnetic layers by a continuous uniform distribution of spins with the same spin density, i.e. we perform in Eq. (4) the substitution

$$\sum_{\mathrm{F2}} \quad \longrightarrow \quad \frac{d}{V_0} \int_{\mathrm{F2}} d^2\mathbf{R}_{\|}\ . \tag{7}$$

In the above equation, $\mathbf{R}_{\parallel}$ is the in-plane projection of $\mathbf{R}_{Oj}$. The interlayer coupling is then given as a function of the distance z by

$$I_{1,2}(z) \sim \frac{d^2}{z^2} \sin(2k_F z) \qquad \text{for } z \to \infty \ , \tag{8}$$

There is a single oscillation period $\Lambda = \lambda_F/2$, and the coupling decays as z^{-2}; this result was first obtained by Yafet [3].

At first sight, the prediction of a period $\lambda_F/2$ much shorter than any observed period might seem to invalidate the RKKY mechanism. Actually, it has been pointed out by several authors [4] that this apparent discrepancy can be removed by a very simple argument: the spacer thickness $z = (N+1)d$ (with N integer) is not a continuous variable, so that an *effective period* much larger than $\lambda_F/2$ may result from this discrete sampling: this effect is known as *aliasing*. Thus the *effective period* Λ associated to a given q vector is given by

$$\frac{2\pi}{\Lambda} = \left| q - n\frac{2\pi}{d} \right| \ , \tag{9}$$

where n is chosen such that $\Lambda > 2d$ (periods smaller than $2d$ would be meaningless for a discrete variable).

We want now to examine the validity of the continuous approximation (7) which was used in early papers on RKKY interlayer coupling [3, 4]. When performing the continuous integration (7) over F2, the integrand is a function of $R_{\parallel}$ which oscillates with a period of the order of $\lambda_F/2$; thus if the interatomic distance b within the plane is smaller than $\lambda_F/2$ we may expect the continuous approximation (7) to be valid; on the other hand, if b large as compared to $\lambda_F/2$, it is clear that approximation (7) must break down.

In order to develop this argument in a more quantitative fashion, we perform explicitely the summation (4) without making the continuous approximation (7):

$$I_{1,2}(z) \sim \int_{-\infty}^{+\infty} dq_z \ \exp(iq_z z) \int d^2\mathbf{q}_{\parallel} \ \chi(\mathbf{q}_{\parallel}, q_z) \sum_{\mathbf{R}_{\parallel} \in \mathrm{F2}} \exp(i\mathbf{q}_{\parallel} \cdot \mathbf{R}_{\parallel}) \ . \tag{10}$$

Because of the translational invariance in the layer plane, the last sum in the above equation is zero unless $\mathbf{q}_{\parallel}$ is a vector $\mathbf{G}_{\parallel}$ belonging to the (two-dimensional) reciprocal lattice of F2. Thus the expression of the coupling becomes

$$I_{1,2}(z) \sim \sum_{\mathbf{G}_{\parallel}} (\cdots) \int_{-\infty}^{+\infty} dq_z \ \exp(iq_z z) \ \chi(\mathbf{G}_{\parallel}, q_z) \ . \tag{11}$$

As already mentioned above, the susceptibility $\chi(\mathbf{q})$ for a free-electron gas depends only on q, and has a logarithmic singularity in its derivative at $q = 2k_F$ (Kohn singularity), which is responsible for the long-range oscillatory behavior of the exchange coupling. Thus, the contribution to $I_{1,2}(z)$ corresponding to a given vector $\mathbf{G}_{\parallel}$ gives a long-range oscillatory interlayer coupling if, when integrating over q_z, one crosses a singularity, i.e. if $G_{\parallel} < 2k_F$; the corresponding wave vector is $[(2k_F)^2 - G_{\parallel}^2]^{1/2}$. Thus the expression of the interlayer coupling in the limit of large thicknesses is

$$I_{1,2}(z) \sim \frac{d^2}{z^2} \sum_{G_{\parallel} < 2k_F} (\cdots) \sin\left\{ \left[(2k_F)^2 - G_{\parallel}^2\right]^{1/2} z \right\} \ . \tag{12}$$

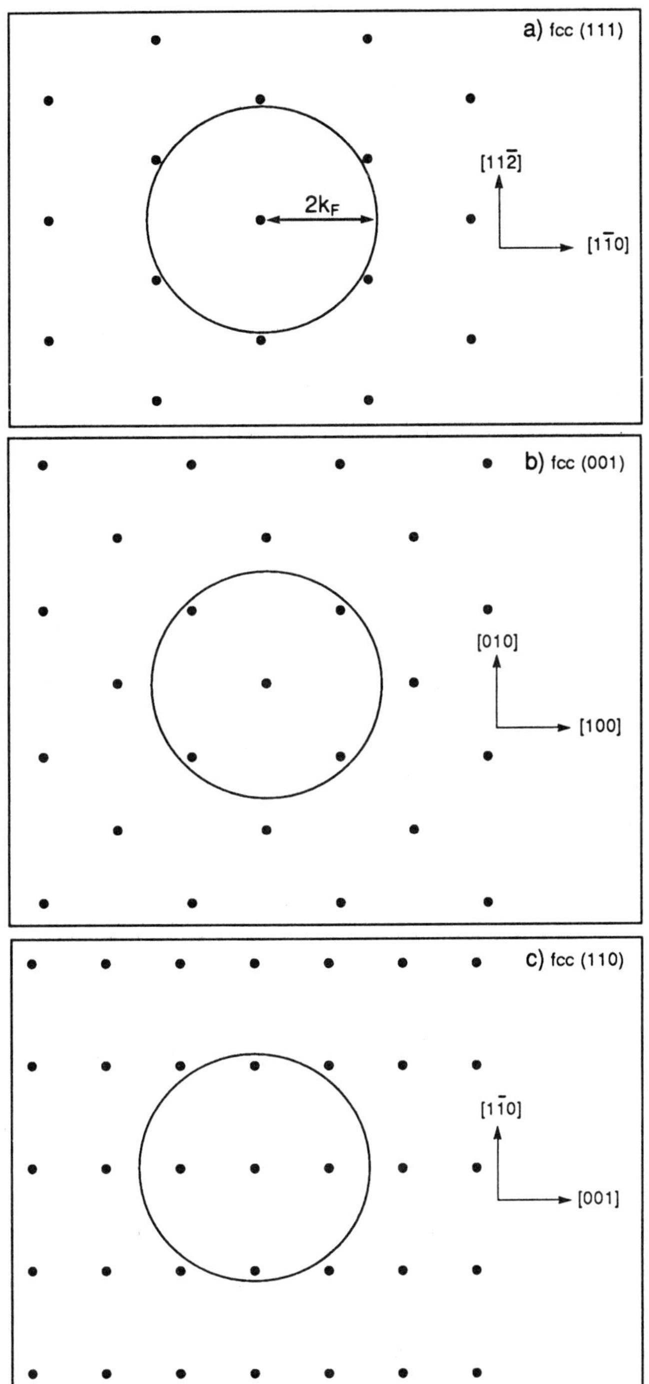

Figure 1. Two-dimensional reciprocal lattice for fcc layers; (a), (b), and (c) correspond, respectively, to the (111), (001), and (110) orientations. The sphere of radius $2k_F$ is the locus of singularities of the susceptibility $\chi(\mathbf{q})$ for a free-electron gas. The unit vectors have a length $2\pi/a$.

It is clear from the above discussion that the multiperiodicity is related to the discrete atomic structure within the layers, and that the number of oscillation periods increases as the in-plane atomic density decreases. This trend is well exemplified for the case of a fcc spacer: as shown on Fig. 1, the number of different oscillation periods for the (111), (001) and (110) orientations is respectively 1, 2 and 3.

2.3 General Theory

The Fermi surface of real metals departs markedly from the spherical one of a free-elctron gas. Thus, in order to be able to make reliable quantitative predictions, one needs to release the free-electron approximation and to formulate a general RKKY theory of interlayer coupling, valid for non-spherical Fermi surfaces. This has been done in Ref. [5], where we showed that the oscillation periods are given by the vectors q_z^α parallel to the z direction, which span the Fermi surface and such that the Fermi velocities at both extremities are antiparallel to each other (see Fig. 2). The period Λ_α corresponding to a given vector q_z^α is $\Lambda_\alpha = 2\pi/q_z^\alpha$ (it is always possible to chose q_z^α such that $\Lambda_\alpha > 2d$).

Since the noble metals have Fermi surfaces which are known experimentally with a very high accuracy from de Haas-van Alphen and cyclotron resonance measurements [12], they appear as ideal candidates to test the predictions of the RKKY theory. Using the Fermi surface data from Ref. [12], we have calculated the oscillation periods of interlayer coupling for Cu, Ag, and Au, in (001), (111), and (110) orientations [5]. Figure 3 shows a $(1\bar{1}0)$-cross section of the Fermi surface of a noble metal, and the vectors giving the oscillation periods. The number of oscillation periods is 1, 2, and 4, respectively, for the (111), (001), and (110) orientations (for the latter, in addition to the period shown in Fig. 3, there are 3 other periods that cannot be seen from the present cross section). Thus the trend obtained within the free-electron approximation, stating that the number of periods increases with decreasing in-plane density, remains valid for noble metals.

The comparison between the periods observed experimentally and those predicted by the RKKY theory is shown in Table 1. For Cu(111), the observed period is somewhat larger than the predicted one; however, the difference is not dramatic and may be attributed to experimental uncertainties, and/or to the influence of internal strains on the Fermi surface. The (001) orientation is of particular interest because the RKKY theory predicts the coexistence of a short and a long period: this has been confirmed subsequently for Cu(001) by Johnson *et al.* [18],for Au(001) by Fuss *et al.* [19], and most recently for Ag(001) by Unguris [20]. This is a major success of the RKKY theory. For Cu(110), the long period observed by Johnson *et al.* [21] is in good agreement with the one predicted by the RKKY theory (it corresponds to the "neck" diameter); the predicted short-period oscillations are not observed: they are possibly smeared out by the roughness.

Note also that the RKKY theory provides a consistent interpretation of the results of first-principles calculations by Herman *et al.* [22], which predict the presence of a short-period oscillation ($\approx$ 2 ML's) for Cu(001), Cu(110), bcc Cu(001), and not for Cu(111).

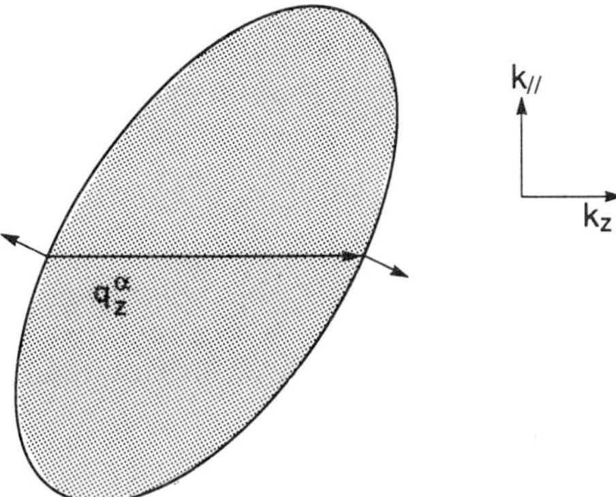

Figure 2. Sketch showing the vector q_z^α which gives the oscillation period for the general case of a non-spherical Fermi surface. The small arrows are the velocity vectors.

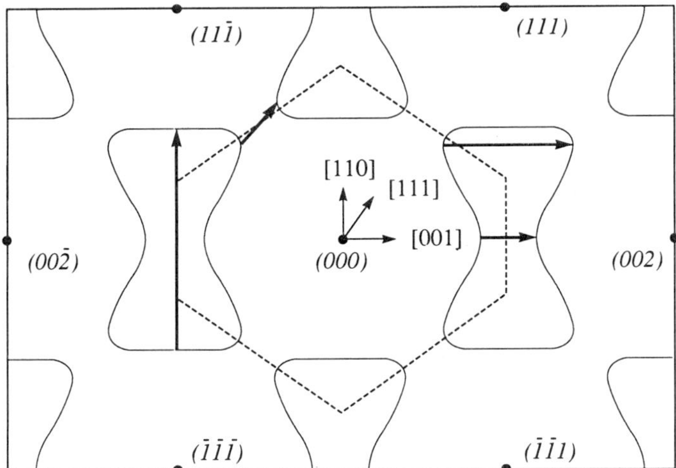

Figure 3. Fermi surface of a (bulk) noble metal: (1$\bar{1}$0) cross-section. The bold points belong to the fcc reciprocal lattice. The first Brillouin zone is indicated by the dashed contour. The horizontal, oblique, and vertical bold arrows are the vectors giving the oscillation periods, respectively, for the (001), (111), and (110) orientations.

Table 1. Comparison between the oscillation periods predicted by the RKKY theory (Ref. [5]) for noble metals and those observed experimentally.

Theory		Experiment		
Spacer	Period(s)	System	Period(s)	Ref.
Cu(111)	$\Lambda = 4.5$ ML's	Co/Cu/Co	$\Lambda \approx 6$ ML's	[13]
		Co/Cu/Co	$\Lambda \approx 5$ ML's	[14]
		Fe/Cu/Fe	$\Lambda \approx 6$ ML's	[15]
Cu (001)	$\Lambda_1 = 2.6$ ML's $\Lambda_2 = 5.9$ ML's	Co/Cu/Co	$\Lambda \approx 6$ ML's	[16]
		Fe/Cu/Fe	$\Lambda \approx 7.5$ ML's	[17]
		Co/Cu/Co	$\Lambda_1 \approx 2.6$ ML's $\Lambda_2 \approx 8$ ML's	[18]
Au (001)	$\Lambda_1 = 2.6$ ML's $\Lambda_2 = 8.6$ ML's	Fe/Au/Fe	$\Lambda_1 \approx 2$ ML's $\Lambda_2 \approx 7$–8 ML's	[19]
Ag (001)	$\Lambda_1 = 2.4$ ML's $\Lambda_2 = 5.6$ ML's	Fe/Ag/Fe	$\Lambda_1 \approx 2.4$ ML's $\Lambda_2 \approx 5.6$ ML's	[20]
Cu (110)	$\Lambda_1 = 2.1$ ML's $\Lambda_2 = 2.5$ ML's $\Lambda_3 = 3.3$ ML's $\Lambda_4 = 9.6$ ML's	Co/Cu/Co	$\Lambda \approx 9.8$ ML's	[21]
bcc Cu(001)[a]	$\Lambda_1 = 2.2$ ML's $\Lambda_2 = 2.6$ ML's	Fe/Cu/Fe	$\Lambda \approx 2$ ML's	[18]

[a]ASW calculation of the bulk Fermi surface of bcc Cu, from Ref. [18].

3 BEYOND THE RKKY THEORY: THE "VIRTUAL BOUND STATE" MODEL

In spite of its considerable success in predicting period(s) of oscillatory coupling, the RKKY theory is, as already mentioned in the Introduction, not capable of describing consistently the *magnitude* of the coupling and the *phase* of the oscillations. This is essentially because the *contact* interaction (1) represents a very rough approximation of the interaction between the (transition metal) ferromagnetic layers and the conduction electrons of the spacer material.

Actually, this problem bears much resemblance with that of magnetic impurities embedded in a non-magnetic host metal, which has attracted considerable attention in the past. In particular, the concept of *virtual bound state*, introduced by Friedel [7], has proved to be extremely useful in explaining the electric and magnetic properties of dilute magnetic alloys. For transition-metal impurities and noble-metal host, the Anderson Hamiltonian [8] provides a modelization of the system which is both computationally convenient and fairly realistic.

In this approach, the impurities are described by localized d-levels, which are coupled to the conduction electrons of the host metal by the so-called s-d hybridization. As a consequence of the latter, the d-levels are broadened into *virtual bound states*. In the case of magnetic impurities, they can be exchange-splitted; the stability of these *localized moments* is governed by a Stoner-like criterion [8]. The first calculation of indirect exchange coupling between two impurities on the basis of the Anderson s-d mixing model is due to Caroli [9].

It is tempting to adapt this approach to the problem of interlayer exchange coupling. The magnetic impurities are thus to be replaced by two-dimensional ferromagnetic layers. An important difference is that the d-electrons are delocalized in the latter case, and have a (two-dimensional) energy band of sizeable width instead of isolated levels as in the former case. This itinerant character is a intrinsic feature of transition metal ferromagnets, and must be taken into account for a realistic description of the interlayer coupling. At first sight, we are thus faced with the difficulty that the effect of the s-d mixing can no longer be described in terms of broadening the isolated energy levels into virtual bound states. Wang *et al.*, and Lacroix and Gavigan, circumvented this difficulty simply by setting to zero the width of the d-electrons energy bands of the ferromagnetic layers [23]; it is obvious that this approximation of neglecting of the itinerant character of the d-electrons is a very unrealistic one.

Actually, as we showed recently [6], the problem can be solved without making use of this unsatisfactory approximation. The clue to the problem is to take advantage of the in-plane translational invariance of the problem, i.e. of the fact that the total Hamiltonian is diagonal in $\mathbf{k}_{||}$: now, for a given $\mathbf{k}_{||}$ the energy levels of the magnetic layers are localized levels and it is clear that the *virtual bound states* are to be defined *locally in the* $\mathbf{k}_{||}$ *plane*. Then, the calculation of the coupling is completely analogous to that of Caroli for impurities [9].

The detailled description of this theory of interlayer coupling is given in Ref. [6], and the results are completely analogous to those obtained for the exchange interaction between magnetic impurities in dilute alloys [9]. The respective influence of the spacer material and of the ferromagnetic layers on the various characteristics of the coupling (oscillation periods, intensity, phase, thermal variation) is displayed transparently:
(i) The *oscillation periods*, the *decay*, and the *thermal variation* of the coupling are the same as from the RKKY theory [5]; they depend only on the Fermi surface characteristics of the spacer material and are independent of the detailled mechanism of

the interaction between the magnetic layers and the conduction electrons of the spacer. This is basically why they are correctly described by the RKKY theory.
(ii) The *magnitude* and the *phase* of the oscillatory coupling depend on the magnetic metal; they are related, respectively, to the *spin polarization* and the *occupancy* of the virtual bound states at specific in-plane vectors $\mathbf{k}_{||}$: those corresponding to stationary values q_z^α of the vector q_z spanning the Fermi surface measure the z direction.

Among the recent experimental results in this field is the observation that (111) Fe/Cu/Fe and Co/Cu/Co multilayers exhibit oscillatory interlayer coupling with same period ($\Lambda \approx 6$ ML's), but with phases differing by about π [15]. The above theory provides a natural explanation for this behavior: because of the valence difference between Fe and Co, one may expect the occupancy of the virtual bound states to be different in both systems.

4 CONCLUSION

We have reviewed the RKKY and "virtual bound state" models of interlayer exchange coupling. A detailled discussion of the RKKY model has shown that, even within the simple free-electron approximation, experimental behaviors which were *a priori* unexpected, such as long periods and multiperiodicity, can be interpreted in a natural manner as a consequence of the crystalline atomic structure of the layers. The general RKKY theory, which accounts for non-spherical Fermi surfaces, has encountered considerable success in predicting the oscillation periods for noble-metal spacers.

The RKKY theory, on the other hand, fails in describing consistently the *magnitude* and the *phase* of the coupling oscillations. In an attempt to improve over the RKKY approach, the "virtual bound state" theory emphasizes the rôle played by the ferromagnetic layers, which may be expressed in terms of virtual bound states at some specific $\mathbf{k}_{||}$ vectors. The analogy with the classical problem of magnetic-impurity solutions, on which this approach relies, proves to be very fruitful.

REFERENCES

[1] S.S.P. Parkin, N. More, and K.P. Roche, Phys. Rev. Lett. **64**, 2304 (1990)

[2] S.S.P. Parkin, Phys. Rev. Lett. **67**, 3598 (1991)

[3] Y. Yafet, Phys. Rev. B **36**, 3948 (1987)

[4] C. Chappert and J.P. Renard, Europhys. Lett. **15**, 553 (1991); R. Coehoorn, Phys. Rev. B **44**, 9331 (1991)

[5] P. Bruno and C. Chappert, Phys. Rev. Lett. **67**, 1602 (1991); ibid. **67**, 2592(Erratum) (1991); Phys. Rev. B **46** (1992), in press

[6] P. Bruno, to be published

[7] J. Friedel, Nuovo Cimento Suppl. **7**, 287 (1958)

[8] P.W. Anderson, Phys. Rev. **124**, 41 (1961); T. Moriya, *Spin Fluctuations in Itinerant Electron Magnetism* (Springer-Verlag, Berlin, 1985), Chap. 6

[9] B. Caroli, J. Phys. Chem. Solids **28**, 1427 (1967)

[10] Y. Wang, P.M. Levy, and J.L. Fry, Phys. Rev. Lett. **65**, 2732 (1990); C. Lacroix and J.P. Gavigan, J. Magn. Magn. Mat. **93**, 413 (1991)

[11] M.A. Ruderman and C. Kittel, Phys. Rev. **96**, 99 (1954); T. Kasuya, Progr. Theor. Phys. **16**, 45 (1956); K. Yosida, Phys. Rev. **106**, 893 (1957)

[12] M.R. Halse, Philos. Trans. R. Soc. London, Ser. A **265**, 507 (1969)

[13] D.H. Mosca, F. Petroff, A. Fert, P.A. Schroeder, W.P. Pratt, Jr., R. Loloee, and S. Lequien, J. Magn. Magn. Mat. **94**, L1 (1991)

[14] S.S.P. Parkin, R. Bhadra, and K.P. Roche, Phys. Rev. Lett. **66**, 2152 (1991)

[15] F. Pétroff, A. Barthélémy, D.H. Mosca, D.K. Lottis, A. Fert, P.A. Schroeder, W.P. Pratt, R. Loloee, and S. Lequien, Phys. Rev. B **44**, 5355 (1991)

[16] J.J. de Miguel, A. Cebollada, J.M. Gallego, R. Miranda, C.M. Schneider, P. Schuster, and J. Kirschner, J. Magn. Magn. Mat. **93**, 1 (1991)

[17] W.R. Bennett, W. Schwarzacher, and W.F. Egelhoff, Jr., Phys. Rev. Lett. **65**, 3169 (1990)

[18] M.T. Johnson, S.T. Purcell, N.W.E. McGee, R. Coehoorn, J. aan de Stegge, and W. Hoving, Phys. Rev. Lett. **68**, 2688 (1992)

[19] A. Fuss, S. Demokritov, P. Grünberg, and W. Zinn, J. Magn. Magn. Mat. **103**, L221 (1992)

[20] J. Unguris, these proceedings

[21] M.T. Johnson, R. Coehoorn, J.J. de Vries, N.W.E. McGee, J. aan de Stegge, and P.J.H. Bloemen, to be published

[22] F. Herman, J. Sticht, and M. van Schilfgaarde, Mat. Res. Soc. Symp. Proc. Vol. **231**, 195 (1992)

[23] Y. Wang, P.M. Levy, and J.L. Fry, Phys. Rev. Lett. **65**, 2732 (1990); C. Lacroix and J.P. Gavigan, J. Magn. Magn. Mat. **93**, 413 (1991).

RECENT DEVELOPMENTS IN THE THEORY OF OSCILLATORY EXCHANGE COUPLING IN MAGNETIC MULTILAYERS

D. M. Edwards[1], J. Mathon[2], R. B. Muniz[3]
Murielle Villeret[2] and J. M. Ward[1]

[1] Department of Mathematics, Imperial College
London SW7 2BZ, UK
[2] Department of Mathematics, City University
London EC1V 0HB, UK
[3] Departamento de Fisica, Universidade Federal
Fluminense, Niteroi, RJ 24020, Brazil

INTRODUCTION

Much of our work has already been published[1-3] so we shall concentrate here on a few recent topics. First we analyse the simplest version of our model in which the electrons (or holes) in the multilayer or sandwich are treated as a Fermi gas with delta-function interactions. This model has an exact Hartree-Fock solution which illustrates the general principles of bilinear and biquadratic exchange. We report the first full treatment of intrinsic biquadratic exchange in sandwich structures, including the temperature dependence. Later in the article we summarize results of our recent calculations using more realistic band structures appropriate to Fe/Cr or Fe/Mo sandwiches.

We consider a sandwich in which two semi-infinite ferromagnetic metals are separated by a non-magnetic spacer layer containing N atomic planes. RKKY-type theories of exchange coupling are most appropriate for simple or noble metal spacers, and Bruno[4] has given a nice theory of the coupling between two magnetic monolayers in such a medium. For transition metal spacers, which are our concern, the difference in energy between different magnetic configurations should largely be determined by the difference in the total one-electron energy of the d-bands. Later in the article we consider real five-orbital d-bands but in the next section holes in the d band are treated as a gas. Bulk Ni and Co have no holes in the majority spin band so that a magnetic layer of one of these metals effectively excludes holes of one spin orientation ("complete hole confinement"). For Fe, however, we speak of partial confinement since holes of both spin orientations penetrate the bulk of the magnetic layer, although at much reduced density in the majority spin band. We assume in all cases that, as in Fe/Cr for example, there is a good match between the spacer d band and the minority spin d band of the magnetic metal so that electrons or holes of this spin orientation move freely across the interface.

If the spacer occupies the region $|z| < \ell/2$ the hole gas Hamiltonian is

$$H = \sum_i [\frac{p_i^2}{2m} + V(\vec{r_i})] + \sum_{i<j} U(\vec{r_i})\delta(\vec{r_i} - \vec{r_j}) \tag{1}$$

with

$$\begin{array}{lll} V(\vec{r}) = V_S, & U(\vec{r}) = 0 & (|z| < \ell/2) \\ V(\vec{r}) = V_M, & U(\vec{r}) = U & (|z| > \ell/2). \end{array} \tag{2}$$

Here $\vec{p_i}$ and $\vec{r_i}$ are the momentum and position of particle i, $\vec{r} = (x, y, z)$, and V_S, V_M and U are constant potentials. If the density of particles of each spin in the bulk spacer is n_S and the densities for each spin in the magnet are n_1 and n_2 ($n_1 > n_2$), the corresponding bulk Hartree-Fock potentials in the magnet are $V_M + Un_2$ and $V_M + Un_1$. The perfect matching in the minority spin band mentioned at the end of the introduction implies

$$n_1 = n_S, \qquad V_M + Un_2 = V_S \tag{3}$$

in our hole gas model. Hathaway and Cullen[5] introduced a model in which the *electrons* are treated as a gas and the same physical matching condition, which they also use, then implies the different relations

$$n_2 = n_S, \qquad V_M + Un_1 = V_S. \tag{4}$$

We shall concentrate on the complete confinement limit of the hole gas model with $n_2 = 0$, $V_M = V_S = 0$ and $U \to \infty$. This simple limit cannot be taken in the Hathaway-Cullen model. It has the advantage that exact Hartree-Fock solutions of the Hamiltonian (1) correspond to potentials which are everywhere either zero or infinite, and the total energy is rigorously just a sum of easily obtainable one-electron energies. The interaction energy is zero because with delta-function interactions only particles of opposite spin interact and they do not coexist in the magnetic regions where $U(\vec{r})$ is different from zero.

To determine the interlayer coupling we need to calculate the thermodynamic potential $\Omega(\theta)$ of the model for arbitrary angle θ between unit vectors $\hat{M}_1$ and $\hat{M}_2$ in the directions of magnetization of the two magnetic layers. It is convenient to define

$$J(\theta) = \Omega(0) - \Omega(\theta) \tag{5}$$

and our usual method[1,2] is easily applied to calculate $J = J(\pi)$, the energy difference per unit area between parallel and antiparallel magnetic configurations. To generalize the result to arbitrary θ it is convenient to use the torque method of Hathaway and Cullen[5], previously introduced by Slonczewski[6] for an insulating spacer. Details of the calculation will be given elsewhere and we just quote the exact Hartree-Fock result for the complete confinement model with arbitrary spacer thickness ℓ, at $T = 0$:

$$J(\theta) = \frac{N_{2d}E_F^2}{\pi k_F^2 \ell^2} \sum_{n=1}^{\infty} \frac{1}{n^3} [1 - cos^{2n}(\theta/2)] \{(1 - \frac{3}{n^2x^2}) sin(nx) + \frac{3}{nx} cos(nx)\}. \tag{6}$$

Here $x = 2k_F\ell$ where k_F is the Fermi wave vector in the spacer, E_F is the corresponding Fermi energy and N_{2d} is the constant density of states per unit area of the gas in two dimensions.

The $n = 1$ term in Eq. (6) gives the fundamental oscillation in $J(\theta)$ as a function of spacer thickness ℓ with period π/k_F. (If J is only observed at discrete thicknesses $\ell = Nd$, where d is the interplane spacing and N is an integer, $sin(2k_F\ell)$ may be replaced by $sin2(k_F - \pi/d)\ell$, yielding a much longer period $\pi/ \mid k_F - \pi/d \mid$ if the Fermi surface approaches the zone boundary (the "aliasing" affect)). Subsequent terms $n = 2, 3, ...$ give higher harmonics with dependences on angle θ of increasing complexity. The angular factor may be written as $(1 - cos\theta)/2$ in the basic $n = 1$ term and as $(3 - 2cos\theta - cos^2\theta)/16$ in the $n = 2$ term. A $cos\theta = \hat{M}_1 \cdot \hat{M}_2$ term corresponds to ordinary bilinear exchange and a $cos^2\theta = (\hat{M}_1 \cdot \hat{M}_2)^2$ term corresponds to biquadratic exchange. Thus, biquadratic exchange first occurs in the $n = 2$ harmonic and generally a term $(\hat{M}_1 \cdot \hat{M}_2)^n$ first occurs in the nth harmonic. It is interesting to compare $J(\theta)$ with RKKY coupling between two planes of spins immersed in the spacer gas a distance ℓ apart. RKKY only gives the $n = 1$ term (bilinear exchange) and $sinx - x^{-1}cosx$ appears[1] instead of $sinx + 3x^{-1}cosx$ with the $x^{-2}sinx$ term missing. Higher harmonics in Eq. (6) distort the basic sine wave so that for large ℓ

$$J(\pi)\ell^2 \sim \sum_{n=1}^{\infty} \frac{1}{n^3} sin(nx) \propto \eta(1-\eta)(2\eta - 1), \qquad 0 \leq \eta \leq 1 \tag{7}$$

with $\eta = x/2\pi = k_F\ell/\pi$. Examination of Eq. (6) shows that the bilinear and biquadratic terms oscillate independently, the biquadratic having half the basic period and, in general, they vanish for different spacer thicknesses ℓ.

It is straightforward to include Fermi factors in the calculation to investigate the temperature dependence of the exchange coupling. The result in the large ℓ limit is

$$J(\theta) = \frac{N_{2d}E_F}{\beta\pi k_F \ell} \sum_{n=1}^{\infty} \frac{1}{n^2} [1 - cos^{2n}(\theta/2)] \frac{sin(2nk_F\ell)}{sinh(n\pi k_F \ell/\beta E_F)} \tag{8}$$

where $\beta = (k_B T)^{-1}$. Clearly higher harmonics, including biquadratic exchange, are suppressed if $T \geq T_0$ where $k_B T_0 = \hbar v_F / 2\pi\ell$, v_F being the Fermi velocity. We have written T_0 in this form because for a general band structure the temperature dependence of the exchange coupling depends on the Fermi velocity normal to the layers in this way. For the actual gas model we can also write $k_B T_0 = E_F/\pi k_F \ell$.

The exchange energy per unit area may be written in the form

$$-J(\theta) = const. - 2A_{12}cos\theta - 2B_{12}cos^2\theta, \tag{9}$$

neglecting higher powers of $cos\theta$, where we have used the notation of the Jülich group with factors of 2 inserted according to their most recent publications[7,8]. If $A_{12} + B_{12} < 0$ and $A_{12} > B_{12}$, which implies a negative biquadratic coupling B_{12}, the stable configuration of $\hat{M}_1$ and $\hat{M}_2$ is a perpendicular one with $\theta = \pi/2$. This is frequently observed and sometimes it is possible to determine the bilinear coupling A_{12} and the biquadratic coupling B_{12} individually from magnetization curves[7]. Slonczewski[9] has proposed a mechanism in which the intrinsic exchange coupling for perfect layered structures is purely bilinear and an effective biquadratic exchange coupling arises as a secondary effect due to interface roughness. Thus, in general, the observed B_{12} is given by

$$B_{12} = B_{12}^{(i)} + B_{12}^{(S)} \tag{10}$$

where $B_{12}^{(i)}$ is the intrinsic biquadratic exchange derived in this paper and $B_{12}^{(S)}$ is due to the secondary Slonczewski effect. It is an experimental challenge to separate these two effects and the following observations may be useful:

1. $B_{12}^{(S)}$ is zero for perfect interfaces but is also small for very rough interfaces with closely-spaced steps.
2. $B_{12}^{(S)}$ only exists if A_{12} has an oscillatory component with short period (~ 2 monolayers (ML)).
3. $B_{12}^{(S)}$ is never positive whereas $B_{12}^{(i)}$ oscillates as a function of spacer thickness.
4. $B_{12}^{(S)}$ is inversely proportional to the thickness of the magnetic layers, assuming this to be smaller than the spacing between steps in the interface, whereas any dependence of $B_{12}^{(i)}$ on this quantity will be complicated and oscillatory.
5. The temperature dependence of $B_{12}^{(i)}$ is characteristic, as discussed below.

Identifying $-2A_{12}$ as the coefficient of the leading $cos\theta$ term in Eq. (8), which occurs in the $n = 1$ term, and $-2B_{12}^{(i)}$ as the coefficient of the leading $cos^2\theta$ term, which occurs in the $n = 2$ term, we have the following dependences on temperature T:

$$A_{12}(T) = A_{12}(0)\frac{T/T_0}{sinh(T/T_0)} = A_{12}(0)[1 - \frac{1}{6}(\frac{T}{T_0})^2 + ...] \tag{11}$$

and

$$B_{12}^{(i)}(T) = B_{12}^{(i)}(0)\frac{2T/T_0}{sinh(2T/T_0)} = B_{12}^{(i)}(0)[1 - \frac{2}{3}(\frac{T}{T_0})^2 + ...] \tag{12}$$

with $k_B T_0 = \hbar v_F / 2\pi\ell$ as before. The temperature dependences exhibited in Eqs. (11) and (12) are those of the first two terms ($n = 1, 2$) of a Fourier-type expansion of the exchange coupling for very general spacer band structure, not just for the gas model[1,2]. Furthermore, we expect it to be a general result that biquadratic exchange first appears in the second term of the expansion. It is therefore likely that Eqs. (11) and (12) are of general applicability and it is interesting to discuss Fe/Al/Fe (001) trilayers where the temperature dependence of B_{12} has been observed[10]. Gutierrez *et al.*[10] find that $B_{12}(T)$ decays exponentially with increasing T and, at least for one of their samples, Eq. (12) fits the data very accurately over the whole temperature range with $T_0 = 122K$[11]. Unfortunately, this value of T_0 is about 10 times smaller than that expected theoretically from the bulk Fermi velocity of Al. This is puzzling because at first sight the Slonczewski mechanism should not operate for an Al spacer and B_{12} should be described by the intrinsic theory. Al is a nearly-free-electron metal and consideration of the Fermi surface (derived from a sphere of radius k_F except very close to zone boundaries) leads us to expect only one period of oscillation in A_{12}. If the spacer thickness ℓ varied *continuously* this period would be $\pi/k_F \simeq a/2 = 1ML$ for an Al(001) spacer, where a is the lattice parameter of the fcc lattice. However, since ℓ changes by discrete steps, $\ell = Na/2$ and the "aliasing effect" yields a long period of $\pi/(k_F - 2\pi/a) = 15.9\mathring{A}$. This is compatible with the observed period[7] and there is no evidence of the additional 2ML period, as found in Cr or Au spacers, normally required for the Slonczewski mechanism. However, there is an indication of the underlying $\pi/k_F \simeq 1ML$ period in the Jülich data[7] which suggests that these measurements, and perhaps the Slonczewski mechanism, sense to some extent an averaged continuous, rather than stepwise, variation of ℓ in the wedge structure used. The Slonczewski mechanism could then operate. For $\ell = 6\mathring{A}$, where A_{12} is large, Fuss *et al.*[7] find $B_{12} \simeq A_{12}/2$ which, as discussed in the next paragraph, is a much larger value of B_{12} than expected from the intrinsic mechanism. However, the rapid temperature dependence of B_{12} remains a puzzle.

We deduce from Eq. (6) that in the present model, even at $T = 0$, the amplitude of the oscillating biquadratic exchange $B_{12}^{(i)}$ is only one-sixteenth of the amplitude of oscillations in A_{12}. Thus, for most spacer thicknesses, except near a zero of A_{12}, it is

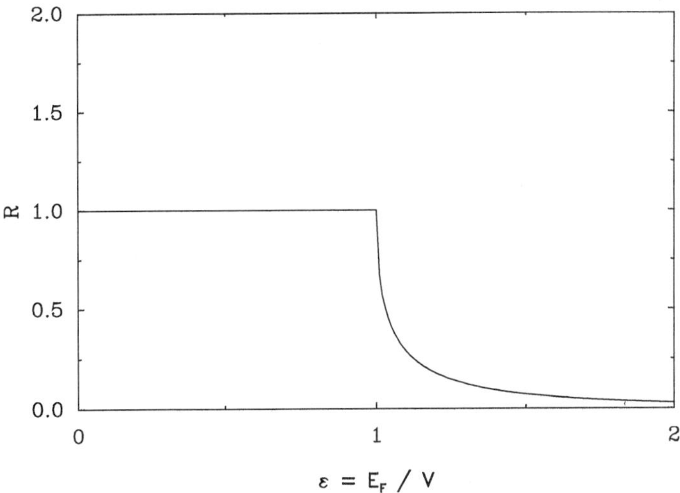

Figure 1. The amplitude factor R as a function of $\epsilon = E_F/V$.

expected that $B_{12}^{(i)}$ will be small compared with A_{12}. It is probable that Slonczewski's $B_{12}^{(S)}$ is often a major part of B_{12} in systems where short period oscillations of A_{12} occur. In connection with their measurements on a bcc Fe/Cu 12ML/Fe trilayer Heinrich *et al.*[12] propose that our $B_{12}^{(i)}$ might account for half the observed B_{12} so that $B_{12}^{(i)} \simeq 0.05A_{12}$.

So far we have considered the case of complete confinement with infinite exchange splitting $V = U(n_1 - n_2)$ in the ferromagnets. We now report recent results[13] on the effect of finite exchange splitting V, and the situation $E_F > V$ is of particular interest where partial confinement occurs with holes of both spin present in the bulk of the magnetic layers. The main result is that an additional factor R^n occurs in the nth harmonic term of asymptotic formulae such as Eq. (8) where R is given by

$$R = 1 + 8\epsilon(\epsilon - 1) - 4(2\epsilon - 1)\sqrt{\epsilon(\epsilon - 1)}, \qquad \epsilon > 1 \tag{13}$$

and

$$R = 1 \qquad \epsilon < 1$$

with $\epsilon = E_F/V$. This amplitude factor R is plotted in Fig.1. Partial confinement leads to a strong reduction in the overall amplitude of the exchange coupling and to a suppression of higher harmonics, including the $n = 2$ term containing biquadratic exchange. Numerical calculations for a one-band tight-binding model [3,13] confirm the strong reduction in amplitude of the oscillatory exchange coupling due to partial confinement and show that in general there is also a phase shift in the oscillations relative to the complete confinement case.

TIGHT-BINDING d-BAND MODEL: Cr OR Mo SPACER

We recently generalized[3] our earlier work[1,2] to a five-orbital d band containing 2.7 electrons of each spin per atom, as is appropriate for paramagnetic Cr or Mo.

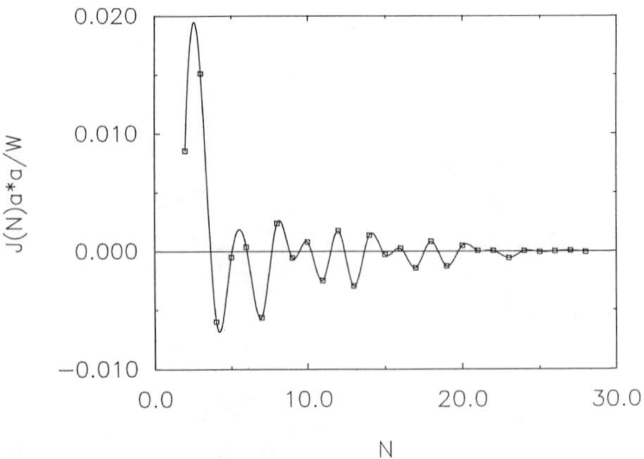

Figure 2. Calculated values of $J(N)a^2/W$ for a Cr(001) spacer in the complete confinement model. Here, a is the bcc lattice constant, W is the canonical d band width parameter, and N is the number of atomic planes in the spacer.

The calculated exchange coupling $J(N)$, with N atomic planes in the spacer, is the energy difference per unit area between parallel and antiparallel magnetic configurations ($J(\pi)$ in the notation of the previous section). The method of calculation is described elsewhere[3] and corresponds to the case of complete confinement. This is unrealistic for a system such as Fe/Cr and grossly overestimates the strength of the exchange coupling as expected from the considerations at the end of the previous section. Even for systems containing Co or Ni, it is likely that inclusion of hybridization with sp electrons will lead to effective partial confinement of d electrons. Nevertheless these calculations are an important step towards our ultimate goal of a realistic fully quantitative theory. An important feature of the work is the need to include about 10^6 points in the summation over $k_{||}$, the wave vector parallel to the layers which classifies the electron states. This is essential[1] to obtain $J(N)$ reliably for N as large as 25 and it puts this type of calculation outside the range of first principles band structure methods.

The calculated $J(N)$ is plotted in Fig. 2 but the analysis is not simple.

We first take second differences to suppress slowly-varying (long period) components. Thus, in Fig. 3 we plot

$$\Delta^2 J = J(N+1) - 2J(N) + J(N-1). \tag{14}$$

A beat period of 6 ML is evident which could arise from the superposition of two oscillations of periods 2 ML and 3 ML. It does not correspond to a single incommensurate oscillation, involving a factor such as $cos N(\pi - \delta) = (-1)^N cos(N\delta)$ with $\delta = \pm\pi/6$, since the sign of $\Delta^2 J$ in Fig. 3 is just given by $(-1)^N$ for $N \geq 7$.

To search for a long-period component in the exchange coupling we simulate experiments on samples with variable spacer thickness. Thus we define a smoothed coupling

$$\overline{J}(N) = \frac{1}{2}J(N) + \frac{1}{4}[J(N+1) + J(N-1)] \tag{15}$$

and this is plotted in Fig. 4.

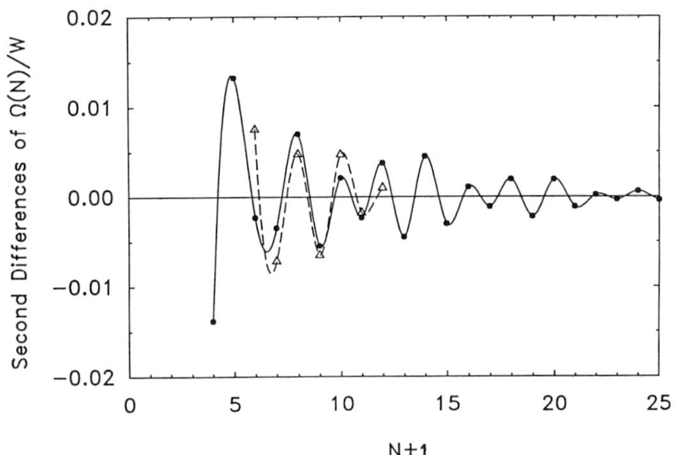

Figure 3. Second differences of the calculated exchange coupling $\Delta^2 J = \Delta^2 \Omega$ (solid line) compared with scaled second differences of the observed saturation field[16] (broken line).

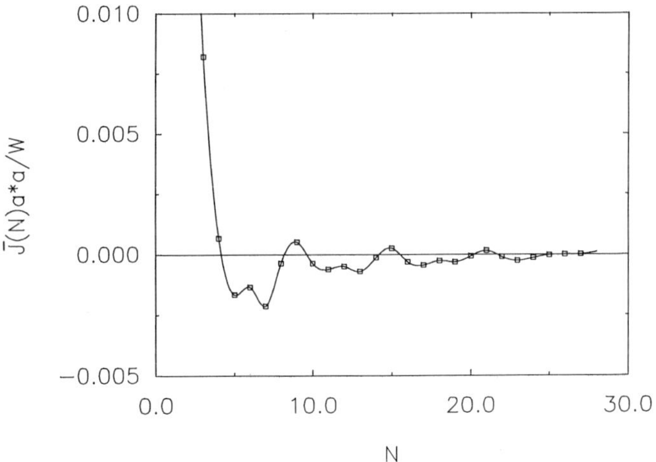

Figure 4. The calculated exchange coupling smoothed to simulate variable spacer thickness (See Eq. (15)).

Figure 5. Cross-section of the Fermi surface for the d band model of Cr or Mo with extremal dimensions in the (001) direction labelled with corresponding expected oscillation periods.

The 2 ML component tends to be filtered out and $\overline{J}(N)$ clearly has components of periods 3 ML and 6 ML. It seems unlikely that the strong 6 ML component arises from beats between the 3 ML component and the much reduced 2 ML component, so it may be that an actual 6 ML period is present. However, this is shorter than the observed long-period of 10 ML[14]. Oscillations with a 3 ML period have been observed in Fe/Mo/Fe (100) sandwiches[15]. We may try to identify these periods with extremal dimensions of the Fermi surface for our d-band model. These are indicated on the cross-section of the surface in Fig. 5. The 2 ML period corresponds to the usual Cr nesting vector and the 2.8 ML spanning vector presumably corresponds to our 3 ML period. The several longer periods indicated in Fig. 5 may relate to the apparent 6 ML component.

It is interesting to compare our calculations quantitatively with experiment in the thickness range $5 \leq N \leq 12$ where measurements of the exchange coupling[16,17] show a short-period oscillation superposed on the antiferromagnetic part ($J > 0$) of a long-period oscillation. Our results (Fig. 2) show short-period oscillations on a ferromagnetic background for this range of N and the short-period oscillations also have their sign changed. This is clear from Fig. 3 where the second differences of our calculated coupling are compared with second differences of the saturation field of Ref. 16, multiplied by a scaling factor and with the sign reversed. Investigation of the scaling factor shows that the calculated amplitude of the short-period oscillation is almost 100 times larger than the observed one. Part of this factor can be attributed to a reduction in the experimental amplitude due to the averaging effect of variable spacer thickness and to the effect of taking measurements at room temperature instead of $T = 0$. However, much of the discrepancy must arise from the unrealistic assumption of complete electron confinement. We saw in the previous section that a strong reduction in amplitude and a phase shift are to be expected on introducing partial confinement.

CONCLUSIONS

One-band models, including the hole-gas and tight-binding models, illustrate many general principles of bilinear and biquadratic exchange in magnetic multilayers and sandwiches. Periods of oscillations in the exchange coupling, considered as a function of spacer thickness, are characteristic of the spacer metal. However, the amplitude and phase of the oscillations depend strongly on the nature of both spacer and ferromagnet which determines the degree of electron, or hole, confinement in the spacer layer. These conclusions are quite similar to Bruno's[4] for coupling between magnetic monolayers immersed in a simple or noble metal. Intrinsic biquadratic exchange is associated with the $n = 2$ harmonic in a Fourier-type expansion of the oscillatory exchange coupling and ways of distinguishing it from Slonczewski's secondary effect are discussed. The intrinsic biquadratic exchange has a characteristic temperature dependence.

Tight-binding calculations of exchange coupling across a Cr or Mo spacer, using five d-orbitals, exhibit oscillations with several different periods. It seems possible to assign these periods to different extremal dimensions of the Fermi surface, as in previous one band calculations. Since complete confinement is assumed, the magnitude of the coupling is grossly overestimated. We hope to remedy this in future work and include sp-d hybridization so that our method of calculation can be applied to both transition and noble metal spacers.

REFERENCES

1. D. M. Edwards, J. Mathon, R. B. Muniz, and M. S. Phan, J. Phys.: Condens. Matter **3**, 4941 (1991).
2. D. M. Edwards, J. Mathon, R. B. Muniz, and M. S. Phan, Phys. Rev. Lett. **67**, 493 (1991); **67**, 1476 (E) (1991).
3. D. M. Edwards, J. Mathon, and R. B. Muniz, Physica Scripta (in press).
4. P. Bruno, preprint and in this volume.
5. K. B. Hathaway and J. R. Cullen, J. Magn. Magn. Mat. **104-107**, 1840 (1992).
6. J. C. Slonczewski, Phys. Rev. B **39**, 6995 (1989).
7. A. Fuss, S. Demokritov, P. Grünberg, and W. Zinn, J. Magn. Magn. Mat. **103**, L221 (1992).
8. A. Fuss, J. A. Wolf, and P. Grünberg, Physica Scripta (in press).
9. J. C. Slonczewski, Phys. Rev. Lett. **67**, 3172 (1991).
10. C. J. Gutierrez, J. J. Krebs, M. E. Filipkowski, and G. A. Prinz, J. Magn. Magn. Mat. (to be published).
11. G. A. Prinz, private communication.
12. B. Heinrich, Z. Celinski, J. F. Cochran, A. S. Arrott, and K. Myrtle, submitted to Phys. Rev. B.

13. J. Mathon, M. Villeret, and D. M. Edwards, submitted to J. Phys.: Condens. Matter.

14. S. S. P. Parkin, N. More, and K. P. Roche, Phys. Rev. Lett. **64**, 2304 (1990).

15. Z. Q. Qiu, J. Pearson, A. Berger, and S. D. Bader, Phys. Rev. Lett. **68**, 1398 (1992).

16. S. T. Purcell, W. Folkerts, M. T. Johnson, N. W. E. McGee, K. Jager, J. aan de Stegge, W. B. Zeper, W. Hoving, and P. Grünberg, Phys. Rev. Lett. **67**, 903 (1991).

17. S. Demokritov, J. A. Wolf, and P. Grünberg, Europhys. Lett. **15**, 881 (1991).

MAGNETIC PROPERTIES OF FERRO/ANTIFERRO-MAGNETIC - BASED SANDWICHES

Daniel Stoeffler and François Gautier

I.P.C.M.S. - GEMME (U.M.R. 46)
Université Louis Pasteur
4 Rue Blaise Pascal
67070 Strasbourg, France

ABSTRACT

In this paper, we point out the importance of the layers's magnetic order on the properties of ferromagnetic/antiferro-magnetic (F/AF) based-sandwiches and more especially on the interlayer magnetic couplings (IMC). We study the magnetic properties of sandwiches based on F (Fe) and AF (Cr and Mn) layers in the itinerant magnetism framework and we use a real space tight binding method to determine the electronic structure at T = 0 K. The amplitudes *and* the orientations of the magnetic moments of the AF layers are determined, the structure of the magnetic walls and their pinning at the interfaces are studied for perfect and diffuse interfaces. The Fe-Cr exchange anisotropy and the IMC are related to the magnetic structure, those couplings resulting from intrinsic quadratic and biquadratic contributions for perfect interfaces.

1. INTRODUCTION

A lot of experimental and theoretical studies have been recently devoted to the magnetic, transport and optical properties of metallic multilayers and sandwiches. From a fundamental viewpoint, one of the most interesting features of such systems is the existence of strong and long ranged interlayer magnetic couplings (IMC) between two successive ferromagnetic (F) layers separated by non magnetic or antiferromagnetic (AF) spacers. These IMC have been first observed by Grünberg *et al* [1] and by Carbone and Alvarado[2] on Fe/Cr/Fe(001) sandwiches. They were found to be AF, the magnetization of the two successive Fe layers being aligned antiparallel. Later on, in Fe/Cr(001) superlattices, these AF IMC, which have been observed for a wide range of Cr thicknesses, have been shown to be related to giant magneto-resistances.[3] More recently, the IMC have been found to oscillate from AF to F with the Cr thickness up to 50 Å with a period of about 18 Å (the magnetization of the two successive Fe layers are alternatively aligned parallel and antiparallel).[4] Such long periods (~ 10 Å) have been observed for a large variety of multilayers.[5]

These observations have given rise to various theoretical studies using schematically two types of approaches based on (i) RKKY-like schemes[6,7] and (ii) band structure calculations.[8-11]

- The *RKKY-like* models relate the period of the IMC to the topology of the Fermi surface for large spacer thicknesses, the corresponding periods being consistent with the experimental results of Fe/Cu and Co/Cu multilayers.[7] However, this theory does not take into account the inter- and intra-plane multiple scatterings so that it cannot determine quantitatively, in its present formulation, the phase and the amplitude of the IMC. Finally, it cannot be applied to systems with magnetically ordered spacers such as those we consider here for which the perturbation is *extended* in the whole spacer.

- The second method uses *ab initio* [8-10] or tight binding[11] (TB) descriptions of the electronic structure. The determination of the IMC is realized by calculating the total energy difference $\Delta E_{F\text{-}AF} = E_F - E_{AF}$ between the two interlayer arrangements (F and AF coupled). For superlattices with noble metal spacers, the periods of the IMC have also been qualitatively related to the shape of the Fermi surface.[10] For superlattices with Cr spacers, the IMC have been first found to oscillate with a period of two monolayers (ML) by a TB method.[11] This has been related to the existence of a central magnetic defect in the Cr layer induced by the strong antiparallel interfacial Fe-Cr coupling. Therefore, the corresponding perturbing potential being extended in the whole spacer, such a situation *cannot* be determined by a RKKY approach. These results have been confirmed by LSDA *ab initio* calculations,[9] the *ab initio* and the TB results agreeing both qualitatively and quantitatively in the considered range of thicknesses and for all investigated superlattices (Fe/Cr, Fe/Mn, Co/Ru). However, the amplitude and the period of these IMC do not agree with the amplitude (they are approximatively 50 times smaller) and the long period (18 Å) of the oscillations found in most superlattices.

More recently, an experimental study[12] has exhibited the important role played by the interfacial roughness on the IMC and a theoretical study[13] has confirmed that the IMC can be strongly modified by introducing interfacial ordered compounds (IOC). From an experimental viewpoint, Unguris *et al* [12] realized a direct visualization of the sign of the IMC as a function of the Cr thickness by imaging with a SEMPA technique a Fe top layer deposited on a Cr wedge previously deposited on a Fe whisker. Such a technique allows to get good and reproducible growth conditions. This study exhibits an IMC with the two ML period found theoretically[9,11] and shows that this short period is obtained only for a high interfacial crystalline quality. Moreover, this study shows that, for high quality sandwiches, the chromium layers are magnetically ordered for temperatures much higher than the bulk Néel temperature. Theoretically, the results we obtain[13] for "imperfect" superlattices agree with these experiments: the IOC pin interfacial magnetic defects at the interfaces and reduce strongly the "interfacial" coupling between the Fe and Cr crystals when the IOC introduce a sufficient number of interfacial frustrated magnetic links. In such cases and for small Cr thicknesses ($n<12$) the magnetic moments are very small in the whole Cr layer and the central magnetic defects disappear. This gives small IMC without short period oscillations.[13]

The previous theoretical results show that the sensitivity of the magnetic moments to the magnetic orientation of their (interfacial) neighbourhood determines the magnetic state of the whole layer. For example, in order to determine, in the itinerant magnetism framework, the sensitivity of the amplitude of the magnetic moments to local disorientations, let us assume that in the bulk (bcc Cr and Fe and bct Mn) the magnetic moment distribution is helicoidal from (001) plane to (001) plane with a tilt angle θ (all moments in the same (001) plane are equivalent) (figure 1-a). In such a case, the magnetic moment of the i-th plane is characterized by its direction $\theta_i = i.\theta$ and by its amplitude $M_i = M_0$. If $\theta = 0$ (π) we have (respectively) a F (AF) ordering. The figure 1-b shows the results obtained for Fe, Mn and Cr. On this figure, we note that (i) the amplitude of the Fe and Mn magnetic moment is nearly constant in a wide range of θ ($[0, \pi/2]$ for Fe and $[3\pi/5, \pi]$ for Mn), (ii) the amplitude of the Cr magnetic moment is strongly sensitive to a small tilt $\Delta\theta$ with respect to its AF ground state: it decreases rapidly and becomes equal to zero for $\theta \le 19\pi/20$. Finally, the energy associated with the disappearance of the Fe, Mn and Cr magnetic moments is

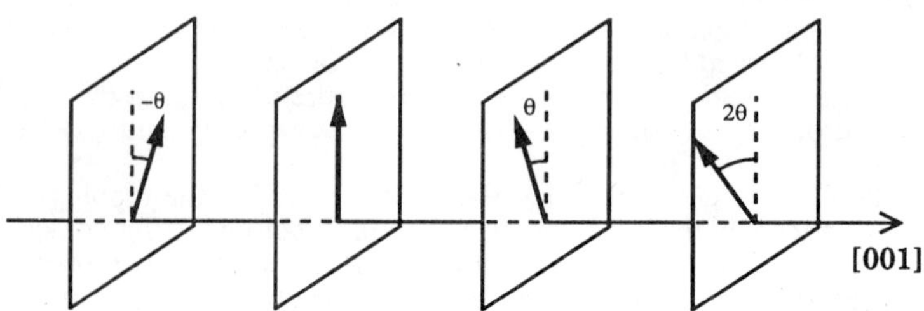

Figure 1-a. Schematic representation of the (001) helicoidal magnetic state.

respectively equal to 200, 80 and 0.8 meV/atom. This shows that the Cr magnetic moment is highly sensitive to a local disorientation whereas Mn and Fe moments remain unmodified for large θ values.

In this paper, we study the magnetic properties of sandwiches based on F (Fe) and AF (Cr and Mn) layers in the itinerant magnetism framework. We use the recursion method and the tight binding approximation (TBA) to determine self-consistently the electronic structure at T=0 K of the considered systems: details of these band structure calculations are reported elsewhere.[11,14] We discuss (i) the relation between the interfacial, interlayer and intralayer couplings and the magnetic moments distributions (ii) the role of the magnetic frustrations induced by an external magnetic field or by an interfacial disorder. We consider here sandwiches made of thin (001) B_n layers between two semi infinite crystals A_∞, all atoms being located on a perfect bcc or bct lattice. In the first section, we consider sandwiches for which B is the F element and we discuss the interfacial couplings and the origin of the exchange anisotropy as a function of the crystalline quality of the interfaces. In the second section, we consider $Fe_\infty/Cr_n/Fe_\infty$ sandwiches: the non colinear magnetic structure of the n Cr ML is determined as a function of the relative orientation of the magnetizations of the two Fe_∞ crystals and is related to the IMC. These IMC are found to oscillate with the parity of n as found previously for superlattices[15] and an intrinsic biquadratic contribution is exhibited for perfect sandwiches.

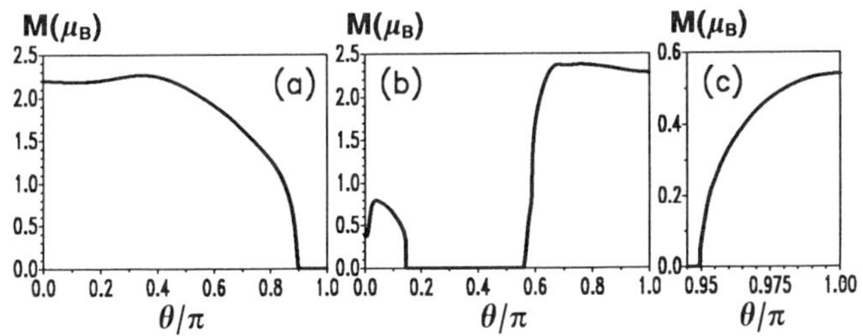

Figure 1-b. Amplitudes of the magnetic moment M as a function of the tilt angle θ of the helicoidal magnetic state for: (a) bcc iron, (b) bct with c/a = 1.15 manganese and (c) bcc chromium.

2. $AF_\infty/F_n/AF_\infty$ SANDWICHES: INTERFACIAL DEFECTS AND INTERFACIAL COUPLINGS

The electronic structure and the magnetic properties of such sandwiches cannot be determined using *ab initio* band structure methods, the size of the super-cell required to get a self-consistent solution being too large (see below). This is why we use the recursion method and we consider a real space cell including the n F monolayers and the m first monolayers α of both semi-infinite AF_∞ crystals. A self-consistent calculation is done versus m for the F and α layers assuming that the AF monolayers which are further apart from the interfaces (β) keep their bulk properties. A satisfactory ground state solution is obtained for $m=m_c$ when the distribution of the magnetic moments is continuous across the fictive α/β interface and remains unmodified for $m>m_c$. Usually, such a solution is obtained for $m_c=10$ but, the Cr moment being highly sensitive to its local environment, we will see that in this case even $m_c=20$ is not sufficient to obtain a realistic magnetic moments distribution.

2.-1. Colinear interfacial arrangement

It is now well known that the Fe-Cr interfacial coupling is strongly antiparallel (AP): this has been shown experimentally[16,17] and theoretically both for films of Fe deposited on Cr(001), of Cr deposited on Fe(001)[18] and for Fe/Cr(001) superlattices.[8,9,11] In the case of $Cr_{\infty}/Fe_5/Cr_{\infty}$ sandwiches, we obtain effectively only a self-consistent solution for an AP Fe/Cr interfacial coupling (figure 2): when the calculation is done starting from a P interfacial arrangement, the Cr moment is progressively returned during the iterations of the self-consistent procedure and a final AP state is always obtained (see § 2-2). For the Fe-Mn interface in $Mn_{\infty}/Fe_5/Mn_{\infty}$ sandwiches, two solutions (parallel P and AP) are found (figure 2) but the solution with P coupled interfacial moments is energetically more stable than the solution with AP coupled interfacial moments: E(AP)-E(P) = 67 meV per Mn interfacial atom. This is in agreement with *ab initio* results of Fe_5Mn_n superlattices.[19] These behaviours are directly related to the sensitivity of the interfacial moments to their local environment. The Cr moment being strongly sensitive to a local magnetic desorientation, it cannot be coupled P with Fe: this is not the case for Fe/Mn interfaces.

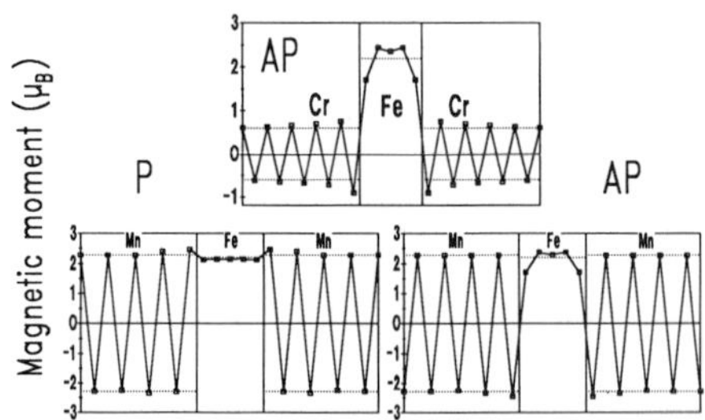

Figure 2. Magnetic moments distributions for $Cr_{\infty}/Fe_5/Cr_{\infty}$ and $Mn_{\infty}/Fe_5/Mn_{\infty}$ for a colinear structure.

2.-2. Non colinear interfacial arrangements

Let us now assume, for simplicity, that the only effect of an external applied field consists in changing the direction of the magnetization of the Fe layer without modifying those of the Cr crystals which remain blocked in a given direction by anisotropy effects. Here the moments of all the Fe atoms are assumed to remain parallel. Two examples of magnetic moments distributions of $Cr_{\infty}/Fe_5/Cr_{\infty}$ sandwiches are shown in figure 3 for (a) the ground state we discussed previously and (b) the first step of the iterations we obtain reversing only the Fe magnetization and therefore introducing a P interfacial coupling. As mentioned previously, the second case is not a self-consistent solution of the problem: the interfacial magnetic moment is progressively turned back during the self-consistent calculation and a magnetic defect appears in the m α Cr layers. Figure 4 shows the progressive apparition of this defect when we vary the angle θ of the Fe magnetization with respect to its ground state value from 0 to π: for $\theta \geq 3\pi/5$ the interfacial Cr magnetic moment

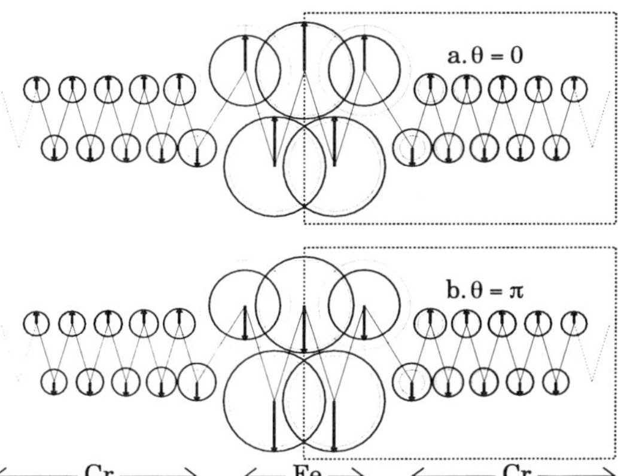

Figure 3. Magnetic moments distributions for $Cr_\infty/Fe_5/Cr_\infty$ sandwiches: (a) ground state with an AP interfacial coupling (b) first step of the iteration procedure obtained reversing the magnetization of the Fe layer. Only one site per (001) plane is represented, the inequivalent atoms for which the calculation is done being inside the dashed line rectangle (m=10). The size of the arrows is proportionnal to the amplitude of the magnetic moment on each site, the dotted line circles giving the amplitude of the bulk magnetic moment.

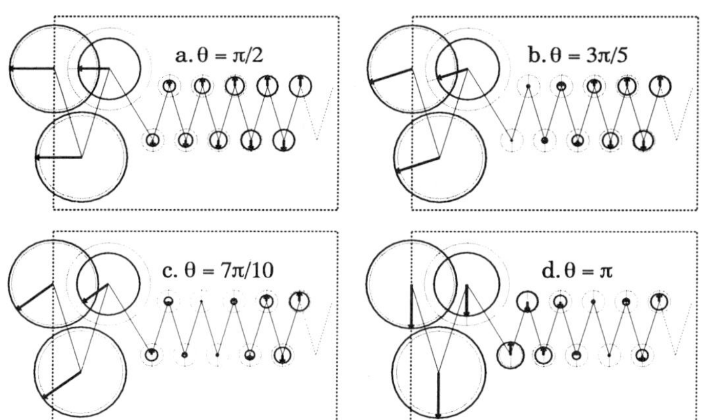

Figure 4. Distributions of the magnetic moments of $Cr_\infty/Fe_5/Cr_\infty$ sandwiches in the real space cell (m=10) for various values of the disorientation θ of the magnetizations of the Fe layers with respect to its ground state direction. Note the appearance of the magnetic defect for $\theta \geq 3\pi/5$.

is opposite to the one of the ground state and the defect, which is located at the interface for $\theta < 3\pi/5$, tends to move away from the interface. These results have been obtained with m=10. For m=20, the result is qualitatively similar except that the defect moves more away from the interface but this value of m is still too small to get the lowest energy solution. On the contrary, for $Mn_\infty/Fe_5/Mn_\infty$ sandwiches, the perturbations remain always located on the Fe and Mn interfacial planes so that a satisfactory self-consistent state is obtained for m_c=10. This is easily understood from the fact that in this case the two interfacial couplings (P and AP) are allowed.

These calculations suggest strongly that if for $Mn_\infty/Fe_5/Mn_\infty$ sandwiches the magnetic wall induced by the Fe disorientation is pinned at the perfect Mn/Fe interfaces, it remains unpinned for $Cr_\infty/Fe_5/Cr_\infty$ sandwiches. Unhappily, the computing time is too large to get the lowest energy state in $Cr_\infty/Fe_5/Cr_\infty$ sandwiches for $\theta \neq 0$ and π, the extension of this unpinned colinear wall being too large in Cr (ξ = 50 ML, see § 3-2).This is why we have used the l-J model we develop in §3-1 to determine the lowest energy state for $\theta = \pi$ by the same iteration scheme. As anticipated, this state is characterized by an interfacial perturbation similar to the one obtained for the ground state and by a separated wall extending over 50 planes. This shows that the interfacial Fe-Cr AP coupling is much stronger than the Cr-Cr AF coupling and that it is energetically easier to frustrate the Cr AF order than to frustrate the interfacial AP coupling.

2.-3. Qualitative discussion

Figures 5 and 6 represent the interfacial magnetic moments (m(θ), M(θ)) and the interfacial energies $\gamma(\theta)$ versus θ we obtained for $Cr_\infty/Fe_5/Cr_\infty$ and for $Mn_\infty/Fe_5/Mn_\infty$ sandwiches. On figure 5, (i) the interfacial Fe moment is nearly constant, (ii) the sign of the interfacial Cr moment changes for $\theta = 3\pi/5$, and (iii) the interfacial energy is maximal at this value. On figure 6, (i) the interfacial Mn moment is nearly constant, (ii) the interfacial Fe moment varies more significantly than is the previous case, and (iii) the interfacial energy varies monotonously with θ.

These behaviours can be qualitatively understood using a simple model (figure 7-a). In such a model, all the magnetic moments remain unmodified by the interfacial coupling except those located on one interfacial ML whose moments m(θ) vary with θ. m(θ) is determined by the formation energy l(m) of the interfacial moment and by its coupling with its A and B neighbours we note j and J. The interfacial energy is then equal to:

$$\gamma(m; \theta) = l(m) - J\, M_0\, m \cos\theta - j\, m_0\, m \qquad (1)$$

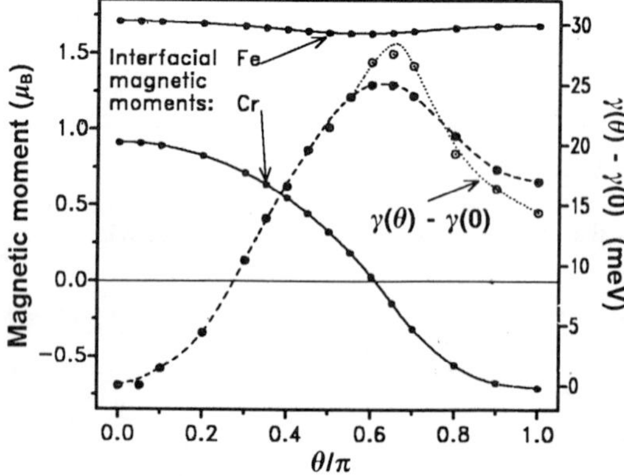

Figure 5. Variations of the Fe and Cr interfacial magnetic moments and of the interfacial energy $\gamma(\theta) - \gamma(0)$ versus θ for $Cr_\infty/Fe_5/Cr_\infty$ sandwiches (TBA). The filled (open) circles and the dashed (dotted) line curve correspond to the results obtained with m=10 (20).

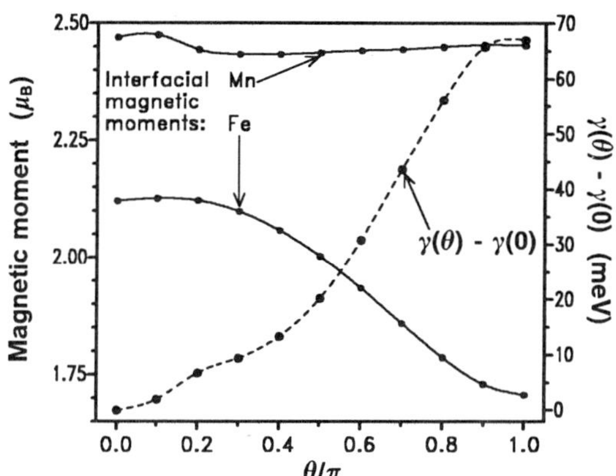

Figure 6. Variations of the Fe and Cr interfacial magnetic moments and of the interfacial energy $\gamma(\theta) - \gamma(0)$ versus θ for $Mn_{\infty}/Fe_5/Mn_{\infty}$ sandwiches (TBA).

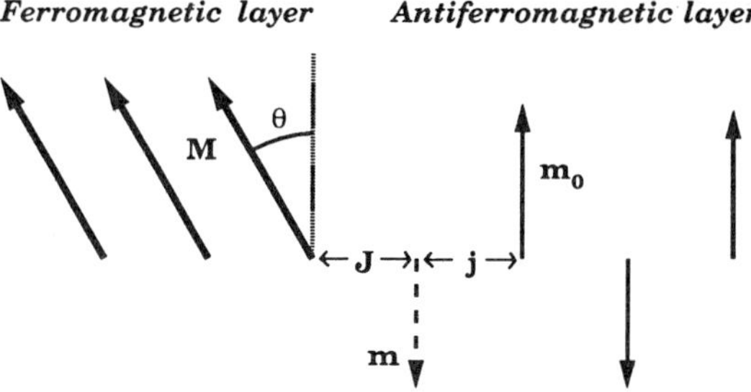

Figure 7-a. Schematic representation of the simple model for Fe/Cr interfaces.

where m_0 and M_0 are the ground state moments. For simplicity, l(m) is taken equal to $(A/2)m^2$. By minimization of this energy with respect to m for a given value of θ, we obtain m(θ) and γ(θ) as a function of $J.M_0$, $j.m_0$ and A. It is easily shown that, according to the value of $\alpha = \frac{J\,M_0}{j\,m_0}$, γ(θ) either inceases (α<1) or presents a maximum for $\theta = \theta_{max}$ (α>1), these behaviours corresponding (respectively) to these obtained for $Mn_\infty/Fe_5/Mn_\infty$ ($Cr_\infty/Fe_5/Cr_\infty$) sandwiches. Figure 7-b shows that the TB results can be qualitatively fitted to this model with the parameters given in the table and with:

$m = m_{Cr}$, $M_0 = m_{Fe} = 1.7\ \mu_B$, $J = J_{Fe\text{-}Cr}$, $j = J_{Cr\text{-}Cr}$, $m_0 = m^0_{Cr}$ for $Cr_\infty/Fe_5/Cr_\infty$ and $m = m_{Fe}$, $M_0 = m_{Mn} = 2.4\ \mu_B$, $J = J_{Fe\text{-}Mn}$, $j = J_{Fe\text{-}Fe}$, $m_0 = m^0_{Fe}$ for $Mn_\infty/Fe_5/Mn_\infty$.

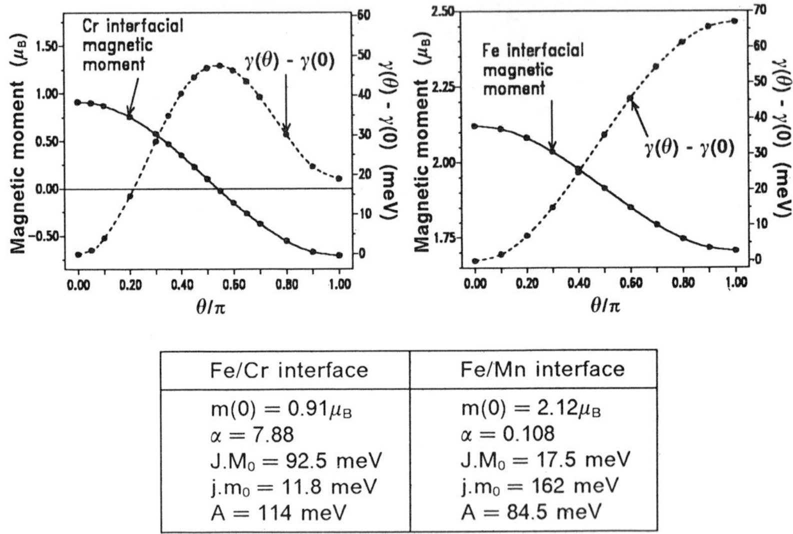

Fe/Cr interface	Fe/Mn interface
$m(0) = 0.91\mu_B$ $\alpha = 7.88$ $J.M_0 = 92.5$ meV $j.m_0 = 11.8$ meV $A = 114$ meV	$m(0) = 2.12\mu_B$ $\alpha = 0.108$ $J.M_0 = 17.5$ meV $j.m_0 = 162$ meV $A = 84.5$ meV

Figure 7-b. Results (m(θ), γ(θ) − γ(0)) obtained after a fit to the tight-binding calculations. The parameters of the fits are given in the table.

2.-4. Discussion: diffuse interfaces and exchange anisotropy

The previous model confirms that the Fe-Cr coupling is much stronger than the Cr-Cr one (approximatively 8 times), the Fe-Fe coupling being approximatively 9 times stronger than the Fe-Mn one. However, the numbers we get for j and J must not be taken too seriously since we assumed that the frustration of the interfacial couplings introduces a localized magnetic perturbation on one interfacial ML whereas the TB calculations show that these perturbations are extended.

The wall induced in the Cr crystal for θ = π is due to (i) the strong interfacial AP coupling, (ii) the high sensitivity of the Cr magnetic moment which propagates the perturbation on a large number of Cr planes and (iii) to the small energy lost when a Cr magnetic moment vanishes. If we introduce now interfacial ordered compounds (IOC) at the Fe/Cr interfaces, to modelize the interdiffusion, the range of the interfacial perturbation and the strength of the interfacial coupling can be strongly modified. This has been done in the TBA[14] with (a) a ($Fe_{0,5}Cr_{0,5}$) IOC extending over one plane and (b) a ($Fe_{0,75}Cr_{0,25}$)/($Fe_{0,25}Cr_{0,75}$) IOC extending over two planes. The interfacial perturbations are localized for both cases. For the IOC (a), the interfacial defect extends only over 10 planes for θ = π but its energy is similar to the one obtained for "perfect" interfaces (for θ = 0, no defect is obtained) so that the exchange anisotropy is large. On the contrary, for the IOC (b), the situations for θ = π and θ = 0 are found to be very similar: an interfacial defect extending over 5 planes is obtained for both θ values: in this case the exchange anisotropy is much smaller (less than 1 meV per interfacial atom) than the coupling energy.[14] This physical origin of the exchange anisotropy will be discussed in more details in a forthcoming paper.[20]

3. $Fe_\infty Cr_n/Fe_\infty$ SANDWICHES: CONSTRAINED WALLS AND INTERLAYER MAGNETIC COUPLINGS

In such sandwiches, the magnetic order in the thin Cr layer is constrained by the strong AP interfacial Fe-Cr couplings and by the relative orientation of the magnetizations of the Fe crystals. For example, if n=4, the ground state is obtained when the interlayer magnetic arrangement (IMA) between the Fe crystals is AF (the magnetization of the two Fe crystals are opposite) as shown in figure 8. For this IMA, the AF order and the interfacial AP couplings are not frustrated. If a strong external magnetic field is applied, it alignes the magnetizations of the Fe crystals and a defect is generated in the Cr layer as obtained in §2-2 for colinear orders, this defect being located at the center of the Cr layer (figure 8).

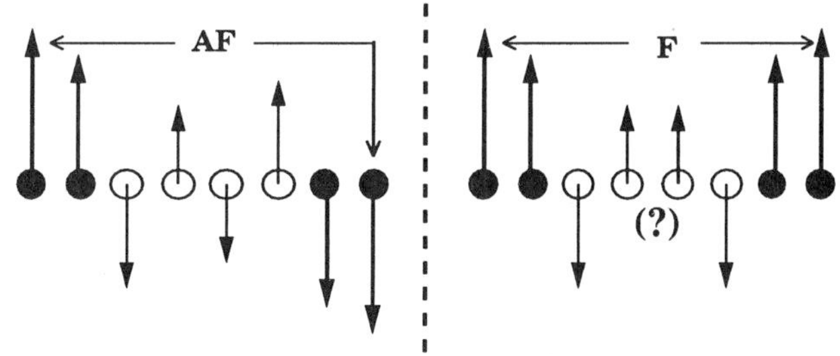

Figure 8. Schematic representation of the magnetic moments distribution in $Fe_\infty/Cr_n/Fe_\infty$ sandwiches for:

- an antiferromagnetic (AF) interlayer magnetic arrangement (left),
- a ferromagnetic (F) interlayer magnetic arrangement giving rise to a frustration of the AF order of Cr (right).

The filled (open) circles correspond (respectively) to Fe (Cr) atoms.

The magnetic moments distribution of the ground state is not necessarily colinear. Moreover, when the angle θ between the magnetizations of the Fe crystals is non zero, the magnetic structure must be not colinear so that we must study the stability of such states: this is the aim of this section 3. However, even if the TB calculations are much less computer time consuming than *ab initio* calculations, they do not allow to get a self-consistent determination of the amplitudes <u>and</u> orientations of the magnetic moments. This is why, (i) we restrict here our study to Cr_n thin films made of n (001) ML, the orientations of the surface magnetic moments, $\mathbf{M}_1$ and $\mathbf{M}_n$, being assuned to be fixed and directed along the strong exchange fields of the neighbouring Fe crystals, (ii) we develop a simple model to describe semi-quantitatively the magnetic defects in such Cr films. In the § 3-1, we define this l-J model and we show that it is sufficient to describe magnetic walls in AF bulk chromium. Then we study in §3-2 the magnetic moments distribution in the Cr constrained thin films we introduced previously.

3-1. l-J model and magnetic walls in bulk chromium

This model allows to perform a complete energy minimization with respect to the amplitudes and the orientations of the magnetic moments. It is based on the assumption[21] that the total energy $E(\{ \mathbf{M}_i \})$ results from the formation energy of the magnetic moments $l(\mathbf{M}_i)$ and from the interlayer coupling energy we assume to be of the Heisenberg form $J_{i,j}\, \mathbf{M}_i.\mathbf{M}_j$.

$$E(\{ \mathbf{M}_i \}) = \sum_i l(\mathbf{M}_i) + \frac{1}{2} \sum_{i,j} J_{i,j}\, \mathbf{M}_i.\mathbf{M}_j \qquad (2)$$

In this section, all the atoms of the i-th monolayer are assumed to have the same magnetization $\mathbf{M}_i$. The interlayer coupling energies $J_{i,j} = J$ are restricted to nearest neighbours planes and the formation energy is given by $l(\mathbf{M}_i) = \frac{A}{2} \mathbf{M}_i^2 + \frac{B}{4} \mathbf{M}_i^4$. We first determine the parameters A,B and J for the bulk chromium (i) calculating the distribution of the magnetic moments for a colinear wall in the TBA and (ii) fitting these results to those of the l-J model (figure 9). The parameters of the l-J model are then given by: A = 608 meV/atom, B=28 meV/atom, $J_{Cr,Cr}$ = 309 meV/atom.

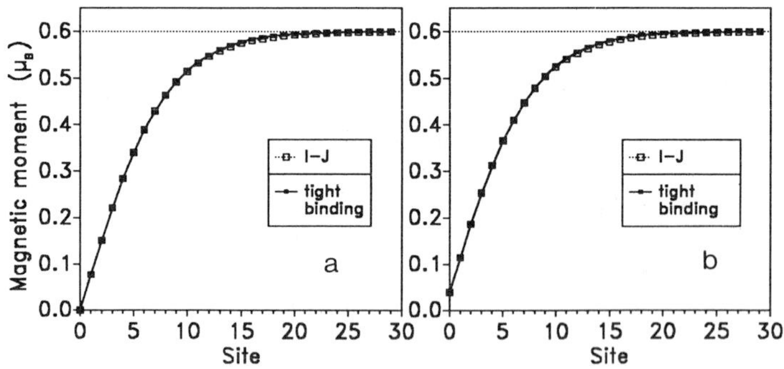

Figure 9. Amplitude of the magnetic moments obtained in the TBA for (a) odd and (b) even colinear walls: only half of the symmetric wall is represented. The TB calculation has been done on an infinite Cr crystal requiring a magnetic constraint to create a wall: for example, if we assume that $\mathbf{M}_{-i} = \mathbf{M}_i$ ($\mathbf{M}_{-i-1} = \mathbf{M}_i$) we create (respectively) an odd (even) wall The filled (open) symbols correspond (respectively) to the distributions obtained with the TBA (the l-J model).

In order to test the validity of this model for bulk Cr, we determined within the l-J model the magnetic moment and the energy of helicoidal states we discussed in the introduction. The agreement between the results of the l-J model and the TB calculation is nearly perfect.

3-2. Modification of the structure of the wall with the thickness n of the Cr film

Let us now consider a Cr_n film whose surface magnetizations $\mathbf{M}_1$ and $\mathbf{M}_n$ are opposite (or parallel), n being odd (or even), so that the Cr AF order and the boundary conditions introduce necessarily a defect. The magnetic structure of this defect is determined in the l-J model by a complete energy minimization with respect to the amplitudes and orientations of the moments $\mathbf{M}_i$. It can be summarized as follows. When this thickness is small (n<20), the wall is colinear (figures 10 and 11): this is due to (i) the strong AP Fe-Cr interfacial coupling

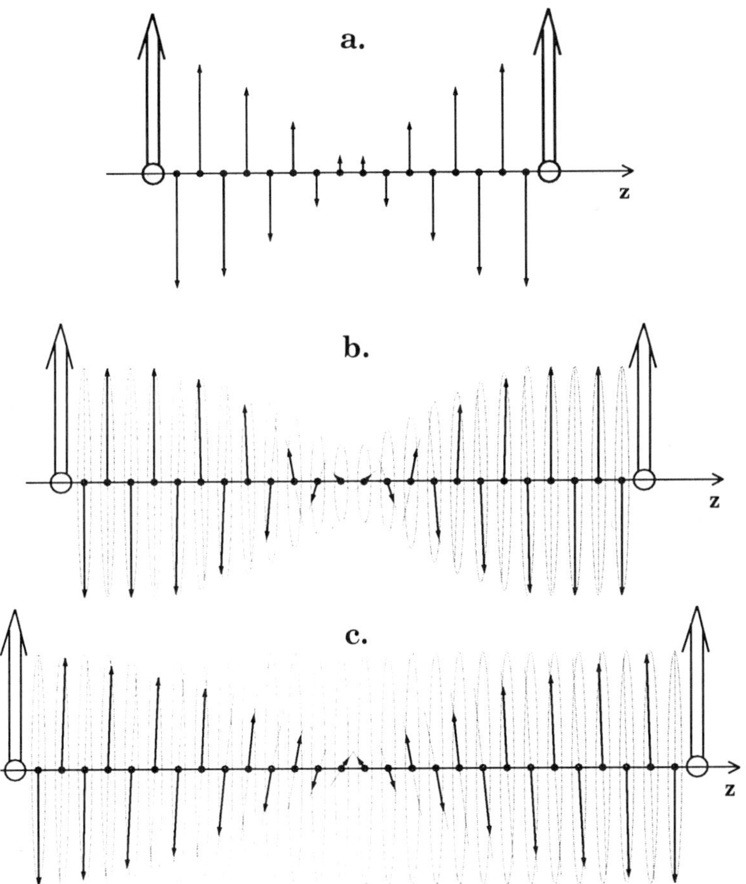

Figure 10. Schematic representation of the magnetic order in an AF thin film. We have assumed that (i) the magnetizations at the interfaces are directed along the exchange field due to the neighbouring ferromagnetic crystals (ii) the parity of the number of planes of the AF thin film and the previous interfacial constraints introduce a defect:

a.- colinear magnetic wall,

b.- non colinear magnetic wall with a strong reduction of the moments at the center of the film,

c.- helicoidal magnetic wall for which all moments keep the same amplitude.

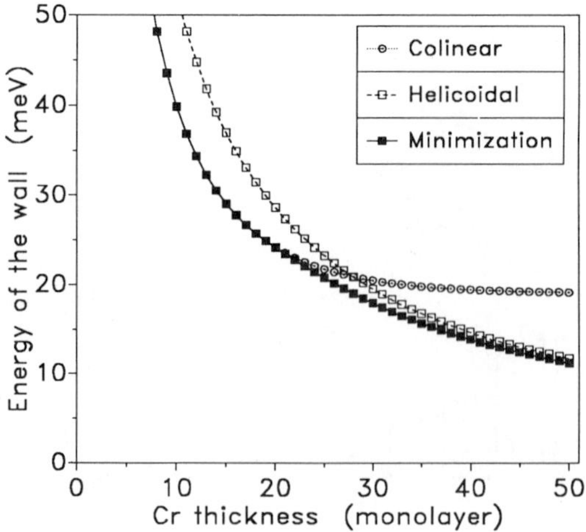

Figure 11. Energy (per atom) of the colinear (open circles), helicoidal (open squares) and non colinear (filled squares) walls obtained with the l-J model in $Fe_{\infty}/Cr_n/Fe_{\infty}$ versus n. Note that the cross-over between colinear and helicoidal walls occurs for n=27.

which tends to align the direction of the magnetic moment in the neigbourhood of the interfaces and (ii) to the high sensitivity of the Cr moment to the disorientation with respect to the colinear AF order (see introduction). For larger Cr thicknesses, n ≥ 20, it is possible to distribute the frustrations introduced at the surfaces of the film without introducing too large disorientations with respect to the colinear AF order: therefore, the wall is non colinear, the amplitudes and orientations of the magnetic moments $\mathbf{M}_i$ varying both with i (figures 10 and 11). Finally, for larger thicknesses, the disorientation of the direction of the magnetic moments from plane to plane becomes very small and the non colinear wall tends towards an helicoidal one (figures 10 and 11).

This behaviour has been confirmed by TB calculations performed on Fe/Cr superlattices[15]: the cross over between colinear and helicoidal walls occurs in such a case for n=24.

3-3. Interlayer magnetic coupling: intrinsic biquadratic contribution

The IMC $\Delta E(\theta) = E(\theta) - E(0 \text{ or } \pi)$ is obtained by determining the total energy $E(\theta)$, for a given angle θ between the directions of the magnetization in the Fe crystals (figure 12), the energy minimization being done with respect to the Cr magnetic moments $\mathbf{M}_i$. $E(0 \text{ or } \pi)$ is the energy of the ground state.

If we assume first that the magnetic order in the Cr layer is helicoidal, the TB calculation shows that $\Delta E(\theta)$ varies as $\Delta E_0 + \alpha.\frac{\mathbf{M}_1.\mathbf{M}_2}{|\mathbf{M}_1||\mathbf{M}_2|}$: $\mathbf{M}_1$ and $\mathbf{M}_2$ are the magnetizations of the two Fe crystals. If we determine the magnetic order with the l-J model, we obtain a magnetic moments distribution in which the disorientation is more pronounced at the center of the Cr layer than at the interfaces where the order remains nearly colinear[14,20]. Consequently, the IMC varies as (figure 13):

$$\Delta E(\theta) = \Delta E_0 + \alpha.\cos\theta - \beta.\sin^2\theta \qquad \text{with } \beta > 0 \tag{3}$$

The second contribution corresponds to a biquadratic term proportionnal to $(\mathbf{M}_1.\mathbf{M}_2)^2$.

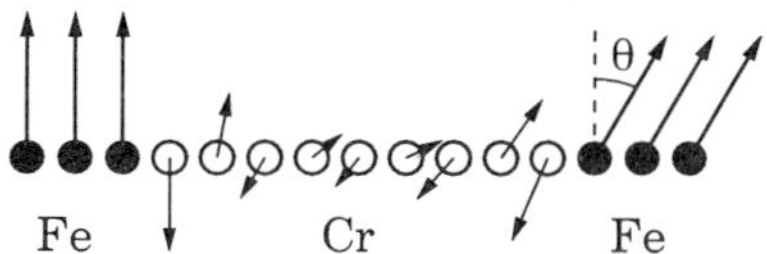

Figure 12. Schematic representation of the magnetic moments distribution when the magnetizations of the Fe crystals are disoriented with respect to each other ($Fe_\infty/Cr_n/Fe_\infty$ sandwiches).

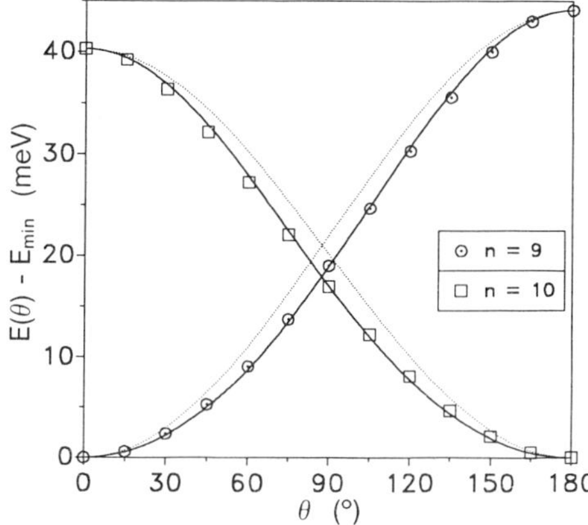

Figure 13. Coupling energies obtained with the l-J model: the dotted line curve corresponds to a fit with a quadratic IMC and the full line curve corresponds to a fit including the biquadratic term.

For example, we can determine the amplitude of the moment $|\mathbf{M}_i|$ in the TBA assuming that these moments are directed as determined in the l-J model. This state is not the lowest TB energy state but (i) it is more stable than the other states (colinear, helicoidal, ...) we considered and (ii) its energy E(θ) is nearly equal to the one calculated in the l-J model and it can be described from quadratic and biquadratic couplings whose ratio $\beta/|\alpha| \approx 15\%$; this value is in qualitative agreement with recent experimental data.[22]

3-4. Discussion

These studies show clearly that the magnetic structure of the Cr spacer is strongly modified by varying the layer thickness and the interfacial boundary conditions. We have shown that the magnetic structure of the wall changes for n=19 (from colinear it becomes non colinear). Similarly, we have shown that, for a given thickness, the IMC contains an intrinsic biquadratic coupling which is due here to the heterogeneous magnetic structure of the spacer and not to the interfacial roughness.[23] Note that if the sign of α varies with the parity of n, the sign of β is always negative: if we assume that some roughness reduce strongly the amplitude of the quadratic contribution, the biquadratic contribution will remain qualitatively the same.

4. CONCLUSION

In this paper, we have determined the magnetic states of various sandwiches based on F and AF layers and we have shown that they are determined by the strong interfacial couplings and by the natural magnetic order of the AF layers. These states are expected to be obtained in sandwiches elaborated recently for which the Cr layers have been shown to be antiferromagnetically ordered[16,17] at room temperature or even above. For non zero external magnetic fields, the strong exchange fields induced by the F layers on the boundaries of the constrained AF layers are modified. They introduce magnetic defects which can be either pinned at the interface or extended far away in the AF layers.

Therefore, the physical origin of the magnetic couplings in such magnetic ordered F/AF sandwiches and superlattices can be understood from pinned or unpinned magnetic walls. It is different from the RKKY mechanism which is thought to explain the couplings when the spacer is non magnetic.

REFERENCES

1. Grünberg P., Schreiber F., Pang Y., *Phys. Rev. Lett.* **57** (1986), 2442.
2. Carbone C., Alvarado S.F., *Phys. Rev. B* **36** (1987), 2433.
3. Baibich M.N., Broto J.M., Fert A., Nguyen Van Dau F., Petroff F., Etienne P., Creuzet G., Friederich A., Chazelas J., *Phys. Rev. Lett.* **61** (1988), 2472.
4. Parkin S.S.P., More N., Roche K.P., *Phys. Rev. Lett.* **64** (1990), 2304.
5. Parkin S.S.P.,Bhadra R., Roche K.P., *Phys. Rev. Lett.* **66** (1991), 2152.
6. Y. Wang, P.M. Levy, *Phys. Rev. Lett.* **65** (1990), 2732.
7. Bruno P., Chappert C., *Phys. Rev. Lett.* **67** (1991), 1602; P. Bruno, these proceedings
8. P.M. Levy, K. Ounadjela, S. Zhang, Y. Wang, C.B. Sommers, A. Fert, *J. Appl. Phys.* **67** (1991), 5914; Ounadjela K., Sommers C.B., Fert A., Stoeffler D., Gautier F., Moruzzi V.L., *Europhys. Lett.* **15** (1991), 975.
9. Herman F., Sticht J., van Schilfgaarde M., *J. Appl. Phys.* **69** (1991), 4783.
10. Herman F., Sticht J., van Schilfgaarde M., *Mat. Res. Soc. Symp.* Vol. **231** (1992), 195.
11. Stoeffler D., Gautier F., *Prog. Theor. Phys.* Suppl. **101** (1990), 139.
12. Unguris J., Celotta R.J., Pierce D.T., *Phys. Rev. Lett.* **67** (1991), 140.
13. Stoeffler D., Gautier F., *Phys. Rev. B.* **44** (1991), 10389.
14. Stoeffler D., Thesis University Louis Pasteur (Strasbourg, 1992) unpublished.
15. Stoeffler D. and Gautier F., to be published in the proceedings of E-MRS (Lyon, september 1992), *J. Magn. Magn. Mater.*
16. Unguris J., Celotta R.J., Pierce D.T., *Phys. Rev. Lett.* **69** (1992), 1125.
17. Walker T.G., Pang A.W., Hopster H., Alvarado S.F, *Phys. Rev. Lett.* **69** (1992), 1121.
18. Fu C.L., Freeman A.J., *J. Magn. Magn. Mater.* **54-57** (1986), 777.
19. Purcell S.T., Johnson M.T., McGee N.W.E., Coehorn R., Hoving W., *Phys. Rev. B* **45** (1992), 13064.
20. Stoeffler D. and Gautier F., to be published.
21. Cyrot M., *Phil. Mag.* **25** (1972), 1031; Lacroix C., Thesis (Grenoble, 1979).
22. Rührig M., Schäfer R., Hubert A., Mosler R., Wolf J.A., Demokritov S., Grünberg P., *Phys. stat. sol. (a)* **125** (1991), 635.
23. Slonczewski J.C., *Phys. Rev. Lett.* **67** (1992), 3172.

STRUCTURE, DEFECTS, GROWTH MODES AND MAGNETIC INTERLAYER COUPLING IN Ni-BASED MULTILAYERS

N.K. Flevaris

Department of Physics, Aristotle University of Thessaloniki
540 06, Thessaloniki, Greece

INTRODUCTION

It has been commonly recognised and noted, for the past decade, that the structural characterisation of the crystalline metallic compositionally modulated multilayers (CMM) is of primary importance both in the improvement and perfection of the growth methodologies as well as in the evaluation of the various physical properties and especially the anisotropic ones. The need for a thorough structural characterization is very often realised, especially when results of different works are to be compared or combined. Namely, when the modification by the multilayering (or modulation) process of some property or bĕhavior is found to depend on the structure of the materials, the necessity for a clear knowledge of the structural properties is more evident. For example, the dependence of the magnetic or magnetotransport properties on the modulation parameters or the growth processes have been recognised for some time and the direct comparison of experimental results is not always possible. However, with the rapid advances on the molecular-beam epitaxy (MBE) methodologies, with the capability for in-situ monitoring of several critical parameters during growth, it becomes easier to evaluate and compare results of different groups and, more so, when the works involve CMM with a total thickness less than about 100 nm for which, as we will discuss later, no significant structural defects are introduced by the coherency-strain effects which is the one most important source of structural (and other) modifications. Another example is the evaluation of the different studies on the so-called "supermodulus effect". In this case, difficulties (or misunderstandings) in a direct comparison is possible to have their origin in the fact that structurally different CMM systems are compared which should not and more importantly in the case of elastic (mechanical) properties as it will be discussed later. Examples of such reviews give Refs. 1-5.

In this contribution we will present examples of structural studies of crystalline metallic CMM. From these studies it will be attempted to develop a unified model for the growth properties of such systems where the primary importançe will be on the coherency strains. Also based on these strains will be a discussion of a possible relation between the structural characteristics, the localised or extended defects and the elastic properties (mainly the supermodulus effect) as well as the various, especially anisotropic ones, other properties.

Magnetism and Structure in Systems of Reduced Dimension
Edited by R.F.C. Farrow *et al.*, Plenum Press, New York, 1993

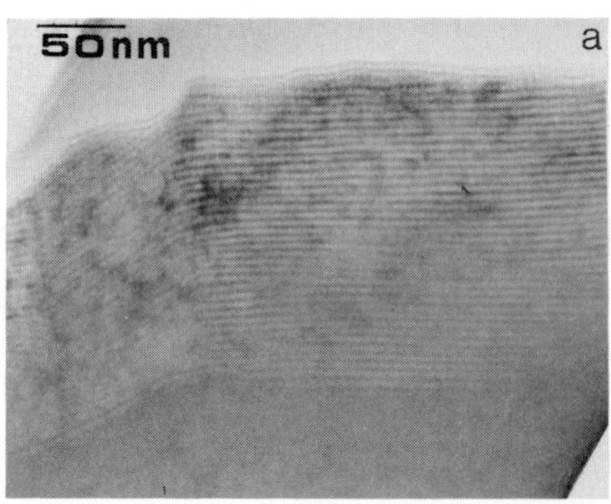

Figure 1. XTEM micrographs for the modulations of (a) Cu_7-Ni_7 and (b) Cu_{33}-Ni_{22} multilayers.

Finally, the magnetic and magneto-optic properties of these CMM systems will be presented and discussed in relation to a modulation-induced indirect magnetic interlayer coupling.

EXPERIMENTAL DETAILS

The CMM thin films were deposited onto heated mica in a dual e-gun evaporator system with a base pressure about $2x10^{-7}$ Torr. Separate monitors and controllers, employing quartz-crystal deposition readings, were used during growth to keep at the predetermined evaporation rates; at about 0.1 - 0.6 nms^{-1}. The desired deposition rates were calculated so that the modulation periods would be comprised of an integral number of atomic planes as it had been found earlier[6] that, especially when the modulation period (λ) consists of very thin constituent layers (of only one or a few atomic planes thick), it is important for the structural quality of the multilayer that this period, λ, is commensurate with the lattice, after care has been taken for the coherency strains. Thus, we will use the notation $\mathbf{A_m}$-$\mathbf{B_n}$ to denote a CMM having a modulation period (λ) consisting of **m** and **n** atomic planes rich in A and B constituent, respectively. The numbers m and n were determined by θ-2θ x-ray diffraction which, for several samples, were later verified by the electron microscopy studies. The latter included transmission, both planar (TEM) and cross-section (XTEM), reflection high-energy electron diffraction (RHEED) and surface morphology studies by scanning electron microscopy (SEM). For the XTEM (and the accompanying diffraction, XTED) studies the film specimen was sandwiched in the Dupuy[7] technique and thinned, initially by mechanical means and subsequently by ion-beam etching. For the TEM and XTEM studies voltages 120 and 200 kV were used while for RHEED 20-100 kV ones. The magnetic studies were performed by vibrating-sample (VSM), SQUID and torque magnetometry (TM) techniques.

RESULTS AND DISCUSSION

Electron Microscopy - Growth Modes

Figure 1(a), a XTEM micrograph for a Cu_7-Ni_7 CMM sample, will show an example of the multilayered structure. The main part of the structure is seen to be very smoothly and evenly layered. However, it will also be observed that, as usually seen for different CMM systems in the literature, the modulation sequences occasionally take an inclined form. Such inclination was not observed for long-λ modulations or for all CMM systems in the same manner. As it will be seen in Fig. 1(b) for a long-λ sample, Cu_{33}-Ni_{22},

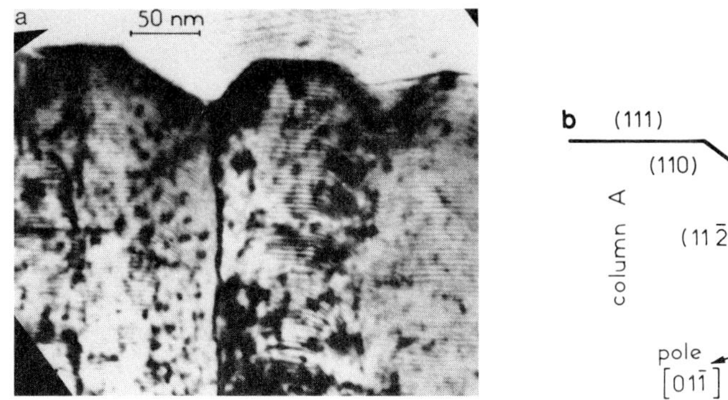

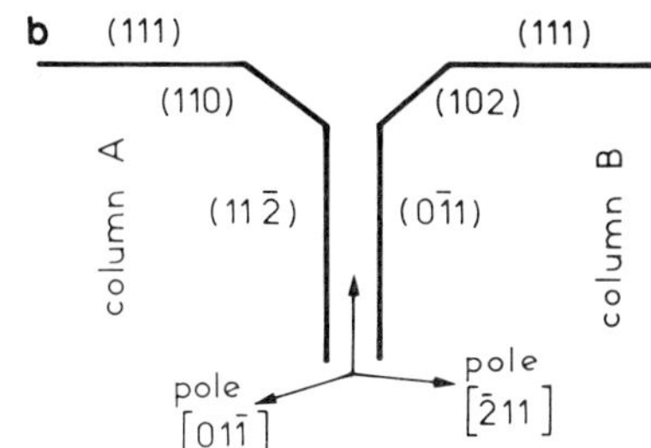

Figure 2. (a) XTEM micrograph for a $Cu_{6.8}$-$Ni_{7.0}$ CMM. (b) The diagram of the columnar growth.

no evidence for such inclined modulations was found; similarly, as it will be discussed later, no such inclined growth was observed for the Pd-Ni or some other systems since all λs studied were too long for the coherency-strain effects to be active. However, several structural properties are hidden in micrographs such as those of Fig. 1; namely, the extended twinning, the dislocations, the small- and high-angle grain boundaries, the possible columnar growth, the microtwinning etc. The main defect of the crystalline CMMs, as it has been studied in detail elsewhere[8,9], is found to be the double-positioning twinning which extends, in a puzzle-like labyrinthic form, over large areas of micrometer dimensions. Examples of the double-positioning twinning as well as average-lattice growth twins will be shown later.

Columnar Growth - Coherency Strains. Let us examine further the "inclined" modulations with the occasionally corresponding columnar-growth mode. The example seen in Fig. 1(a) is best illustrated in the XTEM micrograph seen in Fig. 2(a) for a $Cu_{6.8}$-$Ni_{7.0}$ specimen. It has been demonstrated earlier[8] that this pattern represents a columnar growth with high-angle columnar grain boundaries which, as also analysed earlier, result in a tooth-like faceted surface with crystallografically well defined facet planes. This structure will be shown schematically in Fig. 2(b). It is intriguing that the inclined facets are of the {110}- and the {210}-type for the two columns (A and B, respectively). However, this is expected when the coherency-strain effects are considered[9]. The growth kinetics will be driven by the anisotropic biaxial elastic modulus Y[hkl]. (The strain-dependent thermodynamics, related to Y[hkl], for the general type of dilatational distortion of cubic lattices has been presented earlier[10].) The anisotropy of the relevant elastic modulus for Cu and Ni will be seen in Fig. 3. It will be observed that on planes {100} the directions with the minimum or maximum Y are perpendicular to those of the same type, again with minimum or maximum Y, correspondingly; i.e. the <100> and <110> types. On planes {111} the directions <211> of Y_{min} are perpendicular to those, <110>-type, of Y_{max}. However, the most interesting behavior is observed for planes {110} and {211}, which are of interest for the fcc-on-mica CMM films with the [111] texture. Namely, for both the planes {110} and {211}, Y_{max} is along <111> which is normal to <211> and <110>, respectively. Correspondingly, Y_{min} is along <100> and the high-index <521> which are normal to < 110> and <201>. With the kinetics being treated as for the spinodal decomposition theory, it can be shown that the directions perpendicular to those with Y_{max} and Y_{min} will designate the directions of slowest and fastest growth, respectively. Thus, for example for the crystallite (column A) with the pole [011], in Fig. 2(b), the fastest growth will favour the formation of the "inclined" (112) facet and, correspondingly, for the grain B the (102) one. Similarly, one can assign the formation of the high-angle (columnar) grain boundaries (being (112) and (011) for grains A and B,

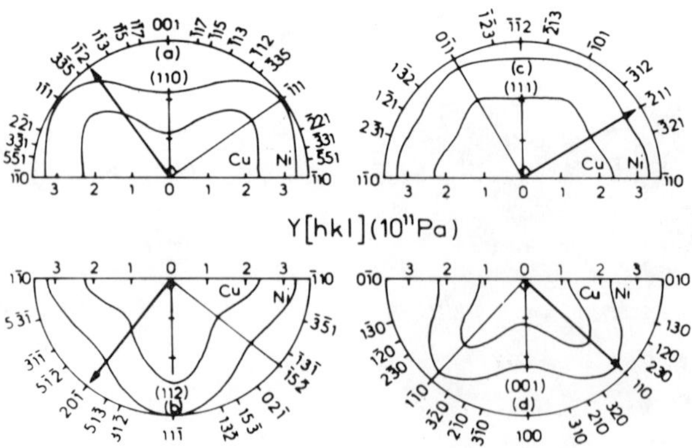

Figure 3. Polar diagrams, for Cu and Ni, of the Y distribution on {110}, {112}, {111} and {001}.

respectively) to the slowest growth as outlined before. Therefore, by considering the appropriate (anisotropic) biaxial elastic modulus Y[hkl], one could argue about the anisotropic growth of crystalline CMM and the formation of the inclined modulation sequences and the corresponding surface facets. Nevertheless, this will be based on the existence of the necessary coherency strains and their effects whose strength (and its dependence on λ) will be discussed further in the remainder. The overgrowth of such columns has been shown in SEM studies[11].

Early Stages of Growth - Thickness Effects. For now we will only comment briefly on another effect of these strains **during** the growth of a coherent crystalline CMM. In Fig. 4(a) and (b) the initial stages of growth will be seen for a Cu-Ni and Pd-Ni, respectively. In both cases the modulation was of roughly corresponding λ (about 2.5 and 1.5 nm) and there was a buffer layer of (about 50 and 30 nm thick) Cu and Pd, respectively, deposited on the mica substrate. It will be seen on both XTEM micrographs that, whilst the very first multilayer periods were deposited and grown smoothly and nearly defect free, the subsequent growth (after a certain thickness for each system) exhibited a variety of defects, for each system differently, as usually observed in our as well as the other structural studies of metallic CMMs. The main defects observed were the threading dislocation and strain fields (significantly more densed for Pd-Ni) and the initiation of the columnar growth (but only in the case) of the Cu-Ni CMM. (Another difference, also commonly observed but not seen with the experimental conditions used for the micrographs of Fig. 4, was the sharpness, size and density of the twins at the early and late stages of growth for the two systems.) The growth properties discussed here can be associated with the strain energy introduced by (and depended on) the lattice mismatch of the two constituents and the elastic properties of the materials. Namely, the coherency-strain induced free energy, within a continuum theory of elasticity, should be λ-independent and only the total excess energy of the system would increase as the CMM thickness increases progressively. On the contrary, the energy due to the formation of threading dislocations (or a dislocation network) would be expected to be inversely proportional to the film thickness. Therefore, two things should be taken into account: First, the dislocation generation should be favoured (for a particular system) for periods longer than a certain λ (without considering the energetics of the twin formation that will be discussed later). Second, after the fully coherent deposition of a certain thickness of the CMM, the total excess energy overcomes that which can be accommodated by the necessary lattice distortion and subsequent growth would result in a symmetry-lowering (or symmetry-breaking) process and - in cases of excessively large lattice mismatches - the generation of dislocations. In the case of Fig. 4(a), the beginning of the columnar growth and the "inclination" of modulation fringes is observed to occur on an otherwise smooth background of flat CMM growth. Correspondingly, threading dislocations together with extensive strain fields will be observed to form after some initial thickness of the CMM film. (A similar observation will be discussed further later.) The initial stages of growth for other CMM systems (Cu-Co and Fe-Ni) have been found to exhibit similar behavior[12]. Thus, it is not always appropriate to compare CMM studies without having an extensive structural study and, more importantly, to compare results from the same CMM system but of different total thickness; it is customary in recent studies on MBE-grown multilayers to restrict the total CMM growth thickness in just a few tens of nanometers which refer only to the early stages of growth.

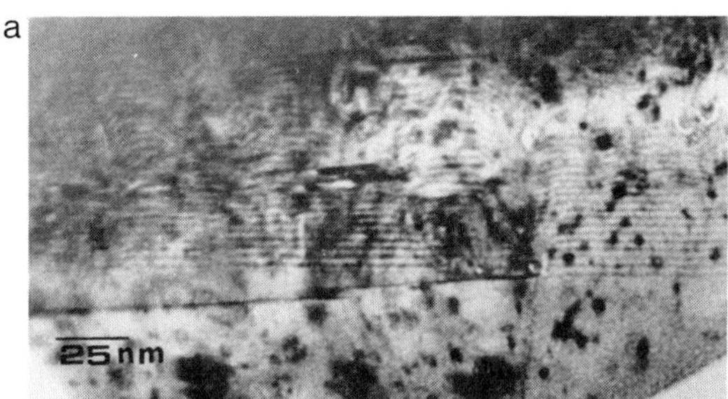

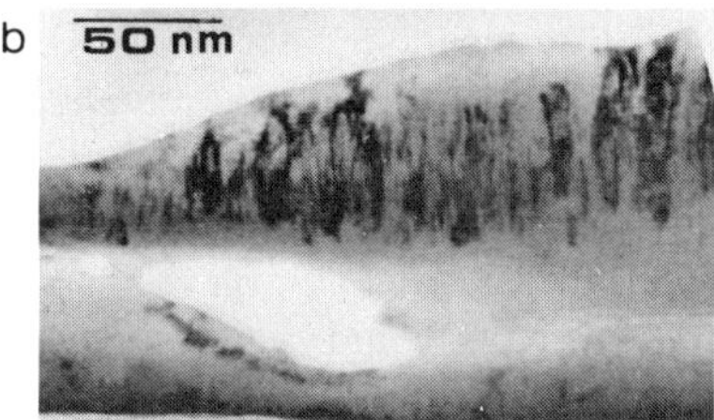

Figure 4. XTEM micrographs showing the early stages of growth of (a) Cu-Ni and (b) Pd-Ni together with the subsequent defect generation (and beginning of columnar growth for Cu-Ni).

Twinning. The twinning behaviour of the textured (metallic) CMM thin films was initially studied by electron microscopy[13] and later by x-ray diffraction[14]. For our CMM systems it has been shown[12,15] that the primary structural defect is the twinning; namely, the double-positioning (D-P) embedded twins have been found to form an interpenetrating puzzle-like structure which usually extends over very large areas. The analysis of this type of twinning, as well as the comparison between different CMM systems, has been published earlier[12]. The basic points can be summarised as follows: For all relatively short-λ modulations the majority of the twins is of the D-P type with occasional formation of microtwins. The 111-type of the twin interfaces were observed to be coherent -very sharp- for the Cu-Ni system but with considerable roughness for the Pd-Ni system; similarly, the vertical 110- and 211-type ones. Whenever the embedded twins penetrated grains with modulations of the "inclined" type, the modulation-fringe sequences were observed to cross the twin interfaces perfectly undisturbed. The density, size and spacing of the embedded twins were observed in XTEM to vary with λ and be different for each system. A typical example of such a twin will be seen in Fig. 5 for a Cu-Ni multilayer. The two TEM micrographs of Fig. 5 depict the bright- and dark-field pictures, respectively, in (a) and (b). Together with the D-P twinned structure one can also see the columnar grains and some microtwins along with the embedded twins. It is interesting to note that, in the micrograph of Fig. 5(a), one can see two D-P twins, connected by a columnar twin interface, only one of which is seen in the dark-field micrograph of Fig. 5(b). This suggests that the boundary will be of 110-type for one and 211-type for the adjacent grain, while both grains have a twin relationship with the matrix.

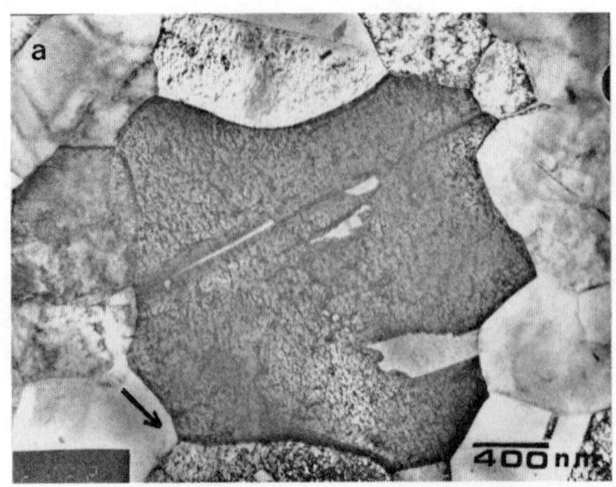

Figure 5. (a) Bright- and (b) dark-field TEM micrographs of columns and the embedded twinning.

As mentioned above, the D-P twinning is characteristic of our fcc CMMs but it also depends on modulation parameters, elastic properties of the constituents and their lattice mismatch. As it was shown earlier[12] the fcc character of the average lattice of a $Cu_{5.8}$-$Co_{3.5}$ multilayer was evidenced by the existence, and observation in XTEM, of a 111-type growth twin which would not be expected for a structure containing hexagonal Co. This type of (growth) twins could also be observed in the Cu-Ni system but it was λ-dependent. In Fig. 6 we will see this type of twinning in the (a) bright- and (b) dark-field XTEM micrographs for a Cu_{33}-Ni_{22} sample. It is clear that the texture and the modulation sequence, as seen before in Fig. 1(b), was [111]. However, the observed twinning, with the very sharp twin interfaces, reveals the equivalence of all <111> directions while for all other, short-λ modulations, the importance of the other directions had been suppressed in favour of the texture direction [111]. As mentioned above, for the case of the D-P embedded twins, we also see in Fig. 6(a) how the modulation-fringe sequence is not interrupted by the twin interfaces. In addition, in Fig. 7 we will see how the modulation process has been left unaffected within such <111> twins. The bright-field XTEM micrograph of Fig. 7(a) shows the modulation in the matrix and the

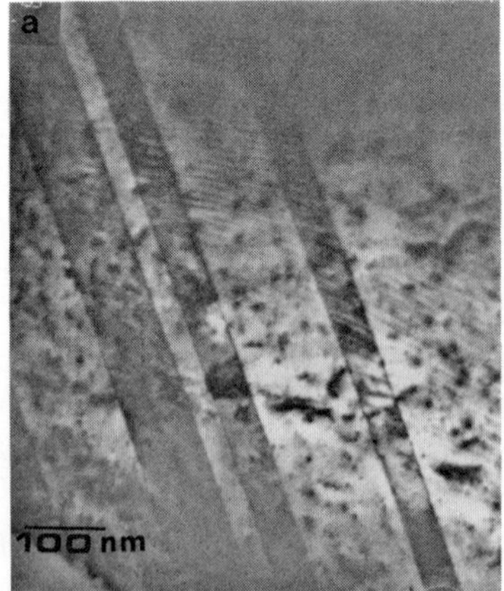

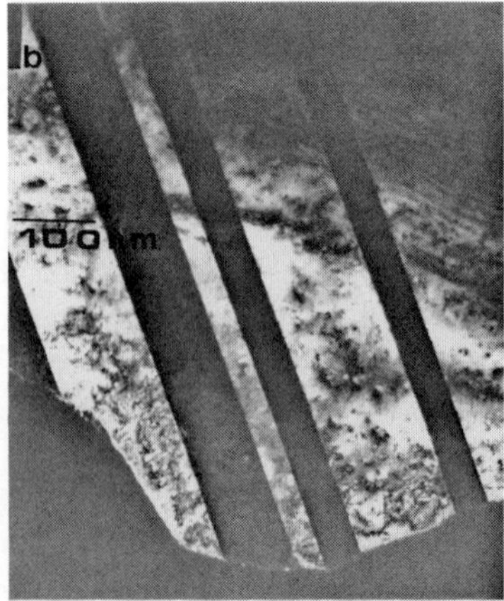

Figure 6. (a) Bright- and (b) Dark-field XTEM micrographs showing <111>-type growth twins.

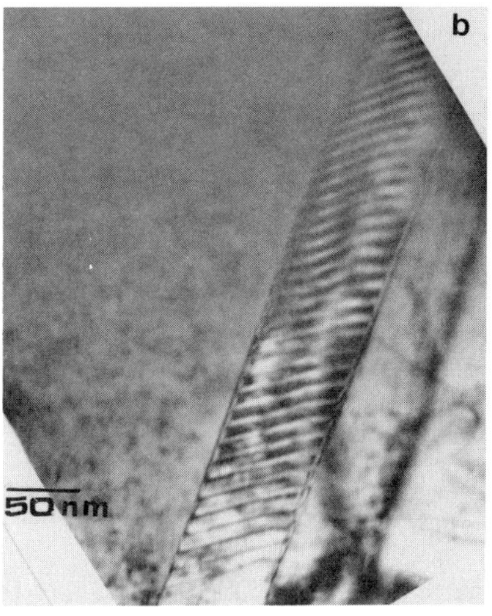

Figure 7. (a) <111> growth twinning and (b) the modulation fringes within this <111>-type twin.

formation of the twin while in the (also bright-field) corresponding appropriate reflection shown in Fig. 7(b) we can see the modulation fringes inside this <111>-type growth twin. Finally, concerning again these growth twins, let us see one more detail of another such twin depicted in the micrograph of Fig. 8. We can observe that, in addition to the very sharp <111>- type boundaries, the system has form the unique -as much as rear- <211>-type twin interface. Such an observation should not be expected (and it was not observed) in a reduced-

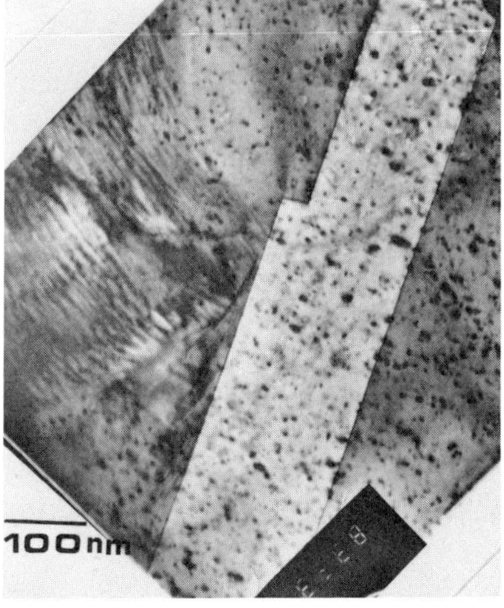

Figure 8. A coherent 211-type twin boundary characteristic of the fcc average lattice.

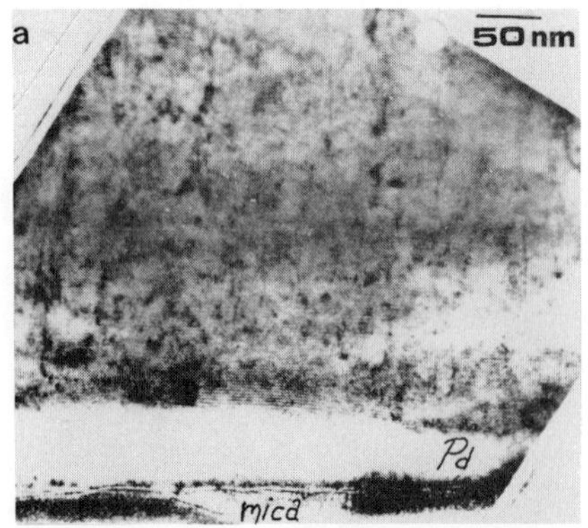

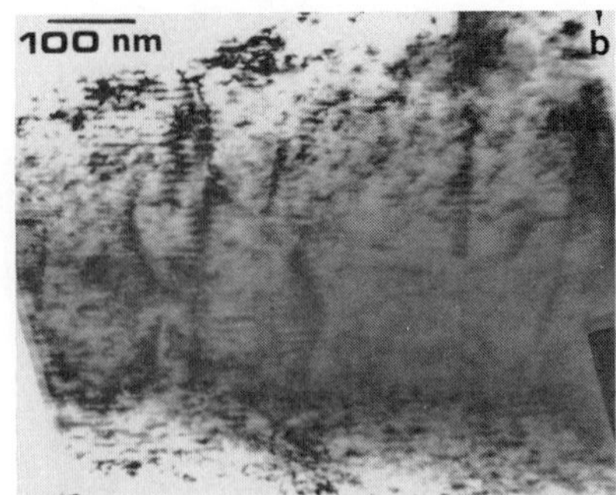

Figure 9. The average-lattice dislocation patterns for (a) Pd_3-Ni_3 and (b) Cu_{33}-Ni_{22} multilayers.

symmetry lattice as the case of the short-λ modulations of this CMM system. As it will be discussed later, this observation suggests that the overall (**average**) lattice, for long modulation periods, retains the fcc characteristics exhibiting all the effects of perfect-symmetry relations.

To conclude, let us investigate this via a study of the dislocation generation processes. In Fig. 9(a) we will see a XTEM micrograph for a Pd_3-Ni_3 multilayer. As also shown and discussed for Fig. 4, one can see the very smooth multilayering process for the early stages of growth to be followed by a dislocated structure and strain fields. Similarly, for a Cu_{33}-Ni_{22} sample, threading dislocations can be seen clearly in Fig. 9(b) while no such observations were possible for any of the short-λ modulations studied. It can be argued that the modulation period of six atomic planes is long enough for the Pd-Ni system (with a lattice mismatch of about 10 %) to make a symmetry-breaking process necessary while for the Cu-Ni system (with the mismatch being about 2.5 %) no such observations were made until the period reached the very high value of 50 atomic planes. (We must distinguish here between the regularly oriented dislocations seen in Fig. 9 and the random ones observed occasionally for other modulations. No evidence was found in our work for any of the so-called superlattice-dislocations[16].)

Discussion of Results. In the remainder of this section we will discuss the electron microscopy observations, we will attempt to associate the observed structural behavior of our crystalline CMM systems to the modulation parameters and, finally, to suggest a model for the structure (and growth mechanisms) in conjunction with the λ-dependent coherency-strain effects within the continuum elasticity theory. The primary defect of our CMM structures was found to be the double-positioning (embedded) twinning. The size, density and coherence of these twins varied with constituent lattice mismatch and modulation period, λ. Also, other kinds of defects were found to depend on the above parameters and this will be the subject of the classification scheme that will follow. We will distinguish different regimes, according to the effects of coherency strains. Namely, we consider the structure to be influenced (and characterised) by the superposition and interplay of two signals: the *carrier* signal (designated by the average CMM lattice) and the *modulating* signal (designating and weighed by the strain effects and being λ-dependent).

First, in the **very-short-λ** regime: The modulating signal dominates over the carrier signal, the effects of the coherency strains are very strong and the system suffers a first symmetry-lowering process; in the case of our fcc CMM structures with the 111 texture, this process results in a 60° equivalence rotation operation. Thus, the result will be the possibility of low-angle grain boundaries and double-positioning twinning and the labyrinthic puzzle-like twinned structure. In this operation, it is significant that the twinning works as a strain-energy sink and considerable amount of the coherency-strain excess energy is accommodated by the twin formation. As a result, new anisotropic behaviours may be observed while the elastic properties would be expected to change considerably. (For example, if[17] the so-called "supermodulus effect" were to be investigated it should be expected to be active mainly in this regime of modulations. But, nevertheless, one should be cautious and examine the structural characteristics of the CMMs to be compared, especially for such strongly structure-dependent properties before conclude about the comparability of their mechanical behavior.) The resulting structure is coherent with minor microstructural incoherence basically in the twin structure. All other defects were observed to be of atomistic dimensions while, because of the symmetry changes, no perfect-symmetry relations were found in the structural formation of the structural defects.

Second, in the **intermediate-λs** regime: As the period λ increases the coherency effects are lowered and the lattice is less vulnerable to structural transformation and more inclined towards various defects. The effect of the (strain-driven) modulating signal is lowered and weighed (or balanced) by the carrier-signal (average lattice) effects in an continuous interplay of the two signals, as the period, λ, increases. The extent of both regimes can be approximated by considering the structural and elastic properties of the constituents. For the second regime (and some part of the first) one is allowed to use the theory of continuum elasticity. (For example, this second regime extends up to about λ~18 and 10 atomic planes for Cu-Ni and Pd-Ni, respectively; the corresponding λs for the first regime would be 9 and only 3-4 atomic planes, where one should actually use microscopic elasticity theories.) In this (second) regime the coherency strains are partly still important for some λs. The incoherence is now wider and the lattice undergoes a second symmetry-lowering (or -breaking) process. The result, in our case, would be 30° rotational relationships and the high-angle grain boundaries with the possibility of well established columnar growth and tooth-like surfaces for some λs until the modulating signal is outweighed by the average lattice carrier signal. (Thus, it could be in this regime that the modification of the elastic properties, and the "supermodulus effect", would be diminished.)

Third, in the **long-λ** regime: This regime is characterised by the effects of only the carrier signal and the CMM behavior is dominated by the average-lattice effects. There are only weak strain effects and no symmetry changes occur. Only perfect symmetry relations are observed and, in our case of 111-textured fcc films, all <111> directions are found to be equivalent. As a result, the growth twins are observed, contrary to the cases of shorter λs. Also, since the average fcc lattice prevails, coherent and perfect-symmetry twin interfaces are observed as shown before. One can estimate a critical λ_c for each CMM system with a certain texture; for our systems one can[18] find $\lambda_c \sim 60$ and 20 atomic planes for Cu-Ni and Pd-Ni. Lastly, for systems with periods **longer than λ_c**, one can no longer talk about composition modulations or superlattices (or multilayers in the current sense); such systems would be incoherently laminated structures and possible effects would be restricted at the interfaces.

The above outlined model contains all observations from the structural investigations and suggests a possible scheme for the classification of structural properties of crystalline CMM. Finally, we will remind the effects of the total thickness and the model of columnar growth. It is important to keep in mind, when designing the preparation of CMM system or analysing various structure-sensitive properties, that the structure of a crystalline coherent CMM can be altered after the deposition of some thickness because of the accumulation of excessive strain energy. Thus, for the preparation of, as much as possible, defect-free CMMs one should limit the total film thickness differently for different constituents or, in case a certain total thickness is needed for some study, the appropriate CMM thickness could be repeatedly grown on successively deposited buffer layers to accommodate the strain energy.

Interlayer Magnetic Coupling - Magnetization Enhancement

The exchange-induced interlayer magnetic coupling has attracted considerable attention recently, mainly because of the very interesting observations of antifferomagnetic-like coupling in Fe- and Co-based multilayers and also the resulting technologically interesting giant magnetoresistance effect. Review articles on this subject can be found in this volume. The observation of such phenomena in Ag/Ni multilayers was also published recently[19]. In this section we will treat the subject of the interlayer coupling, both via magnetic and magneto-optic studies, from a point of view slightly different from the customary one.

It is customary to consider as an indication for the interlayer magnetic coupling the dependence of the magnetization remanence (M_{rem}) or the in-plane saturation magnetic field (H_S) on the thickness of the non-magnetic constituent in each modulation period, λ, when the corresponding thickness of the magnetic layer is kept constant. Occasionally, such a dependence, for several systems, has been observed to exhibit an oscillatory manner. However, as it will be discussed later, the possibility for an oscillatory variation of the interlayer coupling was suggested earlier[20] based on magnetic studies of short-λ Cu_m-Ni_n CMM and their metastability and the spin-glass-like behavior. In that model it was concluded that such a behavior would depend on both *m and n*. The difference between that and the recent studies is that, since the magnetization depends on the magnetic layer thickness, since the latter use Fe or Co (with high magnetization) the minor effects of the coupling are masked by the strong ferromagnetism and one needs to use much thicker non-magnetic spacers while, in the case of Ni-based CMMs, the overall ferromagnetism is considerably weaker allowing the observation of the weak superimposed coupling effects. In addition, as revealed by the x-ray diffraction studies, it was understood that when the ratio (m/n) is much different than unity the coherency-strain effects did not allow the symmetric structural distortion and, therefore, the such asymmetric-λ samples were not structurally comparable and were not used. The magnetic studies to be discussed, involved only samples with m=n or nearly similar.

In Fig. 10(a) the ratio (M_{rem}/M_S) will be shown, for Cu_n-Ni_n at T=5 K, as a function of n and in Fig. 10(b) the corresponding dependence of the in-plane saturation field, H_S. It is seen that the oscillations present the corresponding extrema with a period (in n) of about four atomic planes which, doubled, corresponds to modulation periods (λ) of about 1.6 nm. (The triangles in Fig. 10 refer to different thinner samples of the appropriate modulations.) We observe the variation, in the plots of Fig. 10, to extend from 0.4 to 0.7 in (a) and from 2 to 20 kOe in (b), respectively. As mentioned above, the CMM samples represented in Fig. 10 had both layer thicknesses varying in a way similar to what will also be seen in Fig. 11, for Pd_n-Ni_n samples while in recent literature the magnetic-layer thickness is kept constant. The oscillation period is seen to be similar to that for the Cu-Ni system in Fig. 10. The corresponding quantities, in (a) and (b), vary between 0.2 - 0.9 and 50 Oe - 6 kOe; the high in-plane H_S for the n=2 sample is due to the development of perpendicular anisotropy. The results shown in Fig. 11 depict the room-temperature study; very similar results were obtained at low temperatures also. The dependence of the magnetization of Cu_m-Ni_n and Pd_m-Ni_n on m for constant n, has been established earlier[21,22] for low enough m and n. This represented one of the reasons for not keeping n constant and vary only m which, for thin layers could cause shadowing of the weak effects of the coupling if n was not thin enough.

The notion and the various effects of the possible interlayer magnetic coupling onto the magnetic properties of this kind of CMM thin films has been suggested before concerning the interpretation of either magnetic[21,22] or magneto-optic[23] properties. However, it will be interesting to examine further the possible such effect for the region of the strongest coupling, as suggested from Fig. 10, for the Cu_7-Ni_7 sample. In Fig.12 (a) the temperature dependence of the magnetization will be seen for two such CMM samples (as studied by VSM, TM and SQUID techniques). The magnetization per unit volume of the Ni content is presented as

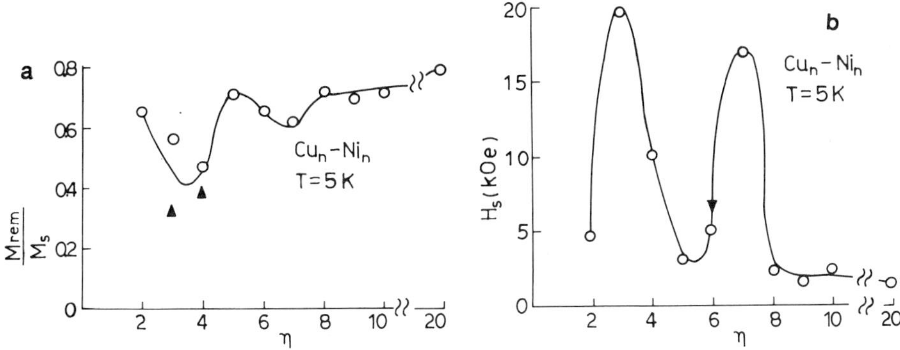

Figure 10. The oscillatory dependence of (a) M_{rem}/M_S and (b) H_S on n for Cu_n-Ni_n, at T = 5 K.

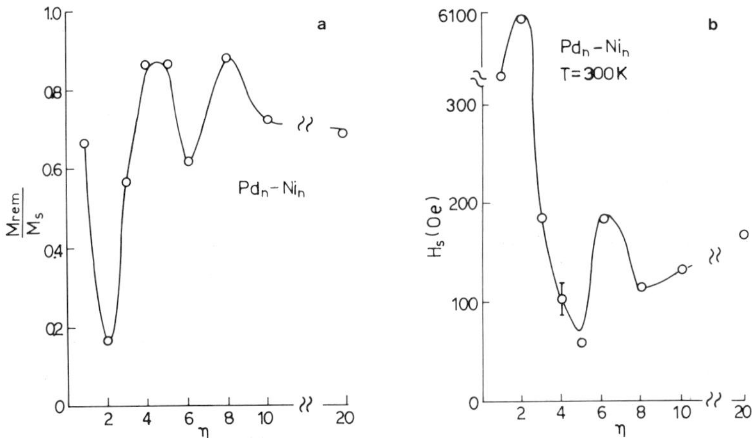

Figure 11. The oscillatory dependence of (a) M_{rem}/M_S and (b) H_S on n for Pd_n-Ni_n at T=300 K.

normalized to that of the pure Ni. It is seen that while M for the Cu_7-Ni_7 sample reached that of pure Ni, the magnetization for $Cu_{6.8}$-$Ni_{7.0}$ is observed to exceed the pure-Ni value by about 10 %. The obvious question of the accurate determination of the total thickness of Ni has been considered in the structural investigations by x-ray diffraction and XTEM and XTED studies. The former two methods reveal the thickness in absolute units (i.e. nm) while the latter in atomic planes (m+n) and they all were in perfect agreement. Therefore, the values shown in Fig. 12(a) may involve only the experimental uncertainty which suggests that the lowest M would be at least equal to that of Ni which, for this kind of thicknesses, also should be considered as a significant enhancement; it must be noted that the previously published M for the Cu_7-Ni_7 sample, and shown in Fig. 12(a), was the highest M among all values for the Cu-Ni multilayers reported in the literature. The difference between the two Cu-Ni samples reported here was that the $Cu_{6.8}$-$Ni_{7.0}$ sample exhibited higher relative modulation amplitude than the Cu_7-Ni_7 one as determined by the x-ray diffraction; the former had a smaller total thickness and the effects of the thickness on the structure were discussed earlier. The structure of these samples was studied in detail as presented in the previous Sections. Also, it must be clarified that this enhancement must not be associated to that published[24] in the beginning of the work on magnetic CMM which was due to erroneous (or incomplete) interpretation of a ferromagnetic resonance study. Thus, it will be rather safe to consider this result as evidence of magnetization enhancement; however, we must remind that this study considers M as (moment-per-unit volume) and not atomic moment. The latter, being the subject of a recent study[25] of surface magnetism has also found evidence for magnetization enhancement. (If one, in comparing the two studies, tried to examine the accuracy in the calculation of the Ni volume in view of the possibility of not representing really the exact amount of Ni because of

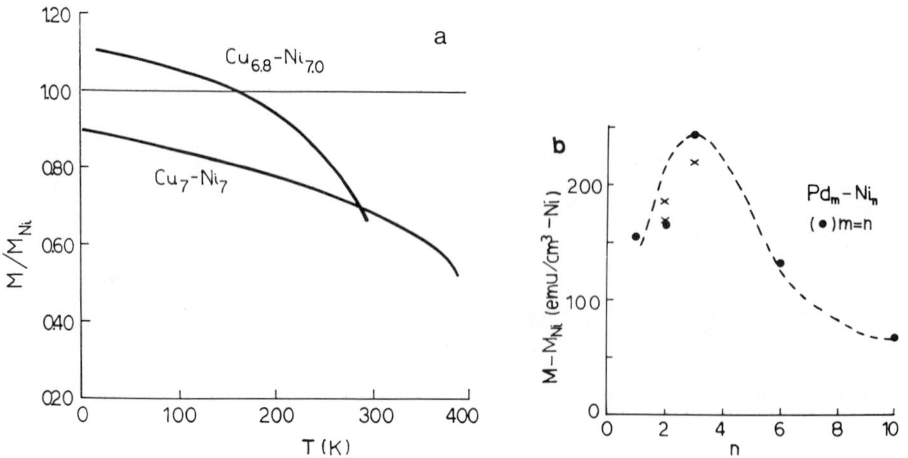

Figure 12. Magnetization enhancement, over that of Ni, for (a) Cu_7-Ni_7 samples and (b) Pd_n-Ni_n.

the possible existence of voids etc., the answer would be that, in that case, the calculated Ni-volume would overestimate the real material contained and, therefcre, underestimate the magnetization density.) Finally, it is interesting to note that the apparent Curie temperature, as suggested by the plot of Fig. 12 (a), is lowered for the enhanced-M sample.

The magnetization enhancement would naturally be expected if the second constituent were the easily polarizable Pd. However, it will interesting to see how this effect is combined with the interlayer magnetic coupling in low-λ Pd_n-Ni_n samples. In Fig. 12 (b), the net magnetization (per unit volume of Ni) over that of the pure Ni, will be shown for the Pd_n-Ni_n samples; with the crosses at n=2 and n=3 we refer to samples Pd_3-Ni_2, Pd_4-Ni_2 and Pd_2-Ni_3, respectively. It must be noted that the enhancement seen in Fig. 12 (b), for room-temperature M, is seen to be up to about 50 % and at low temperatures reached 80 % while the corresponding enhancements reported in the literature for Co-Pd, Fe-Pd, Co-Pt etc., is < 25 % which has its origin in the difference in the modulation sharpness; the latter, for our Pd_m-Ni_n CMM was also revealed by the presence of perpendicular anisotropy exclusively for samples with n=2 while for poor-quality modulations the appearance of such effects spreads over significantly broader modulation thicknesses.

The effects of the coupling to the magneto-optic properties was shown recently[23] for Pd_m-Ni_n multilayers; a review of recent such studies on alloys and multilayers can be found in Ref.[26]. It was observed that the polar Kerr rotation (θ_K) depended not only on the Ni thickness (n), as reasonably expected, for constant m but also on m for constant n. The outcome of such an interplay will be seen in Fig. 13 (a), for Pd_n-Ni_n with thin enough ns. It is seen that the saturation rotation (for light at Λ = 633 nm) increases with n between 1 and 3 (the circle and cross symbols at n = 3 refer to two multilayers with the same modulation λs) and subsequently decreases for n = 4 and 5, while it increased again, as expected for higher ns. (Similar behavior has been observed and reported for the magnetization of Cu_m-Ni_n and Pd_m-Ni_n.) The corresponding behavior for the spectroscopic Kerr rotation is seen in the examples depicted in Fig. 13 (b) for three Pd_m-Ni_2 samples; the rhombic symbols refer to m = 4, the closed triangles to m = 3 and the open ones to m = 2. The data of Fig. 13 (b) suggest that the increase of m, for n = 2, not only modified the appearance of the double peak which is related to Pd content but it also reduced dramatically the amplitude of the rotation enhancement. Some of the effects may be attributed to the modification of the optical properties of the multilayers via the chemical and the strain modulations as well as the reduced dimensionality (which could result in a line broadening such as that observed).

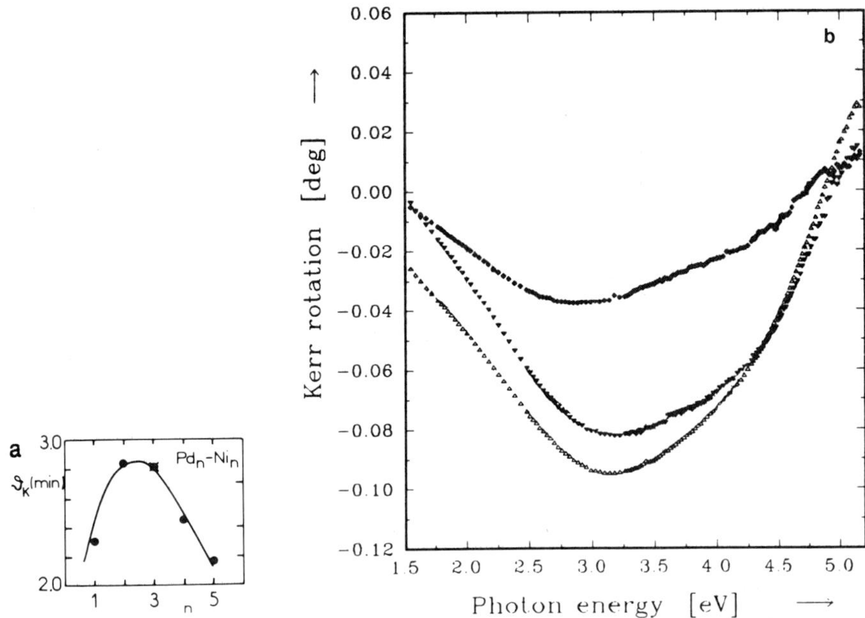

Figure 13. (a) The polar-Kerr- rotation (θ_K) dependence on the layer thickness (n) for short-λ Pd_n-Ni_n. (b) The dependence of spectroscopic polar Kerr rotation on the Pd-layer thickness (m) for Pd_m-Ni_2; rhombic symbols (m=4), closed triangles (m=3) and open triangles (m=2).

Conclusions and Summary. Briefly, the results presented in this Section are to be considered only as a means to discuss the possibility for an interlayer magnetic coupling from a point of view slightly different from that established in the recent literature. We have shown the existence of kinds of oscillatory dependence of magnetic properties on modulation-layer thicknesses while care was taken for the manipulation of the different magnetic contributions, by the variation of both constituent thicknesses, so that these effects would not be masked by the strongly ferromagnetic character of thick magnetic layers. It is shown that the various coupling effects can be studied in this manner when very-short-λ CMM samples are available with excellent structural and modulation characteristics. Nevertheless, in our study we had developed some picture of the structure of our materials by extensive structural investigations. Sometimes it becomes dangerous to misjudge possible influence of the structural effects on the magnetic -or other- properties. It was observed that the magnetic coupling affected strongly the in-plane remanence ratio (M_{rem}/M_S) and saturation field H_S, the polar-Kerr-rotation (θ_K) (at one light frequency as well as spectroscopically) and, finally, it may be responsible for the creation of (not only the λ-dependent magnetic property modification, as it is well established already but also) *magnetization enhancement* at the appropriate modulations even for Cu-Ni. This study of the magnetic properties agrees and complements the model[20] suggested in 1982: *" Therefore, we will interpret the observed phenomena in terms of an oscillatory indirect-exchange-induced distribution which, depending on **n** and **m** as discussed before, may have an atiferromagnetic-like character."*

Acknowledgement. The author is indebted to Prof. J. Stoemenos and Prof. Th. Karakostas for their contributions in the electron microscopy work, to Dr. R. Krishnan for his collaboration in the Kerr-effect, VSM and TM studies and to Dr. S. Visnovski for the spectroscopic-Kerr-effect experiments. The work was supported in part by EEC (BREU-153).

References

1. B.C. Giessen et al., eds., "Synthetic Modulated Structures", Academic, New York (1985).
2. R.C. Cammarata, Scripte Metall. **20**:479 (1986).
3. T. Tsakalakos, ed., "Modulated Structure Materials", Martinus Nijhoff, Dordrecht (1984).
4. B.Y. Jin and J.B. Ketterson, Adv. in Phys., **38**:189 (1989).
5. R.F.C. Farrow et al., eds., "Magnetism and Structure in Systems of Reduced Dimension", this **Volume**.
6. N.K. Flevaris, D. Baral, J.E. Hilliard and J.B. Ketterson, Appl. Phys. Lett. **38**:992 (1981).
7. M. Dupuy, J. Microsc. Spectrosc. Electr. **9**:163 (1984).
8. N.K. Flevaris and Th. Karakostas, J. Appl. Phys. **63**:1228 (1988).
9. N.K. Flevaris and Th. Karakostas, Mater. Lett. **12**:7 (1991).
10. N.K. Flevaris, J. Mater. Sci. **24**:313 (1989).
11. N.K. Flevaris and Th. Karakostas, in: "Alloy Phase Stability", G.M. Stocks and A. Gonis, eds., Kluwer Academic Publs., Dordrecht, (1989) p. 591.
12. N.K. Flevaris, J. Spectrosc. Microsc. Electr. **14**:387 (1989).
13. A. Purdes, Ph.D. thesis (unpublished), Northwestern University, Evanston, USA (1976).
14. E.M. Gyorgy, D.B. McWhan, J.F. Dillon, Jr., L.R. Walker and J.V. Waszcak, Phys. Rev. **B25**: 6739 (1982); A. Segmuller and A.E. Blakeslee, J. Appl. Cryst. **6**:19 (1973); ibid **6**:413 (1973).
15. Th. Karakostas and N.K. Flevaris, J. Mater. Sci. Lett. **5**:1235 (1986).
16. C.S. Baxter and W.M. Stobbs, Appl. Phys. Lett. **48**:1202 (1986); W.M. Stobbs, J. de Phys. **C5**:48 (1987).
17. B.M. Davis, D.N. Seidman, A. moreau, J.B. Ketterson, J. Mattson and M. Grimsditch, Phys. Rev. **B43**:9304 (1991);
    A. Fartash, E.E. Fullerton, I.K. Schuller, S.E. Bobbin, J. W. Wagner, R.C. Cammarata, S. Kumar and M. Grimsditch, Phys. Rev. **B44**:13760 (1991).
18. B. Ditchek, Scripta Metall. **11**:207 (1977) and references therein.
19. C.A. dos Santos, B. Rodmacq, M. Vaezzadeh and B. George, Appl. Phys. Lett. **59**:126 (1991).
20. N.K. Flevaris, Ph.D. thesis (unpublished), Northwestern Univers., Evanston, USA (1983).
21. N.K. Flevaris, J.B. Ketterson and J.B. Hilliard, J. Appl. Phys. **53**:8046 (1982); ibid **53**:2439 (1982).
22. N.K. Flevaris and R. Krishnan, J. Magn. Magn. Mater. **104-107**:1760 (1991).
23. N.K. Flevaris, Appl. Phys. Lett. **58**:2177 (1991).
24. B. Thaler, J.B. Ketterson and J.E. Hilliard, Phys. Rev. Lett. **41**:336 (1978).
25. G. Lugert, G. Bayreuther, S. Lehner, G. Gruber and P. Bruno, Mat. Res. Soc. Symp. Proc. **232**:97 (1991) and references therein.
26. D. Weller and W. Reim, Mat. Res. Soc. Symp. Proc. **232**:71 (1991) and references therein.

ON THE STRUCTURAL QUALITY OF Co/Cu TRILAYERS AND SUPERLATTICES: THE INFLUENCE OF THE TEMPLATE LAYER

A. Cebollada, J. de la Figuera, A.L. Vázquez de Parga, C. Ocal, and R. Miranda

Dpto. de Física de la Materia Condensada
Universidad Autónoma de Madrid, Cantoblanco
E-28049-Madrid, Spain

Introduction

Structural imperfections and chemical intermixing can strongly influence the magnetic properties of metallic superlattices. Co/Cu multilayers grown by sputtering on Si(100) and Si(111) substrates with intermediate seed layers (Cu, Fe, Ru) have been found to display oscillatory MR effects. It has been claimed that the structural quality of multilayers grown on seed layers is high and, in particular, that a 50 Å thick Fe buffer layer provides a template for growth as good as a Cu(100) single crystal [1]. In this work we report on extensive characterization by several diffraction techniques of the detailed crystallography, degree of intermixing and interfacial roughness of Co/Cu trilayers and superlattices grown by MBE on Cu(100) substrates under UHV conditions. Neutron and X-ray diffraction (for superlattices) and LEED (for trilayers) experimental data are compared to detailed calculations (kinematic and dynamic) of the diffracted intensities. On the other hand, the degree of perfection of Fe overlayers deposited on Si wafers and the extent of chemical reaction that takes place at the Fe/Si are also characterized. The data demonstrate that the buffer layers deposited on Si that we have studied represent a template of much lower crystalline quality than Cu single crystals for further growth of Co/Cu superlattices.

Co/Cu superlattices

Several Co/Cu superlattices were grown at room temperature on Cu(100) single crystal substrates by MBE under UHV conditions. Evaporation rates were ≈ 0.01Å/sec. Previous studies on the Co/Cu(100) system [2] make these growth conditions very favourable to obtain flat surfaces from the point of view of density of defects and degree of interdiffusion, that begin to be important at 500 K. Thickness of the Co and Cu layers were measured with a quartz balance previously calibrated with TEAS (Thermal Energy Atom Scattering) oscillations which also allows the determination of density of defects [2]. After deposition of the superlattice, a 200 ML Cu layer was deposited to avoid surface oxidation after exposure to air.

Magnetism and Structure in Systems of Reduced Dimension
Edited by R.F.C. Farrow *et al.*, Plenum Press, New York, 1993

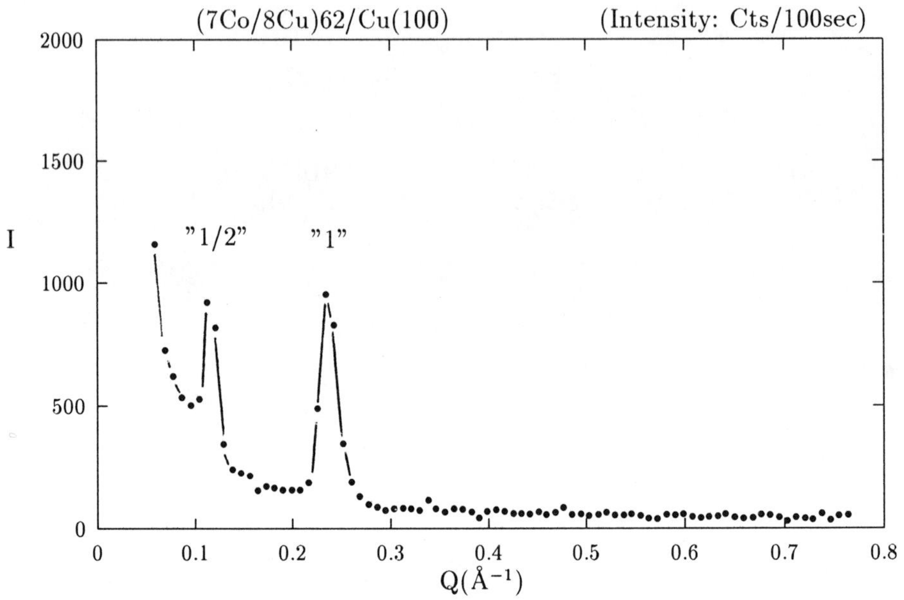

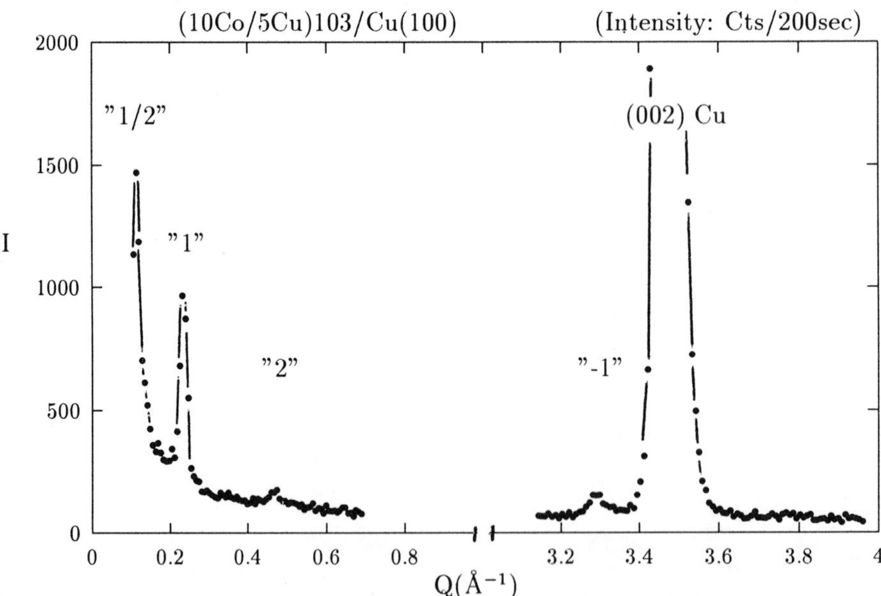

Fig.1 Neutron diffraction measurements for two MBE grown superlattices.

Neutron diffraction spectra for two superlattices, namely (7ML Co/8ML Cu)62/Cu(100) and (10ML Co/5ML Cu)103/Cu(100), are shown in figure 1.

For the superlattice (7ML Co/8ML Cu)62/Cu(100), only low Q diffraction peaks are shown. The peak labeled "1" corresponds to the first superlattice diffraction peak, while the peak labeled "1/2" corresponds to the double periodicity caused by the antiferromagnetic coupling of the Co films through the Cu layers. For the superlattice (10ML Co/5ML Cu)103/Cu(100), both low and high Q diffraction peaks are shown. In the low Q range, two superlattice peaks ("1" and "2") are observed, as well as a "1/2" peak due to the antiferromagnetic arrangement. In the high Q range, the (002) Bragg peak from the Cu(100) substrate overlaps the average lattice peak of the superlattice, and a superlattice peak in the low Q side ("-1") is observed. No superlattice peak is observed in the high Q side due to defocusing efects in the neutron beam at the required scattering angles ($2\theta \approx 90°$). For (7ML Co/8ML Cu)62/Cu(100) superlattice, "-1" satellite has been also observed with polarized neutron diffraction [3].

Finally, both low and high Q satellites, even high Q (002) side ones, are observed in measurements of X ray diffraction for both superlattices.

From these results, a first intuitive analysis shows that the presence of low Q superlattice peaks is due to the existence of long range order in the superlattices, that means sharp Co/Cu interfaces. On the other hand, the high Q superlattice peaks indicate the crystalline character of the Co and Cu layers (short range order).

For a more detailed evaluation of these results, a kinematical theory for X Rays Diffraction can be used [4]. In this approach, the intensity of diffracted peaks from a superlattice composed by N bilayers of Co and Cu is given by:

$$I(Q) = \pounds_N(Q)|f(Q)|^2 \tag{1}$$

where Q is the scattering vector and

$$\pounds_N(Q) = \left|\frac{sen(N\Lambda Q/2)}{sen(\Lambda Q/2)}\right|^2 ;\Lambda = \text{bilayer thickness} \tag{2}$$

is the interference or Laue function, and the form factor is given by

$$\begin{aligned}|f(Q)|^2 = {} & f_{Cu}^2 L_{Cu}(Q) + f_{Co}^2 L_{Co}(Q) + p_{Co}^2 L_{Co}(Q) \\ & +2f_{Cu}f_{Co}[L_{Cu}(Q)L_{Co}(Q)]^{1/2}cos(\Lambda Q/2)\end{aligned} \tag{3}$$

for every "unit cell" in the superlattice, containing the interference functions with periodicities from Co and Cu:

$$L_\nu(Q) = \left|\frac{sen(n_\nu d_\nu Q/2)}{sen(d_\nu Q/2)}\right|^2 ;\nu = Co, Cu. \tag{4}$$

For neutron diffraction, f_{Co} y f_{Cu} are the nuclear scattering cross sections for Co and Cu atoms and p_{Co} is the magnetic scattering cross section for Co atoms. Taking $d_{Cu} = 1.8$Å and $d_{Co} = 1.775$Å (interplanar spacing for Cu and fcc Co in the [100] direction), $f_{Cu} = 0.7718 \times 10^{-12}$cm, $f_{Co} = 0.25 \times 10^{-12}$cm [5], $p_{Co} = 0.461 \times 10^{-12}$cm for low Q peaks and $p_{Co} = 0.315 \times 10^{-12}$cm for high Q peaks [6], the diffracted intensity distribution for the two superlattices studied, assuming sharp interfaces and crystalline bilayers, is shown in figure 2.

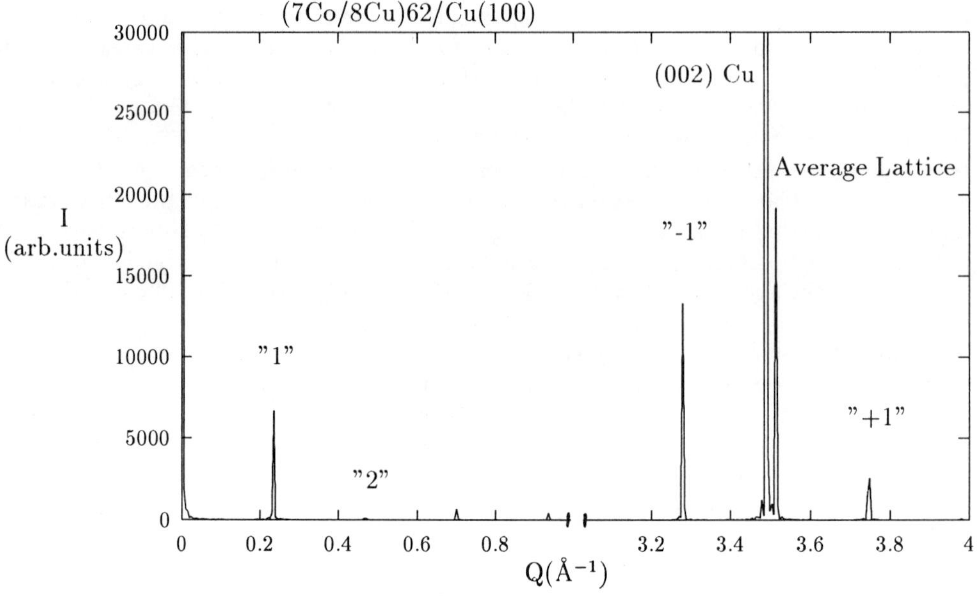

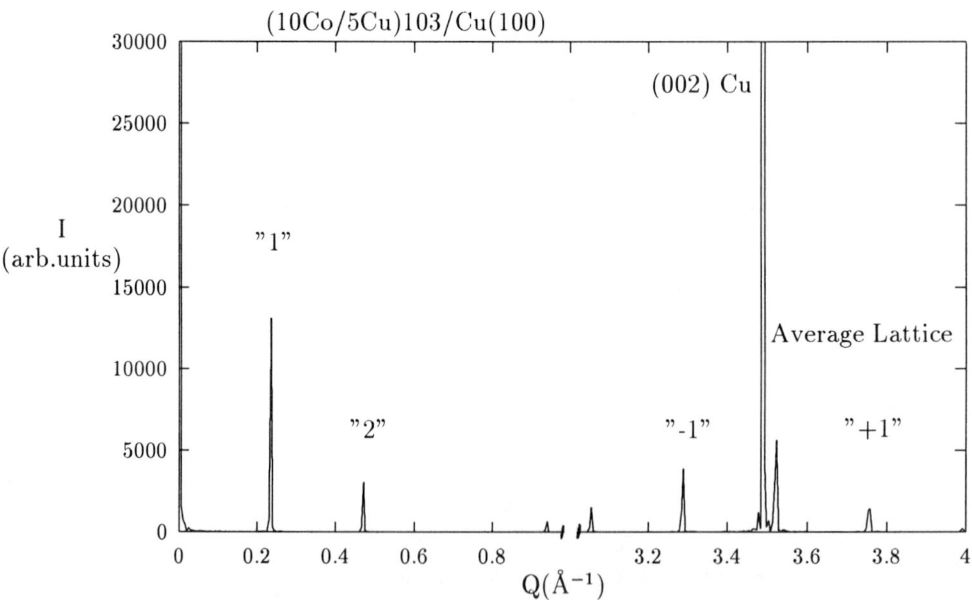

Fig.2 Kinematical calculation of the diffracted intensity assuming crystalline bilayers and sharp interfaces.

It shows the presence of both low and high Q satellites for both samples. In sample (7ML Co/8ML Cu)62/Cu(100), satellite "2" is almost not obsevable, as in fact it happens in the experiment. In sample (10ML Co/5MLCu)103/Cu(100) intensity ratios between low and high Q peaks are also in agreement with the experiment. In the simulation, the average lattice peak is ibserved close to the Cu (002) substrate peak for both samples, while in the experiments it is not observable due to the intrinsic Cu(002) width.

If one of the materials, Co or Cu, is amorphous, the corresponding $L_\nu(Q)$ function is not a Laue Function anymore, changing to

$$L_\nu(Q) \propto \left|\frac{sen\tilde{\Lambda}_\nu Q/2}{Q/2}\right|^2 \tag{5}$$

where $\tilde{\Lambda}_\nu$ is the equivalent thickness to $n_\nu d_\nu$.

In this case, the diffracted intensity distribution is shown in figure 3. The main effects are the dissapearance of the average lattice peak and a drastic reduction in the intensity of the high Q peaks compared to the low ones.

Interdiffusion

A certain degree of interdiffusion can be included in the model via thickness fluctuations, given as a gausian ditribution in the bilayer thickness: $\exp[-(\Lambda-\overline{\Lambda})^2/\sigma^2]$, where $\overline{\Lambda}$ is the most probable Λ and σ is related with the FWHM by $\Delta\Lambda = 2\sqrt{\ln 2}\sigma = 1.67\sigma$.

This thickness fluctuation affects to $\pounds_N$:

$$\pounds_N(Q) = \frac{1+exp(-N\sigma^2Q^2/2) - 2exp(-N\sigma^2Q^2/4)cos(N\overline{\Lambda}Q)}{1+exp(-\sigma^2Q^2/2) - 2exp(-\sigma^2Q^2/4)cos(\overline{\Lambda}Q)} \tag{6}$$

This bilayer fluctuation can be expressed in terms of fluctuations in the number of layers of both materials Co and Cu, $\sigma^2 = d_{Co}^2\sigma_{Co}^2 + d_{Cu}^2\sigma_{Cu}^2$, where $\Delta n_\nu = 1.67\sigma_\nu$ (ν=Co, Cu).

The term $|f(Q)|^2$ is also affected:

$$\begin{aligned}|<f(Q)>|^2 = {} & f_{Cu}^2 L_{Cu}(Q)exp(-d_{Cu}^2\sigma_{Cu}^2Q^2/8) + f_{Co}^2 L_{Co}(Q)exp(-d_{Co}^2\sigma_{Co}^2Q^2/8)\\ & +2f_{Cu}f_{Co}[L_{Cu}(Q)L_{Co}(Q)]^{1/2}exp[-(d_{Cu}^2\sigma_{Cu}^2 + d_{Co}^2\sigma_{Co}^2)Q^2/16]cos(\overline{\Lambda}Q/2)\\ & +p_{Co}^2 L_{Co}(Q)exp(-d_{Co}^2\sigma_{Co}^2Q^2/8)\end{aligned} \tag{7}$$

As an example, the calculated diffracted intensity distribution for $\Delta n_{Co} = \Delta n_{Cu} =$ 0.5ML is shown in figure 4 for both superlattices. The most important feature, apart from the disappearance of the high Q satellites, is an important reduction in the intensity of low Q satellites.

From this analysis, the existence of both low and high Q diffraction peaks in both samples considered leads to the conclusion of sharp interfaces between the cystalline bilayes of Co and Cu.

The influence in the diffraction intensities of substrate quality and interdiffusion is illustrated in figure 5 for two different superlattices. The first one, (5ML Co/9ML Cu)86, was grown on a polydomain Cu substrate, i.e. a substrate with the surface oriented normal to the $< 100 >$ direction and many crystallites randomly oriented within

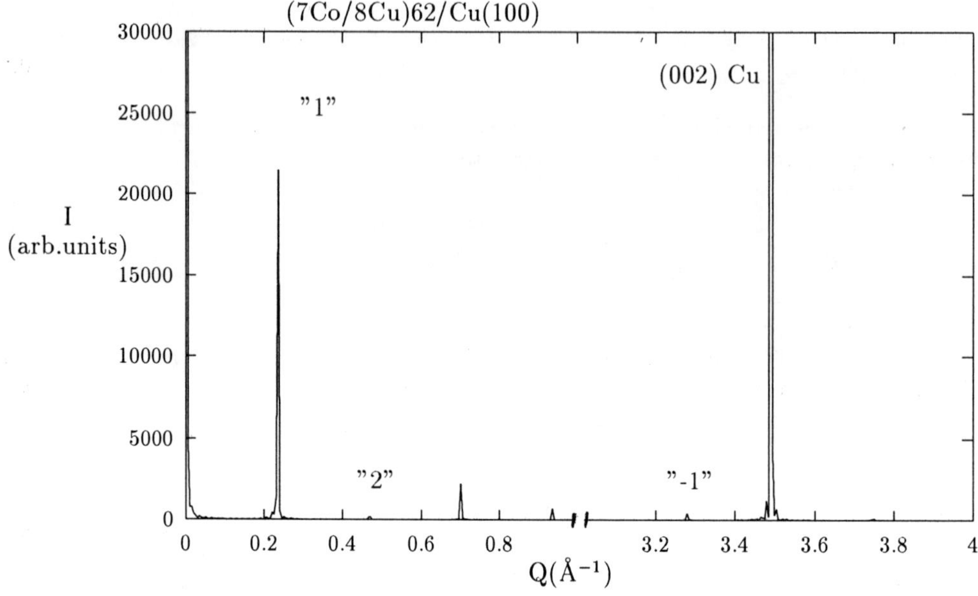

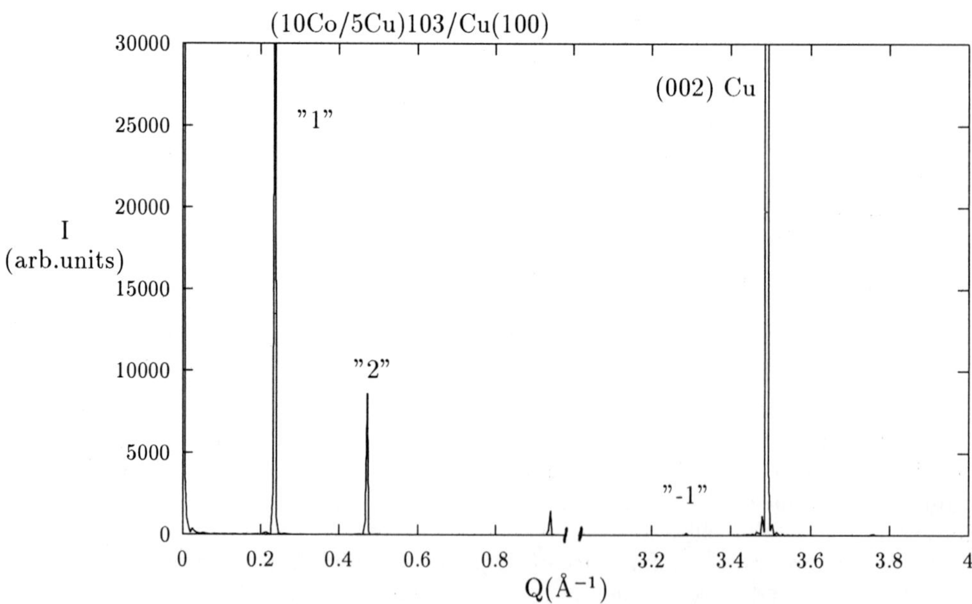

Fig.3 Kinematical calculation of the diffracted intensity assuming amorphous bilayers and sharp interfaces.

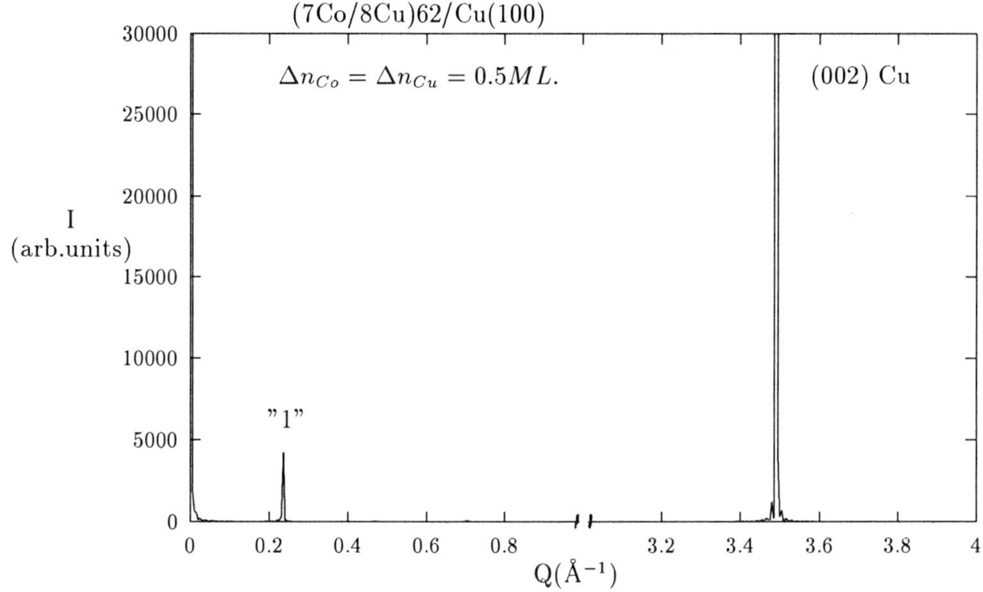

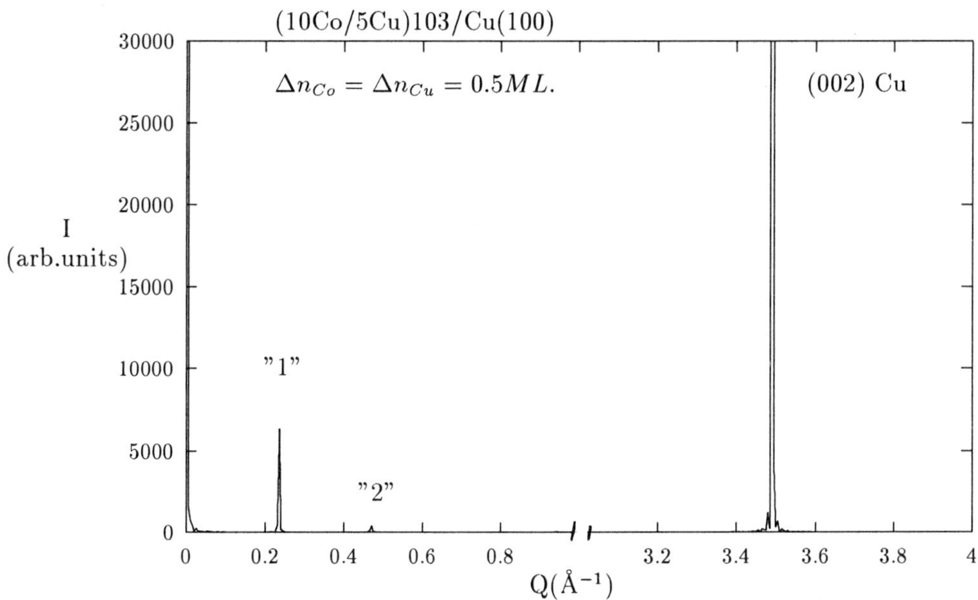

Fig.4 Kinematical calculation of the diffracted intensity assuming interdiffused crystalline bilayers.

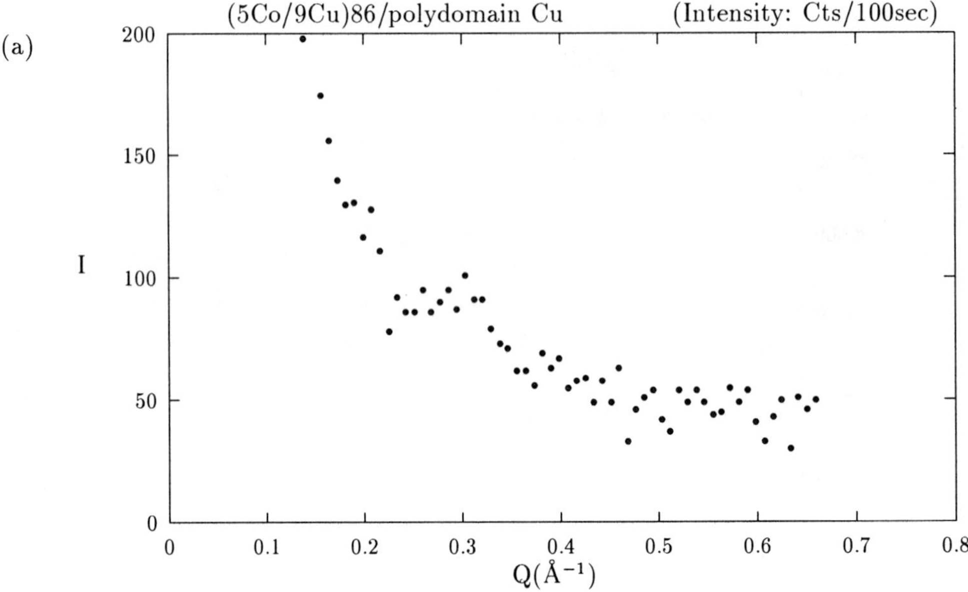

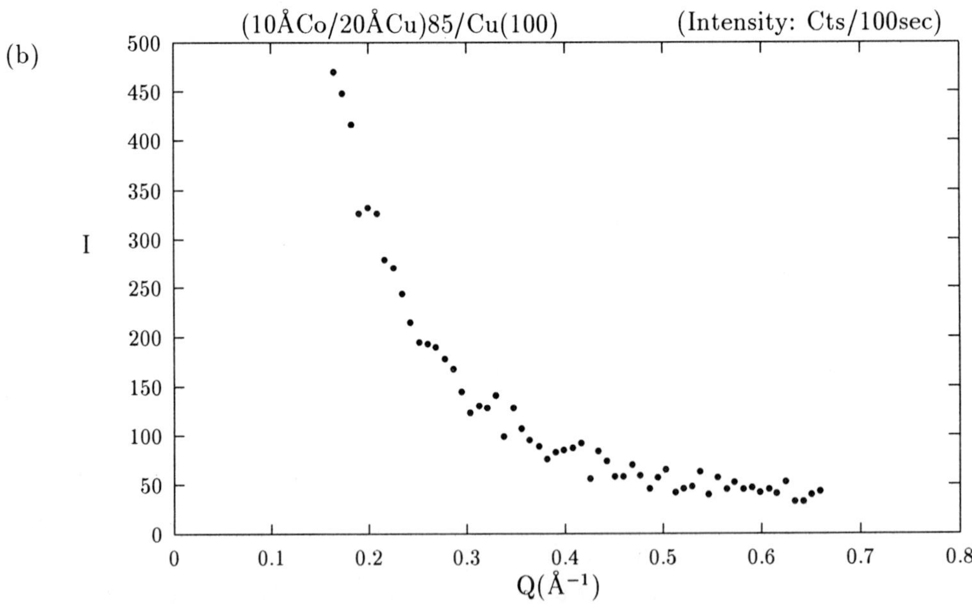

Fig.5 Neutron diffraction measurements for superlattices grown (a) on a polydomain Cu substrate at RT and (b) on a single crystal Cu(100) substrate at 425 K.

the surface plane. This kind of substrate mimics the "textured" polycrystalline substrates used in sputtering growth. In this case, the neutron diffraction data show only a weak superlattice peak at low Q. The second sample, (10ÅCo/20ÅCu)85; was grown on a Cu(100) substrate maintained at 425 K. The absence of superlattice diffraction peaks confirms in this case the interdiffusion between Co and Cu layers.

Co/Cu Trilayers

In the first section, a structural characterization of Co/Cu superlattices grown on Cu(100) single crystals has been performed, but the quality of growth and interfaces in superlattices is determined from the first stages of growth. In this section, a structural characterization of the very first stages of growth of Co/Cu superlattices is performed by LEED Intensity/Voltage (I/V) measurements on Cu(100), 2ML Co/Cu(100), 5MLCo/Cu(100) and 5ML Cu/5ML Co/Cu(100). Samples were grown at RT under UHV conditions and (0,0), (1,1) and (1,0) diffraction beams were recorded. Figure 6 shows as reference I/V curves for (1,1) beam for the samples studied.

Dynamical LEED Calculations using the Renormalized Forward Scattering Method [7], following Van-Hove and Tong's code, produce theoretical I/V curves, that can be compared with experiments via Pendry Reliability Factor (R-Factor) [8]. From these results, both for 2 and 5 ML of Co, a tetragonal distortion from the cubic structure is found in the Co layers. The Co overlayer adopts the lateral spacing of Cu (with a 2% in plane expansion with respect to fcc Co) and exhibits a vertical contraction in an attempt to minimize the strain. In this way, for 2ML Co/Cu(100), the vertical spacing obtained for the Co layers is $d_{12} = 1.77 \pm 0.03$Å and $d_{23} = 1.73 \pm 0.03$Å respectively. For 5ML Co/Cu(100), the vertical spacing for the three last Co layers is $d_{12} = 1.70 \pm 0.02$Å, $d_{23} = 1.74 \pm 0.02$Å and $d_{34} = 1.73 \pm 0.03$Å.

Recent LEED measurements [9] confirm that the Co film maintains the tetragonal distorsion and the lateral spacing imposed by the substrate at least up to 10ML thickness. On the other hand, EXAFS experiments [10], also report the same behaviour for lateral Co distances up to Co thicknesses of 8 ML. Therefore, the Co film grown on Cu(100) does not relax by introducing a lattice of misfit dislocations, but grows up to the thicknesses studied under strain. The deposition of a 5ML Cu film on a 5ML Co/Cu(100) substrate produces identical LEED I/V curves as for a clean, well ordered Cu(100) surface. The horizontal Cu spacing is maintained in the first Co and Cu depositions, and subsequent depositions of both materials will show therefore the same behaviour up to the completion of a superlattice structure.

Fe seed layers on Si(111)

Sputtered metallic multilayers are extensively used for basic an applied science investigations. The use of sputtering techniques allows high deposition rates and the simultaneous fabrication of large quantities of samples on easy achievable high vacuum ($\approx 10^{-8}$torr) conditions. By seed layer epitaxy, the substrate lattice parameter (commonly Si, GaAs or sapphire single crystals) is changed to the parameter of a metallic intermediate layer (Fe, Ru, Ag) that fits better the spacings of the materials

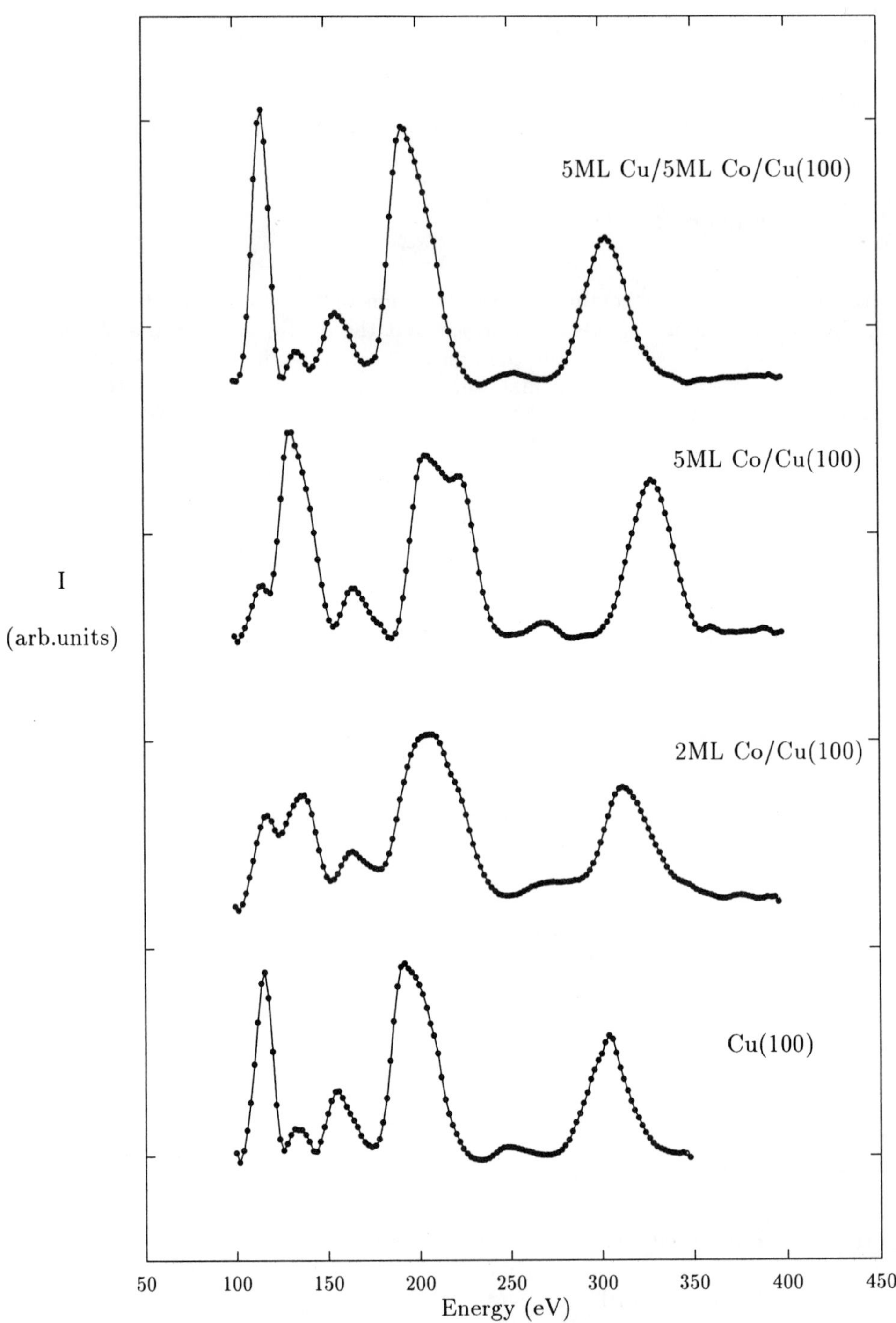

Fig.6 LEED I/V spectra: (1,1) Beams at normal incidence.

to form the multilayer. The multilayers obtained with this method are polycrystalline, textured with a preferent direction in the best situation. Recently, Co/Cu sputtered multilayers grown on Fe/Si(100) seeded surfaces have shown extraordinary large values of magnetorresistance (65 % at RT). Furthermore, it has been claimed [1] that they have better structural perfection than Co/Cu superlattices grown by MBE on Cu(100) single crystals [3].

In the two previous sections, a structural analysis of Co/Cu trilayers and superlattices grown on Cu(100) single crystals under UHV conditions has been performed by inspection of the reciprocal space. From this analysis, the presence of sharp interfaces and crystalline bilayers is concluded from the very first stages of growth up to the completion of a multilayered structure. In this section, results on the structure of a Fe film grown on a Si(111) substrate with a real space technique such as STM will be shown.

Substrates prepared by Shiraki method [11] were introduced in a UHV system, where the thin, uniform SiO_2 layer was desorbed. Afterwards, a thick Si film was evaporated onto the surface at 700°C, showing a characteristic 7 × 7 RHEED pattern. Finally, a 60 Å Fe film, of similar thickness than seed layers proposed for sputtering multilayers growth, is evaporated at 60°C on the surface [12]. Figure 7 shows STM images taken in air of the Fe buffer layer at different (and large) magnifications. The structure of the surface is formed by oblong, platelet-like, Fe crystallites aligned with the terraces of the Si substrate. The average size of these crystallites is 6000 Å long, 1000 Å wide and 60 Å high. On top of the crystallites the surface roughness is very small (less than 3 Å), but domains smaller than 100Å are easily resolved. These domains are randomly oriented in the film plane. Furthermore, Auger Electron Spectroscopy and Ion Scattering Spectroscopy measurements [13] clearly show that chemical reaction between Fe and Si takes place at RT from the very first stages of deposition. The surface-sensitive techniques detect Si close to the surface even after deposition of 50 Å of Fe.

In summary, the Fe buffer layer is a polycrystalline, textured and reacted template for growth of Co/Cu multilayers. This has implications on the magnetic microstructure of these samples, whereby the size of the magnetic domains is probably dictated by the lateral coherency of the buffer layer (≈ 100Å) in contrast to single-crystal, MBE-grown Co/Cu superlattices which are single domain as grown [2]. Since a large value of MR appear only in systems that are magnetically inhomogeneous and contains non-aligned entities (i.e. domains) with a size similar to the electron mean free path [14], we suggest that the microstructure of the buffer layer dictates the large value of MR for sputtered versus MBE-grown Co/Cu multilayers [1]

Conclusions

A reciprocal space characterization of the sutructure from the first stages of growth up to the completion of a multilayered structure for Co/Cu superlattices deposited on Cu(100) single crystalline substrates demostrate the crystallinity, presence of sharp interfaces and absence of misfit dislocations in the system. On the other hand, a real space inspection of a 60 Å Fe film grown under UHV conditions on a Si(111) (7 × 7) at RT indicates a non flat surface, with presence of Fe crystallites several thousands Å large. A TEM inspection of one of these Fe crystallites would lead to the wrong conclusion of the presence of a flat surface after Fe deposition.

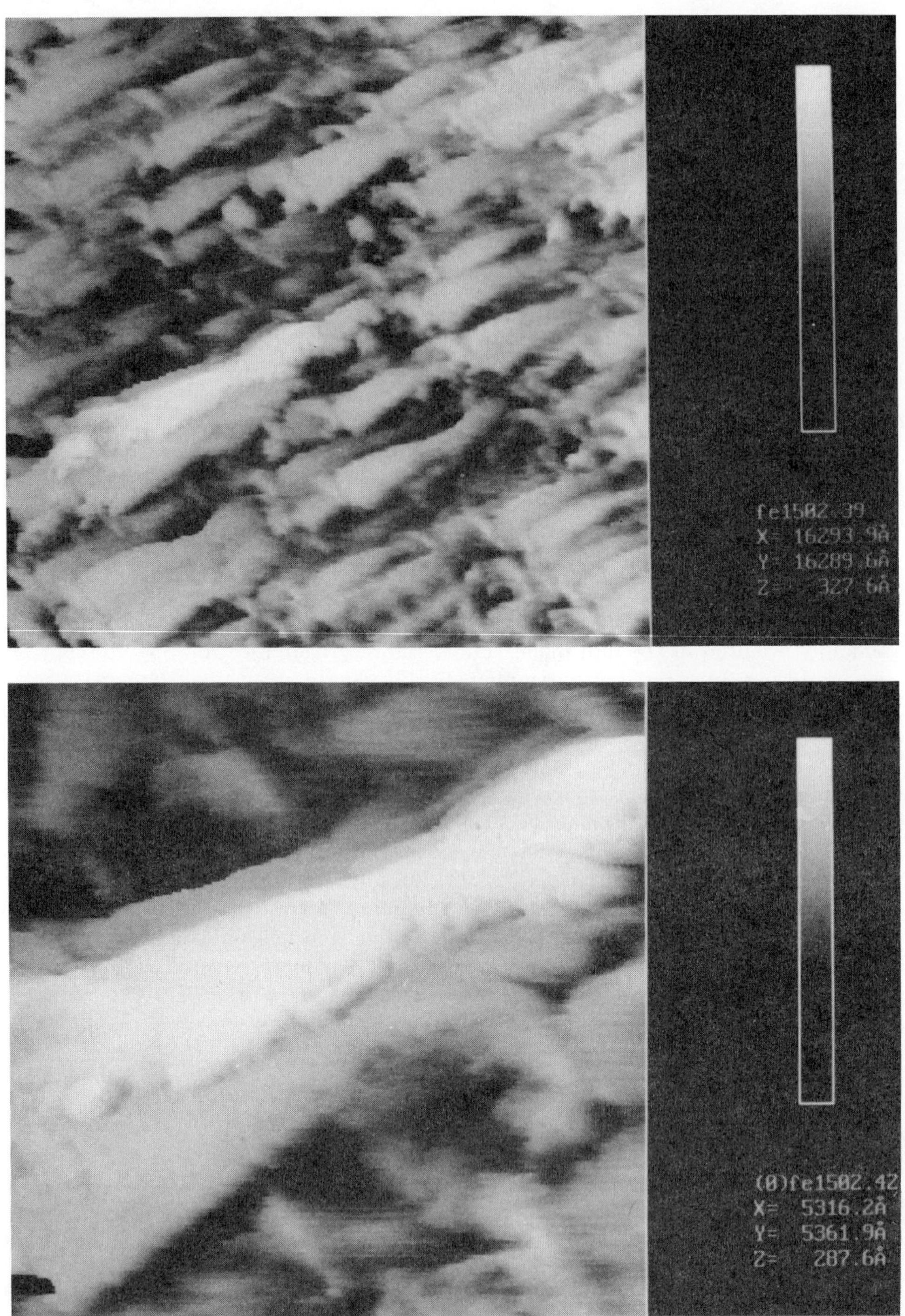

Fig.7 STM images of 60ÅFe/Si(111) deposited in UHV. The images have been recorded in air. The blow up shows an exceptionally large Fe crystallite.

Acknowledgement

We thank P.L de Andrés and J. Cerdá for the LEED calculations. This work has been financed by the CyCIT under contract No 3788 and by the ESPRIT BRA no. P3026 action.

References

[1] S.S.P. Parkin, Z.G. Li and D.L. Smith, *Appl. Phys. Lett.* **58** (1991) 2710.

[2] J.J. de Miguel, A. Cebollada, J.M. Gallego, R. Miranda, C.M. Schneider, P. Schuster and J. Kirschner, *J. Magn. Mag. Mat.* **93** (1991) 1, and references therein.

[3] A. Cebollada, J.L. Martínez, J.M. Gallego, J.J. de Miguel, R. Miranda, S. Ferrer, F. Batallán, G. Fillion and J.P. Rebouillat, *Phys. Rev. B* **39** (1989) 9726.

[4] Y. Fujii, T. Ohmishi, T. Ishihara, Y. Yamada, K. Kawaguchi, N. Nakayama and T. Shinjo, *J. of the Physical Soc. of Japan* **55** (1986) 251.

[5] V.F. Sears, "Thermal Nautron Scattering Lengths and Cross Sections for Condensed Matter Research", Chalk River Nuclear Laboratory, Ontario (1984).

[6] G.E. Bacon, "Neutron Diffraction", Oxford University Press (1975).

[7] See for example M.A. van Hove, W.H. Weinberg and C.M. Chan in "Low Energy Electron Diffraction", edited by G. Ertl and R. Gomer, Springer Verlag (Berlin) 1986.

[8] J.B. Pendry, *J. Phys.* **C13** (1980) 937.

[9] E. Navas et. al. (in preparation).

[10] D. Chandersis et. al. (this issue).

[11] A. Ishizaka and Y. Shiraki, *J. Electrochem. Soc.* **133** (1986) 666.

[12] J. de la Figuera, A.L. Vázquez de Parga, J. Alvarez, J. Ibáñez, C. Ocal and R. Miranda, *Surf. Sci.* **264** (1992) 45.

[13] J.L. Martínez-Albertos et. al., *J. Magn. Mag. Mat.* (submitted).

[14] A.E. Berkowitz et.al., *Phys. Rev. Lett.* **68** (1992) 3745 and J.Q. Xiao et.al., *Phys. Rev. Lett.* **68** (1992) 3749.

INFLUENCE OF GROWTH AND STRUCTURE ON THE MAGNETISM OF EPITAXIAL COBALT FILMS ON CU(001)

C.M. Schneider[1], A.K. Schmid[2], P. Schuster[2], H.P. Oepen[3] and J. Kirschner[1]

[1] MPI für Mikrostrukturphysik, Weinberg 2, O-4050 Halle/Saale
[2] Inst. f. Experimentalphysik, FU Berlin, Arnimallee 14 W-1000 Berlin 33
[3] IGV, Forschungszentrum Jülich GmbH, Postfach 1913 W-5170 Jülich

INTRODUCTION

Artificially layered materials and ultrathin films of ferromagnetic substances currently receive considerable interest in materials science. These systems often show novel magnetic properties which differ dramatically from those of the corresponding bulk ferromagnets, e.g., unexpected magnetic anisotropies or surprisingly low Curie temperatures. Research in this field is stimulated by both scientific curiosity and potential applications in magnetic storage and sensor technology. A successful development of new storage devices and technologies requires a fundamental knowledge of the underlying physical mechanisms, in order to be able to deliberately modify the magnetic properties of a given material or system.

Understanding the magnetic properties turns out to be difficult, since the altered magnetic behavior of films in the monolayer regime must generally be attributed to more than one physical mechanism. Although some of the effects appear due to the reduced dimensionality, it has been realized that crystalline structure and peculiarities in the epitaxial growth process of ultrathin films play a crucial role. This is particularly important in the case of layers built up from metastable crystalline phases, such as face centered cubic iron or cobalt. A thorough understanding of the magnetic properties of epitaxial systems thus requires a broad data base which must include structural aspects and a precise characterization of the growth process in its various stages, as well as the influence of the growth conditions. This "wide band" approach promises new and yet unprecedented insight into the links between structure, topography, morphology and magnetic behavior. As a welcome side aspect, one may expect such an approach to facilitate the comparison of results from different groups on the same system, thus helping to reduce the number of possible inconsistencies. In the following we report on our investigations and results on ultrathin fcc-cobalt films on Cu(001) substrates.

EXPERIMENTAL METHODS AND CRYSTAL PREPARATION

In order to analyze the interrelations between growth, crystalline structure, magnetism and electronic states, we characterized the model system cobalt on Cu(001) by a number of different techniques. We employed in-situ diffraction methods (Low Energy Electron Diffraction, Medium Energy Electron Diffraction) in combination with Auger electron spectroscopy, to monitor the epitaxial growth and to determine precisely the film thickness. Conclusions on the crystalline structure are based on the interpretation of LEED patterns and the analysis of I(E) curves for several diffracted beams. In addition, the surface topology of the substrate and the films has been investigated by real space imaging techniques (Scanning Tunneling Microscopy). The Surface Magneto Optical Kerr effect was employed to study the macroscopic magnetic properties on the basis of hysteresis loops. The electronic structure of the Cu substrate and the Co films has been investigated by means of spin-resolved photoelectron spectroscopy using circularly polarized synchrotron radiation. This gave a detailed insight into the films' electronic structure and the hybridization of film and substrate electronic states. The magnetic microstructure in the Co films has been characterized by imaging magnetic domains with a scanning electron microscope with spin polarization analysis (SEMPA).

The disk-shaped samples of 10 to 12 mm diameter were spark-cut from a high quality copper single crystal and mechanically polished using diamond paste. The crystals were characterized by a mosaic spread of less than 0.1°, the surface normal being within 0.2° along the (001) crystalline axis. After insertion into the ultrahigh vacuum, the surface was prepared by cyclic sputtering and annealing. A well-defined sharp LEED pattern could be observed already after a few cycles, but these surfaces still contained a certain amount of sulphur and carbon and showed a large number of small terraces (size of the order of 100 Å), as judged by the STM. Uncontaminated surfaces with large atomically flat surfaces (average size about 2000 Å) were obtained only after a total cycling time of about 100 hours.

GROWTH OF CO ON CU: RECIPROCAL VS. REAL SPACE APPROACHES

During the deposition of Co onto the Cu(001) substrate, the diffracted beams in a MEED experiment show pronounced intensity oscillations over a wide range of substrate temperatures and deposition rates[1]. An example is given in figure 1a for the intensity variation of the specular beam at room temperature growth. This result may be routinely achieved only on carefully cleaned and annealed copper substrates. The following conclusions can be drawn from this curve. First, the presence of MEED intensity oscillations indicates a nearly perfect layer-by-layer growth of Co on Cu(001). Secondly, the film thickness can be evaluated from the number of periods, since each relative intensity maximum is usually interpreted as the completion of a monoatomic layer. The method there-

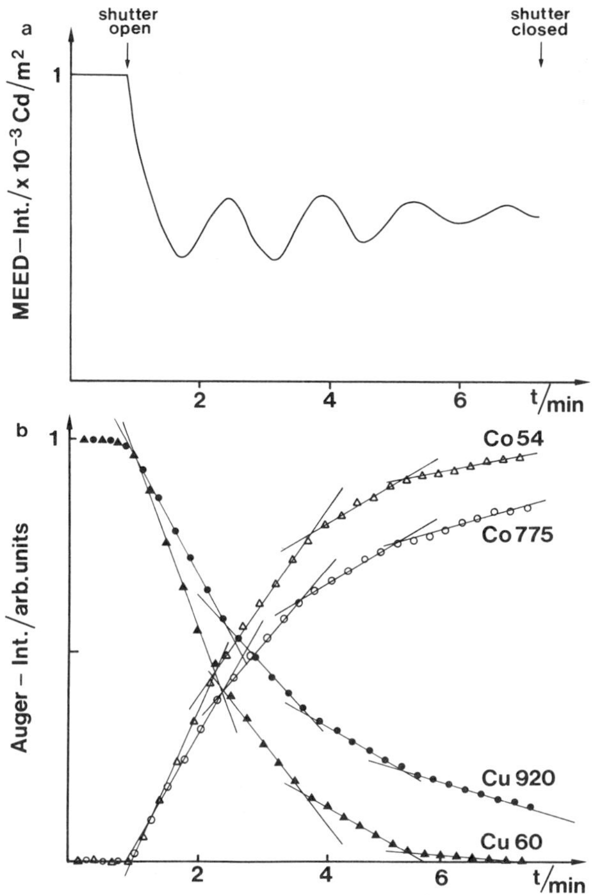

Fig. 1. Evolution of the MEED specular intensity (a), and the Auger lines of Co (54 and 775 eV) and Cu (60 and 920 eV) (b). The data have been recorded simultaneously.

fore yields a direct thickness calibration in units of a monolayer (ML). At room temperature, up to several tens of periods in the oscillations have been observed, indicating that rather thick films of fcc cobalt may be grown[2].

The growth mode is often deduced from the Auger signal vs. deposition time curves. A layer-by-layer growth is concluded if these I(t) curves can be fitted by a series of straight lines. Each intersection between neighboring lines then marks the completion of a monolayer. This is indeed found in Co/Cu(100) for the Auger lines of cobalt and copper (fig. 1b). It is instructive to compare the results from AES and MEED investigations. This can be done best by performing an experiment, which simultaneously measures the intensity of the diffracted MEED beams and the Auger electrons excited by the primary electrons. The procedure gives a direct correlation between the MEED oscillations and the behavior of the Auger intensities without the ambiguity introduced by different samples. This is actually the way the data in figure 1 have been taken. The direct comparison shows that each maximum in the MEED intensity curve closely corresponds to a

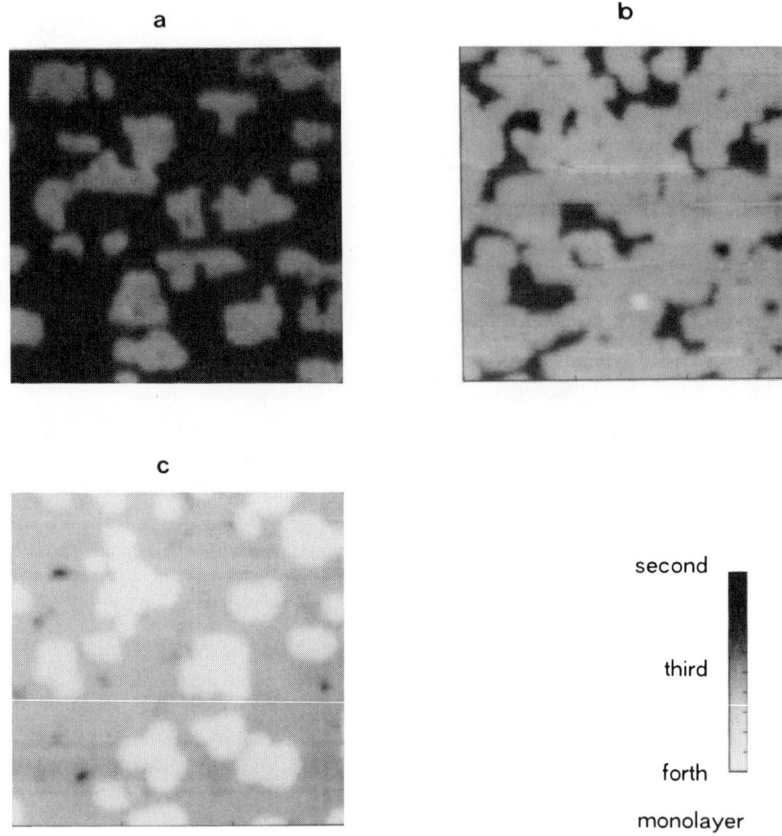

Fig. 2. STM micrographs of the growth of Co on Cu(100) subsequently taken from the same spot of the sample surface at coverages of 2.5 (a), 2.9 (b) and 3.3 monolayers. Scale: 30 × 30 nm

kink in the Auger uptake curves. Thus both methods independently and in combination yield the same result and confirm essentially a layer growth of Co on Cu.

A closer inspection of the MEED intensity oscillations, however, suggests that some deviations from the ideal layer-by-layer growth mode must be present. First, after depositing Co, the average intensity of the diffracted beam drops significantly and never comes back to the initial value of the clean Cu surface. Secondly, the oscillations follow a sinusoidal curve rather than a sequence of parabolas expected for the ideal case[3]. These findings can be interpreted by assuming the film to accumulate a certain amount of defects during growth. Further studies showed the average intensity level to be higher during deposition at elevated temperatures up to 450 K, indicating less defects and slightly better growth conditions. Kinetic processes, i.e., surface diffusion, clearly influence the quality of film growth in the Co/Cu system. If only kinetic arguments are considered, a sufficiently high substrate temperature and thus a high adatom mobility should promote the nearly ideal case of a step-flow growth, as seen for example in the homoepitaxy of copper[4]. The onset of interdiffusion in the Co/Cu system at about 500 K unfortunately sets narrow limits to the choice of substrate temperatures in the growth of Co overlayers.

In order to characterize the role of defects and film topography in more detail we performed in situ STM studies at room temperature. For this purpose, sequences of images were recorded from the same section of the surface after subsequent Co depositions and compared. Figure 2 gives a set of images taken on a 30 × 30 nm scale for Co coverages of 2.5, 2.9 and 3.3 ML. Three discrete height levels corresponding to the second, third and fourth monolayer are transposed into a gray scale. The vertical distance of the neighboring height levels has been determined to 1.8 Å, which agrees with the interlayer distance of fcc-Co. This results justifies the designation "monolayer" for the discrete levels observed in the STM images. Furthermore, it independently confirms the assumption of the model of layer growth, that the islands during growth should have monolayer height. Starting with the 2.5 ML image, one notices an almost closed 2 ML level with very little defects, which is half-covered by islands of the third monolayer. At a coverage of 2.9 ML, the islands have grown in size and most of them have coalesced, but the 3 ML level is not yet completely closed. At a coverage of 3.3 ML, islands of the fourth monolayer have formed, but a few holes in the 3 ML level remain. Similar results have been found at higher thicknesses. These observations lead to the conclusion that coverages corresponding to an

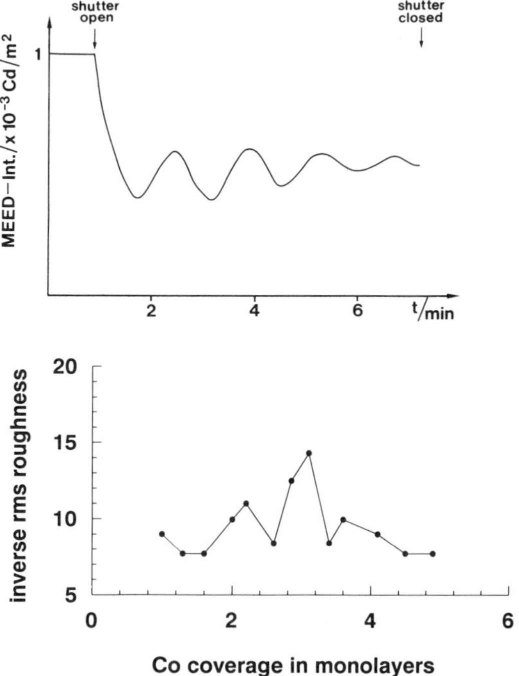

Fig. 3. Comparison of the inverse rms roughness deduced from a sequence of STM micrographs during the growth of a 5 ML fcc-Co film (b) to the MEED specular intensity (a) reproduced from Fig. 1a

integral number of layers still contain a certain number of voids and additional islands, resulting in a deviation from the ideal layer-by-layer growth. This interpretation is also consistent with the MEED data. The roughness at integral coverages is confined to the uppermost layer and is of the order of ±1 ML.

On the other hand, by analyzing a set of images over a larger range of coverages, one can establish a more quantitative relation between the results of diffraction methods and real space techniques. Evaluating the roughness in each image of the set finally yields a dependence of the surface roughness on the film thickness. This is shown in figure 3a which displays the development of the inverse rms roughness during the growth of a 5 ML film, as obtained by STM. Maxima in this function correspond to minimum roughness and thus the completion of integral layers, minima to half-integral layers. This result is compared to the MEED intensity oscillations reproduced from figure 1a, by scaling the x-axis via Auger calibrations taken for both data sets (see, e.g., the calibration curve in ref. 23). The comparison shows clearly that a reasonable agreement between minima in the surface roughness and maxima in the MEED specular intensity exists. The agreement is not perfect, but this should not be expected for two films grown under different circumstances and in different chambers. In conclusion, these STM results independently warrant the interpretation of MEED intensity oscillations in terms of growth induced changes in the surface topography. Based on the picture of a periodically varying surface roughness, MEED experiments can provide a reliable and precise thickness calibration.

IDEAL VS. REAL STRUCTURE: THE INFLUENCE OF TETRAGONALITY

The structure of the films can be deduced from electron diffraction experiments. During the growth of the films one notices intensity oscillations in all diffracted beams, but essentially no deterioration of the diffraction pattern. MEED patterns taken before and after Co deposition are practically indistinguishable. Similar observations have been made with LEED. Experiments on the clean Cu substrate and on carefully prepared Co films yield nearly identical p(1×1) diffraction patterns over a wide range of coverages. Therefore the lateral lattice mesh of the Co overlayer corresponds to that of the Cu substrate. According to this qualitative analysis, cobalt forms rather thick well-ordered pseudomorphic overlayers on the Cu(100) surface. As the films adopt the fcc lattice of the substrate, one stabilizes the metastable fcc crystal structure of cobalt. In bulk cobalt, the stable structure at room temperature is hcp, a phase transition to the fcc lattice occuring at temperatures above 750 K. In Co films on Cu(100), the fcc modification becomes accessible to experiments at and below room temperature.

In a simplified picture of the crystalline structure, the atoms of the first Co layer occupy the four-fold hollow sites in the Cu surface and the adjacent layers adopt the registry of the fcc substrate lattice. This view assumes the same lattices for cobalt and copper. A comparison of the bulk lattice constants of fcc-Co (3.57 Å) and Cu (3.68 Å) shows that this assumption is at least doubtful. On the basis of these bulk values one has to consider a small lattice mismatch. This will give rise to strain in the film, which must be accomodated either by misfit dislocations or structural deviations from the ideal fcc lattice.

A large data set consisting of LEED I(V) curves for various diffracted beams and several film thicknesses has been compared to dynamical LEED calculations[5]. The analysis leads to the conclusion that the films are indeed subject to a notable tetragonal distortion normal to the film plane. Due to the lattice mismatch, the surface unit cell of the films is slightly expanded, followed by a compression of the interlayer distance normal to the film plane. This distortion is thickness dependent and relaxes with increasing coverage, where the mean interlayer distance approaches the value of 1.74 Å. This is still somewhat smaller than the corresponding interlayer distance of bulk fcc-Co. The above findings are consistent with an earlier LEED study[6]. The conclusions of a tetragonal distortion of the lattice agree nicely with the results of recent EXAFS experiments on fcc-Co films on Cu(100)[7].

Electronic States

The majority and minority spin bands of the Co films have been mapped out along the (001) direction by means of spin- and angle-resolved photoemission. The experimental dispersion of the electronic states was compared to fully relativistic bulk band structure calculations for fcc-Cobalt[8]. During the first few monolayers there is a transition from a two-dimensional to a three-dimensional electronic structure, the latter being almost completely established at a thickness of 5ML. The electronic states at this coverage agree well with the theoretical predictions, except for systematic deviations of the minority bands near the

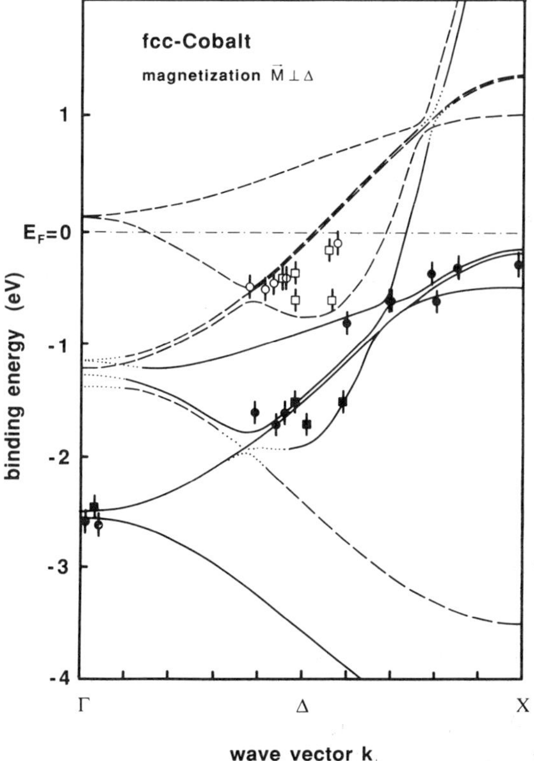

Fig. 4. Full relativistic band structure calculation for fcc-Co. Comparison of experimental data for majority (full) and minority states (open symbols).

Fermi level (fig. 4). The experiment finds the minority states at higher binding energies as predicted by the band structure calculation. It should be kept in mind that photoelectron spectroscopy probes the excited state of an electronic systems. Therefore a perfect agreement with ground state calculations should no be expected. On the other hand, calculations for simple cubic iron[9] and fcc copper[10] have shown that a tetragonal compression or expansion of the lattice leads to systematic changes in the electronic states as compared to the perfect cubic case. In particular, bands shift their energetic position with respect to the Fermi level. We therefore tentatively attribute the above mentioned deviation to a certain extent to a tetragonal lattice distortion. In fact, the energy position of these minority bands slightly shifts with decreasing Co coverage[11], thus reflecting thickness dependent changes in the lattice distortion, which are also observed in the LEED experiments mentioned above.

It is interesting to note in this context that Strange et al.'s work on the tetragonal distortion in sc iron[9] showed the magneto-crystalline anisotropy to be strongly influenced by electronic bands in the vicinity of the Fermi level E_F. A certain band which shifted its energy position and thus the crossing point with E_F upon the degree of tetragonal distortion, led to a distinct change in the magnetic anisotropy energy. The authors suggested this to be the origin of a flip in the direction of the magnetization from in-plane to out-of-plane. A change of the easy axis of magnetization as a function of the overlayer coverage is indeed known to occur in Fe/Cu(100)[12,13]. Although such a spectacular behavior has not been observed in Co films on Cu(100), a variation of the magnetic anisotropy with the film thickness should not be excluded.

The origin of the magneto-crystalline anisotropy is the spin-orbit interaction in the valence bands, which couples the electron spin to the orbital part of the electronic wave function. Investigations of the electronic structure of ferromagnets often neglect the influence of spin-orbit coupling, because it is considered to be small, at least in the case of 3d metals. This view is contrasted by the results of spin-resolved photoemission experiments from the clean Cu substrate using circularly polarized light[14]. These studies revealed strong spin polarization effects in the 3d bands due to spin-orbit interaction and allowed an experimental determination of the spin-orbit splitting along the (100) direction. It was found to be of the order of 100 meV. One is thus forced to assume that the strength of spin-orbit interaction in the 3d ferromagnets is very similar to the value in Cu. In photoemission experiments from fcc-Co films using circularly polarized light, one indeed observes a phenomenon called magnetic circular dichroism (MCD)[15]. This effect occurs due to the interplay of exchange and spin-orbit interaction and thus directly proved the presence of spin-orbit coupling in the Co 3d valence bands.

Magnetic Anisotropies

The remanent magnetization of the Co films is always found to lie within the film plane, independent of the film thickness. As revealed by an azimuthal scan in the SMOKE experiments, the films exhibit a pronounced in-plane four-fold anisotropy, the [110] directions being the easy axes of magnetization[16]. This is directly visible in the domain patterns taken with SEMPA, which show only four different directions of the remanent magnetization, which agree with the orthogonal [110] axes[17]. This result is somewhat unexpected, as the easy axis in fcc materials is known to be the <111> direction if the anisotropy constants K_1, $K_2<0$. Negative values for K_1, K_2 are known for Ni single crystals and have been reported for fcc Co thick films, too[18]. Within a {100} crystallographic plane, the only low-indexed directions left are <110> and <100>, with the latter being the energetically least favorable one (hard axis). In a first crude interpretation, the above finding might therefore be attributed to the demagnetizing field, which forces the magnetization into the plane along the easier <110> axis. In FMR measurements, however, very strong effective demagnetizing fields $4\pi M_{eff}$ of the order of 40 kG have been found[19]. Thus, $4\pi M_{eff}$ in ultrathin fcc Co(100) films is about twice as large as the saturation magnetization $4\pi M_s$. This gives rise to a negative value of the anisotropy constant K_u, corresponding to a large perpendicular uniaxial anisotropy with the hard axis along the surface normal. The observed in-plane magnetization in Co(100) films is therefore only partly due to the demagnetizing field. Recalling the discussion in the previous paragraph, this strong uniaxial anisotropy may be a consequence of the tetragonal distortion in the films' structure.

THE INITIAL STATE OF GROWTH AND THE ONSET OF FERROMAGNETIC ORDER

Monolayer coverages of ferromagnetic materials often show a thickness dependence of the ferromagnetic ordering temperature, which is attributed to the reduced dimensionality in the films. The most striking effect has been reported for fcc cobalt films on Cu(100)[20,21]. According to the results from SMOKE measurements, the Curie temperature T_c of a 2 ML film is slightly above room temperature. For a single cobalt monolayer, an upper limit of 50 K for T_c has been found. The thickness dependence of T_c reported on in ref. 4 is reproduced in Fig. 4 and compared to a recent calculation by P. Bruno[22]. In contrast to the assumptions in the Mermin-Wagner theorem[23], the calculation considers the influence of long-range dipolar interactions and magnetic anisotropies. It therefore predicts a single monolayer to be ferromagnetically ordered well above 0 K. The reasonable agreement with the experimental data at coverages at and above 2 ML indicates the observed thickness dependence to be partly due to the reduced dimensionality. Below a thickness of 2ML the agreement becomes worse with decreasing Co coverage. Finally, the experimental values for T_c of 1 ML films are significantly lower than the theoretical predictions (T_c(theor) ~ 150 K). This discrepancy should not be misinterpreted as an experimental evidence for the validity of the Mermin-Wagner theorem, for two reasons. First, with our present experimental set-up we only can give an upper limit for the ordering temperature (T_c(exp) <50 K). The monolayer may therefore show ferromagnetic order below this value. Second, Bruno's calculation does not take into account two important effects, which are immanent to thin film systems. This is, on the one hand, the electronic hybridization between the Co film and the Cu substrate, which should become more important at low film thicknesses. Some indications for such a hybridization are found in spin-resolved photoemission experiments. Furthermore, covering Co films in the monolayer regime with an additional Cu overlayer, i.e. introducing two Co/Cu interfaces, leads to a sizable reduction of the Curie temperature[20,21]. The electronic hybridization obviously tends to lower T_c and this process gains importance with decreasing film thickness. This effect, however, is not strong enough to explain the extremely low T_c in the case of the 1 ML film.

The second mechanism may be sought in particularities of the surface topography of the cobalt layers. During the growth of the first monolayer (ML) one indeed observes some irregularities in the MEED and TEAS intensity oscillations, which suggest a deviation from the behavior found at higher coverages[23]. A TEAS analysis of step heights in the monolayer and submonolayer regime also indicates differences with respect to thicker layers. These questions have been answered by extensive in-situ STM investigations, which revealed the first Co layer on Cu to grow in a particular manner indeed[24]: contrary to the nearly perfect layer-by-layer growth found at

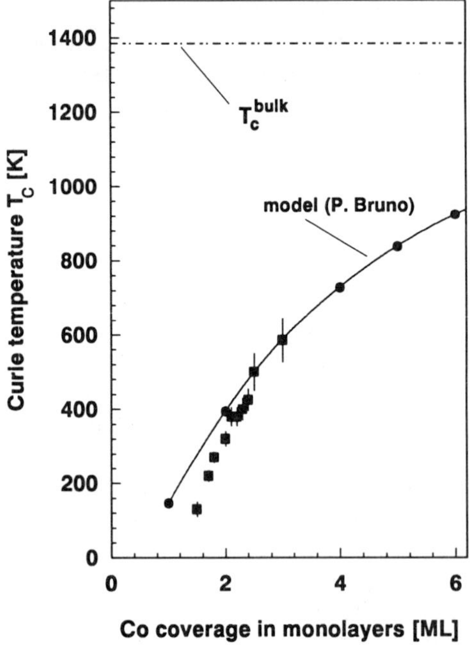

Fig. 5. Thickness dependence of the Curie temperature of fcc-Co films grown on Cu(100). The experimental data points (squares) are compared to a calculation of P. Bruno (circles). The line serves as a guide to the eye.

higher coverages, islands of the second layer nucleate long before the first layer is completed. This is depicted in Fig. 6 showing a comparison of characteristic STM micrographs taken at coverages close to 1 ML (fig. 6a) and 2 ML (fig. 6b). Inspection of fig. 4a prooves clearly that the first monolayer does not grow uniformly. About 20% of the substrate surface are yet uncovered, whereas about 20% of the second layer have already formed on top of the first layer islands. As a consequence a Co deposit equivalent to 1ML yields a significant amount of bilayer islands resulting in a strong deviation from the ideal layer growth. After the deposition of an additional monolayer, the situation is different (fig. 6b). Nucleation of islands of the third layer does not become significant before the second layer is almost completed. At the equivalent of 2ML and higher, the growth mode obviously changes and becomes more perfect layer-by-layer, the surface roughness at half-integral layers being ±1ML. These findings clearly point out a difference in the growth mechanisms of Co/Cu vs. Co/Co, as can be expected from thermodynamical considerations[25].

Coming back to the above question of the discrepancy between the experimental and theoretical T_c dependence, the results suggest, that the surface topography of the film plays an important role, in particular in the monolayer range. Since the monolayer of Co on Cu(100) contains a rather large fraction of isolated islands, the ferromagnetic coupling may be much weaker than in a filled layer, thus leading to a considerably smaller value of T_c.

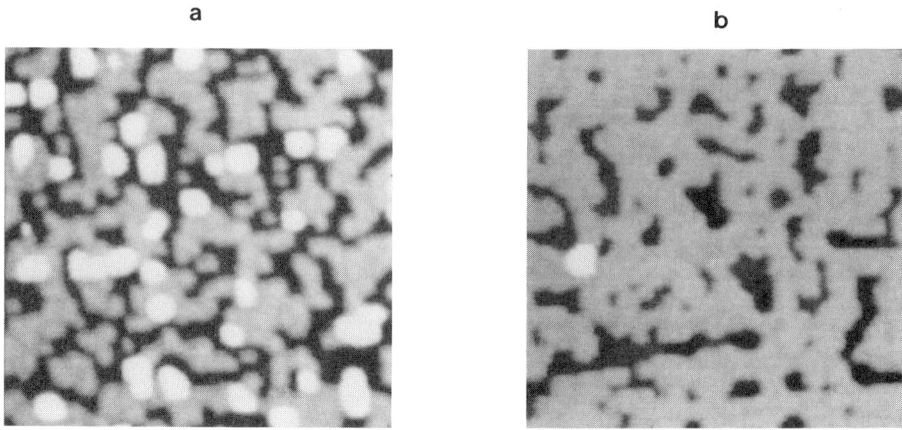

Fig. 6. STM micrographs taken from the same spot on the sample surface at a coverage of close to 1 ML (a), and after the deposition of an additional monolayer (b). Scale: 30 × 30 nm

STEPS AND MAGNETIC ANISOTROPIES

The magnetic properties of a film depend significantly on the structural and topographical quality of the template. Large differences are found, for instance, for films grown on a substrate with large terraces or small terraces. In order to study the influence of atomic steps in a more controlled manner, we prepared a Cu(1 1 13) surface. This surface is vicinal to Cu(001), the surface normal being tilt away from the <001> direction by an angle of 6.2° towards the <110> axis. Cu(1 1 13) has been subject to recent STM investigations by Frohn et al.[26]. These studies revealed the surface to consist of (001) terraces with an average width of 7 atoms in the <110> direction. The terraces are separated by monoatomic steps with the step edges running essentially along {110}.

The growth of Co films on this vicinal surface as seen in MEED experiments proceeds layer-by-layer, but shows a striking difference as compared to the (100) oriented template. The intensity of the specular beam maintained a high level and only very few and weak oscillation periods are visible[27]. This suggests that the growth takes place predominantly via a step flow mechanism. Because of the small width of the terraces, the adatoms can easily reach the next terrace edge, and the step density at the surface remains nearly unaltered. Thus, no major drop in the MEED specular intensity will occur.

The magnetic properties of these Co(1 1 13) films are even more striking. Investigations with SEMPA reveal that only two directions of the remanent magnetization in the domains appear[27]. This is in sharp contrast to the situation in Co(100) films. The magnetization within the domains in Co(1 1 13), and thus the easy axis, is oriented always along the step edges. Despite this surprising difference in the anisotropy, the domain patterns observed in Co(100) and Co(1 1 13) are very similar.

The above SEMPA results have been confirmed by measuring the azimuthal dependence of the hysteresis loops with SMOKE. Results for a 10 ML film are shown in figure 7. An angle $\Phi = 90^\circ$ (0°) corresponds to the external field H applied parallel (perpendicular) to the step edges. In the case where H is almost

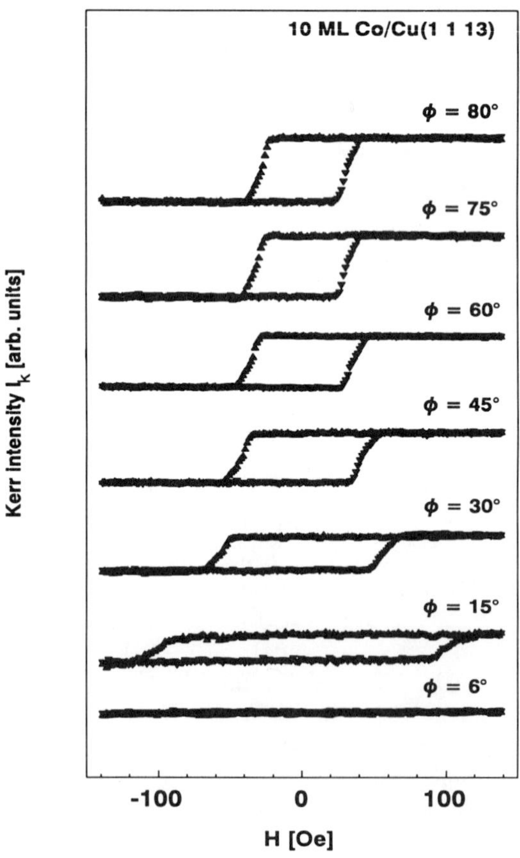

Fig. 7. SMOKE hystereses loops from Cobalt films on Cu(1 1 13) as a function of the azimuthal angle Φ.

normal to the steps ($\Phi = 6^{\circ}$), the coercive field H_c of the film exceeds a value of 150 G, which is the maximum field we can achieve with the present set-up. The angular variation of the coercive force follows closely a $\sin(\Phi)$ behavior. This is a consequence of the fact that the magnetization reversal occurs along the easy axis, being induced only if the geometrical projection of the external field along this direction exceeds the easy axis coercive field.

One might be tempted to think of the steps at the surface being responsible for this behavior. In a film of 10 ML thickness, however, the magnetic properties should be bulk-dominated, the surface having only minor influence. It should also be mentioned that very similar Φ-dependences have been found for Co coverages between 2 and 10 ML[28]. This suggests that a different mechanism must be employed, which is due to the structural differences of the very films. From the crystallographic point of view, the Co(1113) film has a significantly lower rotational symmetry (two-fold) than the low-indexed (100) surface (four-fold). In fact, one may regard Co(1113) films as a new artificial type of solid with a lowered symmetry, the latter being reflected by the uniaxial magnetic anisotropy. This finding points out a very interesting link between structural symmetries and magnetic anisotropies. Similar observations have been reported for Fe films on vicinal tungsten surfaces[29]. In contrast to the above findings, these systems have the easy axis normal to the step edges. The reason for the different behavior of Fe and Co films is yet unknown. Clearly more work is needed to be able to explain the magnetic properties of these novel systems.

SUMMARY AND CONCLUDING REMARKS

The above compilation of results gives an overview of the complex interrelations between structure, electronic states and magnetic properties in the model system Co/Cu(100). One is led to the conclusion that details of the crystalline structure and film topography do indeed strongly influence certain magnetic properties, i.e., the anisotropy and ferromagnetic ordering temperature. Since the effects are already very pronounced in single layer films, one must expect an increase in complexity, when going to sandwiches or multilayers. This is not only due to the larger number of interfaces and films involved, but also caused by the appearance of new magnetic effects in these systems, such as oscillatory interlayer coupling and giant magnetoresistance. It can be suspected that discrepancies like the one reported for the absence[30] or presence[31] of the oscillatory interlayer coupling in Co/Cu(111) multilayers, originate from structural or morphological differences in the samples. Clearly, future investigations on magnetic ultrathin film systems must pay attention to both macroscopic and microscopic aspects, in order to yield a better understanding of the magnetic properties.

ACKNOWLEDGEMENTS

This project has been supported by the Bundesminister für Forschung und Technologie through grant No. 05 413 AXI 03 and the Deutsche Forschungsgemeinschaft through SFB 6. C.M.S, P.S. and J. K. appreciate a fruitful cooperation with the group of R. Miranda at the Universidad Autonoma de Madrid.

REFERENCES

1. J.J. de Miguel, A. Cebollada, J.M. Gallego, S. Ferrer, R. Miranda, C.M. Schneider, P. Bressler, J. Garbe, K. Bethke, and J. Kirschner, *Surf. Science* **211/212**: 732 (1989)
2. H.P. Oepen, priv. communication
3. B. Poelsema, R. Kunkel, N. Nagel, A.F. Becker, G. Rosenfeld, L.K. Verheij, and G. Comsa, *Appl. Phys. A* **53**: 369 (1991)
4. J.J. de Miguel, A. Cebollada, J. M. Gallego, J. Ferron, and S. Ferrer, *J. Cryst. Growth* **88**: 442 (1988)
5. A. Cebollada et al., see these proceedings
6. A. Clarke, G. Jennings, R.F. Willis, P.J. Rous, and J.B. Pendry, *Surf. Science* **187**: 327 (1987)
7. D. Chandesris et al., see these proceedings
8. C.M. Schneider, P. Schuster, M. Hammond, H. Ebert, J. Noffke, and J. Kirschner, *J. Phys.: Condens. Matter* **3**: 4349 (1991)
9. P. Strange, J.B. Staunton, and H. Ebert, *Europhys. Lett.* **9**: 169 (1991)
10. F. Maca and J. Koukal, *Surf. Science* **260**: 323 (1992)
11. C.M. Schneider, J.J. de Miguel, P. Schuster, R. Miranda, B. Heinrich, and J. Kirschner, in: *Science and Technology of Nanostructured Magnetic Materials*, eds. G.C Hadjipnanayis and G.A. Prinz (Plenum Press, New York, 1991)
12. C. Liu, E.R. Moog, and S.D. Bader, *Phys. Rev. Lett.* **60**: 2422 (1988)
13. D. P. Pappas, K.-P. Kämper, and H. Hopster, *Phys. Rev. Lett.* **64**: 3179 (1990)
14. C.M. Schneider, J.J. de Miguel, P. Bressler, P. Schuster, R. Miranda, and J. Kirschner, *J. Electron Spec. Rel. Phen.* **51**: 263 (1990)
15. C.M. Schneider, D. Venus, and J. Kirschner, *Phys. Rev. B* **44**: 12066 (1991)
16. C. M. Schneider, P. Bressler, P. Schuster, J. Kirschner, J.J. de Miguel, R. Miranda and S. Ferrer, *Vacuum* **41**: 503 (1990)
17. H.P. Oepen, *J. Magn. Magn. Mat.* **93**: 116 (1991)
18. J. Goddart and J.G. Wright, *Br. J. Appl. Phys.* **16**: 1251 (1965)
19. B. Heinrich, J.F. Cochran, M. Kowalewski, J. Kirschner, Z. Celinski, A.S. Arrott, and K. Myrtle, *Phys. Rev. B* **44**: 9348 (1992)
20. C.M. Schneider, P. Bressler, P. Schuster, J. Kirschner, J.J. de Miguel, and R. Miranda, *Phys. Rev. Lett.* **64**: 1059 (1990)
21. M.T. Kief, G.J. Mankey, and R.F. Willis, *J. Appl. Phys.* **69**: 5000 (1991)
22. P. Bruno, in: *"Magnetic Thin Films, Multilayers and Surfaces"*, Mat. Res. Soc. Symp. Proc. Vol. 231 (Materials Research Society, Pittsburgh, 1991)
23. J.J. de Miguel, A. Cebollada, J.M. Gallego, R. Miranda, C.M. Schneider, P. Schuster and J. Kirschner, *J. Magn. Magn. Mat.* **93**: 1 (1991)
24. A.K. Schmid and J. Kirschner, *Ultramicroscopy* (1992), in press
25. F. v.d. Merwe and E. Bauer, *Phys. Rev. B* **39**: 3632 (1989)
26. J. Frohn, M. Giesen, M. Poensgen, J.F. Wolf and H. Ibach, *Phys. Rev. Lett.* **67**: 3543 (1991)
27. A. Berger, U. Linke, and H.P. Oepen, *Phys. Rev. Lett.* **68**: 839 (1992)
28. H.P. Oepen, J. Kirschner, T. Reul, and C.M. Schneider, to be published
29. J. Chen and J.L. Erskine, *Phys. Rev. Lett.* **68**: 1212 (1992)
30. W.L. Egelhoff, jr. and M.T. Kief, *Phys. Rev. B* **45**: 7795 (1992)
31. R. Coehoorn et al., see these proceedings

ATOMIC MOMENTS, ANISOTROPY AND EXCHANGE COUPLING IN ULTRATHIN FILMS: MAGNETIC ORDER AND STRUCTURE

G. Bayreuther

Institut für Angewandte Physik
Universität Regensburg
8400 Regensburg, F.R.G.

INTRODUCTION

In recent years a series of new phenomena related to magnetic properties of ultrathin films and multilayers have been theoretically predicted or discovered empirically, e.g. atomic magnetic moments which are enhanced compared to bulk material, magnetic surface and interface anisotropies which often have an easy axis of magnetization perpendicular to the film plane, exchange coupling between ferromagnetic layers across non-magnetic spacer layers which is observed frequently to oscillate between ferro- and antiferromagnetic as a function of spacer layer thickness, spin dependent scattering of conduction electrons in multilayers which leads to a giant magnetoresistance if neighbouring ferromagnetic layers have antiparallel magnetization. However, experimental data often deviate considerably from one laboratory to another. This fact has led to certain controversies but, indeed, it most certainly indicates a strong dependence of the magnetic behaviour on film structure and morphology. These in turn are strongly influenced by the growth conditions for the films which are not perfectly controlled in many experiments.

More generally, when the magnetism of (nearly) two-dimensional systems is investigated by studying ultrathin films it is not *a priori* obvious how far their specific properties which distinguish them from bulk material are an intrinsic effect of reduced dimensionality and how much they are affected by structural defects which in general become more abundant when the film thickness approaches the monolayer range. In this contribution two examples are presented where a close relation between magnetic order and film structure was found:

- ultrathin Fe single films and multilayers grown on Au(111) by MBE ($p < 5 \cdot 10^{-10}$ mbar) or by dc magnetron sputtering;
- Fe/Tb and Fe/Y sputtered multilayered films.

The magnetic properties of the films were studied by SQUID, vibrating sample (VSM), alternating gradient (AGM) and torque magnetometers and by conversion electron Mössbauer spectroscopy (CEMS). A series of different techniques (electron and X-ray diffraction, X-ray fluorescence spectroscopy, AES, LEED, RHEED, STM, CEMS and high resolution electron microscopy) were used for the structural characterization of the films.

Fe on Au(111)

Epitaxial growth of bcc Fe on Au(111) is well known for films thicker than 10 ML to 20 ML: either the so-called Nishiyama-Wassermann orientation [1] is observed with the Fe(110) plane parallel to Au(111) and one of three equivalent axis orientations (Fe[100]||Au[1$\bar{1}$0] or Au[01$\bar{1}$] or Au[$\bar{1}$01]; Fe[110]||Au[11$\bar{2}$] etc.) or the Kurdjumov-Sachs orientation which results from the Nishiyama-Wassermann orientation by rotating the Fe lattice by 5.3°. In both cases the lattice misfit is accommodated by misfit dislocations. For the early growth stages of Fe on Au(111) ($t_{Fe} < 10$ ML), however, contradictory results have been reported in the literature: while in some cases perfectly pseudomorphic growth was observed for Fe films thinner than 4-5 ML [2,3] and a transition from the coherent fcc(111) to a bcc(110) Fe lattice above this thickness, other authors found perfectly coherent growth only for $t_{Fe} \leq 3\,\mathring{A}$ ($\approx$1.5 ML), an increasing density of misfit dislocations above 3 ML and a transition to a (110)bcc structure above 7-8 ML [4], or a bcc(110) Fe phase even for the thinnest films with no indication for a pseudomorphic fcc Fe phase at all [5].

A clear picture for the initial growth of Fe on Au(111), therefore, does not exist on the basis of these studies which are based on electron microscopy or electron diffraction. The reported discrepancies might be explained by subtle differencies in film structure caused by different growth conditions, or by the different sensitivity of the applied analytical tools to details of the film structure.

Quite recently a new wealth of information about the initial stages of metal on metal epitaxy has been provided by *in situ* STM studies [6,7,8]. The epitaxial growth of Fe on Au(111) has been studied by Chambliss et al. [9] and in even more detail by Stroscio et al. [6]: the nucleation of Fe occurs at the corners of the Au(111) 'herringbone' reconstruction with triangular growth shape for the first monolayer islands. Atomic resolution measurements [6] showed a pseudomorphic growth which is nearly layer-by-layer up to 3-4 ML where a gradual transition to rectangular growth shape is observed which may be an indication for the formation of the stable bcc α-Fe phase. The transition itself may be accompanied or triggered by the formation of misfit dislocations as a consequence of the large misfit between the bcc Fe[011] and the fcc Au[11$\bar{2}$] lattice spacing of 18.7%.

It seems natural to expect that these different possible structures and morphologies of ultrathin Fe films on Au(111) will have a considerable influence on their magnetic behaviour. In particular, a change from a pseudomorphic metastable fcc phase to the equilibrium bcc phase should manifest itself clearly in the fundamental magnetic film properties like ground state magnetic moments, magnetic anisotropy, temperature dependence of magnetization and Curie temperature. In the following, measurements of these quantities in Fe films on Au(111) from the submonolayer range to $t_{\mathrm{Fe}} > 10\,\mathrm{ML}$ will be reported[10].

Single Fe films (partly enriched to more than 95 % in ^{57}Fe) were prepared by MBE ($p \leq 5 \cdot 10^{-10}$ mbar) on Au(111) substrate films (t = 400 ML) grown on fused quartz and covered by a protective layer of Au. Scanning tunneling microscopy (STM) and electron microscopy showed that Au films grown under appropriate conditions (substrate temperature $T_S = 300...350\,\mathrm{K}$, $p \leq 5 \cdot 10^{-10}$ mbar) consist of (111)-oriented crystallites with about 100 nm lateral diameter and atomically flat terraces up to 30 nm wide which are separated by mono-atomic steps.

Fe/Au multilayer films were prepared* by DC magnetron sputtering on fused quartz with a Au substrate film of 175 ML. The thickness of the Au layers was chosen large enough (i.e. 33 ML) to avoid direct exchange coupling between neighbouring Fe layers. The Au substrate layers show a pronounced (111) texture along the film normal. The crystallites have a diameter of around 40 nm, i.e. somewhat smaller than for the MBE grown films and they also have extended flat terraces.

Quartz crystal monitors and X-ray fluorescence analysis were used to determine the mass coverage of the layers (in μg/cm^2 or atoms/cm^2; error $< 3\,\%$ for $t_{\mathrm{Fe}} = 1\,\mathrm{ML}$). Defining a geometrical thickness of the films would require the exact knowledge of the density. This is very difficult because the precise structure and defect density are not sufficiently well known for ultrathin films. Therefore, we only consider the film mass and use the term 'thickness' in monolayers (ML) as an equivalent value corresponding to bulk bcc-Fe in (110) orientation.

Conversion electron Mössbauer spectroscopy (CEMS) gave the sharp sixline spectra of bcc-Fe down to 3-4 ML . Some line broadening observed in films thinner than 4 ML can be explained by a certain disorder at the interfaces. The Mössbauer data agree with electron diffraction which showed typical α-Fe patterns for Fe films of 5 ML or more.

Fig. 1 shows magnetization curves of ultrathin Fe films grown on Au(111) by MBE with the field applied parallel($\|$) and perpendicular ($\perp$) to the film plane. It seems remarkable that ferromagnetic order characterized by a spontaneous magnetization (remanence), hysteresis and strong anisotropy exists down to a nominal Fe film thickness of 0.5 ML. This is a strong indication for a mainly 2-dimensional growth for the first few ML which leads to the formation of extended monolayer patches for a nominal thickness below 1 ML.

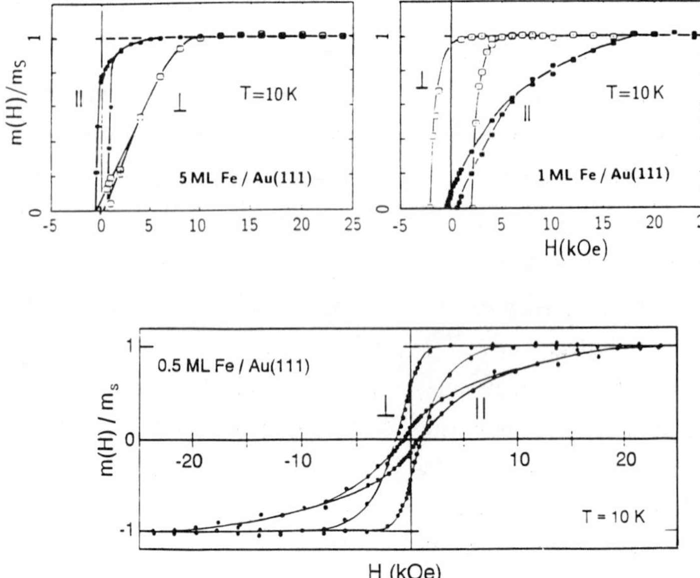

Fig. 1. SQUID magnetization curves of ultrathin Fe films grown by MBE on Au(111) with the field applied parallel (||) and perpendicular (⊥) to the film plane; nominal Fe thickness of 5 ML, 1 ML and 0.5 ML.

The average spontaneous ferromagnetic moment per Fe atom, $\langle\mu\rangle$, is obtained by extrapolating the magnetic moment of the sample, $m(H)$, from the saturation at high fields to $H = 0$ and dividing by the number of Fe atoms in the film. The latter quantity was determined with an error below 3 % of a monolayer by X-ray fluorescence analysis (secondary target excitation and energy dispersive detection of photons).

In fig. 2 the average spontaneous moment, $\langle\mu\rangle$, is plotted as a function of temperature for a 4 ML and a 1 ML Fe film in comparison to bulk α-Fe. For all films which are flat and continuous over distances of several 100 Å the temperature dependence of the magnetization could well be fitted by a $T^{3/2}$ law according to

$$\frac{M_0 - M_S(T)}{M_0} = B_N \cdot T^{3/2} \tag{1}$$

From the fit we get the ground state magnetization, M_0, and the average ground state atomic moment, $\langle\mu_{\text{Fe}}\rangle$, respectively. This is plotted versus Fe film thickness in fig. 3. The experimental data show a considerable enhancement of the atomic moment if the Fe films become thinner than 10 ML. The maximum value is measured for an average film thickness of 1 ML with a moment which is 25 % larger than in bulk α-Fe. Below 1 ML $\langle\mu\rangle$ stays constant within the experimental uncertainty. These results suggest that the atomic moment is determined by the number of monolayers in the film and that submonolayer films indeed consist of patches one monolayer high. Also, it should be noted that no discontinuity or break in $\langle\mu\rangle$ versus Fe film thickness can be detected.

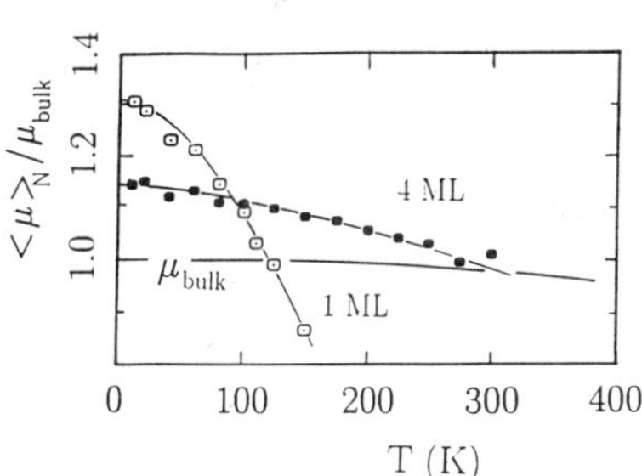

Fig. 2. Average magnetic moment per atom for Fe films on Au(111) as a function of temperature normalized to the bulk moment at $T = 0$.

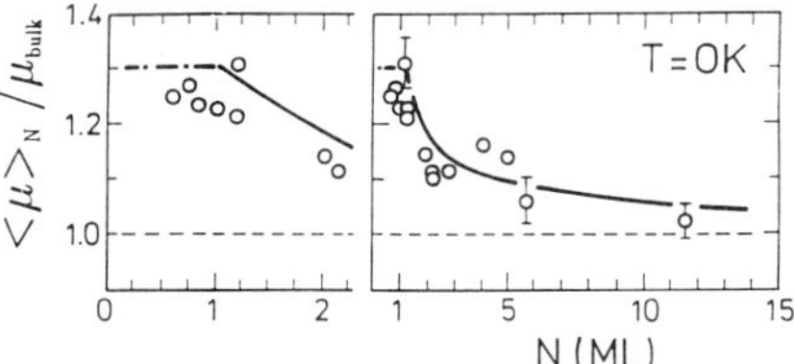

Fig. 3. Average ground state moment per atom of Fe films on Au(111) normalized to the bulk moment ($2.2\,\mu_B$/per atom) as a function of nominal Fe film thickness in monolayers, N; the solid line represents estimated values derived from band calculations (see text).

For the interpretation of this behaviour we compare the atomic moments with theoretical values which have been estimated[11] for (110) oriented bcc Fe films on Au(111) on the basis of existing *ab initio* band calculations and which are represented in fig. 3 by a solid line. Unfortunately, there are no direct calculations for Fe(110) on Au; the estimated values were obtained using general trends in existing band calculations[12,13,14]: i) the surface moment of a (110) oriented Fe film is enhanced by about 28 % compared to the 35 % enhancement found for Fe(100) films; ii) the surface moment is reduced by ≈ 15 % when Fe(100) films are covered with gold; iii) the depth profile of the atomic moment has a very similar shape for all Fe films.

Within the uncertainty of the experimental data we observe an agreement with these estimated values. The estimate, however, is based on calculations which assumed a bcc Fe lattice with the bulk lattice constant. The conclusion therefore is that either in our experiment Fe grows incoherently on Au(111) in a bcc structure from the beginning, or the Fe first grows in a strained fcc lattice and gradually changes to a (110) bcc lattice. This last interpretation then requires that the Fe moment in an fcc Fe(111) layer strained by ≈12 % is not significantly different from the one calculated for ultrathin bcc Fe films. Band calculations for such an fcc Fe structure which might support this interpretation to our knowledge are not available at this time.

On the other hand, there is evidence that enhanced ground state moments are indeed caused by the modification of the band structure at the interface as predicted by band calculations and not primarily by deviations from the bulk lattice structure: in Fe(100) ultrathin films grown on Au(100) substrate films a similar moment enhancement has been found[15]. Here the Fe is known to grow in a nearly unstrained bcc lattice due to the very small lattice mismatch. The respective measurements, however, are not yet as precise as the present experiment for technical reasons and need further improvement.

Next, we discuss the magnetic anisotropy of the films. The hysteresis loops in fig.1 show a switching of the easy axis from the film plane to the film normal when the Fe thickness is reduced from 5 ML to 1 ML[16]. This is expressed quantitatively by the thickness dependence of the effective anisotropy field, H_K^{eff}, relative to the film normal. H_K^{eff} is determined from the area between the easy and hard axis magnetization loops (corrected for hysteresis), from the saturation field in the hard axis and from torque curves. All three methods yield the same values within an experimental uncertainty of about ± 4 kOe except for small absolute values of H_K^{eff} where the saturation method cannot be applied.

The effective anisotropy field of a thin magnetic film with respect to the film normal can be expressed as the sum of the demagnetizing field perpendicular to the film plane, $4\pi M_S$, a volume anisotropy, H_K^V, and the interface term which is proportional to the inverse number of Fe monolayers, N:

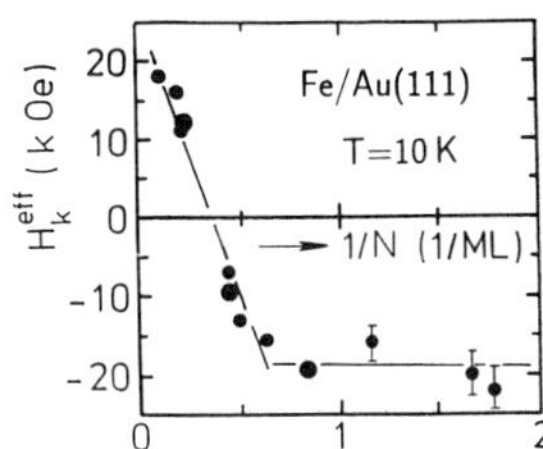

Fig. 4. Effective anisotropy field relative to the film normal, H_K^{eff}, at $T = 10\,\text{K}$ versus inverse number of atomic layers, $1/N$, of Fe films MBE grown on Au(111) at a substrate temperature of $T_S = 310\,\text{K}$.

$$H_K^{\text{eff}} = 4\pi M_S + H_K^V + \frac{1}{N}(H_S^{(1)} + H_S^{(2)}) \tag{2}$$

Here we use Néel's convention that H_K^{eff} is negative for the easy axis perpendicular to the plane. The interface anisotropy field values, $H_S^{(1)}$ and $H_S^{(2)}$, are not necessarily equal because of a potentially different structure at the Au/Fe and Fe/Au interfaces. Here we discuss the average value, $H_S = \frac{1}{2}(H_S^{(1)} + H_S^{(2)})$.

The data points in fig. 4 which correspond to films with the most perfect morphology grown at $T_S = 310\,\text{K}$ indeed follow a straight line as expected from equ.(2) over a large thickness range . This is consistent with the existence of a magnetic interface anisotropy. From the slope we determine the value of the corresponding anisotropy field as $H_S = (-40 \pm 3)\,\text{kOe}$.

Below 2 ML the data deviate from the straight line. It is not surprising that equ.(2) is no longer valid in this thickness range: for $t_{\text{Fe}} < 2\,\text{ML}$ a part of the film is only 1 ML thick. Here, the two interfaces coincide on the same Fe atomic plane, and it can be understood that here the resulting anisotropy field, H_S, is smaller than $\frac{1}{2}(H_S^{(1)} + H_S^{(2)})$.

If for $t_{\text{Fe}} < 1\,\text{ML}$ the films consist of extended monolayer patches H_S should be independent of the nominal thickness because $N = 1$ for each individual ferromagnetic island (s. eq. 2). This is exactly what is observed between 1 ML and 0.5 ML. It again confirms the essentially 2-dimensional growth of the Fe films in the early stage.

By extrapolating to bulk material ($\frac{1}{N} \to 0$) we obtain a H_K^{eff} value of 26 kOe, i.e. slightly above $4\pi M_S$ of α-Fe. This means that any volume anisotropy, e.g. a magnetostrictive component due a uniform lattice distortion perpendicular to the plane is very small.

When the growth temperature of the Fe films is increased to 450 K the absolute value of the interface anisotropy field decreases to (13±5)kOe as seen from the slope of the corresponding data points in fig.5 (open circles). This raises the question whether the reason is an enhanced interdiffusion at the interfaces or a more 3-dimensional growth due to the increased surface mobility of the Fe atoms.

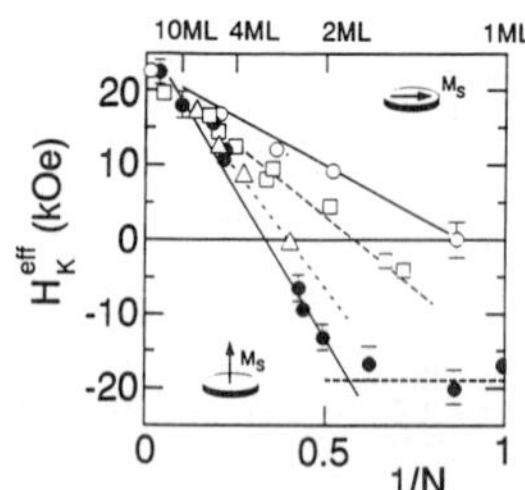

Fig. 5. Effective anisotropy field, H_K^{eff}, relative to the film normal at $T = 10\,\text{K}$ as a function of the inverse number of Fe atomic layers, $1/N$, for Fe films on Au(111): (•) MBE films grown at T_S=310 K; (○) MBE fiims grown at T_S=450 K; (□) MBE films with 0.5 and 2 ML interfacial alloy zone, T_S=310 K; (△) sputtered multilayer films (t_{Au}=33 ML).

In order to determine the influence of intermixing on the interface anisotropy two series of epitaxial films were prepared with thin alloy zones at the interfaces: alloy layers with a composition of $Fe_{60}Au_{40}$ and a nominal thickness of 0.5 ML and 2 ML, respectively, were deposited by controlled co-evaporation at both interfaces. The resulting anisotropy fields are marked in fig.5 by open squares both for t_{alloy}=0.5 ML and 2 ML. This result means that already a very small amount of alloy reduces the value of $|H_S|$ to (17±5) kOe, i.e. by more than 50 %. Further, it means that a thicker alloy zone does not reduce H_S any more. Mössbauer spectra help us to understand this behaviour: 0.5 ML of interface alloy cause a much stronger change in the spectra than would correspond to the nominal amount of ^{57}Fe contained in the alloy. We interpret this result as an indication that the interface alloy not only shows up as an additional Fe phase but drastically changes the structure of the subsequently grown pure Fe film. Deposition of a Fe-Au alloy in sub-monolayer quantity might affect the surface reconstruction of the Au substrate film mentioned above thus changing the conditions for the growth of the Fe.

Next, we discuss the behaviour of sputtered Fe/Au multilayered films. Their anisotropy fields are marked by triangles in fig.5. Although the resulting interface anisotropy field $H_S = (-33 \pm 8)$ has only a slightly smaller value than the best MBE grown films there is a marked difference: below 3 ML the sputtered films exhibit an increasing tendency towards superparamagnetic relaxation even at 10 K. Consequently, no remanence and magnetic anisotropy are observed in the thickness range where a perpendicular spontaneous magnetization would be expected from the thickness dependence of H_K^{eff}.

In conclusion, epitaxial Fe films on Au(111) which are grown under appropriate conditions in a mainly 2-dimensional mode show the largest value of the interface anisotropy field, $|H_S| = 40\,\mathrm{kOe}$. Any deviation from the optimized structure leads to a reduction of $|H_S|$:

- an interfacial alloy zone of 0.5 ML drastically reduces $|H_S|$;
- MBE growth at elevated temperature (450 K instead of 310 K) produces the lowest values of $|H_S|$. In addition to some interdiffusion a more 3-dimensional growth is the main reason for this effect. This can be inferred from the Mössbauer spectra;
- sputtered multilayered Fe films show an $|H_S|$ value nearly as high as the best MBE grown films, but they are less continuous with a pronounced island structure below 3 ML. Sputtering also produces smaller crystallites in the Au substrate layer and an enhanced mesoscopic roughness of the films.

These results mean that the interfaces of the best MBE films are indeed very sharp. The introduction of an artificial alloy zone at the interface strongly affects the growth of the following Fe layer.

Finally, we discuss spin wave excitations in MBE grown Fe films on Au(111). The temperature dependence of the magnetization for all films which are flat and continuous over distances of several 100 Å can well be fitted with a $T^{3/2}$ law according to equ. (1) as illustrated in fig. 2.

In addition, for a series of films CEMS spectra were measured in the temperature range from 77 K to 300 K. The spectrum of a 2 ML Fe film on Au(111) grown at 310 K in fig. 6 shows the typical sixline pattern with a slight line broadening. The small intensity

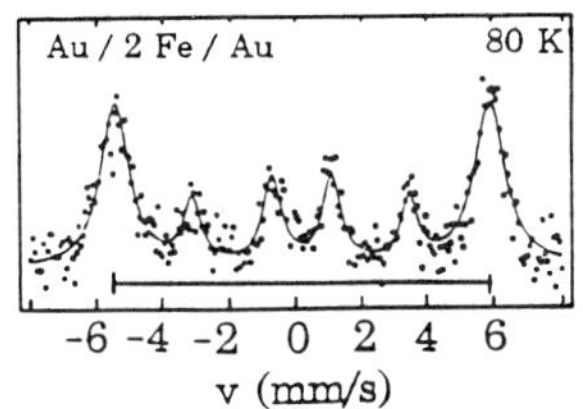

Fig. 6. Conversion electron Mössbauer spectrum of a 2 ML Fe film MBE grown on Au(111) ($T_S = 310\,\mathrm{K}$) measured at 80 K.

of lines 2 and 5 indicates that the magnetization is oriented mostly perpendicular to the film plane in agreement with the magnetization data in fig. 4. The fact that these lines are not completely absent can easily be explained by assuming that for an average film thickness of 2 ML some parts will exist with a local thickness of 3 ML which causes an in-plane orientation of the magnetization according to fig. 4.

The effective field at the ^{57}Fe nucleus , H_{eff}, derived from the splitting of the Mössbauer resonance lines is plotted versus $T^{3/2}$ in fig. 7. The resulting straight line shows that H_{eff} has the same temperature dependence as the magnetization (see eq. 1). This is important because it rules out the possibility that the external field applied for the magnetization measurements on which the data in fig. 2 are based has a decisive consequence on the observed $T^{3/2}$ behaviour.

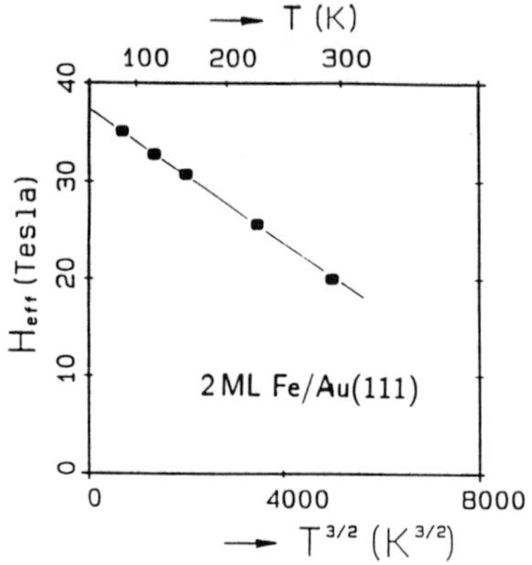

Fig. 7. Hyperfine field, H_{eff}, for an MBE grown 2 ML Fe film on Au(111) (T_S = 310 K) at variable temperature.

The spin wave parameter B_N (s. equ. 1) which is directly related to the average exchange energy per Fe atom turns out to vary linearly with the inverse number of atomic layers, $^1/_N$, for each series of films prepared under identical conditions. In addition, B_N is very strongly affected by any imperfection of the film structure: substrate and interface roughness, holes etc. all enhance the thermal spin excitations at a given temperature and, hence, lead to an increase of B_N.

From this study we conclude that the magnetic properties considered here strongly indicate that Fe on Au(111) grows in the bcc like structure from the beginning under the conditions of our experiment. There is no evidence for a transition from fcc to bcc around 5 ML as concluded from recent STM studies [6]: within the experimental uncertainty we do not observe any discontinuity or break neither in the atomic moments per Fe atom (s. fig. 3) nor in the effective anisotropy field relative to the film normal (s. fig. 4) nor in the spin wave parameter B_N. For this quantity a linear dependence on the inverse film thickness from $N = 1$ to $N \geq 100$ would not be expected if a phase change from fcc to bcc Fe occurred at $N \approx 5$ because the resulting exchange interaction would most probably be different in both structures. A direct correlation between STM observations with atomic resolution, LEED data and the magnetic properties of the identical samples should be helpful to answer remaining questions like the effect of substrate properties and a particular reconstruction of the Au surface on the growth of Fe on Au(111).

3. Fe/Tb MULTILAYERS

Fe/Tb multilayers prepared by rf or dc magnetron sputtering show a number of peculiarities of their magnetic behaviour which seem to be strongly correlated with their structure [17].

Small angle X-ray diffraction and high resolution electron micrographs of cross sections proof that a pronounced layer structure is present if the single layer thickness of Fe and Tb are thicker than 5-7 Å [18]. Mössbauer spectra show that the Fe is in an amorphous state for a layer thickness of $t_{\mathrm{Fe}} < 20..24$ Å [17]. The exact value of the critical thickness above which a crystallization into polycrystalline $\alpha-$Fe takes place depends on the growth conditions. For thicker Fe layers always an amorphous component equivalent to 2-3 ML remains. It is not definitely clear whether an amorphous alloy zone has formed or the Fe lattice is only strongly distorted close to the interface without considerable intermixing.

The structure of the multilayers has a decisive influence on their magnetic behaviour:

i) for a layer thickness of $t_{\mathrm{Fe}} \leq 20-22$ Å and $t_{\mathrm{Tb}} \leq 30$ Å the film systems behave like usual ferrimagnets with a Curie temperature of $T_C \approx 390$ K and a compensation temperature, T_{comp}, equal to the value in a homogeneous amorphous alloy with the same nominal composition. This is illustrated in fig. 8 for a film consisting of 75 periods of 10 ÅFe/7.5 ÅTb. The Curie temperature is independent of the layer thickness for a fixed ratio of Fe to Tb thickness ($t_{\mathrm{Fe}} \leq 20$ Å) and only weakly dependent on the net composition if $t_{\mathrm{Fe}}/t_{\mathrm{Tb}}$ is varied;

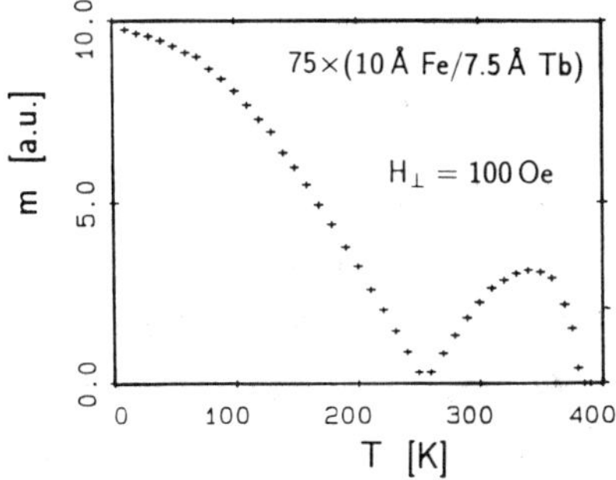

Fig. 8. Temperature dependence of the film moment (absolute value) in an applied field of 100 Oe for a multilayer with 75 periods of (10 Å Fe/7.5 Å Tb).

ii) for thicker Fe layers ($T_{Fe} \geq 25$ Å) T_C is much higher ($> 500\,^\circ$C) in agreement with the dominant α-Fe component seen in the Mössbauer spectra;

iii) by variation of the Tb layer thickness it is found that at room temperature the magnetic polarization of the Tb by interaction with the Fe only takes place over a distance shorter than ≈ 15 Å;

iv) a strong perpendicular magnetic anisotropy is observed with the effective easy axis along the film normal in the entire temperature range up to T_C if the layers are thin enough ($t_{Fe} \leq 15$ Å). As an example fig. 9 shows a square hysteresis loop perpendicular to the film plane for a 64×(14 ÅFe/12 ÅTb) film with 100 % remanence and a coercive field of 27 kOe at T = 100 K. For $t_{Fe} > 30$ Å the easy axis is always in the film plane. In the transition region the easy axis switches from perpendicular to in plane with increasing temperature; close to the switching temperature the magnetic order is inhomogeneous resulting in an apparent biaxial anisotropy due to non-saturation in fields up to 25 kOe.

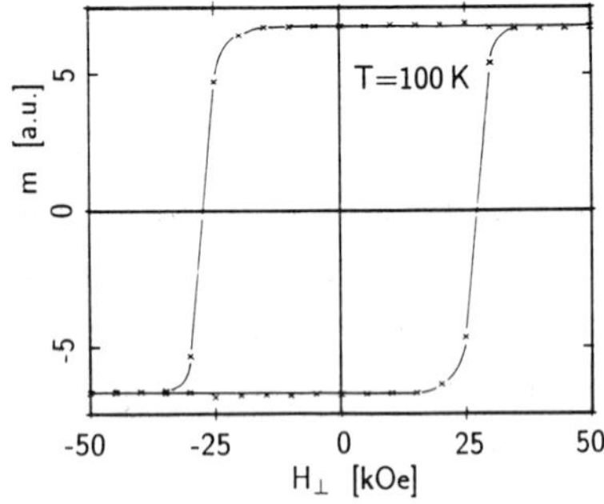

Fig. 9. Hysteresis loop of a multilayer with 64 periods of (14 Å Fe/12 Å Tb) with the field applied perpendicular to the film plane.

As a conclusion, the existence of a magnetic phase transition with a Curie temperature around 390 K is clearly connected with the amorphous state of the Fe layers. This is confirmed by a comparative study of Fe/Y multilayers [19] where the exchange coupling between neighbouring layers is absent due to the lack of the rare earth moment. Here a transition from amorphous to α-Fe takes place at $t_{Fe} \approx 23$ Åwhich is accompanied by a change of the magnetic order from a spin glass like character where the average local moment vanishes around 80-100 K to ferromagnetic with a high Curie temperature (>600 K). A similar behaviour is observed [19] on thin Fe films between Y layers grown by e-beam evaporation in UHV by Landes et al. [20].

We therefore observe the fact that two materials with a magnetic ordering temperature of $\approx$80 K (a-Fe) and $\leq$ 220 K (Tb) exhibit ferrimagnetic order with a Curie temperature of 390 K if they are exchange coupled through the interface. The existence of this phase transition is closely connected with the amorphous structure of the Fe layers. The role of the amorphous phase and a certain intermixing at the interfaces for the nature and interaction range of the exchange coupling and for the observed anisotropy is not yet clear. The experimental data available at this time suggest that the peculiar magnetic properties of Fe/Tb multilayer films are the combined effect of the transition from amorphous to α-Fe and the finite range of the exchange interaction.

Acknowledgement

The author is most indebted to his collaborators on whose experimental work this report is based: G. Lugert, W. Robl, M. Haidl, S. Lehner, G. Gruber, L. Pfau, M. Brockmann, and E. Arenholz.
This work was supported by the Deutsche Forschungsgemeinschaft and the European Community (Contract SC1*/0106-C (MB), CAMST project).

* Preparation of sputtered multilayer films by R. Pollard and P. G. Grundy (University of Salford) is gratefully acknowledged.

References

1. E. F. Wassermann and H. P. Jablonski, Surf.Sci. 22 (1970) 69
2. P. Gueguen, M. Cahoreau and M. Gillet, Thin Solid Films 16 (1973) 27
   P. Gueguen, S. Camoin and M. Gillet, Thin Solid Films 26 (1975) 107
3. S. Araki, T. Takahata, H. Dohnomae, T. Okuyama and T. Shinjo, Mat.Res.Soc.Symp.Proc. 51 (1989) 123
4. G. Honjo, K. Takayanagi, K. Kobayashi and K. Yagi, J.Cryst.Growth 42 (1977) 98
5. C. Marlière, J.,P.,Chauvineau and D. Renard, Thin Solid Films 189 (1990) 359
6. J. A. Stroscio, D. T. Pierce, R. A. Dragoset and P. N. First, J.Vac.Sci.Technol. A 10 (1992) 1981
7. D. D. Chambliss, K. E. Johnson, R. J. Wilson and S. Chiang, to appear in J.Magn.Magn.Mat. (March 1993)
8. R. Q. Hwang, C. Günther, J. Schröder, S. Günther, E. Kopatzki and R. J. Behm, J.Vac.Sci.Technol. A 10 (1992) 1970
9. D. D. Chambliss, R. J. Wilson and S. Chiang, in: "Structure /Property Relationships for Metal/Metal Interfaces", A. D. Romig, D. E. Fowler, and P. D. Bristowe eds., Materials Research Society, Pittsburgh, 1991, p. 15
10. G. Lugert, thesis, Universität Regensburg 1992
    G. Lugert, W. Robl, L. Pfau, M. Brockmann and G. Bayreuther, to appear in J.Magn.Magn.Mat. (March 1993)

11. G. Lugert and G. Bayreuther, to be published
12. C. L. Fu, A. J. Freeman and T. Oguchi, Phys.Rev.Lett. 54 (1985) 2700
13. A. J. Freeman and C. L. Fu, J.Appl.Phys. 61 (1987) 3356
14. S. Blügel, S. Drittler, R. Zeller and P. H. Dederichs, Appl.Phys. A 49 (1989) 547
15. C. Turtur and G. Bayreuther, to be published
16. G. Lugert and G. Bayreuther, Thin Solid Films, 175 (1989) 311
17. Z. S. Shan and D. J. Sellmyer, Phys.Rev. B 42 (1990) 10433
    G. Bayreuther, G. Lugert, S. Lehner, P. Bruno, G. Endl and M. Haidl, Proc. of the IBM Workshop "Fundamentals of Magnetic Materials for Storage", St. Paul-de-Vence, 1990
18. G. Endl, thesis, Universität Regensburg, 1991
19. E. Arenholz and G. Bayreuther, to be published
20. S. Handschuh, J. Landes, U. Köbler, Ch. Sauer, G. Kisters, A. Fuss and W. Zinn J.Magn.Magn.Mat., in press

MAGNETISM AND STRUCTURE OF Gd MONOLAYERS AN AC SUSCEPTIBILITY INVESTIGATION

A. Aspelmeier, U. Stetter*, and K. Baberschke

Institut für Experimentalphysik
Freie Universität Berlin, Arnimallee 14, D–1000 Berlin 33, Germany

INTRODUCTION

One of the most classical experimental techniques to investigate bulk magnetism are ac susceptibility measurements in a mutual induction bridge with low frequency lock–in technique. Curie–Weiss temperatures, the paramagnetic $\chi(T)$ and its critical behavior, as well as the ferromagnetic susceptibility have been determined in metallic and nonconducting magnets. Recently this technique became compatible with UHV. It was used to study in–situ a 300 Å Gd film evaporated on a W(110) substrate. $\chi(T)$ at the paramagnetic site (at T_C^+) was analyzed[1]. We could show the effect of film annealing on the sharpness of the Curie–Weiss peak, and for the magnetically most homogeneous film the critical exponent γ was determined. In a second publication we studied $\chi(T)$ in the ferromagnetic regime (at T_C^-), a full theoretical analysis for $T<T_C$ was given, which fits the experiments within error bars and clarifies some dubious discussions in the literature on the Hopkinson maximum[2]. An effective anisotropy energy and a small distribution of T_C values for the macroscopic film (5 mm diameter) were determined. However, all this has been measured at films with nm thickness. It would be of some interest if this technique for UHV in–situ measurements is sensitive enough for metallic films of monolayer (ML) thickness. In this communication we present the first results of thickness d=7 to 100 ML Gd(0001) epitaxial films. T_C varies from T_C(bulk)=292.5 K down to $T_C \approx$

* Present address: Department of Material Science and Engineering, Stanford University, Stanford, CA 94309–2205

150 K. This will be analyzed and compared to previous measurements[3] at which T_c remained almost constant even for films with a mass–equivalent thickness of 1 ML only.

GROWTH MODI AND FILM STRUCTURE

The films were prepared in an UHV.chamber with a base pressure of ≈ 1.5 x 10^{-11} mbar on an Auger–clean W crystal showing a sharp bcc (110) LEED pattern. During evaporation the pressure never exceeded 1x10^{-10} mbar, and the W crystal was kept at a substrate temperature T_s = 300 K. The evaporation rate is ≈ 1 ML in 3 min. Note this is different from the technique used by Farle et al.[3]. In these previous experiments the substrate was kept at T_s = 720 K, during evaporation. In the literature[4] a Stranski–Krastanov (SK) growth modus is discussed. Strictly speaking one would expect a formation of the first monolayer and islands on top of this. For the present evaporation conditions a layer–by–layer growth is very likely[4], for details see Stetter[5]. Fig. 1 shows χ(T) for a 9 ML film measured instantaneously after evaporation and 1 to 2 hours later. It is obvious that the Curie–Weiss peak is reduced after 1 1/2 hours. We conclude that the contamination of the residual gas even at a pressure of ≈ 2x10^{-11} mbar destroys the magnetism of the clean metallic Rare Earth film after few hours. Assuming a sticking coefficient of 1, the relevant gas coverage is in the order of 1/10 of a ML. This effect shows up only for the 5 to

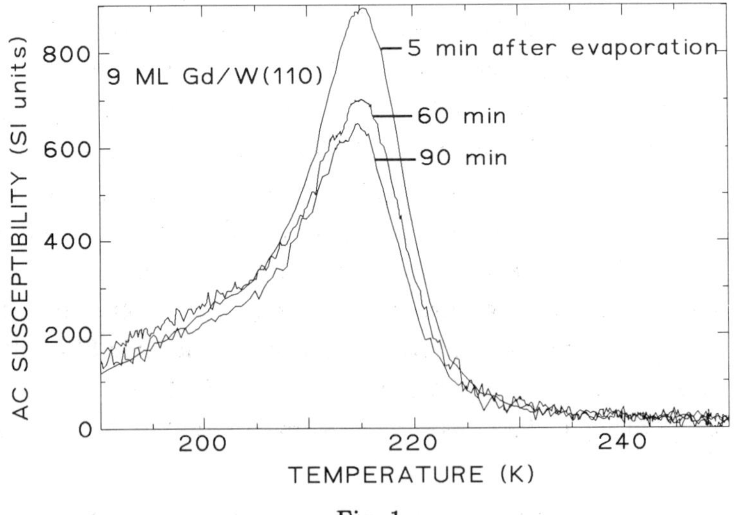

Fig. 1

12 ML films. For thicker films of d $\geq$ 30 ML the $\chi(T)$ remains unchanged for several hours.

The misfit of the lateral lattice constant for hcp (0001) of Gd and for bcc (110) of W produces some in–plane strain. To reduce and homogenize this, all freshly evaporated films were annealed for $\approx$ 1 min at 50 to 150 ^{0}C. Only after such a treatment sharp peaks (Fig. 1 and 2) were obtained. In the next paragraphs we will discuss the experiments with the underlying picture of a layer–by–layer grown film. However, we do not have a real geometrical image on an atomic scale of our films. The ac field ($H_{pp} \lesssim 1$ G) oscillates in the film plane. Our technique has no spatial resolution, like a focussed electron or LASER beam. The χ_{ac} signal is the response of the whole macroscopic specimen, including edges, etc. It might well be that the films consist of several macroscopic (several μm) slabs. One outcome of the present investigation will be that the magnetic measurements themselves give information on the film structure.

CURIE–WEISS TEMPERATURE AND FINITE SIZE SCALING

Fig. 2 shows $\chi(T)$ for various film thickness d. Within $\pm$ 1 K the maximum in χ determines T_c[1,2]. These values are plotted in Fig. 3 ( ). Here the vertical error bar is very small, while horizontal relative error bars are large in particular for small d. A film thickness given in ML is not an integer number, the uncertainty is at least $\pm$ 1 ML. For 7 ML this yields a variation of T_c of about 7%. Both figures show a drastic reduction of T_c for few atomic layers. As discussed elsewhere for Ni(111) films[6], finite size scaling

$$\frac{\Delta T_c(d)}{T_c(\infty)} = \frac{T_c(\infty) - T_c(d)}{T_c(\infty)} = C_0 \cdot d^{-1/\nu} \qquad (1)$$

determines $T_c(d)$. For an arbitrary factor C_0 (variation over orders of magnitude) power law fits through many data sets are possible. However theory predicts a physically meaningful $C_0 \approx 2$ to 10. The dashed line in Fig. 3 is drawn with $C_0 = 7.0$ and $1/\nu = 1/0.63$, the 3D Ising value. We do not vary and fit the exponent, rather we take it as given and use eq. (1) for a qualitative discussion of $T_c(d)$. We have prepared several films with the same nominal thickness d=7, 9, 15 etc. ML, and found within error bars good agreement for $T_c(d)$. One might turn the argument around and deduce from $T_c(d)$ the film thickness. Under the error bars given and the discussions above this seems to be a fairly safe procedure. Depending on the assumption of a Heisenberg or Ising interaction T_c reduces to $\approx$ 50% of its

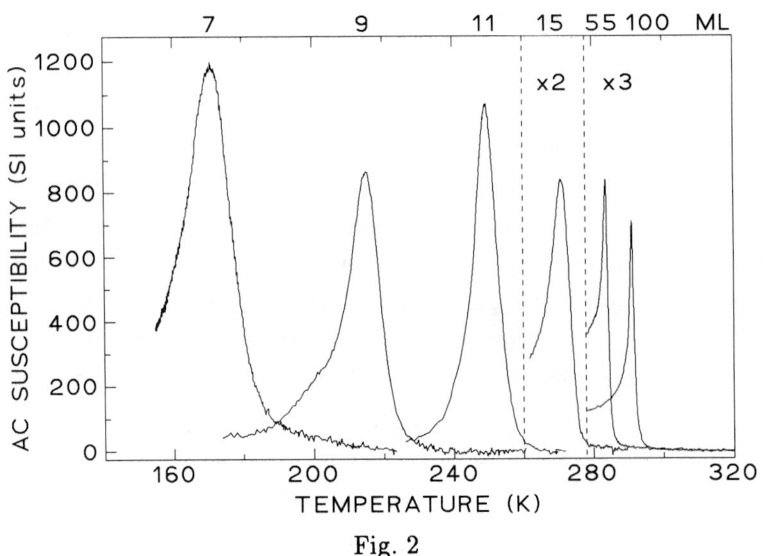

Fig. 2

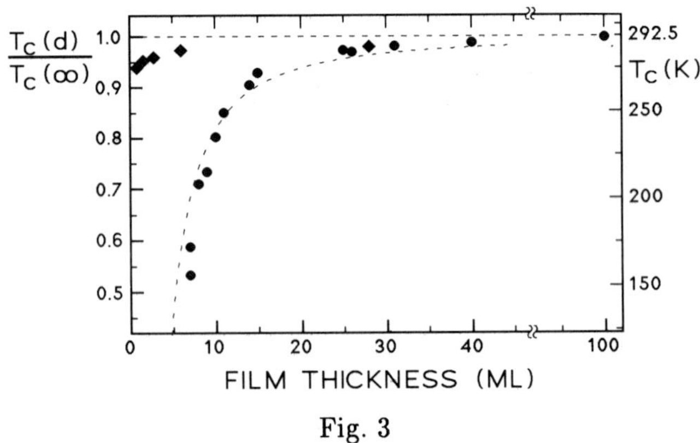

Fig. 3

bulk value for 3 to 5 ML. There is no example known to us in the literature which proofs this "rule" to be wrong. It is interesting to note that for example Fe and Co films are very impracticable for such an investigation. The bulk T_Cs of both metallic ferromagnets are so high that films of 5 to 20 ML already have a $T_C(d)$ too high to avoid interdiffusion with the substrate surface. Fe/Pd(100) for example was investigated up to 3 ML only ($T_C \approx 600$ K)[7]. On the other hand thicknesses of 1 or 3

ML only is a completely different subject to discuss. First of all, eq. (1) is the asymptotic limit for 3 D (bulk) films and yields T_cs without physical meaning for d = 1 or 2 ML. Second, Schneider et al.[8] have shown by means of real–space STM investigation that a 2 ML Co structure, yielding nice layer–by–layer MEED oscillations during evaporation, has a fairly rough morphology.

In Fig. 3 we added our previous data[3] ( ). Within the present frame of analysis and having in mind the discussion above, we have to conclude that the previous 0.8 and 1.6 ML with a T_c of approx. 270 K were not grown layer–by–layer. These data can be explained if we assume a formation of islands about 15 ML high and lateral dimensions on a μm scale (SK growth). The old data point for d = 28 ML (≈ 100 Å) is in agreement with the present result.

Finally we comment on the width (δT_c) of the Curie–Weiss peaks. Fig. 2 shows that δT_c is larger for very thin films than for 40 and 100 ML. δT_c ranges from 15 K for the former to 1 K for the latter (the film analyzed by Stetter et al.[1]). This will be discussed in detail elsewhere, we attribute this to larger lateral strain in thin films. However one should note that the χ_{max} values for 7 to 11 ML range up to 1200 whereas χ_{max}(15 ML) ≈ 400 and χ_{max}(100 ML) ≈ 200, only (note the gain factors in Fig. 2). At first glance this seems to be contradictious in a naive picture of an inhomogeneous superposition. This will be discussed in the following Section.

GIANT χ_{max} VALUES, DEMAGNETIZATION FACTOR N, AND THE CRITICAL EXPONENT γ

Since this is to our knowledge the first publication where the ac susceptibility is employed to study the thickness dependence of the Curie–Weiss temperature of metallic films grown in UHV and measured in–situ, we would like to draw the reader's attention to some points. First, despite the fact that χ_{ac} measurements are quite common in bulk magnetism investigation, the majority of published data is given in uncalibrated units, i.e., in microvoltage of the bridge unbalance, etc. This makes a theoretical analysis and comparison to different experiments difficult. Second, we have not just used a scale factor between various films, but we used an absolute calibration with a Gd salt[1,5]. This enables us to relate our signal to the number of spins (the magnetic moment, respectively) to an accuracy of ± 5%. To stress this we give χ in "absolute SI units" (Fig. 1,2 and Ref. 1). This should not be confused with the fact that one measures the relative susceptibility – but calibrated in SI units.

Now we will discuss $\chi(T)$ quantitatively at the paramagnetic site ($T>T_c$). In

Ref. 1 we have shown how to analyze $\chi(T>T_c)$ including a small distribution of δT_c for the macroscopy specimen, and including small but finite demagnetizing factors $N_{\parallel}$ for the in plane susceptibility. For a flat disk $N_{\perp}$ can be approximated by

$$N_{\perp} \approx 1 - \pi\gamma/2 + 2\gamma^2, \qquad \gamma = c/a$$

and consequently

$$N_{\parallel} \approx \pi\gamma/4 - \gamma^2 \tag{2a}$$

For macroscopic samples it is very hard to reach $N_{\parallel}<10^{-2}$. Only for thin films on an atomic scale, $\chi(T)$ curves as in Fig. 2 can be recorded. With the layer spacing along the c–axis for Gd (2.9Å) we rewrite eq. (2a) in

$$N_{\parallel} \approx 4.5\cdot 10^{-8}\, d, \qquad d \text{ in ML} \tag{2b}$$

For an ideal flat 10 ML film $N_{\parallel} \approx 5\cdot 10^{-7}$. N can only increase in real films with some roughness. Smaller values have no physical meaning.

Having fixed the range of values for $N_{\parallel}$ we will now estimate the area under $\chi(T)$ on the paramagnetic $(T>T_C)$ site. For a rough estimate we neglect the convolution with a T_C–distribution as discussed in Ref. 1,2. For the present experiments this is a minor effect. The area under the ideal susceptibility is a measure for the number of atoms and their magnetic moment, it is given by

$$A = \int_0^{\infty} \frac{\chi_0^{+} \cdot t^{-\gamma}}{1 + N\chi_0^{+}\cdot t^{-\gamma}}\, dt = \frac{\pi(\chi_0^{+})^{1/\gamma} \cdot N^{-1+1/\gamma}}{\gamma\cdot \sin(\pi/\gamma)} \tag{3}$$

t is the reduced temperature[1,2]. 97% of the full area lie between t=0 and t=0.1. This is the range of temperature where the signal is seen. The full analysis will be given elsewhere, we refer here to the results only:

For films of thickness d ≈ 15 to 100 and thicker eq. 3 can be fitted consistently using γ=1.24 and $N \approx 10^{-5}$ (see eq. 2b). For the thin films of 7 to 11 ML this is impossible. The experimental area is about 100 times larger than the calculated one (eq. 3). Agreement is achieved only if we take $N<10^{-11}$. This is unrealistic as discussed above. As a conclusion we were focussed to take larger γ values. Using γ = 1.75, N can be kept in the 10^{-6} range. This is a strong indication for a dimensional crossover below ≈ 10 ML. Similar results have been observed for thin Ni films[6].

SUMMARY

The ac susceptibility set–up shown here is sensitive to detect monolayer magnetism. Provided conventional UHV equipment is available, set–up is very simple[5] and inexpensive (one Lock–In–Amplifier). The technique and its results will give complementary informations to those experimental set–ups that dominate the research on magnetic thin films at the time.

The present results show that finite size scaling hold for Gd as it does for Ni[6]. It is very important to know the real structure of a film. In some of our previous[3] and other experiments the given number of monolayers can only be regarded as a mass equivalent.

In the last Section we gave a strong indication that Gd films below 10 ML show a γ of $\approx$ 1.75, this corresponds to a 2D–Ising behavior.

REFERENCES

1. U. Stetter, M. Farle, K. Baberschke, W.G. Clark, Phys. Rev. B 45, 503 (1992)
2. U. Stetter, A. Aspelmeier, K. Baberschke, J. Magn. Magn. Mat., July 1992
3. M. Farle, K. Baberschke, Phys. Rev. Lett. 58, 511 (1987) and M. Farle, Ph.D. Thesis (1989) FU Berlin (unpublished)
4. J. Kolaczkiewicz, E. Bauer, Surf. Sci. 175, 487 (1986)
5. U. Stetter, Ph.D. Thesis (Febr. 1992) FU Berlin (unpublished)
6. Yi Li, K. Baberschke, Phys. Rev. Lett, 68, 1208 (1992)
7. C. Liu, S.D. Bader in: Magnetic Properties of Low–Dimensional Systems II, Eds. L.M. Falicov, F. Mejia–Lira, Y.L. Moran–Lopez, Springer Proc. in Physics 50, 22 (Springer, Berlin, 1990)
8. C.M. Schneider, A.K. Schmid, H.P. Oepen, J. Kirschner, these proceedings

X-RAY STUDIES OF MAGNETIC SUPERLATTICES

Roy Clarke, Francisco J. Lamelas, and Sezai Elagoz

Department of Physics
Randall Laboratory
University of Michigan
Ann Arbor, MI 48109-1120

INTRODUCTION

It is becoming clear that many of the unusual properties of magnetic multilayers are directly related to their microstructure. Interface roughness[1], stacking faults[2], lattice strains[3], and islanding[4], are just some of the imperfections that are known to influence the magnetic behavior strongly, and, in some cases, to produce entirely new effects. It is important therefore to characterize these structures, especially the interfaces, as completely as possible in order to understand their magnetic properties.

In this contribution we describe x-ray diffuse scattering methods[5] as they are applied to magnetic superlattices. We will emphasize the diverse types of information that x-ray scattering can provide with appropriate modeling, giving examples based on our recent studies of Co-Cu and Co-Au superlattices grown by molecular beam epitaxy (MBE)[6]. With the continuing improvement of x-ray sources, particularly the introduction of new synchrotron facilities, x-ray techniques have emerged as a powerful non-destructive probe of interface structure[7]. In many ways the information provided by x-ray scattering is complementary to that given by transmission electron microscopy (TEM). At the highest spatial resolution, however, many concerns relating to interpretation of the data are common to both x-ray and TEM approaches.

At one level x-ray diffraction is regarded as a routine structural characterization method for analyzing multilayer structures, since basic features of the scattering (e.g. the positions of the x-ray peaks) can be interpreted by inspection. However, for a more complete analysis of x-ray data, it is useful to separate the various aspects of an epitaxial layer structure into several distinct components:

Layering, refers to the chemical composition of the individual layers, their thickness and periodicity and the abruptness of the interfaces between the different layers. The key question here is what type of roughness is present: steps and terraces, or continuous interdiffusion?

In-plane epitaxy describes the crystallographic relationships of succeeding layers of the structure, both with respect to the substrate and with respect to each other. Important issues to be taken into account here include relative mismatch of the lattices

Magnetism and Structure in Systems of Reduced Dimension
Edited by R.F.C. Farrow *et al.*, Plenum Press, New York, 1993

corresponding to the different layers, the strains arising in response to this mismatch, and the reduction in coherence that might result from strain-relieving dislocations at the epitaxial interface.

Stacking refers to the relative crystallographic arrangement of neighboring atomic planes. A simple example of particular relevance to transition metal magnetic structures is the stacking of close-packed planes of atoms in either the hexagonal close-packed (hcp) sequence: ABABAB..., or in the face centered cubic (fcc) sequence: ABCABC... . An important point here is that the internal energies of the different stacking sequences are very similar and so stacking faults are to be expected. This particular defect has been linked by some authors to a possible mechanism for perpendicular anisotropy[8].

X-RAY DIFFRACTOMETER SCANS

Our x-ray measurements are performed using a 12 kW Rigaku rotating anode x-ray generator in conjunction with a four-circle diffractometer. It is advantageous to use a Mo target which produces relatively energetic x-rays [17.5 keV; $\lambda \sim 0.71$ Å]. This results in a penetrating x-ray beam which is useful in performing scans in transmission geometry, and also when the sample is mounted in a partially absorbing enclosure such as a cryostat or an annealing furnace.

The most commonly employed x-ray technique for layered materials is the *perpendicular* scan where intensities are measured along the specular rod. During this scan [Fig. 1(a)] the scattering vector is perpendicular to the superlattice layers and serves as a probe of layer structure in the growth direction. At relatively large wavevectors, on the order of the reciprocal lattice spacings of the bulk lattice, this scan is sensitive to a combination of the local atomic ordering and the elemental modulation throughout the superlattice. At small wavevectors, of order $2\pi/\lambda_{SL}$ (where λ_{SL} is the superlattice period), the scattering is sensitive to modulation in elemental composition only.

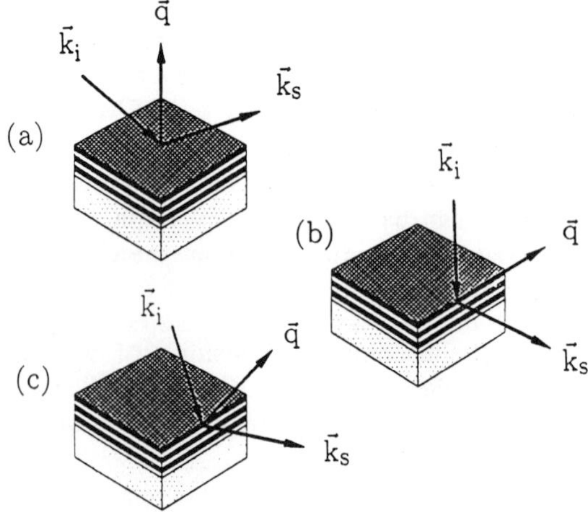

Figure 1. Types of diffractometer scan for x-ray diffuse scattering measurements on ultrathin layered structures: (a) perpendicular scan (b) in-plane scan (c) c^* scan.

In a second type of scan, the *in-plane* scan, the scattering vector lies in the plane of the superlattice [Fig. 1(b)]. This scan probes the atomic arrangement within the growth plane and is useful, for example, in determining epitaxial relationships between the film and the substrate and in estimating in-plane coherence lengths. The in-plane scans which we will describe were all performed in transmission geometry. In this geometry the entire superlattice and the substrate are sampled by the beam and scattering intensities with a rotating anode source are acceptable.

The third important type of x-ray scan for our purposes is the c^* *scan* [Fig. 1(c)], where the scattering vector has both in-plane and out-of-plane components. This scan, which is performed in transmission geometry along a non-specular rod, is sensitive to the lateral stacking coherence of the atomic layers.

LAYERING CHARACTERIZATION

The simplest approach to analysis of x-ray scattering data is through the so-called kinematical approximation[9]. This assumes that contributions from multiple scattering are negligible, it neglects interference between the scattered and incident radiation, and assumes that the reduction in intensity of the scattered wave inside the crystal lattice (extinction) is small. For metallic superlattices, which tend to be far from perfect crystals, these approximations are well obeyed except at very low diffraction angles close to where total external reflection of the x-ray occurs. In this case it is necessary to perform a more detailed calculation of the x-ray reflectivity, one which includes dynamical scattering effects (see following section).

In the kinematical approximation one may calculate the scattering amplitude from a layer structure as

$$A(q) = \sum_{n=1}^{M} f_n e^{iqr_n} \tag{1}$$

where the sum is over each of the M monolayers in the superlattice, f_n is the layer scattering factor (atomic scattering factor multiplied by the atom density), q is the scattering vector amplitude, and r_n is the position of the nth monolayer.

Segmüller and Blakeslee[10] have obtained a closed-form expression for the scattering intensity from an ideal superlattice with atomically abrupt interfaces. As we shall see, real samples often deviate significantly from this ideal case; nonetheless the model is useful as a guide. The simplest deviation from an ideal superlattice that one might consider is the occurrence of diffusion at the interfaces between neighboring layers. A damping of the superlattice satellites is to be expected since these correspond to the higher-order Fourier coefficients that are necessary in reproducing compositionally abrupt profiles. On the other hand, since interfacial diffusion does not diminish the long-range order of the superlattice, it does not lead to peak broadening. McWhan et al.[11] have developed a closed-form expression for satellite intensities for the case of trapezoidal composition profiles.

A second type of imperfection is the occurrence of irregularities in the superlattice layer thickness. These might arise, for example, in the growth of multilayers while the deposition rates are fluctuating. In the case of a crystallographically ordered superlattice, one may consider *discrete* fluctuations in layer thickness. In the case when the superlattice has amorphous or randomly oriented polycrystalline layers, which includes a large fraction of samples being prepared currently, the layer thickness fluctuations can be *continuous* rather than discrete. Both of these types of thickness fluctuations destroy the long range order of the superlattice, resulting in a finite structural coherence length and diffraction peak broadening. However, the effects turn out to be less severe in the case of discrete fluctuations, as one might expect.

Sevenhans et al.[12] have derived an analytical expression for the scattering intensity from an amorphous-crystalline superlattice with a continuous Gaussian distribution of amorphous layer thicknesses. Clemens and Gay[13] later extended the analysis of cumulative disorder to the case of discrete fluctuations in layer thickness, using a generalized Patterson function method. In a recent article Locquet et al.[14] combine features of these earlier models and derive an expression for the scattering intensity which simultaneously includes both continuous and discrete fluctuations in layer thickness.

Computer Simulations of the Scattering Intensity

As modeling of superlattices becomes more involved, simultaneously including structural disorder of various kinds, the derivation of closed form expressions for the scattering intensity at some point becomes intractable. We have recently described a numerical approach to modeling the intensities obtained in a perpendicular scan[15].

In our model the calculation of the scattering intensity is carried out by applying Equation 1 to a stack of multilayers that contain both layer thickness fluctuations and interface roughness. We begin with an ideal superlattice where each layer has a given chemical composition and in which the lattice parameter changes abruptly across the interfaces between layers [Fig 2(a)]. Next we introduce disorder in the layer thicknesses [Fig. 2(b)] using a random number generator to chose the number of atomic planes in each layer from a parabolic layer-thickness distribution. Interface roughness effects (Fig. 2(c)] are now introduced by allowing the scattering factors and layer spacings to vary exponentially according to the distance r', from the nearest interface. For example,

$$f_n = f_A + \frac{1}{2}(f_B - f_A)\exp(-2r'/t_d) \tag{2}$$

within the A layer where t_d is the interfacial diffusion width; the atomic plane spacing at the AB interface is assumed to be $\frac{1}{2}(d_A + d_B)$.

Figure 2. Illustrations of superlattice disorder introduced into numerical simulations. (a) Perfect superlattice; (b) Layer thickness disorder; (c) Simultaneous layer thickness disorder and interfacial diffusion.

Once the values d_n and f_n are determined for each atomic plane at a given scattering angle, the scattering amplitude is calculated using Eqn. 1 and squared to give the scattering intensity. Lastly, this intensity is multiplied by the Lorentz and polarization factors, and instrumental broadening is introduced by convoluting with a Gaussian resolution function.

Layer thickness fluctuations destroy the superlattice periodicity. Higher order superlattice peaks may appear split and/or shifted as shown in Fig. 3. During a particular execution of the program, the fine structure in the scattering intensity is sensitive to the actual sequence of numbers of atomic planes N^i_A and N^i_B that is used in the scattering calculation. N^i_A and N^i_B in turn are determined by the sequence of values returned by the random number generator. On the other hand, a real sample is composed of many local domains within the area sampled by the x-ray beam, each of which contains a different sequence of layer thicknesses. Thus the measured superlattice peaks are broadened rather than split into a set of sharp peaks. We simulate this effect in our calculation by averaging intensities over structures which are generated by different sequences of random numbers [Fig. 3(c)]. An example of the numerical simulation of Co-Au superlattices is shown in Fig. 4 for both low angle and high angle satellites. Typical values of t_d extracted from these fits are 1.3-1.5 monolayers. Fluctuation amplitudes for the layer thicknesses are ~ 2 monolayers and lateral coherence lengths of 0.5 to 2μm are inferred from the widths of low-angle rocking curves. From these numbers we concluded that our Co-Au superlattices have essentially atomically abrupt interfaces.

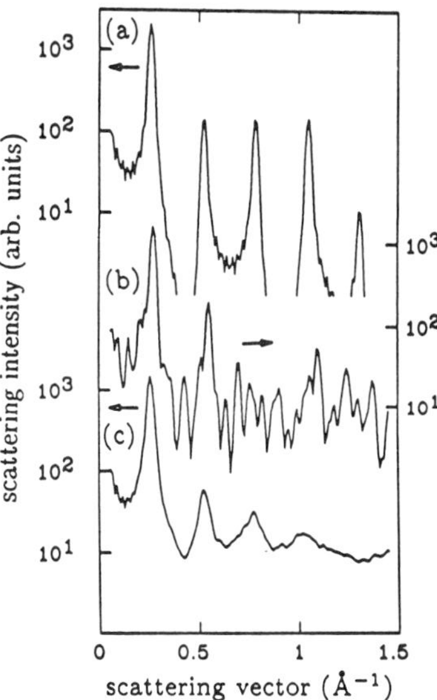

Figure 3. Effect of layer-thickness disorder on low-angle superlattice scattering intensities.

(a) Perfect superlattice with parameters $N_A = 5$, $N_B = 7$, $d_A = d_B = 2.0$ Å, $f_A = 0$, $f_B = 1$.

(b) Same parameters as (a), but with layer thickness fluctuations of $\pm$ 1 monolayer.

(c) Same as (b), but with intensities averaged over 100 calculations.

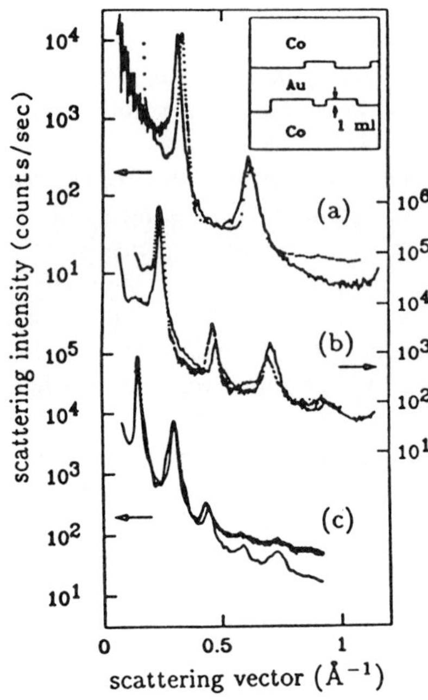

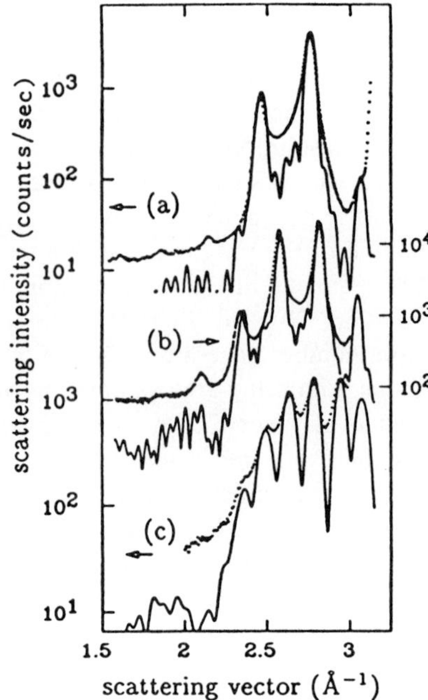

Figure 4. Measured (points) and calculated (solid curves) low-angle and high-angle scattering intensities for Co-Au superlattices. (a) 5 Å Co - 16 Å Au; (b) 10 Å Co - 16 Å Au; (c) 30 Å Co - 16 Å Au; Inset: Interface profile inferred from simulations.

Small-angle Scattering

Miceli et al.[16] have pointed out that refraction effects can noticeably shift the low-angle $00l$ Bragg peaks away from the positions given by the simple kinematical expression $\lambda = 2d\sin\theta$. The effect is to underestimate, by a few percent, the multilayer periodicity based on a simple measurement of the $00l$ peak positions. Dynamical effects can also lead, in general, to asymmetries in the low-angle peaks shapes. In order to analyze the low-angle x-ray data correctly it is necessary to include these optical effects explicitly in a calculation of the reflectivity as a function of the angle of incidence. A powerful technique has been developed for the multilayers by Barbee et al.[17]. The calculation uses an iterative Fresnel approach, where the reflection coefficients are determined at each interface in the sample. For example, the reflection coefficient at the interface between layers j and $j+1$ is given by

$$R_{j,j+1} = a_j^4 \left[\frac{R_{j+1,j+2} + \mathcal{F}_{j,j+1}}{R_{j+1,j+2}\mathcal{F}_{j,j+1} + 1} \right] \tag{3}$$

where

$$\mathcal{F}^{\sigma}_{j,j+1} = \left[\frac{E_j^R}{E_j} \right]^{\sigma} = \frac{g_j - g_{j+1}}{g_j + g_{j+1}} \tag{4}$$

and

$$\mathcal{F}^{\pi}_{j,j+1} = \left[\frac{E^R_j}{E_j}\right]^{\pi} = \frac{g_j/\tilde{n}_j^2 - g_{j+1}/\tilde{n}_{j+1}^2}{g_j/\tilde{n}_j^2 + g_{j+1}/\tilde{n}_{j+1}^2} \tag{5}$$

are the Fresnel coefficients for reflection of the σ and π components of polarization from this interface. Here $g_j = (\tilde{n}_j^2 - cos^2\theta)^{\frac{1}{2}}$ and a_j is an amplitude factor, $e^{-i\frac{\pi}{\lambda}g_j d_j}$. The computation of the reflectivity starts at the substrate and works recursively back to the surface. The resulting reflectivity is related to the intensity by $I(\theta)/I_o = R_{12}^2$.

An example of the use of low-angle x-ray reflectivity measurements for the analysis of ultrathin magnetic layers is shown in Fig. 5. Here we compare data on one of our MBE-grown Co-Cu superlattices, measured using a high-resolution x-ray diffractometer on a synchrotron radiation source, with reflectivity curves calculated using the optical model described above.

The calculated reflectivity curve shown in Fig. 5(a) is for an ideal stack of 26 bilayers of Co(14.0 Å/Cu (11.4 Å) grown on Ge-buffered (110) GaAs. The model includes the additional buffer layers of bcc-Co and Au required to initiate coherent growth of the superlattice in a (111) orientation; also included is a (15 Å) Au cap layer. The prominent peak at ~ 0.25 $Å^{-1}$ corresponds to the first (*l*=1) superlattice Bragg peak. The series of regularly spaced fringes arises from the finite thickness of the thin film superlattice. These fringes are sensitive to roughness on a scale shorter than the longitudinal coherence length of the x-ray beam (about 1000 Å in our measurements).

In order to achieve a reasonable fit to the data [Fig. 5(b)] we found it necessary to introduce *steps* at the substrate interface. The effect of the steps, which we estimate to be ~ 2 ML in height, is to reduce the amplitude of the finite thickness fringes by introducing phase shifts between rays reflected from different terraces. The origin of the interface steps is in a small amount of miscut (typically ~0.1°) of the GaAs substrate which produces a stepped 'vicinal' surface. The presence of these interface steps, and their characteristic length scale (100 - 1000 Å), leads us to believe that they may play a very important role in the unusual interlayer coupling behavior of these particular superlattices. While the terraced interface model fits our data quite well there are significant discrepancies especially in the phase of the interference fringes immediately below the *l*=1 Bragg peak. This may be due to anomalous scattering contributions arising from an absorption edge in the Cu layers close to the energy of the x-ray beam (~8 keV). These are not included in the model calculation at this point.

The measured intensity shown in Fig. 5(b) includes only the specular component of the reflectivity. A broad diffuse component, due to short-range correlations of the interface steps from layer to layer, has been subtracted from the raw data. Analysis of the diffuse scattering in terms of interface roughness remains a significant challenge for the detailed characterization of the interfaces in multilayer samples. However, as this example illustrates, there is a great deal of useful information to be gained from the specular reflectivity.

IN-PLANE SCANS

Fig. 6(a) is a contour plot of the measured in-plane scattering intensity for a Co-Au superlattice with nominal layer thicknesses of 20 Å for Co and 10 Å for Au. The

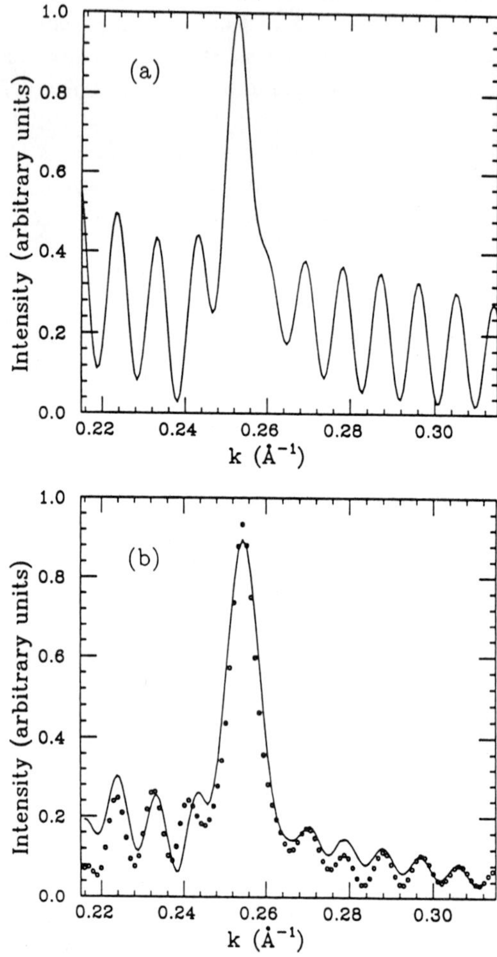

Figure 5. Low-angle x-ray reflectivity from a cobalt-copper superlattice [Co(14.0 Å) - Cu(11.4 Å)]
(a) calculated reflectivity for ideal case with atomically abrupt interfaces;
(b) calculated curve including interface steps. Open circles: synchrotron data.

presence of the Au $2\bar{2}0$ peak and the Co $11\bar{2}0$ peak shows that both Co and Au layers are oriented within the film plane. The metal peak positions relative to those of the GaAs substrate peaks confirm the epitaxial relationships given earlier[18] by RHEED: Au[$20\bar{2}$]: GaAs [001] and Co[$11\bar{2}0$]: Au[$2\bar{2}0$]. The splitting of the Co and Au peaks shows that the in-plane structure is incoherent from layer to layer. That is, even though there is sufficient coherence at the interface to *orient* each layer with respect to the previous one, the average lattice parameters within the layers have relaxed towards their bulk values.

In contrast, if the interface lattice mismatch is not too large (< 2%) the interface can have a coherent epitaxial structure. Fig. 6(b) is a scattering intensity contour plot for a sample with nominal layer thicknesses of 40 Å Co and 20 Å Cu. Our conclusions with respect to orientation of the layers within the film plane are identical to those in the Co-Au case. However there is no evidence of any peak splitting at the cubic $2\bar{2}0$ position. This indicates that the entire superlattice structure occurs with the same in-plane lattice spacing. Note that in the case of the Co-Cu superlattice the mismatch is considerably smaller (~ 2%) than for Co-Au (~13%). An interesting sidelight on this comparison is that the coherent epitaxial growth in the Co-Cu system stabilizes an fcc phase of cobalt, which has a much smaller magnetocrystalline anisotropy energy than the bulk hcp phase. This has important implications for the strength of the perpendicular anisotropy in cobalt-based superlattices.[3]

STACKING STRUCTURE

Many of the ultrathin magnetic structures of current interest are formed by the stacking of layers of close-packed metal atoms. At typical deposition rates (0.1 - 1.0 Å/sec) and substrate temperatures (50 - 200°C), the growth of such structures cannot be considered an equilibrium process. As a result, faulting of the stacking sequence is highly likely during growth. Moreover, the presence of interfaces in the structure can

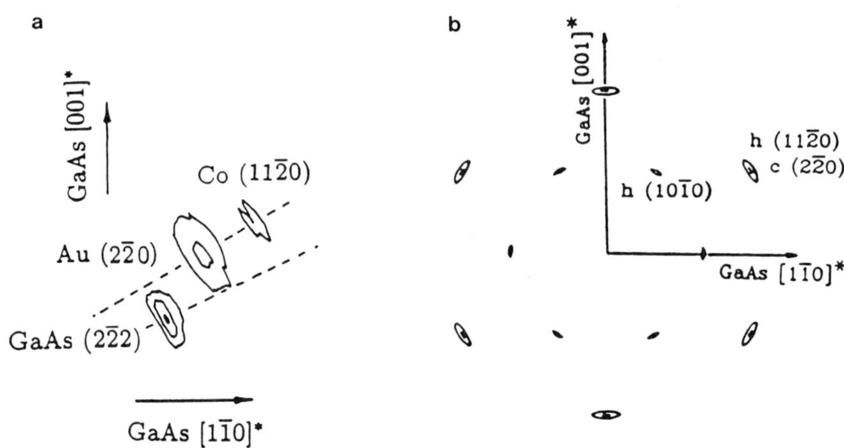

Figure 6. In-plane scattering intensity contour plots for two cobalt-based superlattices.
(a) 20 Å Co - 16 Å Au superlattice.
(b) 40 Å Co - 20 Å Cu superlattice.

strongly influence the stacking sequences, for example through epitaxial strains, or *via* effects arising from the electronic band structure. That is, apart from producing faults in a superlattice, such effects during an epitaxial growth process can induce the formation of stacking sequences corresponding to metastable phases.

We have analyzed diffuse x-ray scattering data from several ultrathin magnetic structures by comparing the measured intensities in a c*-scan with those given by Jagodzinski's two-parameter model[19] for growth faulting. This model may be described by a growth rule for close-packed structures, where the probability of cubic (c) or hexagonal (h) site occupancy by a given layer depends on the previous three layers. Specifically, if the previous three layers are stacked according to the hcp sequence (e.g. ABA), the succeeding layer will occupy the fcc (C) sites with a probability α and the hcp (B) sites with a probability $1 - \alpha$. That is, $P(h \rightarrow c) = \alpha$ and $P(h \rightarrow h) = 1 - \alpha$. Similarly $P(c \rightarrow c) = \beta$ and $P(c \rightarrow h) = 1 - \beta$. A perfect fcc structure corresponds to $\alpha,\beta = 1$ and an hcp structure correspondence to $\alpha,\beta = 0$.

Holloway[20] has presented an analytical solution of the scattering intensity in the Jagodzinski model. In Fig. 7 we plot calculated intensities generated using this solution along with the measured intensities for a Co-Cu superlattice with 10 Å Co layers. The two peaks in this plot arise from the cubic (fcc) stacking symmetry in this sample, while those corresponding to hcp Co (which would occur at $q_{c^*} = 0$ and $q_{c^*} = \pm 1.54$ Å^{-1}) are entirely absent. We find that the measured profiles correspond to $\alpha = 0.99$ and $\beta = 0.90$. This value of β implies that stacking faults in the predominantly cubic structure will occur with a 10% probability; the value of α indicates that whenever a stacking fault does occur, the succeeding layer will occupy fcc sites with a probability of 99%. That is, there are almost no hcp regions in the crystal other than those associated with isolated faults; e.g. ...cchcccchcccccccccccccchcccccccchccchccccccccccccccccccchcccccc... .

Another approach reported in the literature is to employ analytic solutions to more sophisticated models, such as the three-parameter model of Sebastian and Krishna[21], which has been developed in the context of hcp-fcc phase transformations in the bulk. In this model the parameters describing the stacking sequence are: γ, the probability of random growth faults in an hcp phase; α, the probability of nucleation of fcc growth sites; and β, the probability of continued fcc growth at the nucleation sites. This model therefore lends itself well to the quantitative analysis of stacking transitions where *correlated* regions of one type of stacking (e.g. fcc) coexist with regions of another (e.g. hcp). This is often found to be the case in metallic superlattices grown in the (111) orientation.

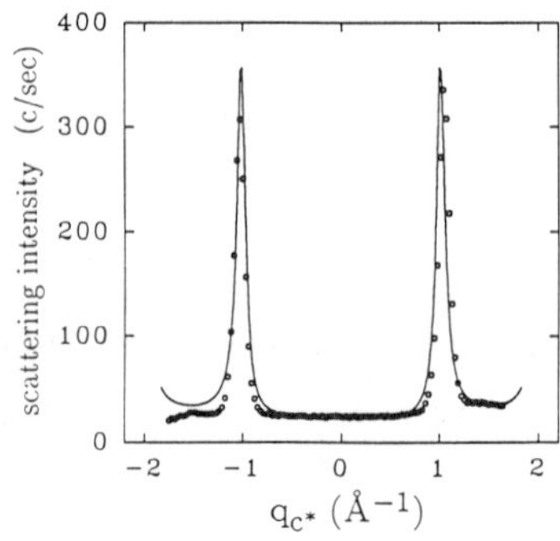

Figure 7. X-ray scattering intensities along $(10\bar{1}l)$ for Co-Cu superlattice with 10 Å Co layers and 20 Å Cu layers. The solid line is a fit to the model described in the text.

CONCLUSIONS

We have emphasized how important it is to compare measured x-ray intensity profiles with those calculated from realistic models of the structure. At present the quality of most metallic heterostructures does not warrant the use of scattering approaches beyond the kinematical approximation. However, as growth mechanisms become better understood and more control over the growth parameters is achieved, the crystalline perfection will improve to the point where dynamical scattering methods are necessary. At the point where dynamical diffraction approaches become worthwhile for these systems, it should be relatively straightforward to implement techniques which are already in routine use for analysis of epitaxial semiconductor materials.

In the context of epitaxial growth, current detailed work on ultrathin layers such as those described here suggests that defects intrinsic to the growth may be responsible for significant lattice distortions. These may have important consequences for the magnetic behavior, including magnetoresistance effects and interlayer coupling.

In the future we will undoubtedly see much more structural work on these systems being performed at synchrotron radiation sources, especially measurements which require tunable x-ray beams (e.g. anomalous scattering performed near an absorption edge). *In-situ* synchrotron measurements, performed in a time-resolved mode (for example during growth) will also find increasing applications in studies of the interface structure.

ACKNOWLEDGEMENTS

We gratefully acknowledge the contributions of W. Vavra, D. Barlett, C. H. Lee and H. He to Molecular Beam Epitaxy growth of samples described in this work. This work was supported in part by ONR Grant N00014-92-J-1335. The synchrotron studies were performed at the National Synchrotron Light Source on the AT&T Bell Laboratories beam line X-16B. The NSLS is supported by the U.S. DOE, Basic Energy Sciences, Materials and Chemical Sciences under Contract No. DE-AC02-76CA0016.

REFERENCES

1. J. C. Slonczewski, Phys. Rev. Lett. **67**, 3172 (1991).
2. S. Chikazumi, *Physics of Magnetism* (Wiley, New York, 1964) p 359.
3. C.H. Lee, H. He, F. J. Lamelas, W. Vavra, C. Uher and R. Clarke, Phys. Rev. **B42**, 1066 (1990).
4. S. Iwasaki and K. Ouchi, IEEE Trans. Magn. **MAG-14** 849 (1978).
5. R. Clarke and F. J. Lamelas, *Ultrathin Magnetic Structures*, eds. J.A.C. Bland and B. Heinrich (Springer, Berlin) in press.
6. C. H. Lee, H. He., F. Lamelas, W. Vavra, C. Uher and R. Clarke, Phys. Rev. Lett. **62**, 653 (1989).
7. See, for example, *Handbook on Synchrotron Radiation*, Vol. 3, eds. G. S. Brown and D.E. Moncton (North Holland, Amsterdam, 1991).
8. C. J. Chien, R.F.C. Farrow, C. H. Lee, C. J. Lin and E. E. Marinero, J. Magn. Mag. Mater. **93**, 47 (1991).

9. A Guinier, *X-ray Diffraction in Crystals, Imperfect Crystals, and Amorphous Bodies* (W. H. Freeman, San Francisco, 1963).

10. A. Segmüller and A. E. Blakeslee, J. Appl. Cryst. **6** 19 (1973).

11. D. B. McWhan, M. Gurvitch, J. M. Rowell and L. R. Walker, J. Appl. Phys. **54**, 3886 (1983).

12. W. Sevenhans, M. Gijs, Y. Bruynseraede, H. Homma, and I. K. Schuller, Phys. Rev. **B34**, 5955 (1986).

13. B. M. Clemens and J. G. Gay, Phys. Rev. **B35**, 9337 (1987).

14. J.-P. Loquet, D. Neerinck, L. Stockman, Y. Bruynseraede and I. K. Schuller, Phys. Rev. **B39**, 13338 (1989).

15. F. J. Lamelas, H. D. He and R. Clarke, Phys Rev. **B43**, 12296 (1991).

16. P. F. Miceli, D. A. Neumann and H. Zabel, Appl. Phys. Lett. **48**, 24 (1986).

17. T. W. Barbee Jr., W. K. Warburton and J. H. Underwood, J. Opt. Soc. Am. **B1**, 691 (1984).

18. H. He, C. H. Lee, F. J. Lamelas, W. Vavra, D. Barlett and R. Clarke, J. Appl. Phys. **67**, 5412 (1990).

19. H. Jagodzinski, Acta Cryst. **2**, 208 (1949).

20. H. Holloway, J. Appl. Phys. **40**, 4313 (1969).

21. M. T. Sebastian and P. Krishna, Physica Status Solidi (a) **101**, 329 (1987).

ADVANCED RESEARCH WORKSHOP ON MAGNETISM AND STRUCTURE IN SYSTEMS OF REDUCED DIMENSIONS
June 15-19, 1992
Cargese, Corsica, France

CONTRIBUTORS

INDEX